ESTATES GAZETTE LAW REPORTS

Estates Gazette Law Reports

2010

Volume 1

Edited by
HH Judge Hazel Marshall QC

Routledge
Taylor & Francis Group
LONDON AND NEW YORK

EG Books

EG Books is an imprint of Routledge
2 Park Square, Milton Park, Abingdon, Oxon OX14 4RN
711 Third Avenue, New York, NY 10017, USA
Routledge is an imprint of the Taylor & Francis Group, an informa business

First published 2010

British Library Cataloguing-in-Publication Data
A catalogue record for this book is available from the British Library

Library of Congress Cataloging-in-Publication Data
A catalog record for this book is available from the Library of Congress

ISBN for this volume: 978-0-7282-0580-2
ISSN: 0951-9289

For information on all EG Books publications
visit our website at www.routledge.com

Transferred to Digital Print 2011

CONTENTS

Cases reported in "Estates Gazette" [2010] issues 1-16

SUPPLEMENT

Cases not reported in "Estates Gazette"

TABLE OF CASES

Titles of cases shown in bold type
Names of parties reversed shown in ordinary type

INDEX OF SUBJECT MATTER

ADJUDICATION

Shaw and another *v* MPF Foundations & Piling Ltd

CHANCERY DIVISION

[2010] EWHC 9 (Ch)

6 January 2010

HH JUDGE STEPHEN DAVIES, SITTING AS A HIGH COURT JUDGE

Estates Gazette 20 March 2010

[2010] 11 EG 122

Construction — Adjudication — Statutory demand — Cross-claim — Insolvency Rules 1986 — Adjudicator's award enforced by Technology and Construction Court — Statutory demand in respect of sum awarded by adjudicator — Application to set aside statutory demand on ground of cross-claim equalling or exceeding debt specified in statutory demand — R 6.5(4)(a) of 1986 Rules — Whether discretion to set aside exercisable where demand founded on judgment enforcing adjudicator's award and debtor having ability to pay

The appellants engaged the respondent to carry out building works on the terms of a written minor works contract at a contract price of £168,253. The appellants became dissatisfied with the progress and quality of the works and did not honour certain interim certificates issued in accordance with the contract. The respondent left the site. Each party contended that the other had repudiated the contract. The respondent submitted its final account valuation, which the appellants disputed.

The respondent referred the dispute to adjudication under the Housing Grants, Construction and Regeneration Act 1996. The appellants declined to participate, having been advised, erroneously, that the adjudicator had no jurisdiction. The adjudicator decided the dispute and concluded that the respondent's employment had been wrongly terminated and that it was entitled to damages of £80,954.55. The Technology and Construction Court (TCC) allowed the respondent's claim to enforce the adjudicator's award.

The appellants served notice of arbitration on the respondent, claiming the repayment of any sums due or paid as a result of the adjudicator's decision that exceeded the proper evaluation of the works. The respondent served statutory demands on the appellants for the amount of the adjudicator's award and subsequently issued bankruptcy petitions against the appellants.

The appellants applied to set aside the statutory demands, under r 6.5(4)(a) of the Insolvency Rules 1986, on the ground that they had a cross-claim that equalled or exceeded the amount of the debt claimed. The district judge found that the appellants had a genuine and substantial cross-claim but upheld the statutory demands on the ground that because the appellants could afford to pay, there was no risk of them being made bankrupt and thus unable to pursue arbitration proceedings to have the adjudicator's decision set aside. The appellants appealed.

***Held*: The appeal was allowed. When determining an application to set aside a statutory demand, in respect of a sum awarded in an adjudicator's decision and enforced by a judgment of the TCC, the bankruptcy court should exercise its decision with regard to all the relevant circumstances. The authorities do not establish any rule that prevents a creditor from pursuing insolvency proceedings in the event of a genuine cross-claim. Nor, on the other hand, should the court start from a presumption that a debtor is not entitled to rely on a substantial and genuine cross-claim save where it would be oppressive to prevent it from doing so. Where such a cross-claim exists, the existence of a TCC judgment enforcing the adjudicator's decision does not materially affect the position, since the grounds for opposing enforcement are circumscribed and do not permit reliance on a cross-claim. The "pay now, litigate later" philosophy of the 1996 Act does not displace the position applied to personal insolvency by r 6.5(4) of the 1986 Rules or to corporate insolvency by case law. Where a statutory demand is founded on an adjudicator's decision and the debtor can show that it has a substantial cross-claim, it should not be prevented from raising those matters in opposition to bankruptcy proceedings simply because it could have raised, or even unsuccessfully did raise, those matters before the adjudicator. Further, the fact that the debtor is able to pay the sums due under the statutory demand is not a factor of any great weight and should not be regarded as a decisive or principal reason for refusing to set aside the demand. The district judge had erred in law in that respect and his decision was therefore set aside. Exercising the discretion afresh, it was appropriate to set aside the statutory demands in the instant case.**

The following cases are referred to in this report.

AEI Rediffusion Music Ltd v *Phonographic Performance Ltd (Costs)*; *sub nom Phonographic Performance Ltd* v *AEI Rediffusion Music Ltd (Costs)* [1999] 1 WLR 1507; [1999] 2 All ER 299, CA

Bayoil SA v *Seawind Tankers Corporation (The Leonidas)* [1999] 1 WLR 147; [1999] 1 All ER 374; [1999] 1 Lloyd's Rep 211, Comm

Dennis Rye Ltd v *Bolsover District Council* [2009] EWCA Civ 372; [2009] 4 All ER 1140

Garrow v *Society of Lloyd's* [1999] BPIR 668

Guardi Shoes Ltd v *Datum Contracts* [2002] Adj LR 10/28

Parke v *Fenton Gretton Partnership* [2000] Adj LR 08/02

Popely v *Popely*; *sub nom Popeley* v *Popeley* [2004] EWCA Civ 463; [2004] BPIR 778

Remblance v *Octagon Assets Ltd*; *sub nom Octagon Assets Ltd* v *Remblance* [2009] EWCA Civ 581; [2010] Bus LR 119

Shaw v *MFP Foundations & Piling Ltd* [2009] EWHC 493 (TCC)

This was an appeal by the appellants, Christopher Shaw and Gabrielle Shaw, from a decision of District Judge Smith refusing an application to set aside statutory demands served by the respondent, MPF Foundations & Piling Ltd, in respect of sums awarded in an adjudicator's decision enforced by a judgment of the Technology and Construction Court.

Gabrielle Shaw appeared in person for the appellants; Richard Bradley (instructed by CE Law, of Birkenhead) represented the respondent.

Giving judgment, **HH JUDGE STEPHEN DAVIES** said:

Introduction

[1] This is an appeal from a decision of District Judge Smith made on 30 June 2009, in which he refused to set aside statutory demands served by the respondent on the appellants relating to a sum that an adjudicator appointed under the Housing Grants, Construction and

Regeneration Act 1996 (the HGCRA) had awarded to them and the subject of a judgment in enforcement proceedings in the Technology and Construction Court (TCC). Although the district judge refused permission to appeal, the appellants subsequently successfully obtained permission to appeal on 25 September 2009 on one ground, which was that:

> The District Judge was wrong in law on the facts as he found them when he decided that although he was bound by the decision in *George Parke* v *The Fenton Gretton Partnership* (which was a case admitted by the District Judge as being "on all fours" with the present case) he would nevertheless exercise his discretion in the Respondent's favour.

[2] In summary, what the appellants say is that although the district judge, as they say rightly, decided that they were entitled in law to assert a genuine and substantial cross-claim that equalled or exceeded the amount claimed in the statutory demands, he had erred in exercising his discretion not to set aside the statutory demands on the basis that, because they could afford to pay that amount, there was no risk of them being made bankrupt and thus being unable to pursue arbitration proceedings to have the adjudicator's decision set aside.

[3] The appellants were also given permission to rely on further evidence on appeal, namely the first arbitral award of Mr Anthony Bingham published on 31 July 2009.

[4] On 16 October 2009, the respondent filed a respondent's notice contending that:

> 1. A statutory demand should not be set aside on the grounds of a counter-claim, set-off or cross demand equalling or exceeding the amount of the debt specified in the statutory demand where the debt or debts in the statutory demand are based upon a judgement enforcing an adjudication award and bankruptcy proceedings are not being used in order to stifle a genuine counterclaim, set-off or cross demand.
>
> 2. The Appeal Court is not bound by the decision in *George Parke* v *The Fenton Gretton Partnership*.

[5] It will be apparent to those familiar with the scheme for construction adjudications established by Part II of the HGCRA that the cross-appeal raises the question as to whether a party to a relevant construction contract that has obtained an adjudicator's decision in its favour and also obtained a judgment in the TCC enforcing that decision is entitled to pursue the losing party to insolvency even where the losing party has a substantial cross-claim against the sum awarded by the adjudicator. The respondent says that the losing party should not be entitled to rely on such a cross-claim; the appellants say that they should.

[6] It is convenient to begin by setting out the relevant facts as they were put before the district judge and as to which there was little or no substantial dispute so far as the hearing below and this appeal are concerned.

Relevant facts

[7] The appellants are the owners of a property known as East Lodge, Great Moreton Hall, Cheshire; the respondent is a building company. By a written contract dated 27 April 2007, made in the form of the minor works contract issued by the Architects and Surveyors Institute (August 1990 ed), the respondent agreed to carry out building works to the East Lodge for a contract price of £168,253. The "Adviser" appointed under the form of contract was a Mr Carter, the architect instructed by the appellants. So far as the conditions of contract are concerned, I need refer only to clause 6, which provided for interim payments and a final ("single") payment on "proper completion of the work and defects and when the contractor's account is approved and certified by the Adviser", and clause 10, which provided for disputes to be put to the adviser in writing for his decision and if not resolved by that decision referred to arbitration. There was no contractual provision for adjudication, not surprisingly because the standard form used predated the HGCRA.

[8] Works were undertaken and the appellants made interim payments of £102,421. However, by early 2008, the works were overrunning, the appellants were dissatisfied with progress and with the quality of some of the works and interim certificates issued by the adviser had not been honoured. In February 2008, the respondent left site in disputed circumstances, each party contending that the other had repudiated the contract and that they had accepted the other's repudiation. The respondent submitted its final account valuation, which was disputed by the appellants.

[9] On 27 August 2008, the respondent served a notice of adjudication on the appellants. Having referred to the contract, it stated:

> On 11 February 2008 the Referring Party accepted your repudiation of the contract.
>
> On 23 June 2008 the Referring Party submitted as its claim for damages arising from the Responding Party's breach of contract its assessment of the value of the works it had completed at the date of your repudiation of the contract in the sum of £189,134.69 and indicating that it was due a further payment of £86,730.24.
>
> You have failed to respond to the Referring Party's assessment of the value of the works it has completed and have not made any payment in respect of the same. In the circumstances a dispute exists between the Referring Party and you as to the sums to be paid in respect of the work the Referring Party has undertaken at your instruction.
>
> The Referring Party seeks an award of damages for your breach of contract and an adjudicator will be asked to determine what further sums if any are to be paid as damages to the Referring Party in respect of the works it carried out at the property and will be asked to make an award that:
>
> 1. That the Responding Party pay the Referring Party damages of £86,713.24 or such sum as the adjudicator shall decide being the unpaid value of the works, pleaded by the Referring Party at the date of the Responding Party's breach of contract.
>
> 2. …

[10] I have set out the notice of adjudication at some length because its precise terms are relevant to a question canvassed during the course of the appeal in the light of the subsequent first arbitral award, namely the nature of the dispute referred to adjudication.

[11] The appellants, on advice, decided not to participate in the adjudication on the ground, subsequently held to be erroneous, that the adjudicator had no jurisdiction because the contract fell within the exception in section 106 of the HGCRA relating to construction contracts with residential occupiers. It appears that they were not advised to participate in the adjudication and thus to respond to the claim on the merits without prejudice to that jurisdictional objection, which undoubtedly would have been a course open to them. The adjudicator proceeded to decide the dispute and, by his decision dated 25 September 2008, decided that the respondent's employment under the contract had been wrongly determined and that it was entitled to the payment of damages in the sum of £80,954.55.

[12] On 29 September 2008, Mr Wilson, the construction consultant then advising the appellants, wrote to the respondent's solicitor, enclosing his evaluation of the respondent's claim. This was a 25-page document that set out in some detail Mr Wilson's evaluation of the value of the works, taking into account matters such as architect's instructions, other additional works, works the subject of provisional sums, the cost of completing incomplete works and the cost of rectifying defective works. In summary, it valued the works at £113,619.45, which, having regard to payments already made, left a balance of £11,199.47 to pay. No payment was made by the appellants, not even the balance accepted due.

[13] On 10 October 2008, the respondent brought proceedings in Liverpool County Court, TCC list, to enforce the adjudicator's decision. The appellants sought to resist enforcement on the ground that the adjudicator had no jurisdiction, but that argument was rejected by the TCC judge in Liverpool and again by Coulson J on appeal.

[14] In the meantime, on 13 January 2009, the respondent issued and served statutory demands on the appellants, claiming that they owed it the amount decided by the adjudicator they should pay, together with costs and judgment interest. A day earlier, on 12 January 2009, Mr Wilson, on behalf of the appellants, served a notice of arbitration on the respondent, claiming the "repayment of the sums due or paid in excess of the proper evaluation of the works as a result of the adjudicator's decision". An arbitrator was appointed but the respondent

took the point that the notice was premature because the appellants had not first submitted the dispute to the adviser for his decision, as was required by the contract, and the arbitrator agreed. It appears that the appellants then sought to refer the dispute to the adviser for his decision, but without success.

[15] The appellants also applied to set aside the statutory demands, but this process was delayed owing to various procedural mishaps that it is not necessary for me to detail, and, in March 2009, the respondent issued bankruptcy petitions against the appellants. On 7 April 2009, the appellants put in witness statements in support of their application to set aside the statutory demands, referring to the history of events and setting out in some detail their case as to the termination of the contract, the defects and their valuation of what was properly due under the contract as *per* the Wilson evaluation, which they suggested established a cross-claim in the sum of £74,798.82. The hearing of their application to set aside the statutory demands was listed for 30 June 2009.

[16] On 25 June 2009, the appellants paid the respondent a sum that their solicitor explained, in a subsequent witness statement, was intended to discharge the amount admitted due as *per* the Wilson evaluation, together with their liability for interest and costs. Some further payments were made in the days leading up to the hearing to ensure that all such undisputed sums were paid.

[17] In the meantime, on 19 June 2009, the appellants served a second notice of arbitration, the dispute referred being the proper valuation of the final account, including eight specified issues, one being the dispute concerning the "termination of the contract". Acting with commendable speed, Mr Bingham was appointed on 25 June 2009 and, on 29 June 2009, gave directions with a view to making an initial decision about how the arbitration should proceed.

Hearing of 30 June 2009 and judgment of district judge

[18] As is apparent from the transcript of the hearing with which I have been provided, two points were taken by the appellants in support of their application. The first, in respect of which the district judge found against the appellants and in respect of which permission to appeal was refused, was an argument that the statutory demands were perplexing because of a change of name by the respondent and the failure of the statutory demands to make it clear who had issued the demands and on what basis. The second was that the appellants had paid all undisputed sums due to the respondent, and that there was a cross-claim that equalled or exceeded the amount claimed by the respondent. The appellants' counsel referred to *Parke* (being a decision of HH Judge Boggis QC, sitting as a judge of the Chancery Division in Birmingham on 2 August 2000*) and submitted that this was authority for the proposition, contrary to the approach adopted by the TCC in respect of enforcing adjudicators' decisions, that the bankruptcy court would not allow the winning party to bring insolvency proceedings against the losing party where there was a genuine cross-claim. In response, counsel for the respondent submitted, first, that save in exceptional circumstances, as to which there was none here, the existence of a genuine cross-claim could not prevent a party from bringing insolvency proceedings where it had the benefit of an adjudicator's decision and a judgment enforcing that decision. His second submission was that the court had a discretion whether to set aside a statutory demand, that here the evidence showed that the appellants were able to pay but unwilling to do so, so that bringing insolvency proceedings would not prevent them from arbitrating the dispute already referred to adjudication, and that the court should not allow them deliberately to disregard the previous judgment of the TCC in enforcement proceedings.

[19] The district judge proceeded to give an admirable *ex tempore* judgment. He considered the authorities referred to by the respondent's counsel that confirmed the approach taken by the TCC, and then considered *Parke*. He accepted that there appeared to be some tension between these decisions, but observed that there was a "qualitative difference between bankruptcy or other insolvency process and other mechanisms of enforcement, and that may well be that way in which the decisions can be reconciled". He held that the decision in *Parke*, being a decision of a judge of the High Court directly applicable to the facts before him, was binding on him and, accordingly, he rejected the first submission by counsel for the respondent.

[20] He then, however, went on to consider the second submission of counsel for the respondent. He recorded that it was not disputed that the appellants were able to pay the sums claimed by the respondent. He then held as follows:

> 30. It therefore seems to me that there is no real issue of bankruptcy of Mr and Mrs Shaw. There is no real issue of the right to pursue the arbitration being lost because of a supervening bankruptcy. It seems to me that this is simply a decision for Mr and Mrs Shaw as to whether they choose to pay the amounts due or not. To set aside the statutory demand simply to require the respondent to pursue enforcement by other means of a sum that (it is quite clear both from the facts of the case and all the authorities I have referred to) is undoubtedly due, is a step which, in my judgment, on the facts of this case [is] not one which the court should take. I do have a discretion. For those reasons, I exercise my discretion and decline to set aside the statutory demands.

First arbitral award and subsequent conduct of the arbitration

[21] On 31 July 2009, Mr Bingham published his first arbitral award. This dealt only with the issue as to whether the respondent was or the appellants were entitled to treat the contract as having been wrongfully terminated by the other. Dealing only with a debate about whether the respondent was in breach in respect of the quality of the stonework windows and stonework surrounds to the patio doors, he answered that issue by deciding that it was in breach and that the appellants were entitled to treat the contract as having been terminated wrongly by the respondent.

[22] I should also record that I was informed at the appeal hearing that Mr Bingham has gone on to hear the remaining matters in dispute and has advised the parties that he has signed his award on the substantive issues that may be taken up on payment of his fees. There is, however, a dispute between the parties as to the payment of his fees, which has resulted in an unfortunate situation where the award is not yet available. This is particularly unfortunate because although there is a dispute as to the effect of the first arbitral award, it is common ground that this award, on the substantive issues, will be a final determination of the dispute decided on by the adjudicator and thus, pursuant to section 108(3) of the HGCRA and para 23(2) of the scheme for construction contracts, will effectively supersede the adjudicator's decision. It follows that this award is likely to be determinative of the respondent's ability to proceed with the bankruptcy petitions, irrespective of the decision on this appeal. None the less, it appears that there will be costs and, possibly, other consequences arising out of this appeal that will have to be determined regardless, so that it is common ground that I must determine this appeal on the basis of what is currently before me.

Relevant principles applicable to this appeal

[23] It is common ground that this appeal is against the decision by the district judge of a matter in respect of which he had a discretion, and that it is not a rehearing. As set out by Lord Woolf in *AEI Rediffusion Music Ltd* v *Phonographic Performance Ltd (Costs)* [1999] 1 WLR 1507, at p1523:

> Before the court can interfere it must be shown that the judge has either erred in principle in his approach, or has left out of account, or has taken into account, some feature that he should, or should not, have considered, or that his decision was totally wrong because the court is forced to the conclusion that he has not balanced the various factors fairly in the scale.

Appellants' appeal

[24] The central focus of the appellants' appeal, summarised in paras 30 and 31 of the skeleton argument that accompanied their appeal notice, was that it is wrong and contrary to established law that the court should refuse to set aside a statutory demand simply because the debtor appears to have the means to pay. They placed particular reliance on a decision of the Court of Appeal, given on 17 June 2009 and to which

* Editor's note: Reported at *Parke* v *Fenton Gretton Partnership* [2000] Adj LR 08/02

District Judge Smith was not referred, in *Remblance* v *Octagon Assets Ltd* [2009] EWCA Civ 581*.

[25] In *Remblance*, the facts were that the creditor had served a statutory demand on the debtor, as guarantor, for rent due from his company, which it had covenanted to pay "without deduction or set off". The debtor's company had a counter-claim for damages for breach of covenant of quiet enjoyment, which it was pursuing in the county court. It was common ground that if the creditor had served a statutory demand on the company, it would have been able successfully to apply to set it aside on the ground that, pursuant to r 6.5(4)(a) of the Insolvency Rules 1986 (the 1986 Rules), it appeared to have a counter-claim equalling or exceeding the amount of the debt specified in the statutory demand. Equally, however, it was common ground that the appellant, as guarantor, was unable to set up that counter-claim, with the result that he could not bring himself within that sub-rule. He argued, however, that he was entitled to bring himself within the residual ground provided for by sub-rule 6.5(4)(d), where the court is entitled to set aside the statutory demand if it is satisfied, on other grounds (that is, on grounds other than those appearing in sub-rules (a) to (c) inclusive), that the demand ought to be set aside. The district judge had agreed, but her decision had been overturned on appeal by the High Court judge. He had held that he would have acceded to the appellant's argument that it was unfair to allow the creditor to proceed to bankruptcy against the appellant, as guarantor, when it could not have proceeded to wind up against the company as principal debtor, save for one point, which was that the appellant could pay the amount claimed so that there was no risk that he would be made bankrupt in respect of that amount. Dyson and Ward LJJ allowed the appeal; Mummery LJ would have dismissed the appeal.

[26] Dyson LJ said, in [44] and [45] of his judgment:

> 44. Sub-paragraph (a) provides that, if one of the conditions specified in the sub-paragraph is met, then the court may set aside the demand. I accept that the court retains a discretion not to do so even if one of the conditions is satisfied. But it seems to me that, if one of the conditions is satisfied, then the demand will usually be set aside. That is because it will usually be unjust to require the principal debtor to face the consequences of bankruptcy if he appears to have a counterclaim, set-off or cross demand. There may be circumstances where the principal debtor's ability to pay is a relevant factor. But as Nicholls LJ said [in *Re A Debtor* [1989] 1 WLR 271, at p276D], the theme running through rule 6.5(4) is that a statutory demand will be set aside where it is just to do so. In my judgement, it is difficult to see how it can be just not to set aside a demand where the principal debtor satisfies one of the conditions in sub-paragraph (a) merely because he can afford to pay the debt.
>
> 45. It has not been suggested (rightly in my view) that, even where the principal debtor appears to have a bona fide arguable counterclaim, set-off or cross demand, an application to set aside a statutory demand under sub-paragraph (a) should be dismissed because he has the means to pay the debt that he owes to the creditor. Were the position to be otherwise, it could always be said of a principal debtor who has the means to pay the debt that he has the ability to avoid the consequences of a bankruptcy and that his application to set aside the statutory demand should be dismissed for that reason. Ability to pay would always be a trump card in the hands of the creditor in answer to such an application. I would add that, if ability to pay were fatal to an application to set aside a statutory demand, one would have expected that to be spelt out in the rules.

[27] Accordingly, in [48], Dyson LJ concluded that the judge had been wrong to consider that the fact that the appellant had the means to pay the debt was an important factor that militated against setting aside the statutory demand. He added:

> I would not rule out the possibility that ability to pay may be a relevant factor in certain circumstances. But for the reasons that I have given, I find it difficult to conceive of circumstances where ability to pay can be the sole or principal reason for refusing to set aside a statutory demand. No such circumstances have been identified in the present case.

[28] In this case, it is clear from the argument and from the decision of District Judge Smith, to which I have already referred, that the decisive or principal reason that he gave for exercising his discretion not to set aside the statutory demand was that he was satisfied that the appellants were able to pay the sums due, so that there was no real risk that they would be prevented by bankruptcy from pursuing the arbitration to overturn the adjudicator's decision. In my judgment, it follows from the decision in *Remblance* that it was wrong as a matter of law for the district judge to have regarded this as a decisive or a principal reason for refusing to set aside the statutory demands. Of course, in fairness to District Judge Smith, he can scarcely be criticised for approaching the matter in the way that he did when he was not referred to *Remblance*, which had only recently been decided.

[29] It appears from [30] of his judgment that the only other point taken into account by the district judge was that he thought that the only consequence of setting aside the statutory demand would be that the respondent would be able to enforce the judgment obtained from the TCC by other means, and that this was not a step that the court should, on the facts of this case, require the respondent to have to take. However, it is clear from his decision that this was a supplementary rather than a decisive or principal reason for his decision. Furthermore, it is not immediately apparent why it would be wrong to set aside the statutory demands, so as to prevent the respondent from using bankruptcy as a means of enforcement, simply because the respondent would be entitled to seek to enforce the judgment in other ways. If the respondent here was able to enforce its judgment against the appellants by, for example, obtaining a third-party debt order against funds standing to their credit in their bank account, that is something they would be perfectly entitled to do. It does not seem to me that their ability to do so could be relevant to the exercise of the discretion as to whether to allow the respondent to seek to bankrupt them instead or as well. The point, as Ward LJ observed in *Remblance*, is that bankruptcy is a draconian step, especially when compared to other means of enforcement and, as Dyson LJ observed in [50] of the judgment: "The question whether execution of a judgement should be stayed is distinct from the question whether a statutory demand should be set aside."

[30] In his submissions before me on appeal Mr Richard Bradley, for the respondent, submitted that this case could be distinguished from *Remblance* because the statutory demand was based on an adjudicator's decision that had already been the subject of an enforcement judgment that the court had refused to stay and where the policy of parliament, as shown in the HGCRA, is to require adjudicators' decisions to be paid pending final determination of the dispute regardless of any cross-claim.

[31] It does not seem to me, however, that these arguments can prevail so far as seeking to justify the decision on the basis that the district judge did in this case. It is apparent that the district judge did not purport to decide the case on the basis that these factors, taken together with the appellants' ability to pay, justified him in exercising his discretion against setting aside the statutory demands. To the contrary, his decision was that, following the approach in *Parke*, a debtor was entitled to rely on cross-claims even in cases where the statutory demand was founded on an adjudicator's decision, treated as the equivalent of a judgment.

[32] In my judgment, therefore, the district judge did err in a material respect, so that the appeal should be allowed on this basis and, accordingly, it is necessary for this appeal court to exercise the discretion afresh. I must therefore go on to consider the respondent's cross-appeal and, in the light of my conclusions about that, to consider whether or not the statutory demands should be set aside.

Respondent's cross-appeal

[33] Before the district judge, Mr Bradley submitted that, as a matter of principle, the court should never exercise its discretion under r 6.5(4) to set aside a statutory demand where the creditor has a judgment in its favour enforcing an adjudicator's decision. In his skeleton argument produced for this appeal, adopted in his oral submissions before me, Mr Bradley's submission was more nuanced, and was as follows:

> 11.5.1 Where the court is faced by an application to set aside a statutory demand issued in support of a judgement based upon an adjudication award

* Editor's note: Reported at [2010] Bus LR 119

> it has a discretion as to whether or not to set aside that statutory demand. In exercising that discretion the court should give great weight to the policy of the 1996 Act, namely that adjudicator's awards are to be paid and execution of judgments based upon adjudicator's awards should not be delayed pending final determination of the dispute whether by litigation or arbitration. However, in exercising the discretion the court should be alive to the possibility that bankruptcy proceedings could be used to stifle genuine litigation to finally resolve the dispute provisionally determined by adjudication. Therefore, if the evidence suggests that bankruptcy proceedings are being used to prevent a party pursuing a genuine dispute the statutory demand should be set aside. If however the position is that the court has refused a stay of execution based upon the inability of the claimants to repay an award (as it has in this case) and, if it is clear that the debtor has the funds to pay the judgment sum (as is the case in this matter) then, the discretion should be exercised against the setting aside of the statutory demand.
>
> 11.5.2 A late issue of a statutory demand shortly before the final determination of the dispute by arbitration or litigation would also lead to the exercise of the discretion to set aside the demand on the grounds that it was being used oppressively.

[34] Mr Bradley submitted that in so far as the decision in *Parke* is inconsistent with that approach, it is wrong. He referred me to, and relied on, the subsequent decision of Ferris J in *Guardi Shoes Ltd* v *Datum Contracts* 28 October 2002*. In oral argument, he also submitted that the requirements for relying on a cross-claim, as stated by the Court of Appeal in *Bayoil SA* v *Seawind Tankers Corporation (The Leonidas)* [1999] 1 Lloyds Rep 211 (namely that a cross-claim had to have substance, be in an amount exceeding the amount of the debt and also one that the company had been unable to litigate), were not satisfied in this case because the appellants had been able to raise the arguments that they now advance before the adjudicator but had chosen not to do so.

[35] Mr Bradley also submitted that the first arbitral award of Mr Bingham did not, on proper analysis, finally determine the dispute referred to and decided by the adjudicator and, thus, was irrelevant.

[36] Mrs Gabrielle Shaw, by contrast, submitted that there was no reason to treat the case of a statutory demand founded on a judgment enforcing an adjudicator's award as being any different from any other case in which the debtor could show a genuine cross-claim and, thus, that the approach in *Parke* should be followed. She submitted that there was in law no requirement that the cross-claim had to be one that the debtor had been unable to litigate, and, moreover, that even if there were in this case the district judge had accepted that, from January 2009 onwards, the appellants had indeed been endeavouring to have the dispute referred to the adjudicator finally determined by arbitration. She submitted that the first arbitral award of Mr Bingham did, properly analysed, finally determine the dispute referred to the adjudicator and, on proper analysis, removed the essential foundation of the claim referred to the adjudicator and accepted by him, namely that the respondent was entitled to damages as a result of the appellants' wrongful repudiation of the contract.

Discussion

[37] I should begin by setting out Part 6.5(4) of the 1986 Rules, which provides that on the hearing of an application to set aside a statutory demand:

> The court may grant the application if
>
> (a) the debtor appears to have a counterclaim, set-off or cross demand which equals or exceeds the amount of the debt or debts specified in the statutory demand; or
>
> (b) the debt is disputed on grounds which appear to the court to be substantial; or
>
> (c) it appears that the creditor holds some security in respect of the debt claimed by the demand, and either Rule 6.1(5) is not complied with in respect of it, or the court is satisfied that the value of the security equals or exceeds the full amount of the debt; or
>
> (d) the court is satisfied, on other grounds, that the demand order ought to be set aside.

[38] Next, I should refer to *Bayoil*, which was a company winding-up case, as opposed to a bankruptcy case. The facts were that the petitioner was a shipowner that chartered a ship to the company. It made a claim for freight, but the company asserted a counter-claim for damages for breach of the charter. Both claim and counter-claim were referred to arbitration. The arbitrators, acting in accordance with the well-established common law rule that freight must be paid free of all deductions, made a summary interim award in favour of the petitioner for its claim for freight, leaving the counter-claim to be arbitrated in the normal way. The petitioner then proceeded to seek to wind up the company and, at first instance, the judge, who accepted that the counter-claim was genuine and serious and, moreover, was one that it had been unable to arbitrate to a conclusion, none the less held that his jurisdiction was at large and that he should exercise it in favour of making a winding-up order. The Court of Appeal held that the discretion in such a case ought to be exercised in favour of the company in the absence of special circumstances. Nourse LJ also rejected the petitioner's submission that the "finality and unappealability of the interim award" was a special circumstance.

[39] As has already been noted, Nourse LJ also held that it was a requirement in such a case that the cross-claim had to be one that the company had been unable to litigate. However, Mrs Shaw was able to refer me to a more recent decision of the Court of Appeal in *Dennis Rye Ltd* v *Bolsover District Council* [2009] EWCA Civ 372*, in which Mummery LJ, in [19], in a judgment with which Elias LJ agreed, referred to *Bayoil* and to two subsequent first instance decisions in the Companies Court and said:

> The authorities are illustrations of the well established practice of the Companies Court that, if a company has a genuine and serious cross-claim, which is likely to exceed the petition debt, the court will normally exercise its discretion by dismissing the winding up petition and allowing the Company the opportunity to establish its cross-claim in ordinary civil proceedings. A company is not prevented from raising a cross-claim in winding up proceedings simply because it could have raised or litigated the claim before the presentation of the petition or it has delayed bringing proceedings on the cross-claim. The failure to litigate the cross-claim is not necessarily fatal to a genuine and serious cross-claim defeating a winding up petition. However, in deciding whether it is satisfied that the cross-claim is genuine and serious, the court is entitled to take into account all the relevant circumstances, such as the fact that a company has not even attempted to litigate the cross-claim, or that there are reasons why it has not done so.

[40] Mrs Shaw also pointed out that neither r 6.5(4) nor para 12.4 of the Practice Direction Insolvency Proceedings make any reference to there being any requirement that the debtor had been unable to litigate the counter-claim, set-off or cross-demand.

[41] In my judgment, it is apparent from the decision in *Dennis Rye* that even in corporate insolvency cases it is no longer a requirement that the company has been unable to litigate its counter-claim, set-off or cross-demand; that is something that may be a relevant circumstance in a particular case, but it is not decisive. Furthermore, I accept Mrs Shaw's submission that the law applicable to companies winding up is not necessarily applicable to personal insolvency, where the Insolvency Act 1986 and the 1986 Rules introduced a new code.

[42] The approach that the TCC takes to the enforcement of adjudicators' decisions is also not in dispute. It was succinctly summarised by Coulson J in his decision on appeal in this case, reported at [2009] EWHC 493 (TCC) as follows:

> 15. The law in this area is clear and can be condensed into a few simple propositions:
>
> (a) Absent a want of jurisdiction or a failure to comply with the rules of natural justice, the court will enforce the decision of an adjudicator. The position was summarised by Chadwick LJ in *Carillion Construction Ltd* v *Devonport Royal Dockyard Ltd* [2005] EWCA Civ 1358 in these terms:
>
> 85. The objective which underlines the Act and the statutory scheme requires the courts to respect and enforce the adjudicator's decision unless it is plain that the question which he has decided was not the question referred

---

* Editor's note: Reported at [2002] Adj LR 10/28

* Editor's note: Reported at [2009] 4 All ER 1140

to him or the manner in which he has gone about his task is obviously unfair. It should be only in rare circumstances that the courts will interfere with the decision of an adjudicator. …

86. It is only too easy in a complex case for a party who is dissatisfied with the decision of an adjudicator to comb through the adjudicator's reasons and identify points on which to present a challenge under the labels "excess of jurisdiction" or "breach of natural justice". … The task of the adjudicator is to find an interim solution which meets the needs of the case. Parliament may be taken to have recognised that, in the absence of an interim solution, the contractor (or sub-contractor) or his sub-contractors will be driven into insolvency through a wrongful withholding of payments properly due. The statutory scheme provides a means of meeting the legitimate cash flow requirements of contractors and their sub-contractors. The need to have the "right" answer has been subordinated to the need to have an answer quickly.

…

87. In short, in the overwhelming majority of cases, the proper course for the party who is unsuccessful in an adjudication under this scheme must be to pay the amount that he has been ordered to pay by the adjudicator. If he does not accept the adjudicator's decision is correct (whether on the facts or in law) he can take legal or arbitration proceedings in order to establish the true position. …

(b) The paying party can, as Chadwick LJ indicated, issue a claim form or an arbitration notice in order to obtain a substantive decision on the merits, and that may lead to some or all of the money, originally ordered to be paid by the adjudicator, being repaid to the paying party. But that does not affect the temporary finality of the adjudicator's decision, or the directive that that decision must be complied with and, if necessary, enforced by the court.

(c) The fact that the paying party has commenced substantive arbitration proceedings is usually irrelevant to the successful parties right to enforcement (and indeed execution) of any judgment: see *Harlow & Milner* v *Teasdale (No 2)* [2006] EWHC 535 (TCC).

(d) The only authority of which I am aware in which the existence of ongoing arbitration proceedings affected the mechanics of enforcement is my decision in *Kier Regional Ltd* v *City & General (Holborn) Ltd* [2008] EWHC 2454 (TCC). But that was a case where the enforcement of an adjudicator's decision had in fact been ordered by the court in 2006. The difficulty was that the claimants had done nothing to execute that judgement and only sought charging orders and the like at a time when the substantive arbitration was only a few weeks away. In those wholly exceptional circumstances I granted a stay of execution.

[43] Turning next to the decision in *Parke*, the facts were that the creditor had obtained an adjudicator's decision against the debtor, Mr Parke, in respect of its final account claim, in circumstances where it appears that Mr Parke had received little or no notice of the adjudication. By the time the application to set aside the statutory demand was heard by the district judge, Mr Parke had already commenced proceedings in the TCC, arguing that the true final account showed a balance payable in his favour, but for some reason the district judge was not made aware of this fact. On appeal, the judge considered that this factor justified him in looking at the matter afresh, and he held that although the adjudication did create a debt that failed to be treated in the same way as a judgment or order, none the less Mr Parke had a valid cross-claim. In response to an argument that to allow Mr Parke to rely on the cross-claim would be contrary to the scheme of the HGCRA, the judge said, in [17], that:

In my judgment it cannot be right that an employer or main contractor can be made bankrupt when it is known that he has proper proceedings on foot which if successful will result in a payment to him. I do not accept that the scheme of the 1996 Act is that an adjudication can be pursued to bankruptcy no matter the underlying state of account. The court would be required to close its eyes to the overall position, which in the context of bankruptcy is in my judgment wrong in principle.

[44] It appears clear, therefore, that, in *Parke*, the judge was not prepared to accept the argument that the approach taken by the TCC to enforcement proceedings should also be applied to bankruptcy proceedings.

[45] Turning then to *Guardi*, the facts in that case were that the company sought an injunction restraining the petitioner from advertising its winding-up petition. The petition had been presented some 11 months after the works the subject of the claim were completed, in circumstances in which the company had failed to serve a withholding notice as required under the HGCRA against the petitioner's claim for payment, with the result that the petitioner obtained an adjudicator's decision in its favour, and subsequently a consent judgment in the TCC enforcing that decision. Furthermore, the company made various payments to the petitioner under the judgment. It was not until after the presentation of the petition that the company first produced evidence of a counter-claim for defective work. In those circumstances, although Ferris J, not surprisingly, was unconvinced by the counter-claim, none the less he was prepared to accept that it was "not obviously without substance". He held that it was appropriate to restrain advertisement only if the court were satisfied either that the debt asserted was disputed to the knowledge of the petitioner on substantial grounds and in good faith or if the court could be certain that the petition was bound to be dismissed. He decided that he was satisfied on neither point, saying that: "The whole situation is very unconvincing and to my mind smacks of an attempt to drag together sufficient grounds to get the advertisement of the petition restrained."

[46] He went on to refer to *Bayoil* and the two subsequent first instance decisions referred to in *Dennis Rye* and held that that the circumstances surrounding the supposed inability to litigate were of crucial importance. He observed that, in the case before him, the claim and alleged cross-claim were closely connected, and held that in those circumstances the company's failure to serve the appropriate withholding notice, which if it had been served would have put payment into suspense until both claim and counter-claim were adjudicated on, meant that the company had only itself to blame.

[47] In my judgment, there is no inconsistency between the decisions in *Parke* and in *Guardi*. It seems to me that, in *Guardi*, Ferris J was not purporting to hold that a creditor under an adjudicator's decision, with or without a TCC judgment enforcing that decision, was entitled to pursue insolvency proceedings against the debtor even where there was a genuine counter-claim that the debtor had been unable to litigate. There is nothing in his judgment that suggests to me that he was intending to do anything other than to decide the case in front of him on its own particular facts. It seems to me that, in that case, he was concentrating his focus on the particular circumstances that, in his judgment, did not make it unjust for the petitioner to pursue insolvency proceedings, specifically the unconvincing nature of the company's counter-claim, its failure to serve a withholding notice and its failure to take any steps to refer the dispute to final determination in the considerable period of time from practical completion.

[48] Furthermore, and by the same token, it does not seem to me that, in *Parke*, HH Judge Boggis QC was purporting to hold that there was a rule that would prevent a creditor from pursuing insolvency proceedings in any such case. In my judgment, he was also focusing on the particular circumstances of the case and concluding that the existence of proceedings on foot by the date of the hearing before the district judge to assert what he was satisfied was a genuine and substantial cross-claim made it unjust for the creditor to be permitted to pursue bankruptcy proceedings against Mr Parke.

[49] It follows, in my judgment, that there is no authority, binding or persuasive, that supports either of the extreme propositions contended for by the respective parties in this case. In my judgment, the bankruptcy court, when faced with an application to set aside a statutory demand in such cases, will have regard to all relevant circumstances, and will not be circumscribed in the exercise of its discretion in the manner contended for by either party.

[50] So far as the respondent's position is concerned, in my judgment there is a clear difference between enforcing an adjudicator's decision in the TCC, which itself will provide the platform for the usual panoply of enforcement proceedings, and seeking to use that decision and/or the enforcement judgment to found bankruptcy proceedings even where there is a genuine and substantial cross-claim that the debtor is either actively pursuing or for genuine reasons has been unable to pursue thus far. Although the respondent places considerable emphasis on the policy behind the 1996 Act, that is, the pay-now litigate-later philosophy, there is nothing, in my judgment, either in the Act or in the scheme that indicates that it was intended that this should displace the

position as applied to personal insolvency by r 6.5(4) of the 1986 Rules or, for that matter, to corporate insolvency by case law. The structure of the HGCRA is to require the contract to provide that the decision of the adjudicator is binding until the dispute is finally determined by legal proceedings or by arbitration (section 108(3) of the HGCRA), so that the obligation to pay the amount decided due by the adjudicator is a contractual one. That, it seems to me, is not fundamentally different from the common law rule in respect of payment of freight without deduction, which does not override the normal position in insolvency (see *Bayoil*) nor from the contractual "pay now, sue later" clause unsuccessfully relied on by Lloyds in *Garrow* v *Society of Lloyd's* [1999] BPIR 668. Moreover, it appears that the "pay-as-you-go" ethos of the Civil Procedure Rules did not prevent a debtor from relying on a cross-claim in response to a statutory demand founded on a costs order: *Popely* v *Popely* [2004] EWCA Civ 463; [2004] BPIR 778. In both of the latter cases, the decisions to that effect by the judge at first instance were upheld by the Court of Appeal.

[51] Although I recognise that Mr Bradley's revised formulation of the appropriate test on appeal does give more weight to the individual circumstances of the case, and thus requires the court to be vigilant to prevent abuse of the insolvency procedure from causing injustice to the debtor by preventing it from litigating its claim, I am still not prepared to accept that the court should start from the presumption that a debtor is not entitled to rely on a substantial and genuine cross-claim save where it would be oppressive to prevent it from so doing.

[52] So far as the appellants' argument is concerned, for the reasons that I have given, I am satisfied that the decision in *Parke* is not one that purports to lay down any general principle. It seems to me that cases involving the enforcement of adjudicator's decisions should be approached in the same way as any other case, that is, by having regard to all relevant circumstances. I can well see that in some cases, such as in *Guardi*, the failure to serve a withholding notice, as is required by section 111 of the HGCRA, if coupled with a failure to commence arbitration or litigation within a short compass of the adjudicator's decision and an asserted cross-claim not supported by convincing evidence, would justify a court in refusing to set aside a statutory demand.

Exercise of discretion

[53] In this case, it appears to me to be clear on the evidence that the appellants' cross-claim is one that is both genuine and substantial and that it equals or exceeds the amount of the statutory demand. That is not surprising, since it is the subject of an extremely detailed and comprehensive evaluation by Mr Wilson, as I have already observed.

[54] Although, in para 11.6 of his written submissions on appeal, Mr Bradley suggested that even on the appellants' own case they were still indebted to the respondent in a sum greater than £750, he did not pursue that in oral submissions, nor was it pursued before the district judge. Indeed, it appears inconsistent with the evidence of the appellants put before the district judge, which was in terms that they had paid everything that they did not dispute, and it appears from the items referred to in this subparagraph that Mr Bradley is seeking to include sums not referred to in the statutory demands, which is not permissible.

[55] At one stage in argument, I did raise the question as to whether the appellants' case was, on true analysis, a cross-claim, as opposed to a combination of defence and cross-claim. In many final account claims, it is likely that the grounds for challenging the claim are a combination of arguments concerning either the extent or valuation of the work done or the entitlement to payment for extra works (both of which would appear to be simple matters of defence) and complaints concerning defects and or delays (which could be matters of abatement, set-off, cross-claim or counter-claim). The point may be significant where the creditor has obtained a judgment enforcing the adjudicator's decision because, under para 12.3 of the insolvency practice direction, it is provided that:

> Where the statutory demand is based on a judgment or order, the court will not at this stage go behind the judgment or order and enquire into the validity of the debt nor, as a general rule, will it adjourn the application to await the result of an application to set aside the judgment or order.

[56] This may raise an interesting issue as to the correct approach where the grounds for setting aside the statutory demand could be said to raise a genuine dispute concerning the merit of the claim allowed by the adjudicator as well as counter-claims, set-offs or cross-demands. However, it is not one that I have to deal with because it is quite apparent that the argument before the district judge proceeded on the undisputed basis that the Wilson evaluation provided a cross-claim that, if established, would equal or exceed the claim, and Mr Bradley did not seek to submit to the contrary on appeal. It may be that para 12.3 is intended to apply to the case of judgments only in cases where the debtor has had the opportunity (even if it has not taken it) of setting out the grounds for disputing the debt the subject of the statutory demand. In an enforcement action in the TCC, of course, the debtor is not permitted to defend by arguing that the adjudicator's decision was wrong on the merits, so that the rationale behind para 12.3 should not apply. A similar but connected argument would be that an enforcement judgment in the TCC is intended only to give temporary finality to the adjudicator's award, so that again it should not be regarded in the same way as a judgment in ordinary proceedings that, subject to subsequent setting aside or appeal, is final and conclusive for all purposes.

[57] Mr Bradley did, however, submit that it was a relevant factor that the appellants chose not to participate in the adjudication, where they could have set up the matters raised in the Wilson evaluation. I accept that it is a relevant factor, but it does not seem to me that it has any great weight in this case. First, it is quite apparent that the reason they chose not to participate in the adjudication was not because they did not believe that their case on the merits lacked substance; instead, it was because of advice (subsequently proved to be incorrect) that the adjudicator lacked jurisdiction. Second, the whole essence of adjudication is that it provides a speedy but temporary determination; as has been said on many occasions, it is implicit in the scheme that mistakes will be made and a wrong decision may be arrived at. Parliament must be intended to have contemplated that the injustice that might be caused by an unsuccessful party having to pay up under a wrong decision would be ameliorated by the opportunity to overturn the decision by subsequent litigation or arbitration. It follows, in my judgment, that where a statutory demand is founded on an adjudicator's decision, if the debtor can show that it has a substantial cross-claim, the insolvency regime does not contemplate that it should be shut out from raising those matters in opposition to bankruptcy proceedings simply because it could have, or even unsuccessfully did, also raise those matters before the adjudicator.

[58] The respondent has not sought to suggest that the appellants failed to take prompt steps to litigate or arbitrate the adjudicator's decision after it was promulgated. As is apparent from the chronology, and as the district judge found, the appellants attempted to bring arbitral proceedings in January 2009, but were frustrated by the respondent's reliance on the contractual dispute resolution clause. This is not intended as a criticism of the respondent, which was perfectly entitled to take this point, but it does explain the reason for the delay in commencing the arbitration proceedings in June 2009. Furthermore, the respondent has not sought to criticise the delay between receipt of the adjudicator's decision in late September 2008 and the initial attempt to arbitrate in January 2009. It is apparent from the procedural chronology that, over this period, the appellants had already submitted the Wilson evaluation to the respondent and were awaiting its response and were also engaged in responding to the respondent's commencement of enforcement proceedings in the TCC.

[59] In any event, by the time the application to set aside the statutory demands came on before District Judge Smith, the appellants had commenced arbitration to seek to have the adjudicator's decision finally determined.

[60] Mr Bradley submitted that it is a relevant factor that the adjudicator's decision has been the subject of a TCC judgment to enforce it, which itself has been the subject of an unsuccessful appeal

to Coulson J. He also submitted that it is a relevant factor that the TCC has rejected the appellants' application for a stay of enforcement. In my judgment, these factors are of little or no weight in this case. It does not seem to me that the existence of a judgment enforcing the adjudicator's decision materially affects the position where there is a genuine and substantial cross-claim; not least because, as Coulson J explained in that part of his decision on appeal to which I have already referred, the grounds for opposing enforcement in such circumstances are so circumscribed and, in particular, do not permit reliance on a cross-claim. By the same token, the fact that the TCC has rejected the application for a stay is not surprising, given the extremely limited circumstances in which the TCC would allow a stay in an adjudication enforcement case.

[61] Mr Bradley submitted that the appellants' ability to pay is a relevant factor. For the reasons given by the majority of the Court of Appeal in *Remblance*, it does not seem to me that this is a factor of any great weight. It is quite apparent, in my judgment, that the appellants assert a cross-claim on substantial grounds and take the view that because they are not truly indebted for this amount they should not have to pay it. Indeed, it is also apparent, regardless of whether this is justified, that one reason for their unwillingness to pay is their concern that they will be unable to recover that money if as they expect they are successful in the arbitration. That is tied in, in this case, with the fact that there has undoubtedly been a rather curious state of affairs whereby those who control the respondent have engaged in a number of corporate name changes and other procedures that, rightly or wrongly, have understandably given the appellants cause for concern.

[62] Accordingly, and even before any question of the subsequent first arbitral award falls to be considered, if I had been exercising my discretion on 30 June 2009 I would unhesitatingly have exercised my discretion under r 6.5(4) in favour of granting the application to set aside the statutory demands.

Effect of first arbitral award

[63] I am reinforced in my decision when I come to consider the first arbitral award. It is quite apparent, in my judgment, that the dispute that was referred to the adjudicator was the respondent's claim for damages for what they contended was the appellants' repudiatory breach of contract. Although Mr Bradley has valiantly attempted to argue that what, in substance, was referred to the adjudicator was the respondent's final account claim, either as an alternative to the breach of contract claim or as, in effect, the only claim made, I am unable to accept that submission. It is quite clear from the evidence that it would have been open to the respondent to advance a claim based, for example, on the non-payment of earlier certificates or, it would appear, on the adviser's certification of its final account claim. However, it is quite plain to me that this is not what it did; the wording of the notice of adjudication is quite clear. Although I accept that the claim for "damages of £86,713.24" is qualified by the addition of the words "or such other sum as the adjudicator shall decide being the unpaid value of the works completed by the referring party at the date of the responding party's breach of contract", it is apparent, in my judgment, that this was not intended to set up a separate final account claim, independent of the claim for damages for breach of contract. It is simply a way of saying that the claim for damages is made by reference to the unpaid value of the works as at that date, and that the respondent is inviting the adjudicator either to award the sum that they contend represents that value, or such other sum as he may decide represents that value.

[64] In so finding, I am conscious that one should not seek to construe a notice of adjudication as though it were a statute or a carefully drafted contractual provision, but none the less it seems to me that its meaning is clear. Moreover, although I am also conscious that it is the notice of adjudication, read in its proper factual matrix, that identifies the dispute referred to the adjudicator, none the less when I refer to the subsequent referral notice and, indeed, the adjudicator's decision, they fortify me in my conclusion that what was intended to be referred was the claim for damages, and that is what the adjudicator decided.

[65] It follows, in my judgment, that since the arbitrator in his first arbitral award decided that the appellants were not in repudiatory breach of contract, and instead it was the respondent that was in repudiatory breach of contract, that is a final determination by the arbitrator that the respondent does not have a claim for damages for repudiatory breach of contract by the appellants. It follows, in my judgment, that as from the date of publication of the first arbitral award, the decision of the adjudicator was no longer binding. I appreciate, of course, that the respondent may still establish its entitlement to a final account claim in its favour when the substantive arbitral award is released; however that is an entirely separate matter and an entirely separate claim.

[66] It also follows, in my judgment, that from 31 July 2009 the adjudicator's decision was no longer binding on the appellants. It would appear to follow that the appellants would be entitled to apply to have the enforcement judgment set aside, although I did not hear argument on that point. None the less, it seems to me that it could not possibly be right for the statutory demands not to be set aside in those circumstances. Although Mr Bradley submitted that this was not a point taken by the appellants on this appeal, and although it is true that it is not raised in terms in the appeal notice, it is quite apparent from the skeleton argument subsequently served in support of the appeal, together with the application for permission to rely on this evidence, that the appellants were seeking to rely on the first interim award on that basis, as they said in para 19 of the skeleton, that this has overturned the adjudicator's decision.

[67] For all of those reasons, therefore, I would allow the appeal and set aside the statutory demands.

Appeal allowed.

AGRICULTURAL HOLDINGS

Kemp *v* Fisher and others

CHANCERY DIVISION

[2009] EWHC 3657 (Ch)

24 November 2009

HH JUDGE RAYNOR QC, SITTING AS A HIGH COURT JUDGE

Estates Gazette 13 March 2010

[2010] 10 EG 118

Agricultural Holdings — Agricultural Holdings Act 1986 — Succession — Holding farmed by family for generations — Consensual successions by grant of new tenancy to different family members in 1972 and 1998 — Whether exhausting maximum two statutory successions permitted by 1986 Act — Whether section 37(1) and (2) of 1986 Act applying to consensual tenancy granted prior to coming into force of original statutory succession scheme in 1976

The claimant held a tenancy of an agricultural holding to which Part IV of the Agricultural Holdings Act 1986 applied. The holding formed part of an estate of which the defendant landlords were the trustees. The claimant's family had farmed the land for generations. In 1972, the claimant's father had taken over the holding under a new tenancy agreement. In 1998, the claimant became the tenant by a further tenancy agreement between the trustees, as landlords, the claimant, as tenant, and his father, as outgoing tenant. The claimant wanted his son to take over the farm.

The defendants contended that the statutory succession provisions of the 1986 Act no longer applied to the tenancy since the maximum of two successions permitted by the Act had already been made. They argued that both the 1972 and 1998 tenancies counted as successions falling within section 37(1) of the 1986 Act, since both qualified as occasions on which the consensual grant of a new tenancy to an eligible person, prior to the death of the sole tenant, was to be deemed, under section 37(2), to have been obtained by virtue of a direction of the Agricultural Land Tribunal.

The claimant sought a declaration that the statutory succession provisions continued to apply. He accepted that section 37(1) and (2) applied to the agreement under which he had become tenant, but argued that they could not apply to his father's tenancy since it predated the coming into force of the original statutory succession scheme under the Agriculture (Miscellaneous Provisions) Act 1976 and the creation of the Agricultural Land Tribunal.

***Held*: The claim was allowed. A tenancy granted by agreement prior to the coming into force of the 1976 Act does not take the case out of the statutory succession scheme introduced in 1976. A tenancy should not be deemed, under section 37(2), to have been obtained by a tribunal that did not exist at the relevant date. For an event to be deemed to be the equivalent of an Agricultural Land Tribunal direction, it should have taken place at a time when it was possible to obtain such a direction. Pre-1976 tenancies have nothing to do with the statutory scheme. The purpose of that scheme is to limit the number of statutory successions to two, and there is no reason to assume that parliament had intended pre-1976 consensual acts to restrict the statutory rights of an eligible person. In the light of the presumption against retrospective legislation, had parliament intended section 37(2) to apply to transactions that occurred before 1976 and to confer on them retrospectively a significance that they did not have at the time, clearer words would have been used.**

The following cases are referred to in this report.

Francis Perceval Saunders Dec'd Trustees v *Ralph*; *sub nom Trustees of Saunders* v *Ralph* (1993) 66 P&CR 335; [1993] 2 EGLR 1; [1993] 28 EG 127, QB

Yew Bon Tew v *Kenderaan Bas Mara* [1983] 1 AC 553; [1982] 3 WLR 1026; [1982] 3 All ER 833, PC

This was the hearing of a claim by the claimant, Malcolm Kemp, against the defendants, the Hon Patrick Vavasseur Fisher, the Trustees of the Kilversstone 1989 Settlement and the Trustees of the Blakeney Park Settlement, for declaratory relief as to the right of succession to an agricultural holding under Part IV of the Agricultural Holdings Act 1986.

Martin Rodger QC (instructed by Birketts LLP, of Ipswich) appeared for the claimant; the defendants did not appear and were not represented.

Giving judgment, **HH JUDGE RAYNOR QC** said:

[1] In these proceedings, the claimant seeks a declaration that his tenancy of an agricultural holding at Kilverstone, in Norfolk, is a tenancy to which the statutory succession provisions in Part IV of the Agricultural Holdings Act 1986 continue to apply.

[2] The proceedings are not actively opposed by the defendants, the claimant's landlords, but neither do they admit the claim. They do, though, admit that Part IV of the Act applies to the claimant's tenancy.

[3] This action raises an issue of law that has been the subject of some debate in textbooks relating to agricultural holdings in the light of what is seen as uncertainty created by an *obiter* remark of Jowitt J in *Trustees of Saunders* v *Ralph* (1993) 66 P&CR 335*.

[4] The short point that arises is whether the consensual grant of a new tenancy that takes place prior to 14 November 1976, which was the date of the coming into effect of the Agriculture (Miscellaneous Provisions) Act 1976 (the 1976 Act), can adversely affect the succession rights created by that Act of the tenant of an agricultural holding.

[5] The circumstances that give rise to this claim can be stated shortly. The claimant, Mr Malcolm Anthony Kemp, is the tenant of an agricultural holding in Norfolk, namely Lodge Farm, which is on the Kilverstone Estate. The family of Mr Kemp has farmed that land for generations and the initial tenancy was created as long ago as 1947. A new tenancy agreement was made on 16 October 1973, under which the claimant's father, Mr Roy Kemp, became the tenant of Lodge Farm under a yearly tenancy from 11 October 1972. The current tenancy of Lodge Farm was granted by an agreement made on 25 February 1998 between the trustees of the Kilverstone Settlement, as landlords, the claimant, as tenant, and his father, as outgoing tenant. The term was from 11 October 1996 until October 1997 and thereafter from year to year.

---

* Editor's note: Also reported at *Francis Perceval Saunders Dec'd Trustees* v *Ralph* [1993] 2 EGLR 1

[6] The issue that arises in this action is whether the tenancy granted by agreement on 16 October 1973, before the inception of the statutory succession scheme introduced by the 1976 Act, is deemed by the provisions of section 37(2) of the Agricultural Holdings Act 1986 (the 1986 Act) to be one of the maximum of two statutory succession opportunities permitted by the 1986 Act. If it does count as one, it follows that there can be no further succession rights under the Act and Mr Kemp's son, who farms the land and whom he would want to take over, would not be an eligible person.

[7] Section 18(1) of the 1976 Act provided that where, after the passing of the Act, the sole or sole surviving tenant of an agricultural holding died and was survived by any of a number of defined persons, the succession rights under the Act would apply unless excluded by section 18(4). By section 18(4)(*e*), those rights were excluded if, on each of the last two occasions when there died a sole or sole surviving tenant, there occurred one or other of the following: (i) an eligible person becoming entitled to a tenancy of the holding by direction of the Agricultural Land Tribunal; or (ii) a tenancy being granted by the landlord to a person who, being an eligible person, was or had become the sole or sole remaining applicant for such a direction. This last provision was inserted to make it unnecessary for parties who agreed that a tenancy should be granted to pursue the application to a decision by the tribunal.

[8] The exclusion in section 18(4)(*e*) was stated to be without prejudice to subsection (5), which contained a deeming provision providing (in effect) that the consensual grant of a new tenancy to an eligible person prior to the death of the sole or sole surviving tenant should be deemed to be an occasion such as is mentioned in section 18(4)(*e*)(i), on which a direction was given by the tribunal for the grant of a tenancy. This provision made it unnecessary to wait for the death of the tenant, but avoided disadvantaging a landlord that granted a new tenancy in such circumstances.

[9] The Act was amended in 1984 under the Agricultural Holdings Act of that year. Paragraph 2 of Schedule 1 to the Act amended the provisions of section 18(4) and (5) of the 1976 Act so that, *inter alia*, the assignment of the tenancy to a close relative by agreement of the landlord was now an event deemed to be within section 18(4)(*e*).

[10] The succession provisions of the 1976 Act were repealed by the 1986 Act, which was a consolidating Act; they were re-enacted in Part IV of the 1986 Act. The exclusion provisions in sections 37(1) and (2) provide:

Exclusion of statutory succession where two successions have already occurred.

(1) Section 36(1) above (and section 41 below) shall not apply if on each of the last two occasions when there died a sole (or sole surviving) tenant of the holding or of a related holding there occurred one or other of the following things, namely

(*a*) a tenancy of the holding or of a related holding was obtained by virtue of a direction of the Tribunal under section 39 below, or such a tenancy was granted (following such a direction) in circumstances within section 45(6) below, or

(*b*) a tenancy of the holding or of a related holding was granted by the landlord to a person who, being a close relative of the tenant who died on that occasion, was or had become the sole or sole remaining applicant for such a direction.

(2) If on any occasion prior to the date of death, as a result of an agreement between the landlord and the tenant for the time being of the holding or of a related holding, the holding or a related holding became let

(*a*) under a tenancy granted by the landlord, or

(*b*) by virtue of an assignment of the current tenancy,

to a person who, if the said tenant had died immediately before the grant or assignment would have been his close relative,

that occasion shall for the purposes of subsection (1) above be deemed to be an occasion such as is mentioned in that subsection on which a tenancy of the holding or a related holding was obtained by virtue of a direction of the Tribunal under section 39 below.

[11] In this case, it is common ground that the agreement of 25 February 1998, under which the claimant became the tenant, was one agreement falling within subsection (2). It follows, therefore, that if the agreement of 16 October 1973 also counts, there can be no further succession rights under the Act.

[12] In his very impressive skeleton argument, Mr Martin Rodger QC, for the claimant, set out at length his submissions as to why it cannot have been intended by parliament, when it passed the 1986 Act, that the provisions of section 37(1) and (2) would apply if the agreement under which the tenancy was granted predated the coming into being of the original statutory succession scheme and the Agricultural Land Tribunal in 1976.

[13] As I have indicated, the issue is uncertain because of a passage in the judgment of Jowitt J in *Saunders*. His remarks were clearly *obiter* because he held that there had been no prior statutory succession in the case before him, and so he did not need to consider the retrospectivity issue. However, he chose to do so: at p349, under the heading "Retrospectivity" he stated: "For the sake of completeness I deal with the issue of retrospectivity raised by Mr Denbin for the respondent. He submits that in any event a succession falling within subsection (2) of section 37 cannot be one which occurred before the passing of the 1976 Act. The submission in my judgment is misconceived. Before the passing of that Act there were no rights of succession. The Act created such rights and subsection (2), just as subsection (1) of what is now section 37, merely placed limits upon the availability of those rights."

[14] The passage that I have quoted has been the subject of considerable academic criticism and, in the current edition of *Scammell and Densham's Law of Agricultural Holdings* (9th ed), at pp575-577, the editors set out why, in their view, the judgment of Jowitt J is wrong. A similar conclusion is reached by the editors of *Muir Watt and Moss: Agricultural Holdings* (14th ed), at p428n44.

[15] I have considered those views and the much fuller submissions of Mr Rodger and I have concluded that those submissions are sound. In my judgment, essentially for the reasons advanced by Mr Rodger and the editors of *Scammell and Densham*, and with all due respect of the judgment of Jowitt J, a tenancy granted by agreement prior to 14 November 1976 does not have the effect of taking the case out of the statutory scheme introduced in 1976. This is because:

(a) Subsections (1) and (2) of section 37 should be read together. Subsection (2) deems the consensual grant of a new tenancy (or letting by assignment) to be an occasion on which the tenancy was obtained by virtue of a direction of the Agricultural Land Tribunal. It is an unnatural construction to envisage that something would be deemed to be obtained by direction of a tribunal that did not exist at the relevant date that is, to deem something to have happened that could not possibly have occurred. I accept Mr Rodger's submission that the more natural and, in my view, correct construction is that events that are to be deemed to be the equivalent of a tribunal direction should have occurred at a time when it was possible to obtain a direction.

(b) This construction advances the obvious legislative purpose of sections 37(1) and (2), which is to limit the number of statutory successions to two. Pre-1976 tenancies have nothing whatever to do with the statutory scheme, and there is no reason to think that parliament had intended that pre-1976 consensual acts should restrict the statutory rights of an eligible person. I agree with the view expressed by the editors of *Scammell and Densham* that "it would be very surprising if Parliament intended that landlords could exhume the past in this way".

(c) It is legitimate to go back to the original legislation if it is contended that the words of a consolidating Act are ambiguous. In this case, it seems to me that the editors of *Scammell and Densham* are right in their view that, on their true construction, the succession provisions contained in Part II of the 1976 Act are intended to be governed by the opening words of section 18(1): "Where after the passing of this Act…". This construction accords with the well-known presumption against retrospective legislation as enunciated in the advice of the Privy Council in *Yew Bon Tew* v *Kenderaan Bas Maria* [1983] 1 AC 553, at p558. As submitted by Mr Rodger, the effect of construing section 37(2) as applying to transactions taking place before the commencement of the 1976 Act would be retrospectively conferring on those transactions a significance that they did not have at the time, and I agree that one

A would have expected parliament to make that intention much clearer if that had indeed been its intention.

(d) If the statute is construed so as to apply to pre-1976 consensual grants, its operation would be capricious. Many families will have been entirely excluded from the statutory scheme because two voluntary successions had taken place years before the statutory right of succession had been conceived. It is difficult to see why parliament should have decided to disadvantage these families, compared to the family of a first generation tenant.

(e) If the Act were retrospective in the way suggested by Jowitt J, as the editors of *Scammell and Densham* say, the possible extent of retrospectivity would be limitless, and it would be possible for an B owner of a traditionally managed estate to point back to a succession of handovers from one family member to another stretching back over decades. It would need very clear language to induce me to construe the Act in a way that would have that effect and I do not find that the Act was intended to be so construed.

[16] It follows that, in my judgment, the claimant is entitled to the declaration that he seeks.

Claim allowed.

Allen *v* Royal Bank of Scotland A Group plc

COURT OF APPEAL

[2009] EWCA Civ 1213

11 October 2009

DYSON, HUGHES AND WALL LJJ

Estates Gazette 20 February 2010

B **[2010] 07 EG 102**

Disability discrimination — Service provider — Bank — Lack of wheelchair access to bank branch — Disability Discrimination Act 1995 — Whether bank fulfilling duty to make reasonable adjustments within section 21(2)(a) to (d) of 1995 Act — Nature of service provided by bank — Whether provision of telephone and internet banking facilities amounting to adequate alternative means of making service available to disabled persons — Whether duty requiring implementation of scheme to provide disabled access to branch — Whether discrimination occurring contrary to C ***section 19(1)(b) of 1995 Act***

The respondent suffered from muscular dystrophy and used an electric wheelchair. He opened a bank account with the appellant bank at one of its main branches, which was located in a listed building. Access was by means of steps, which rendered the building inaccessible to wheelchair users. Moreover, the bank's two ATM machines were inaccessible owing to their height above the ground. The appellant offered the respondent what it claimed to be a reasonable alternative, namely a combination of internet and telephone banking and the use of other branches. It rejected a scheme to provide wheelchair access that involved installing a platform lift in part of the banking hall on D the ground floor because it would result in the loss of one of its eight interview rooms.

The respondent brought an action against the appellant, alleging unlawful disability discrimination, contrary to section 19(1)(b) of the Disability Discrimination Act 1995, on the ground that the appellant had failed, without justification, to make reasonable adjustments, under section 21(2)(a) to (d), to enable disabled persons to use its service. Allowing the claim, the county court judge made a declaration that the appellant had discriminated against the respondent contrary to section 19(1)(b) of the 1995 Act, awarded the respondent £6,500 in damages for injury to feelings and ordered the appellant to install a E platform lift. He found that the appellant had not offered reasonable alternative means of providing the relevant service outside the branch and had not shown that all methods of achieving disabled access to the branch were unreasonable.

The appellant appealed. It contended that the judge had misapplied section 21(2) by, *inter alia*: (i) inadequately identifying the relevant service and conflating the service of providing banking facilities with the provision of a building for that purpose; and (ii) asking only if remedial steps were reasonable, rather than applying the stricter statutory test of whether it was reasonable to require the appellant to carry out those steps.

Held: **The appeal was dismissed. The duty under section 21(2) arises when a physical feature makes it impossible or unreasonably** F **difficult for disabled persons to use a service and there is a step that, in all the circumstances of the case, it is reasonable for the service provider to take to achieve one of the goals stated in section 21(2)(a) to (d). The relevant service in the case of the appellant was the provision of banking facilities at the main branch. Although a distinction between the service and the means by which it is delivered might be helpful in some contexts, the means of delivery is often an integral part of the description of the service. The provision of internet or telephone banking facilities involves the provision of different banking services from those offered by traditional, face-to-face means.** G **Unless a platform lift were installed in the appellant's branch, persons whose disability required them to use wheelchairs could not use the facilities that persons who were not disabled had available to them at that branch. The judge had been entitled to conclude that face-to-face facilities were an important element of the service provided by the appellant at the branch and that the alternative means of making banking services available relied on by the appellant were not a reasonable alternative unless there were no reasonable means to afford the respondent physical access to the main branch. To determine whether a step is unreasonable involves an exercise of judgment, taking into account all the circumstances of the case.** H **The judge had objectively addressed the question of whether the reasonable adjustments duty required the appellant to adopt the platform lift scheme and had been entitled to conclude that it did so.**

The following cases are referred to in this report.

Lewisham London Borough Council v *Malcolm* [2008] UKHL 43; [2008] 1 AC 1399; [2008] 3 WLR 194; [2008] 4 All ER 525; [2008] 2 P&CR 18; [2008] HLR 41

Roads v *Central Trains Ltd* [2004] EWCA Civ 1541

This was an appeal by the appellant, Royal Bank of Scotland Group plc, from a decision of HH Judge Dowse, sitting in Sheffield County J Court, allowing a claim by the respondent, David Allen, for damages and injunctive and declaratory relief for disability discrimination contrary to the Disability Discrimination Act 1995.

Richard Lissack QC and Dr Stephen Hardy (instructed by Cobbetts LLP, of Manchester) appeared for the appellant; Robin Allen QC and Catherine Casserley (instructed by the Sheffield Law Centre) represented the respondent.

Giving the first judgment, **DYSON LJ** said:

Introduction K

[1] Mr David Allen is now 18 years of age. He suffers from Duchenne muscular dystrophy and has been using an electric wheelchair since an early age. He opened a bank account with the Royal Bank of Scotland (the bank) at its main branch (the main branch) in Church Street, in the centre of Sheffield.

[2] The main branch is a 19th-century listed building. Access to all the entrances is gained by flights of stone steps. There are two principal customer entrances to the front elevation, situated at the far right and far left. There is also a staff entrance adjacent to the far right entrance door. The front elevation incorporates two automated telling machines

A (ATMs), which are inaccessible to wheelchair users owing to their height above the ground.

[3] The main branch is therefore inaccessible to wheelchair users. Following his attempts to use the main branch and complaints by Mr Allen's mother, the bank suggested that Mr Allen use the branch staff entrance. However, this entrance was also inaccessible to him. The bank also offered what it asserted to be a reasonable alternative, namely a combination of internet banking, telephone banking and the use of branches elsewhere in the city. The bank also suggested that Mr Allen could use NatWest branches. Mr Allen attempted to do this, but was turned away on the ground that NatWest did not permit the use
B of an RBS card at its branches.

[4] On 20 August 2007, Mr Allen issued these proceedings, alleging unlawful disability discrimination on the ground of a breach of section 19(1)(b) of the Disability Discrimination Act 1995 (the DDA) in failing to comply with the duty to make reasonable adjustments in section 21(2) without justification.

[5] The court had before it two potential building solutions to the access difficulties encountered by Mr Allen. One solution (the lobby scheme), which the bank had considered and rejected, involved the installation of a platform lift within the entrance lobby adjacent to the customer entrance. The bank had obtained planning permission for this
C scheme but rejected it on the ground that the turning circle required could not be accommodated and the works would cause severe disruption because they required alterations to incoming gas mains, water mains and internal services. The other solution (the Owen scheme), which was recommended by the single joint expert, Mr Richard Owen, involved the installation of a platform lift in the area that forms part of the existing banking hall. This was rejected because it would require the loss of an interview room on the ground floor. A similar scheme to the Owen scheme had been rejected for the same reason in around 2002. The main branch has eight interview rooms. Four of these are located on the ground floor and four on the first floor.

D [6] By an order made on 16 January 2009, HH Judge Dowse declared that the bank had discriminated against Mr Allen, contrary to section 19(1)(b), awarded damages in the sum of £6,500 for injury to feelings and ordered the bank to install a platform lift in accordance with the Owen solution no later than 30 September 2009.

[7] The bank appeals with the permission of Moses LJ, who also granted a stay pending the outcome of the appeal. It does not seek to disturb the award of damages made by the judge.

Relevant statutory material

[8] So far as material, the DDA provides:

E **19. Discrimination in relation to goods, facilities and services**

(1) It is unlawful for a provider of services to discriminate against a disabled person

(a) in refusing to provide, or deliberately not providing, to the disabled person any service which he provides, or is prepared to provide, to members of the public;

(b) in failing to comply with any duty imposed on him by section 21 in circumstances in which the effect of that failure is to make it impossible or unreasonably difficult for the disabled person to make use of any such service;

(c) in the standard of service which he provides to the disabled person or the manner in which he provides it to him; or

F (d) in the terms on which he provides a service to the disabled person.

(2) For the purposes of this section and sections 20 and 21

(a) the provision of services includes the provision of any goods or facilities;

(b) a person is "a provider of services" if he is concerned with the provision, in the United Kingdom, of services to the public or to a section of the public; and

(c) it is irrelevant whether a service is provided on payment or without payment.

(3) The following are examples of services to which this section and sections 20 and 21 apply

(a) access to and use of any place which members of the public are permitted to enter;

…

(e) facilities by way of banking or insurance or for grants, loans, credit G
or finance;

20. Meaning of "discrimination"

(1) For the purposes of section 19, a provider of services discriminates against a disabled person if

(a) for a reason which relates to the disabled person's disability, he treats him less favourably than he treats or would treat others to whom that reason does not or would not apply; and

(b) he cannot show that the treatment in question is justified.

(2) For the purposes of section 19, a provider of services also discriminates against a disabled person if

(a) he fails to comply with a section 21 duty imposed on him in relation to the disabled person; and H

(b) he cannot show that his failure to comply with that duty is justified.

(3) For the purposes of this section, treatment is justified only if

(a) in the opinion of the provider of services, one or more of the conditions mentioned in subsection (4) are satisfied; and

(b) it is reasonable, in all the circumstances of the case, for him to hold that opinion.

(4) The conditions are that

(a) in any case, the treatment is necessary in order not to endanger the health or safety of any person (which may include that of the disabled person);

21. Duty of providers of services to make adjustments

(1) Where a provider of services has a practice, policy or procedure which makes it impossible or unreasonably difficult for disabled persons to make use of a service which he provides, or is prepared to provide, to other members J of the public, it is his duty to take such steps as it is reasonable, in all the circumstances of the case, for him to have to take in order to change that practice, policy or procedure so that it no longer has that effect.

(2) Where a physical feature (for example, one arising from the design or construction of a building or the approach or access to premises) makes it impossible or unreasonably difficult for disabled persons to make use of such a service, it is the duty of the provider of that service to take such steps as it is reasonable, in all the circumstances of the case, for him to have to take in order to

(a) remove the feature;

(b) alter it so that it no longer has that effect;

(c) provide a reasonable means of avoiding the feature; or K

(d) provide a reasonable alternative method of making the service in question available to disabled persons.

[9] Prior to its repeal by the Equality Act 2006 (the EA), section 53A of the DDA conferred on the Disability Rights Commission the power to prepare and (subject to approval by the Secretary of State and absent any negative resolution of either house of parliament) to issue codes of practice (the codes) giving practical guidance on how to avoid unlawful acts or "on any other matter relating to the operation of any provision" under the relevant parts of the DDA. The codes were admissible in proceedings, and where any provision of a code of practice appeared to a court to be relevant section 53A(8A) provided that it must take L that provision into account. Section 42(3) of the EA, which makes transitional provision for the functions of the previous statutory commissions, provides that "a code of practice issued by a Commission dissolved by virtue of section 36… (a) shall continue to have effect until revoked by the Secretary of State at the request of the Commission for Equality and Human Rights, by order made by statutory instrument". The code of practice *Rights of Access: services to the public, public authority functions, private clubs and premises* (the 2006 Code of Practice) came into force on 4 December 2006: see the Disability Discrimination Code of Practice (Services, Public Functions, Private Clubs and Premises)(Appointed Day) Order 2006 (SI 2006/1967). M

Grounds of appeal

[10] There are five grounds of appeal. Ground 1 is that the judge misunderstood the legal nature of Mr Allen's case. He approached it as though it were an allegation of discrimination by less favourable treatment (section 20(1)), whereas it was in fact a claim that the bank had failed to make reasonable adjustments in accordance with the duty imposed by section 21. Ground 2 is that the judge misapplied the reasoning in *Lewisham London Borough Council* v *Malcolm* [2008] UKHL 43; [2008] 1 AC 1399. Ground 3 is that the judge erred in failing: (i) to identify the relevant service that the bank was providing; (ii) properly to deal with what he referred to as the bank's concerns

A and considerations; and (iii) to apply the objective test. Ground 4 is that the judge erred in finding that the bank had not been justified in its treatment of Mr Allen in accordance with section 20(3) and (4). Ground 5 is that the judge gave too much weight to the evidence called on behalf of Mr Allen and discounted the bank's evidence, in particular that given by Ms Jean Cluness and the report of Mr Owen.

[11] I shall start with ground 3, since that is the principal ground on which Mr Richard Lissack QC relied.

Ground 3

First criticism

B [12] The first criticism made by Mr Lissack under this ground was that the judge had not correctly identified the relevant service provided by the bank. The judge said, in [15]:

> There can be no argument that the Defendant is a service provider within the meaning of Section 19 of the DDA or that its range of services generally offered are those of a normal bank and include the paying in of money, practical matters and advice concerning the opening of accounts including what accounts may be best for the client and the provision of money from cash points as well as obtaining money at the service counter. In my view a client can expect to be offered a facility for confidential discussion and the Bank recognises this in the establishment at Church Street of no less than eight interview rooms.

C [13] Mr Lissack submitted that this statement of the services provided by the bank was insufficient to enable a proper analysis to be undertaken of the section 21(2) duty to make reasonable adjustments. He said that the judge had failed to distinguish between a service (for example, financial service) and the means by which a service is delivered (for example, by telephone or internet banking or face-to-face confidential discussion). He submitted that the judge had failed to identify the particular services that, on Mr Allen's case, could not be used by him at all or without unreasonable difficulty. In the absence of such an analysis, the judge could not properly address the question of what steps (if any) it was reasonable for the bank to have to take under
D section 21(2)(a) to (d). The judge's incorrect approach was illustrated by the way in which he dealt with the question of internet banking. In [50], he said:

> The Bank also sought to assert that David could (and effectively should) use Internet banking. Of course he cannot use that facility for paying money in or drawing money out. Internet banking is a facility provided by the Bank which does not provide equal services to the services provided within the confines of a branch of the Bank.

Mr Lissack made the point that the judge did not identify the particular service that he was saying was denied to Mr Allen or acknowledge
E that internet banking is a means of accessing the principal banking services offered.

[14] Mr Lissack submitted that the judge should have classified the services into six separate elements. Had he done so, he would have realised that there was no breach of section 21(2) in respect of any of them because, with regard to each element, the bank had taken such steps as it was reasonable for it to have taken in order to "provide a reasonable alternative method of making the service in question available": see section 21(2)(d).

[15] The first element is paying money into the account. Mr Lissack submitted that this was not an issue before the judge because Mr Allen's
F regular income comprised payments of disability living allowance that were made directly into his account every four weeks. For any other payments that Mr Allen wished to make, the evidence was that a quick deposit service was available at other branches in Sheffield that had disability access.

[16] The second element is withdrawal of cash. The evidence was that Mr Allen could withdraw cash from ATMs in other banks that had disability access or in some shops that also had that facility.

[17] The third element is obtaining information on bank balances. The evidence was that Mr Allen could obtain this information by using an ATM, a telephone or the internet.

[18] The fourth element is making money transfers. The evidence was that this could be done by telephone.

[19] The fifth element is opening accounts. The evidence was that G
an account could be opened by telephone or via the internet. It was also said by Ms Vigar, a manager at the bank, that home visits could be considered in some circumstances.

[20] The sixth element was obtaining advice. In [51] of his judgment, the judge said that it was possible for arrangements to be made for a customer to be seen at home. He added that it was "notable that no such offer has ever been made in this case".

[21] Mr Lissack submitted that the judge fell into error in failing to consider each of these services separately. He should have asked in respect of which, if any, of these services Mr Allen was disadvantaged by not being able to have physical access to the main branch, H
notwithstanding the availability of the quick deposit service at other branches, the internet, telephone, ATMs and home visits.

[22] In short, Mr Lissack submitted that the judge had wrongly equated access to the premises with access to the services. The service of providing banking facilities is not the provision of a building where the services are provided; it is the provision of the services and, for present purposes, the six elements that he identified.

[23] In my judgment, the relevant service in this case was the provision of banking facilities at the main branch. Section 19(3)(a) of the DDA provides that examples of services to which sections 19 to 21 apply are access to and use of "any place which members of the J
public are permitted to enter". Also included are "facilities by way of banking". It is not in dispute in the present case that, unless a platform lift was installed in the main branch, persons whose disability required them to use wheelchairs could not gain access to the main branch and could not therefore make use of the facilities that were available at that branch to persons who were not disabled. The analysis of the banking facilities in the way for which Mr Lissack contended cannot disguise this fact.

[24] I accept, however, that it was necessary for the judge to examine the six elements identified by Mr Lissack in order to see whether, as the bank contended, it has discharged the duty under section 21(2)(d) K
by providing a reasonable alternative method of making available to Mr Allen the facilities that are available at the main branch to people who are not disabled.

[25] In [15] of his judgment, the judge had correctly identified the services provided by the bank. The distinction made by Mr Lissack between the service and the means by which it is delivered may be helpful in some contexts. I do not find it helpful here. The means by which a service is delivered is often an integral part of the description of the service. In any event, section 19(2)(a) defines the "provision of services" as including "the provision of any goods or facilities". The provision of banking facilities by internet or telephone is the provision L
of different banking facilities (and, therefore, different banking services) from the provision of traditional face-to-face banking facilities.

[26] The judge considered some of the six elements identified by Mr Lissack. He concluded that at least some of the alternative methods of making the facilities available that were relied on by the bank did not amount to reasonable alternative methods of "making the service in question available to disabled persons". Take the giving of advice. In [15], he said that a client could expect to be offered a facility for confidential discussion. In [48], he described the incident, on 31 August 2006, when a member of the bank's staff saw Mr Allen in the street outside the main branch and discussed with him the opening of M
a savings account and helped Mr Allen to complete the paperwork then and there. It is clear (and not surprising) that the judge did not regard this as a reasonable alternative method to the facility of giving advice that is available face to face to those who are able to gain physical access to the main branch. I have already referred to [51], where the judge noted that arrangements for the giving of advice at home had not been offered to Mr Allen. It is clear, therefore, that the judge found that the bank had not provided a reasonable alternative method of providing an important element of banking services (the giving of advice) to Mr Allen.

[27] However, there is a more fundamental point. The judge clearly considered that an important feature of the facilities provided at the

main branch was that they were available to customers in the traditional way across the counter. He did not spell this out, but it emerges most plainly from [54], where he said that "the closest the Bank can get to a reasonable alternative method is its assertion that there are other banks to which David could gain access". He then proceeded to explain why those other banks were not suitable, and there is no challenge to that part of the judgment.

[28] In my view, the judge was entitled to conclude that the fact that face-to-face facilities are available at the main branch was of itself an important element of the service provided by the bank at that branch. Many customers prefer distance banking. However, there are many who prefer the traditional way of doing things. The bank clearly recognises this, since otherwise it would not maintain some 2,300 branches throughout the country to which its customers have physical access. Although the judge did not refer to it in his judgment, Ms Catherine Casserley drew his attention in her closing submissions on behalf of Mr Allen to the 2006 Code of Practice, para 6.4 of which provides:

> The policy of the [DDA] is not a minimalist policy of simply ensuring that some access is available to disabled people; it is, so far as is reasonably practicable, to approximate the access enjoyed by disabled people to that by the rest of the public. Accordingly, the purpose of the duty to make reasonable adjustments is to provide access to a service as close as it is reasonably possible to get to the standard normally offered to the public at large.

[29] The theme of para 6.4 was replicated by Sedley LJ in *Roads v Central Trains Ltd* [2004] EWCA Civ 1541, in [13]:

> the policy of the Act, as I would accept, is what it was held to be by Mynors Ch (albeit by way of restricting the duty) in *In re Holy Cross, Pershore* [2002] Fam 1, para 105: "to provide access to a service as close as it is reasonably possible to get to the standard normally offered to the public at large"

[30] The public at large has physical access to banks in order to make use of traditional counter banking services and the bank's non-disabled customers have physical access to all its branches, including the main branch. The judge was entitled to conclude that the provision of the alternative methods of making those banking services available that were relied on by the bank was not a reasonable alternative, unless there was no reasonable way of affording Mr Allen physical access to the main branch. I shall deal with the question of whether there was a reasonable way of affording Mr Allen such access when I consider the third criticism made by Mr Lissack in [38] to [46] below.

[31] To summarise, I do not consider that the judge was led into error by failing to analyse the nature of the services that were being provided by the bank with sufficient rigour.

Second criticism

[32] The second criticism is directed in [16] of the judgment, where the judge said:

> Has the Bank made it impossible or unreasonably difficult for disabled people to use the service provided at Church Street and has it taken such steps as it would be reasonable to take to remove or to provide a reasonable means of avoiding or provide reasonable alternative methods of service? If it has is that failure justified? Has the failure to take steps made it impossible or unreasonably difficult for David to use the service.

[33] Mr Lissack submitted that this paragraph misstates the relevant test in section 21(2), which includes "it is the duty of the provider of that service to take such steps as it is reasonable, in all the circumstances of the case, for him to *have to* take in order to…." (emphasis added). Mr Lissack submits that: (i) [16] of the judgment ignores the words "have to"; (ii) it may be reasonable for a step to be taken, but this does not mean that it is unreasonable for the step not to be taken; and (iii) the error can be seen when the judge went on to find, in [33], that he did "not accept that [the single joint expert's] report has demonstrated that all methods of achieving disabled access are unreasonable".

[34] As to (i), Mr Lissack said that the words "*have to*" in "such steps as it is reasonable… for him to have to take" in section 21(2) require a strong test to be applied. It is not sufficient to show that it was reasonable for a step to be taken; before the duty can arise, it must be shown that it is reasonable for the step to *have to* be taken.

[35] The language of section 21(2) is plain and not difficult to understand. The duty arises once (in this case) a physical feature makes it impossible or unreasonably difficult for disabled people to use a service and there is a step that, in all the circumstances of the case, it is reasonable for the service provider to have to take to achieve one of the goals stated in paras (a) to (d). I accept that, by omitting the words "have to" in [16], the judge did not state the statutory test strictly accurately. However, for reasons that I set out later in this judgment, I am satisfied that he did in fact apply the correct test. I would add that I find it difficult to think of circumstances in which it would be reasonable for the service provider to take steps to achieve one of the goals stated in paras (a) to (d) but not also reasonable to require it to take those steps. In practice, therefore, it is unlikely to be necessary to apply a two-stage test of asking, first, whether it is a reasonable step and, second, whether it is reasonable to *have to* take it.

[36] As to (ii), I do not find it necessary to decide whether the fact that it is reasonable to require a person to do X necessarily entails the proposition that it is unreasonable not to require him or her to do X. In [16], the judge rightly asked whether the bank had taken such steps as it would be reasonable to take to satisfy section 21(2)(a) to (d). That test accurately reflects the language of section 21(2).

[37] As for (iii), [33] does not indicate that the judge had applied the wrong test. It was clear that Mr Allen's case was that: (a) there was a physical feature that made it impossible/unreasonably difficult for him to use the facilities available at the main branch; and (b) there were reasonable steps that the bank could have taken and had to take to remove that feature (by adopting the Owen scheme). The question of the reasonableness of the Owen scheme was therefore a central issue in the case. It was the bank's case that Mr Owen had demonstrated that "all the methods of achieving disabled access are unreasonable": see para 3.3.4 of the bank's closing submissions to the judge. Paragraph [33] of the judgment was a direct response to and rejection of this submission. It has no more significance than that. It does not reveal an application of the wrong test.

Third criticism

[38] The third criticism is that the judge had failed to apply the correct objective test. In [17] to [25] of his judgment, the judge considered the *attitude* adopted by the bank to the lobby scheme, when he should have applied a strictly objective test to the application of the duty to make adjustments in section 21. The bank was opposed to the lobby scheme. Similarly with regard to the Owen scheme, the judge concentrated on the bank's opinion. He said this in respect of the Owen scheme:

> 28. It appears to me that this proposal has never been considered seriously by the Bank and so much was obvious from the frank oral evidence from Mr Hannah who made it plain that the project engineers were not permitted to fully consider such a proposal since it involved a loss of an interview room which the Bank would not sanction.
>
> 29. It follows, in my judgement, that the Bank has not at any time been open to the proposal of this lift. It has failed to properly consider it. It is therefore impossible to say, as is suggested at paragraph 3.3.2 of the Bank's closing submissions that Mr Fairgrieve's evidence demonstrated that no appropriate solution had been found to resolve the difficulties at Church Street. Since the proposal has never been considered by the Bank it is not possible for the Bank to rely upon questions of structure security or safety. Mr Hannah's observation that "Safety was paramount above all other concerns" can play no part, in my judgement, in a scheme that has not been properly considered by the Bank.
>
> 30. I conclude therefore that the failure to consider the proposal must lead to a finding of discrimination under Section 20(1)(a) and Section 20(2)(a) unless the treatment is justified. The burden of showing justification falls upon the Bank.
>
> 31. The evidence from the Bank and particularly Miss Jean Cluness (and to an extent Miss Sarah Vigar) was that the Bank's facilities at Church Street are fully utilised. The interview rooms are regularly in use and the loss of such an interview room would have a considerable effect upon this busy branch.
>
> 32. In fact no detailed evidence of the use of any of the interview rooms was provided during the course of the hearing, nor is it contained within the substantial body of documentary evidence submitted. The reason for this

A omission is quite straight forward. The Bank simply declines to consider sanctioning the loss of one of its interview rooms. That failure cannot be justified on any evidence adduced before me.

33. I do not accept that Mr Owen's report has demonstrated that all the methods of achieving disabled access are unreasonable.

[39] Mr Lissack submitted that the proper test requires a consideration of all the circumstances, regardless of whether or not they had been considered by the provider of the services. The judge should have conducted a balancing exercise, in which the nature and extent of the difficulty faced by disabled persons in accessing the particular service sets the context. The factors to be taken into account would B include the anticipated cost of any adjustment and the disruption that would be likely to be caused by it; the fact that the bank had made good efforts at a high percentage of its 2,300 UK retail branches to remedy deficiencies in access (see [68] of the judgment); and the availability of other means of removing or reducing the difficulty and/or otherwise making the service available. All these factors needed to be considered together. For example, the availability of internet and telephone banking facilities should have been considered in conjunction with the cost and disruption of the installation of a lift, and not in isolation, as was done by the judge in [50].

[40] I accept that what is reasonable for the purposes of the test C in section 21(2) must be judged objectively. Mr Robin Allen QC did not seek to argue otherwise. In deciding whether it was reasonable for the bank to have to remove the offending feature by adopting the Owen scheme, the judge inevitably had to consider the grounds relied on by the bank to support its case that it was not reasonable for it to adopt the Owen scheme. However, provided that the judge assessed the reasonableness of those grounds objectively, his approach cannot be criticised.

[41] On a fair reading of the judgment, I consider that the judge had applied an objective test. In doing so, it was inevitable that he would make reference to the reasons advanced by the bank for its contention D that the reasonable adjustments duty did not require it to adopt the Owen solution and that the duty was fulfilled by the alternative methods on which it relied. Thus, in [45], he considered that the provision of alternative facilities at NatWest branches was not reasonable because Mr Allen's card had been rejected when he tried to access a NatWest branch. In [47], the judge considered ATMs and accepted that he could access cash at some ATMs. In [48], he referred to the incident when Mr Allen was given advice in the street outside the main branch. The judge considered the bank's resources in [20] and the cost and possible disruption that would be involved in the Owen scheme in [23], [26] and [27]. In [59] and [60], he also considered, but rejected E as unsuitable, other branches of the bank at The Moor and Broomhill. He also assessed the implications for the bank of the loss of an interview room.

[42] In his oral submissions, Mr Lissack did not place much weight on the argument that the judge had failed to apply an objective test. Instead, he concentrated his attack on [30] of the judgment, where the judge set out his conclusion. He submitted that even if the bank had failed to "consider" the Owen scheme, that could not logically lead to the conclusion that there had been discrimination contrary to section 20(2)(a).

[43] If [30] is considered in isolation, I am in no doubt that it is F legally flawed. In the end, Mr Allen was constrained to accept this. The fact that the bank did not consider the Owen scheme could not of itself justify the conclusion that there was unlawful discrimination contrary to sections 19(1)(b) and 21(2)(a). Put simply, it is a *non sequitur*. Does this error invalidate the central part of the judge's reasoning? Mr Lissack submitted that it does. The question that the judge had to decide on this part of the case was whether the bank was in breach of its duty in failing to remove the offending feature by adopting the Owen scheme.

[44] He dealt with this issue in [28] to [33]. He found that the sole reason why the bank objected to the Owen scheme was that it would involve the loss of an interview room. There were no objections of a technical nature or on the grounds that the installation of the lift would create health and safety problems or on the grounds of cost. G Indeed, Mr Hannah, a regional director of Faithful & Gould, which was appointed as project manager by the bank to manage its DDA programme, said in his oral evidence that the Owen scheme "does provide a technically competent solution and it overcomes the physical space constraints".

[45] With regard to the business implications of the loss of a room, the judge referred to the evidence of Ms Cluness, who is employed by the bank as a manager in the operations support team. She said that all eight interview rooms were fully used. However, she could not give any direct evidence on this. She conceded that the bank had not considered the possibility of reducing the size of the rooms so as to compensate H for the loss of one room or finding an additional room by opening up one of the upper floors or in some other way. Nor had there been any costing of the effect of losing one of the interview rooms. In the light of this evidence, the judge had concluded that, despite the loss of one of the existing interview rooms, it was reasonable to require the bank to adopt the Owen scheme. That is why the judge made the statement that he made in [32] and [33].

[46] As I have said, the judge should not have concluded that a breach of sections 19(1)(b) and 21(2)(a) was proved by the bank's failure to consider the Owen scheme. However, it is clear that what he had meant was that, by failing to consider the proposal, the J bank was not in a position to adduce evidence to show that the duty to make adjustments imposed by section 21(2) did not require it to remove the offending physical feature by adopting the Owen scheme. The judge knew that the Owen scheme would remove the offending physical feature and that the only argument advanced against it was that it would involve the loss of one of eight interview rooms. Whether a step is or is not unreasonable involves an exercise of judgment taking into account all the circumstances of the case. This court will interfere only if it is satisfied that the judge misdirected himself or reached a conclusion that no reasonable judge could have reached. Although the language used by the judge was unfortunate, I am not persuaded K that he misdirected himself. As I have said, he addressed the central questions, which were whether the reasonable adjustments duty required the bank to adopt the Owen scheme or whether it was a sufficient discharge of that duty that the bank made available the alternative facilities on which it relies. In my view, the judge addressed these questions objectively and reached a conclusion that was open to him on the evidence.

[47] By way of a postscript to ground 3 as a whole, I should add that it was common ground, for the purposes at any rate of this appeal, that the correct sequence of reasoning that should have been followed by the judge before he could conclude that there had been unlawful L discrimination contrary to sections 19(1)(b) and 21(2), involved asking the following questions:

(i) Was the bank a provider of services within the meaning of section 19(2)?

(ii) Did a duty arise under section 21(2), that is, did the physical feature make it impossible or unreasonably difficult for disabled people, or a class of disabled people, to use the service?

(iii) If so, were there reasonable steps that should be taken to remove, alter, provide a reasonable means of avoiding or a reasonable alternative method of service?

(iv) If so, did the bank take those steps? M

(v) If not, did the failure to take those steps make it impossible or unreasonably difficult for Mr Allen to use the service?

(vi) If the answer to (v) is yes, is the failure to take the steps justified in accordance with section 20(4)?

[48] With regard to (i), I have already dealt with the question of whether the bank was a provider of services and what these were. The judge dealt with this in [15]. With regard to (ii), the judge raised the section 21(2) question in [16], although unhelpfully in that paragraph he rolled it up with other questions. There was no dispute that there was a physical feature of the main branch (no disabled access) that made it impossible or unreasonably difficult for Mr Allen to use the facilities available at that branch. Step (iii) was critical in this case. Was the bank

in breach of the duty to take such steps as were reasonable in all the circumstances of the case for it to have to take in order to remove the feature by installing a lift (as was contended on behalf of Mr Allen) or was the duty discharged by the provision of reasonable alternative methods of making the facilities available (as was contended on behalf of the bank)?

[49] I have already referred to those parts of the judgment in which the judge dealt with the alternative method point and rejected the bank's contention that it had discharged its duty by reference to section 21(2)(d) and to those parts in which he dealt with the question of whether the bank was in breach of section 21(2)(a) in failing to adopt the Owen scheme so as to remove the offending feature.

Ground 1

[50] Mr Lissack drew attention to the fact that there are signs in the judgment that the judge believed this to be a claim on the grounds of less favourable treatment, contrary to section 20(1). He pointed in particular to the fact that, in the first draft of the judgment that he circulated to the parties, [30] included the words "under section 20(1)(a) and" between "discrimination" and "section 20(2)(a)". These words were deleted from the final approved version of the judgment. Mr Lissack also relied on the fact that the judge referred to *Lewisham* in [49]. *Lewisham* was a less favourable treatment case and, it is common ground, had no application to the present claim.

[51] It is rightly conceded by Mr Allen that there is some confusion in the judgment between section 20(1) and (2). However, any consideration of less favourable treatment by the judge cannot detract from his central findings and conclusions in respect of the duty to make reasonable adjustments. Most of the judgment and findings are correctly directed to the allegation of breach of the duty in sections 19(1)(b) and 21(2). I am satisfied that the unfortunate confusion (which, to some extent was contributed to by the presentation by counsel for the bank, who was not Mr Lissack) did not infect the judge's findings on the issue that was before him.

Ground 2

[52] Mr Lissack submitted that if the judge did have jurisdiction to deal with a less favourable treatment claim, he had erred by failing to apply *Lewisham* correctly.

[53] In my judgment, there is nothing in this ground of appeal. Had the claim been a less favourable treatment claim, it would have been necessary to consider *Lewisham*. But it was not.

Ground 4

[54] The complaint here is that the judge had failed to take account, or sufficient account, of the bank's submission that section 20(4)(a) was satisfied on the facts of this case. As we have seen, in [29] of his judgment the judge said: "Since the proposal has never been considered by the Bank it is not possible for the Bank to rely upon questions of structure security or safety. Mr Hannah's observation that 'Safety was paramount above all other concerns' can play no part, in my judgment, in a scheme that has not been properly considered by the Bank."

[55] Mr Lissack submitted that the judge had failed to consider justification at all. He said that the bank demonstrated that it had health and safety concerns in respect of the proposed platform lift and that these gave rise to a sustainable belief that alternatives ought to be considered.

[56] There is some confusion in this ground of appeal. There may have been safety concerns expressed with regard to the lobby scheme. However, it was the Owen scheme that formed the basis of the judge's decision in [28] to [33]. As I have said, the evidence was that there never was any safety or indeed security concern on the part of the bank in respect of the Owen scheme. That is the short and decisive answer to ground 4.

Ground 5

[57] The final ground of appeal is that the judge had generally erred in his approach to important issues of fact. It is said that the judge gave too much weight to the evidence adduced on behalf of Mr Allen and too little weight to the bank's evidence, in particular the evidence of Ms Cluness in respect of the interview rooms and the evidence of Mr Owen, in paras 8.6 and 9.8 of his report, regarding the disruption that would be created by the Owen scheme. It is also said that the judge had wrongly overlooked the substantial efforts undertaken by the bank to provide disabled access to disabled users across the UK.

[58] The judge took the evidence of Ms Cluness into account. He heard her give that evidence. The weight that he gave to it was a matter for him. He was entitled not to be persuaded by it for the reasons mentioned in [44] above. The disruption entailed by the Owen scheme was not a matter on which reliance was placed at the hearing.

[59] With regard to the bank's general record in discharging its obligations under the DDA, the judge said, in [68]:

> I have already made it plain that I find that the Bank has not approached this case and the difficulties which David faces with the zeal that one could have expected. I am prepared to accept that the Bank has about 2,300 branches in the United Kingdom (see the witness statement of Mr Ian Hannah at page 364 paragraph 5). I am also prepared to accept that the Bank had made good efforts at a high percentage of its branches to remedy deficiencies in access to disabled persons. It has not done so for the Church Street branch in Sheffield.

[60] In the light of this paragraph, the criticism that the judge had overlooked the bank's general efforts is unfounded. However, as the judge said, however well the bank may have performed generally in this difficult area, it had not discharged its statutory duty at the main branch in Sheffield.

Overall conclusion

[61] For the reasons that I have given, I would dismiss this appeal.

Hughes LJ said:

[62] I agree that this appeal should be dismissed for the reasons given by Dyson LJ.

Giving the second judgment, **Wall LJ** said:

[63] I agree that this appeal should be dismissed for the reasons given by Dyson LJ, whose judgment I have had the advantage of reading in draft. I add a short judgment of my own out of deference to Mr Lissack's careful and well-structured attack on the judgment.

[64] Like Dyson LJ, however, I do not believe that the judge had misdirected himself in any fundamental way. This was a physical access case. The claimant wanted to use counter facilities that the bank offered as part of its services at its main branch in Sheffield. The claimant could not access those facilities and, in my judgment, a duty on the part of the bank plainly thereby arose under section 21(2) of the DDA.

[65] There were, on the facts as found by the judge, reasonable steps that could be taken by the bank to make disabled access possible. The bank did not take those steps, giving as its reason, not the disproportionate cost of carrying out the work but simply the fact that it would lose the use of an interview room. The failure to take those steps undoubtedly made it unreasonably difficult for the claimant to use the services provided by the bank, and the judge was entitled to find that the failure on the bank's part to take the necessary steps was unjustified. In my judgment, therefore, the claimant was discriminated against by the bank within section 20(2) of the DDA and the judge was entitled so to find, albeit that the structure of his judgment and the means whereby the judge reached his conclusion are open to criticism.

[66] Speaking for myself, I acknowledge that there was a superficial attraction about Mr Lissack's analysis of the individual services that the bank could provide without taking the steps necessary to facilitate the claimant's access to counter services. However, it seems to me that this analysis broke down when applying the words of DDA, particularly section 19(1)(b) and 19(3)(a). In addition, and without in any way wishing to doubt the good faith of the bank, it does seem to me that the structure of Mr Lissack's submissions amounted, in reality, to a means for the bank to escape from the provisions of the Act certainly as a means of avoiding steps that were, in my judgment, reasonable for the bank to have to take in order to facilitate the claimant's access to the services that he wanted to use.

A [67] Like Dyson LJ, I do not give the words "have to" in section 21(2) of the DDA the significance that Mr Lissack would have us give them. In this respect, I agree with Dyson LJ's reasoning in [35] of his judgment. In addition, the words "have to", as it seems to me, apply to the reasonable steps that the service provider is required to take in order to make the necessary adjustment. In other words, if as was the B case here on the facts the steps are reasonable, the service provider is obliged to take them.

Appeal dismissed.

INSOLVENCY

Chen and another *v* Delaney

CHANCERY DIVISION

[2010] EWHC 6 (Ch)

8 January 2010

HH JUDGE CHARLES PURLE QC, SITTING AS A HIGH COURT JUDGE

Estates Gazette 27 March 2010

[2010] 12 EG 101

Insolvency — Sale of property — Sale at undervalue — Insolvency Act 1986 — Sale and leaseback entered into with purpose of putting assets beyond reach of judgment creditors — Sale price less than unencumbered freehold value owing to grant of lease — Section 423 of 1986 Act — Whether lease forming part of consideration Whether transaction at undervalue

The appellant purchased a residential property and simultaneously let it back to the vendors for a 21-year term at a rent of £500 per calendar month. The tenancy was exclusive to the vendors and unassignable. Owing to the leaseback, the sale price of £210,000 was substantially lower than the £275,000 unencumbered freehold value. The day before the sale, charging orders had been granted against the vendors' interest in the property to secure the payment of costs to the respondents, as judgment creditors. However, the transfer of the freehold took place before the charging orders could be protected on the register. The vendors used £160,435 of the sale proceeds to discharge their mortgage and they retained the balance.

The respondents applied to set aside the sale, under section 423 of the Insolvency Act 1986, on the ground that it was a transaction at an undervalue, made with the purpose of putting assets beyond their reach or otherwise harming their interests as judgment creditors. The district judge: (i) held the sale to be a transaction at an undervalue with the requisite purpose under section 423 of the Insolvency Act 1986; (ii) declared it to be void as against the respondents; and (iii) ordered the retransfer of the property to the vendors subject to a charge in the appellant's favour in the sum of £160,435, the amount of the previous mortgage, to have priority over the respondents' charging orders.

The appellant appealed. He contended that the transaction had not been at an undervalue because he had acquired only a freehold reversion subject to the tenancy, rather than the unencumbered freehold interest. Alternatively, the consideration for the sale of the freehold was both the £210,000 purchase price and the grant of the tenancy, which had a value to the vendors of £65,000. The respondents argued that the tenancy did not add to the value of the consideration because it was valueless, being exclusive and unassignable.

***Held*: The appeal was allowed. The impugned transaction was not a sale at an undervalue. Although a sale of the freehold reversion for £210,000 might be viewed as such, in the light of the fact that the vendors had an unencumbered freehold up to the point of sale and could have sold to another purchaser at the full unencumbered value, the consideration that the vendors received was not only the £210,000 but also a tenancy, which was far from valueless. Assuming the unencumbered value of the property to be £275,000, the tenancy had a value of £65,000 to the vendors, and it was irrelevant that the tenancy was unassignable. A transaction that depletes the vendor's assets will not be at an undervalue if the vendor, although receiving nothing saleable in return, none the less receives full value. Had the vendors sold the property to the appellant for £275,000, and then taken a long lease at a low rent of a comparable property at a market premium of £65,000, the payment of that premium would not have been a transaction at an undervalue. The instant transaction, which was equivalent to a sale and leaseback, should be regarded no differently; the premium value of the tenancy made up for any shortfall in the purchase price. The surrender value of a tenancy is measurable in money's worth and is to be regarded as part of the consideration for the purposes of section 423. Further, the fact that the tenancy was not registered did not affect its value at the date of the transaction because it was binding between the vendors and the appellant. Nor did the fact that it was subject to forfeiture on all the usual grounds, since it was not suggested that the tenancy was put in place as a device to engineer a forfeiture and it had the potential to last for the full 21-year term.**

The following cases are referred to in this report.

Abbey National Building Society v *Cann* [1991] 1 AC 56; [1990] 2 WLR 832; [1990] 1 All ER 1085; (1990) 60 P&CR 278; 22 HLR 360, HL
Agricultural Mortgage Corporation plc v *Woodward*, CA (1994) 70 P&CR 53; [1995] 1 EGLR 1; [1995] 04 EG 155; [1995] 1 BCLC 1
Inland Revenue Commissioners v *Hashmi* [2002] 2 BCLC 489
MC Bacon Ltd (No 1), Re [1991] Ch 127; [1990] BCC 78; [1990] BCLC 324, Ch
Moon v *Franklin* [1996] BPIR 196, Ch
National Bank of Kuwait SAK v *Menzies* [1994] 2 BCLC 306, CA
Phillips (liquidator of AJ Bekhor & Co) v *Brewin Dolphin Bell Lawrie Ltd (formerly Brewin Dolphin & Co Ltd)* [2001] UKHL 2; [2001] 1 WLR 143; [2001] 1 All ER 673
Redstone Mortgages plc v *Welch* [2009] 3 EGLR 71; [2009] 36 EG 98, CC

This was an appeal by the appellant, Paul Delaney, from a decision of District Judge Ingram, sitting in the Chancery Division, Birmingham District Registry, allowing an application by the respondents, Can Chen and An Xiang Du, under section 324 of the Insolvency Act 1986, to set aside a sale of residential property as a transaction at an undervalue.

Mark Anderson (instructed by The Wilkes Partnership, of Birmingham) appeared for the appellant; Aubrey Craig (instructed by Elliott & Co, of Birmingham) represented the respondents.

Giving judgment, **HH JUDGE CHARLES PURLE QC** said:

[1] This case concerns the effect of section 423 of the Insolvency Act 1986 (the Act) on sale and leaseback transactions. There are now a number of organisations that will purchase properties from homeowners on terms requiring the buyer to grant a tenancy or other interest back to the original homeowners. The purchase price is reduced to take account of the tenancy or other interest granted back. The original homeowners thus release some of the equity in their house, without having to move. The issue is whether and to what extent such a transaction may be vulnerable to attack as a transaction at an undervalue.

[2] The present case concerns the sale of a house (the property) by Mrs Ge Chiu and Mr Ji Feng Ding (the sellers) to Mr Paul Delaney (the buyer) at a price that was substantially lower than the unencumbered freehold value because of the grant back to the sellers of a tenancy of a day or two over 21 years. The address of the property is

10 Pioneer Close, Simpson Manor, Northampton, and it is registered at HM Land Registry. It was sold to the buyer for £210,000 on 8 May 2008. Of the sale proceeds, £160,435.15 was applied to discharge an existing mortgage taken out by the sellers. The balance (a little under £50,000) was paid to the sellers. The sale was challenged by judgment creditors of the sellers, Mrs Can Chen and Mr An Xiang Du (the victims) who say that a substantial purpose of the sellers was to put assets beyond their reach or otherwise harm their interests as judgment creditors. I emphasise that in describing the judgment creditors as "the victims", I in no way prejudge the outcome of this appeal. I do so because that is (as will appear) the language chosen by parliament to describe persons claiming (as they do) the benefit of section 423. For the avoidance of doubt, the expression "the victims" should be understood whenever it is used in this judgment in respect of the judgment creditors as a reference to their alleged status as such.

[3] The matter came before me by way of an appeal from the order of District Judge Ingram made on 9 July 2009. The district judge decided that the sale was a transaction at an undervalue with the requisite purpose. She declared that the transfer to the buyer of the property was void as against the victims. She also ordered the retransfer of the property to the sellers, although subjecting it to a charge with repayment to the buyer of the sum of £160,435.15 (the amount of the previous mortgage), which was to have priority over two charging orders made in the victims' favour to secure costs orders made on 7 May 2008 by Judge McCahill QC and by the district judge on the day she handed down judgment. The charging orders would, in the events that then ensued, have had no effect on the freehold without a retransfer since they were made against the interests of the sellers in the property, who transferred the freehold to the buyer before the charging order could be protected on the register.

[4] The litigation between the victims and the sellers has a long history. The proceedings (commenced in 2004) did not originally involve the buyer, who is a property investor and long-standing acquaintance of the sellers, although he gave evidence at the trial before Judge McCahill QC. Following that trial, the sellers were, on 22 November 2007, found to hold half of the shares in a company (Herbmagic (UK) Ltd) on trust for the victims, who were awarded their costs to be assessed. The sum of £30,000 (then standing in court) was ordered to be paid to the victims on account of their costs, and was paid. Subsequently, Judge McCahill QC, at a hearing on 7 May 2008 in Bristol, made an order for a further interim payment of £30,000 on account of the victims' costs and made a charging order *nisi* in respect both of the property and another buy-to-let property owned by the sellers. Since the order was not drawn up immediately by the court in Bristol, the charging orders could not then be protected on the register. With regard to the other buy-to-let property, there presently appears to be no, or no significant, equity in it following its remortgage by the sellers.

[5] The impugned transfer was the next day, 8 May 2008. The sellers sold the property to the buyer for £210,000. The sale was not with vacant possession (despite routine completion information suggesting otherwise) although it was with full title guarantee. The intention was that the sellers would continue to live there. In anticipation of the transfer, the buyer had previously, on 1 May 2008, granted a tenancy of the property to the sellers expressed to commence from 8 May 2008 "to and including the ninth day of 2029" at a rent of £500 per calendar month exclusive of council tax and water charges. The argument proceeded before the district judge (and before me) on the assumption that the intended reference was to 9 May 2029. The buyer was responsible for the maintenance of the external areas of the property, and for insuring the building. The sellers in turn covenanted to keep the interior of the property in good repair and condition. There was no provision for rent review during the term granted by the tenancy. The tenancy was also expressed to be an assured shorthold tenancy, as defined by section 19A of the Housing Act 1988 as amended. It was also expressed to be "exclusive" to the sellers and "unassignable". It was common ground before me that the transfer took effect subject to the tenancy, as plainly it did.

[6] Section 423 of the 1986 Act relates to transactions (defined in section 436 as including a gift, agreement or arrangement) entered into at an undervalue. If such a transaction is entered into by a person for the purpose of putting assets beyond the reach of a person (the victim) making a claim against that person or otherwise to prejudice the interests of the victim in respect of his claim, the court may make such order as it thinks fit for restoring the position to what it would have been had the transaction not been entered into. A person enters into a transaction at an undervalue if the value of the consideration, in money or money's worth, provided by the other party to the transaction is significantly less than the value, in money or money's worth, of the consideration provided by himself. This is not a complete or verbatim statement of the section, but a summary of the parts relevant to the present appeal.

[7] The district judge noted that the section required a comparison to be made between the value obtained by and the value of the consideration provided by the sellers. Founding herself on the observations of Millett J (as he then was) in *Re MC Bacon Ltd (No 1)* [1990] BCLC 324 (which related to section 238 of the Act), she said that both values must be considered from the point of view of the sellers. In that case, Millett J held that the granting of a debenture was not a transaction at an undervalue since the granting of the debenture did not deplete or diminish the value of the company's assets. The real complaint was not that the company had entered into the transaction at an undervalue, but that it had entered into it at all.

[8] The Court of Appeal in *National Bank of Kuwait SAK* v *Menzies* [1994] 2 BCLC 306 approved the approach of Millett J towards undervalue and applied it to a section 423 case.

[9] On the evidence before the district judge, the unencumbered freehold value of the property in May 2008 was £275,000. This was based on the buyer's own assessment (as recorded in para 34 of his witness statement) and a valuation of Mr Jonathan Carpenter FRICS obtained for the purpose of these proceedings by the buyer. Although given the opportunity to do so by [6] of Judge McCahill QC's order of 28 July 2008, the victims did not seek permission to adduce valuation evidence of their own, but based their submissions on the buyer's evidence (including Mr Carpenter's valuation, which was admitted into evidence without objection). The district judge held that there was an undervalue of £65,000.

[10] The buyer appeals this ruling on the ground that he did not acquire the unencumbered freehold, but a freehold subject to the tenancy. The tenancy had been granted prior to the 8 May transfer and took effect simultaneously with that transfer, which coincided with the commencement date of the term. Even if the tenancy had been created at the same time as the transfer, the position would have been the same. The transfer and tenancy were indissolubly bound up as part of the same overall transaction. Just as a purchaser dependent on a mortgage never acquires anything other than an equity of redemption (*Abbey National Building Society* v *Cann* [1991] 1 AC 56, in particular Lord Oliver of Aylmerton, at pp92F-93C) so the buyer in this case never acquired anything other than a freehold reversion. The full title guarantee was irrelevant to this conclusion because the tenancy was an incumbrance created by the buyer himself (and therefore well known to him) to take effect on completion as part of the bargain between himself and the sellers: see section 6(2) of the Law of Property (Miscellaneous Provisions) Act 1994.

[11] In *Redstone Mortgages plc* v *Welch* [2009] 36 EG 98*, HH Judge Worster held that the registered purchaser under a similar sale and leaseback transaction acquired nothing more than the freehold reversion, so that the rights of the tenants (who remained in actual occupation throughout) had priority over the rights of the purchaser's mortgagee. Although that decision is not binding on me (Judge Worster was sitting in Birmingham County Court), I find the reasoning compelling on this point, and agree with it.

---

* Editor's note: Also reported at [2009] 3 EGLR 71

[12] It follows, according to Mr Mark Anderson, for the appellant, that what the district judge should have focused on was the value of the freehold reversion acquired by the buyer and compared that with the consideration of £210,000 provided to the sellers. On the evidence, the freehold reversion was worth no more than £210,000, and may have been worth a lot less. According to Mr Carpenter, the value of the freehold reversion as at 16 September 2008 was £115,000. This was, unfortunately, the wrong valuation date for the purpose of the statutory comparison and Mr Carpenter explained that this was merely the investment value, commenting that a prudent investor buyer would probably pay a higher figure (albeit lower than the vacant possession valuation) in the light of the possibility of the tenancy terminating before the end of the term. Even taking that into account, there is nothing in the evidence to justify a drop in value of as much as almost 50% between May and September 2008, although it cannot be doubted that the market was on a downward trend during that period. More importantly, the comparable transactions considered by Mr Carpenter (which were actual quotations obtained by the sellers in April 2008 from UK-Property Buyers, an organisation specialising in sale and leaseback transactions) indicated that the price of £210,000 for the freehold reversion was, in the present case, a fair one. There was therefore, according to Mr Anderson, no transaction at an undervalue.

[13] Mr Anderson, in advancing the appeal in that way, was inviting me to compare the value of what the buyer got with what the buyer paid. For my part, I accept that, thus regarded, the buyer paid full value, treating him as having acquired no more than the freehold reversion; £210,000 was a fair price for that interest.

[14] However, Mr Aubrey Craig, for the victims, urged me to look at the matter, in line with *MC Bacon*, from the sellers' perspective. What the sellers had immediately before the impugned transaction was an unencumbered freehold worth £275,000. What they got in return was £210,000 and a tenancy having no value because it was exclusive and non-assignable. Thus regarded, there was indeed, as the district judge held, a significant undervalue. Mr Anderson countered this by saying that what the sellers transferred was no more than what the buyers received, which was the freehold reversion. Any diminution in value occasioned by their rights under the tenancy was merely incidental and should, consistently with the *MC Bacon* decision, be ignored.

[15] I cannot without more accept Mr Anderson's analysis of the position. The sellers had an unencumbered freehold down to the point of sale to the buyer. They could have sold the property to anyone else at its full unencumbered value, notwithstanding the tenancy agreement put in place on 1 May in anticipation of the subsequent sale to the buyer. Their decision to sell to the buyer had the result that the purchase price was lower than it needed to have been, and the sale of the property for £210,000 was on this footing at an undervalue. The same result would follow had the transaction followed the more conventional course of the transfer and leaseback being executed at the same time. The sellers would (on the approach presently under consideration) have transferred the property at an undervalue by selling on those terms.

[16] However, Mr Anderson had another point. The consideration that the sellers received was £210,000 and a tenancy that was far from valueless. On the footing that the unencumbered value of the property was £275,000, the tenancy must (on my finding that £210,000 was a fair price for the freehold reversion) have had a value of £65,000, since that is effectively what the sellers paid for it by receiving a discount on the price of an unencumbered freehold. The fact that the tenancy was unassignable is irrelevant. The tenancy had a value to the sellers of £65,000, and that is enough.

[17] There are many instances where a transaction may be entered into that depletes the transferor's assets, but that is not sufficient to stigmatise the transaction as being at an undervalue. Thus, a debtor facing bankruptcy may decide to dissipate substantial amounts on food, drink and other aspects of high living. He gets nothing saleable in return, but receives full value. Whatever else may be said concerning such a transaction, it is not at an undervalue. The debtor pays full value for the consumables or services in question. Once consumed or enjoyed, there is nothing left and what he pays is irretrievably lost to his creditors. The complaint would, however, as in *MC Bacon*, be that the transaction was entered into at all, not that it was entered into at an undervalue.

[18] Closer to the present case, had the sellers sold the property to the buyer for £275,000, and then taken a long lease at a low rent of a comparable property, paying a market premium of £65,000, the payment of that premium would not be a transaction at an undervalue. It would make no difference that the tenancy once granted would not be assignable. The premium element would be no different in principle from part-payment in advance for any other services, such as hotel accommodation. The result can be no different in the case, as here, of a sale and leaseback, where the premium value of the tenancy made up for any shortfall in the purchase price. In those circumstances, the sale at £210,000 was (on the present example) the equivalent of a sale at £275,000, with a leaseback at a premium of £65,000.

[19] I note in this connection that the monthly rent of £500 throughout the entire 21-year term was a small yield (a little under 2.2%) for a property valued with vacant possession in May 2008 at £275,000. That alone suggests that the tenancy had a premium value, and this is confirmed by the quotations obtained from UK-Property Buyers. It is, of course, possible that the mere existence of a tenancy, even at a rack-rent, will, in any given case, have a depressive effect on value. However, evidence is needed to establish that fact, and none was forthcoming in this case, despite the opportunity the victims had to seek permission to adduce valuation evidence. The legal burden was on the victims to establish that the transaction in question was at an undervalue, which means that they had to show that the tenancy granted to the sellers had a premium value of less than £65,000. They have not discharged that burden. Had they done so, I might then have needed to consider whether any diminution in value was merely incidental, as that expression is used in *MC Bacon*. The Court of Appeal left open in *Agricultural Mortgage Corporation plc* v *Woodward* [1995] 1 BCLC 1* whether incidental diminution in value takes the case out of section 423. I need not decide whether it does, since the victims have not established even an incidental diminution in value in the present case.

[20] The reference to *Agricultural Mortgage* is instructive. A farmer granted a tenancy of mortgaged property to his wife at what was assumed, for the purpose of summary judgment proceedings, to have been at full market rent. The effect was to diminish the value of the mortgaged property to the detriment of the mortgagee. The fact that the tenancy was at a market rent was not sufficient to save the transaction from being at an undervalue, since the wife also received from the transaction the benefit of the surrender value, which exceeded the value of the consideration provided by her.

[21] That case is therefore authority for the proposition that the surrender value of a tenancy is measurable in money's worth and is to be regarded as part of the consideration for the purposes of section 423. Since, moreover, the tenancy that the district judge was considering in the present case appears to have been for less than the full market rent, the district judge needed to be satisfied that the premium element, as well as any surrender value, was less than £65,000.

[22] The district judge did consider whether a value could be put on the tenancy, and, taking her lead from *Phillips (liquidator of AJ Bekhor & Co)* v *Brewin Dolphin Bell Lawrie Ltd (formerly Brewin Dolphin & Co Ltd)* [2001] UKHL 2; [2001] 1 WLR 143, concluded that the tenancy had only a speculative value, and that meant that the burden was on the buyer to produce evidence establishing that the tenancy had a value to the sellers of at least £65,000.

[23] I do not consider that a proper appreciation of *Phillips* justified this conclusion. That case concerned the sale of a business coupled with a subletting of computer equipment under a collateral agreement in the face of an absolute bar in the headlease. The House of Lords held that the consideration included the collateral agreement, but that, because the collateral agreement was so precarious as to make it worthless and the head lessors had almost immediately declared it to be a repudiatory breach, it was for the party relying on the collateral

---

* Editor's note: Also reported at [1995] 1 EGLR 1

agreement to establish its value. In the present case, there was nothing precarious about the tenancy, and no justification therefore for throwing any burden on the buyer. The burden of proving an undervalue under the section is on the victims. The circumstances in *Phillips* were such as amply to discharge that burden and shift the evidential burden onto the buyer. That is not the case here. The tenancy had the potential to last for at least the whole of the 21-year term. There was no basis for concluding that it had no value to the sellers.

[24] Moreover, the buyer did adduce valuation evidence before the district judge, namely the buyer's own valuation evidence, which I have summarised, and the comparables embodied in the UK-Property Buyers' quotations. The district judge discounted the comparables because she treated the UK-Property Buyers' quotations as being based on a 12-month lease, not a 21-year lease. That would, if anything, suggest a higher premium or surrender value in the case of the tenancy agreement granted to the sellers. However, perusal of the quotations and accompanying correspondence from UK-Property Buyers showed that, although the proposed assured shortholds (which also contained a covenant against subletting or parting with possession) were for 12 months in the first instance, UK-Property Buyers would enter into a binding commitment to allow the tenancy to roll forward indefinitely, for as long as the sellers wanted to live in the property.

[25] The district judge also appears to have accepted the following submissions of the victims:

(i) Since the tenancy was not assignable, it had no value in money or money's worth. I reject this submission for the reasons given above.

(ii) There was no provision for rent review. This, however, must have increased rather than reduced any premium or surrender value.

(iii) The tenancy was subject to forfeiture on what was described as "all the usual grounds". It is not suggested, however, that the tenancy was put in place as a device to engineer a forfeiture. The buyer's valuation evidence took into account the terms of the tenancy, and the UK-Property Buyers' proposed tenancy also contained provisions for forfeiture.

(iv) The tenancy was not registered. This seems to me to be irrelevant because the tenancy was binding between the sellers and the buyer and non-registration did not affect the tenancy's value at the date of the transaction.

[26] In the circumstances, I consider that there was no evidence before the district judge entitling her to conclude (as she did) that the impugned transaction was at an undervalue. It follows from this that section 423 has no application, and the appeal must be allowed.

[27] Having reached that conclusion, it is not strictly necessary for me to express a conclusion on the second aspect of the case, namely whether the requisite statutory purpose was established. I will, however, deal with the matter briefly.

[28] The district judge correctly noted that what mattered in this context was the sellers' purpose (*Moon* v *Franklin* [1996] BPIR 196), which need be only a substantial as opposed to the predominant purpose: *Inland Revenue Commissioners* v *Hashmi* [2002] 2 BCLC 489. As to that, the evidence of one of the sellers (Mr Ding), and the evidence of the buyer, was that they did not know of the charging order made on 7 May 2008, when the transaction was completed on 8 May 2008, and that the reason that date was chosen was because a fixed-rate mortgage was coming to an end. The district judge rejected this evidence, preferring, she said, the victims' evidence. However, the victims gave no evidence that was relevant to the sellers' or buyer's state of mind, and I have considerable doubt as to whether it was open to the district judge to reject the sellers' and buyer's evidence in this respect, since there was no cross-examination (as there would have been had it been sought) on these or any other points.

[29] None the less, I do consider that the evidence before the district judge justified the conclusion that the sellers' purpose was to harm the interests of creditors generally, which would of course include the victims, who were known to be applying for a further interim payment. Mr Ding's witness statement recorded that it became clear after 2007 that the property had to be sold to pay off debts, legal costs and to fund further legal action. He then explained the difficulties that the sellers had encountered in their attempts to sell the property, and how they came to turn their minds to a sale and leaseback. With regard to the sale to the buyer, Mr Ding explained, at the very end of his witness statement, that "the sale was done with the best interest of my family to retain a roof over their heads and to reduce my debts and outgoings; it was not to cheat anyone or deprive any creditors".

[30] The reference to retaining a roof over his family's heads clearly recognised that, absent the sale and leaseback, the roof may have been blown away to satisfy the demands of creditors. The effect of the sale and leaseback was, as was obvious, to make the property, including the leasehold interest, immune from the claims of creditors, including the victims, and this must have been, on a proper understanding of Mr Ding's evidence, one of the sellers' purposes as well as its effect. It is noteworthy also that, despite the concluding words of Mr Ding's witness statement, the sellers have not paid any part of the £30,000 due under the order of 7 May 2008. In those circumstances, the district judge was entitled to conclude that the requisite purpose was made out, whether or not the charging order came to the sellers' attention before completion and even though there may have been other reasons for completing on 8 May 2008. I would, on Mr Ding's evidence, reach the same conclusion myself.

[31] Despite my upholding of the district judge's conclusion on the issue of purpose, the appeal is allowed because I consider that the burden was on the victims to establish the requisite undervalue, which they have not done.

[32] I should finally mention that had I upheld the decision on undervalue, Mr Anderson would have wanted to address me on the appropriate remedy. It will be noted that the district judge, treating the impugned transfer as void (although that word is nowhere to be found in section 423) made no provision for the repayment to the buyer of the purchase price of £210,000, but limited his recovery (by way of subrogated charge) to the mortgage paid off from the proceeds of sale. I have real doubts as to whether it was appropriate for the district judge to go that far. The real complaint with regard to the surplus proceeds is that they have been disbursed or dissipated in some way. I can, as at present advised, see no justification for visiting the consequences of that on the buyer because that would presumably have happened even if the sale had been for £275,000, and he had no control over the destination of the proceeds. Since, however, I have heard no detailed argument, which was deferred pending my decision on the main appeal, it is not necessary for me to rule on the point.

Appeal allowed.

Goldacre (Offices) Ltd *v* Nortel Networks UK Ltd (in administration)

CHANCERY DIVISION: COMPANIES COURT

7 December 2009

[2009] EWHC 3389 (Ch)

HH JUDGE CHARLES PURLE QC, SITTING AS A HIGH COURT JUDGE

Estates Gazette 6 March 2010

[2010] 09 EG 168

Insolvency — Administration — Administration expenses — Payment of rent Insolvency Rules 1986 — Company in administration holding premises on long leases — Administrators occupying part of premises for purposes of administration — Whether full rent continuing to be payable to landlord as expense of administration — Rule 2.67(1) of 1986 Rules — Application of salvage principle

The respondent company held commercial premises on two long leases under which rent was payable to the applicant landlord quarterly in advance. Parts of the property were sublet. After the respondent went into administration, the administrators used part of the premises to carry out the administration and continued to pay rent to the applicant. The right to receive rent from the subtenants was also transferred to the applicant under section 6 of the Law of Distress Amendment Act 1908 and, although the administrators continued to receive some rents, these were passed on to the applicant.

The applicant sought a court ruling that the administrators, having decided to use part of the premises for the beneficial outcome of the administration, were liable to pay the full rent as it fell due as an expense of the administration, pursuant to the Insolvency Rules 1986. The administrators contended that any payments should be tailored to their actual use of the premises, with the proportion of rent payable to be assessed according to the floorspace that they occupied.

***Held*: The application was allowed. (1) The "salvage" or "liquidation expenses" principle, under which liquidators are held liable to pay rent as a liquidation expense where they make use of or retain possession of leasehold premises for the benefit of the liquidation, also applies to administration. The relevant expenditure falls within r 2.67(1) of the 1986 Rules, either as expenses properly incurred by the administrators in performing their functions, within r 2.67(1)(a), or as necessary disbursements by the administrators in the course of the administration, within r 2.67(1)(f). The word "necessary" in the latter provision extends to a payment that ought to be made in fairness and justice and is not confined to a situation in which a specific threat for redelivery has been made; it is therefore apt to include a case where the salvage principle applies. The relevant rules, when properly construed in accordance with the salvage principle, compelled the administrators to pay the rent and the court should direct them to act in accordance with the mandatory requirements of the rules. (2) The administrators were not entitled to tailor their payments to the use they were making of the premises. Where administrators adopt a contract for conducting the administration in a beneficial manner, all liabilities arising under that contract after the date of adoption are entitled to priority. Administrators who elect to hold leasehold premises may do so only on the terms and conditions contained in the lease, such that any liability incurred while the lease is being enjoyed or retained for the benefit of the administration is payable in full as an administration expense. Accordingly, the administrators were obliged to pay the next quarter's rent in full when it fell due, as one of the costs and expenses of the administration, and it did not fall to be apportioned in the event that the administrators vacated the premises during that quarter. The rent would continue to be payable as an administration expense quarterly in advance under the terms of the two leases for so long as the administrators retained or used any part of the premises demised under each lease for the benefit of the administration. The court had no discretion as to whether to declare the rent a liquidation expense, although it might have a discretion as to whether to grant a remedy to obtain payment.**

The following cases are referred to in this report.

ABC Coupler & Engineering Co (No 3), Re [1970] 1 WLR 702; [1970] 1 All ER 650, Ch
Atlantic Computer Systems plc, Re [1992] Ch 505; [1992] 2 WLR 367; [1992] 1 All ER 476; [1991] BCLC 606, CA
Centre Reinsurance International Co v *Curzon Insurance Ltd*; *Centre Reinsurance International Co* v *Freakley* [2006] UKHL 45; [2006] 1 WLR 2863; [2006] 4 All ER 1153
Davis (S) & Co, Re [1945] Ch 402
Ellis v *Rowbotham* [1900] 1 QB 740, CA
Exeter City Council v *Bairstow*; *sub nom Trident Fashions plc, Re* [2007] EWHC 400 (Ch); [2007] 4 All ER 437; [2007] 2 P&CR 8; [2007] BCC 236; [2007] 2 BCLC 455
Lehman Brothers International (Europe) (in administration), Re; *sub nom Lomas* v *Rab Market Cycles (Master) Fund Ltd* [2009] EWHC 2545 (Ch)
Levi & Co Ltd, Re [1919] 1 Ch 416, Ch
Linda Marie Ltd (in liquidation), Re (1988) 4 BCC 463, Ch
Lundy Granite Co, ex parte Heavan, Re (1871) LR 6 Ch App 462
Oak Pits Colliery Co, Re (1882) 21 Ch D 322
Powdrill & Lyle (joint liquidators of Kentish Homes Ltd) v *Tower Hamlets London Borough Council* [1993] BCLC 1375, Ch
Powdrill v *Watson* [1995] 2 AC 394; [1995] 2 WLR 312; [1995] 2 All ER 65, HL; [1994] 2 All ER 513, CA
Shackell & Co v *Chorlton & Sons* [1985] 1 Ch 378, Ch
Spring Valley Properties Ltd v *Harris*; *sub nom Salmet International Ltd (in administration), Re* [2001] BCC 796, Ch
Sunberry Properties Ltd v *Innovate Logistics Ltd (in administration)*; *sub nom Innovate Logistics Ltd (in administration)* v *Sunberry Properties Ltd* [2008] EWCA Civ 1321; [2009] 1 BCLC 145; [2009] BCC 164
Toshoku Finance UK plc (in liquidation), Re; *sub nom Inland Revenue Commissioners* v *Kahn*; *Kahn* v *Inland Revenue Commissioners*; *Khan* v *Inland Revenue Commissioners* [2002] UKHL 6; [2002] 1 WLR 671; [2002] 3 All ER 961

This was the hearing of a claim by the applicant landlord, Goldacre (Offices) Ltd, against the respondent tenant, Nortel Networks UK Ltd, a company in administration, for a ruling as to the liability of the administrators, under the Insolvency Rules 1986, to pay rent on demised premises used for the purposes of the administration.

Stephen Jourdan QC and Blair Leahy (instructed by Olswang LLP) appeared for the applicant; William Trower QC and David Allison (instructed by Herbert Smith LLP) represented the respondent.

Giving judgment, **HH JUDGE CHARLES PURLE QC** said:

[1] In this case, I have to decide whether the rent that is soon to fall due, and future rent thereafter, in respect of property held by the respondent company (now in administration) in Harlow is payable as an expense of the administration. There are two long leases, both predating the date of the administration. The premises in question have, since the date of the administration, been used to an extent by the administrators for the more efficient conduct of the administration and rent has been paid to date, albeit late so far as the last September quarter was concerned. It is accepted on behalf of the administrators that interest in respect of that late payment will be paid. The real issue therefore is a future one, but the problem is a pressing one because the commercial needs of the applicant, which is the landlord of the premises in question, require an immediate decision. Moreover, the administrators wish to know where they stand, since that may inform their future conduct of the administration. I would ideally have wished to take more time to prepare this judgment, but the parties expressed a preference for speed of decision rather than perfection of language. I have nevertheless reached a clear view and have, in approving the transcript, taken the opportunity of refining my reasoning.

[2] Although the administrators are using part of the premises in respect of both of the demises in question, they are using only part and

A it is a relatively small part. There are, in addition, subtenants in respect of other parts. With regard to those subtenants, notices have been served under section 6 of the Law of Distress Amendment Act 1908, having the effect of transferring the company's right to receive rent to the landlord. As it happens, some rents are still coming in to the administrators, as I understand the position, but they are passing those over to the landlord. In those circumstances, it cannot be said that the company in administration is, in any realistic sense, in receipt of rents and profits, so that possession of the premises in that sense can be ignored. What I am concerned with is the relative minor use that the administrators are making of the properties in question.

B [3] Mr Stephen Jourdan QC, for the applicant landlord, submitted that, on established authority, once the administrators decide, as they have done, to continue to use any part of the properties for the beneficial outcome of the administration, they are liable to pay the rent as it falls due in full as an administration expense. He said that this follows nowadays from the insolvency rules that were adopted following the enactment of the Enterprise Act 2002, and that the previous decision of Blackburne J in *Spring Valley Properties Ltd* v *Harris* [2001] BCC 796 should not be followed in respect of this administration, which is governed by the Insolvency Rules 1986 as now amended. He prayed in aid the more recent decision of David Richards J in *Exeter City Council* v *Bairstow*
C [2007] EWHC 400 (Ch); [2007] 4 All ER 437; [2007] 2 BCLC 455; [2007] BCC 236. In that case, David Richards J, after a careful analysis both of the rules and of the previous decision of the House of Lords in *In Re Toshoku Finance UK plc (in liquidation)* [2002] UKHL 6; [2002] 1 WLR 671 (which related to a company in liquidation), held that the expenses regime set out in the rules is mandatory in the case of administrations as well as liquidations. David Richards J was struck in my judgment, rightly so by the closeness of the wording between what is now r 2.67 (applicable to administrations) with the wording in r 4.218, considered by the House of Lords in *Toshoku*. He held, in accordance with his ruling, that rates in an administration are payable
D as an administration expense. He also expressed the view that that was so in respect of unoccupied as well as occupied property, although that is no longer so as a result of subsequent delegated legislation that came into effect on 1 April 2008. The House of Lords in *Toshoku* held that corporation tax was payable in a liquidation as a liquidation expense, since it was made necessary by statutory enactment, notwithstanding that the income that was taxed was only deemed income and not income in fact received by the company in liquidation.

[4] I agree with Mr Jourdan's submission that the matter is now to be considered exclusively by reference to the rules and that if the rental liability falls within the rules, that is payable as a matter of
E mandatory obligation not as a matter of discretion, either on the part of the administrators or on the part of the court.

[5] *Toshoku,* although a case on corporation tax, considered what is referred to sometimes as the salvage principle, sometimes as the *Lundy Granite* principle and sometimes as the liquidation expenses principle. The main speech in the House of Lords was given by Lord Hoffmann, all other four law lords agreeing, although in Lord Hobhouse's case with uncharacteristic diffidence. Lord Hoffmann, in a masterly analysis of the previous law, traced the history of the *Lundy Granite* principle: so named after *Re Lundy Granite Co, ex parte Heavan* (1871) LR 6 Ch App 462. Under that principle, liquidators are held liable to pay rent as
F a liquidation expense where the liquidators make use of or retain, for the benefit of the liquidation, possession of leasehold premises. Express reference was made and approval given to the judgment of Lindley LJ (a judge of at least as great eminence in this field as Lord Hoffmann) in *Re Oak Pits Colliery Co* (1882) 21 Ch D 322. In [26] of his speech, Lord Hoffmann referred to the following citation of Lindley LJ, at p330:

> When the liquidator retains the property for the purpose of advantageously disposing of it, or when he continues to use it, the rent of it ought to be regarded as a debt contracted for the purpose of winding up the company, and ought to be paid in full like any other debt or expense properly incurred by the liquidator for the same purpose…

[6] Lord Hoffmann continued in [27]: G

> My Lords, it is important to notice Lindley LJ was not saying that the liability to pay rent had been incurred as an expense of the winding up. It plainly had not. The liability had been incurred by the company before the winding up for the whole term of the lease. Lindley LJ was saying that it would be just and equitable, in the circumstances to which he refers, to treat the rent liability *as if* it were an expense of the winding up and to accord it the same priority.

[7] That principle (if it applies in administrations at all) plainly applies to the present case and has effectively been acknowledged as applicable by the payment of rent hitherto. As Lord Hoffmann went on to explain, although the rule emerged historically in the context of the
exercise of a discretion (whether or not to allow forfeiture proceedings H to be brought or distress to be levied), it has evolved into a principle that, as he put it in [38], "does not involve an exercise of discretion any more than the application of any other legal principle to the particular facts of the case". He held that the rules and in that case he was concerned with r 4.218 applicable to liquidations were to be construed to include debts that, under the *Lundy Granite* principle, were deemed to be expenses of the liquidation, and concluded that, ordinarily, this meant that debts, such as rent under a lease, would be treated as coming within para (a) of r 4.218(1), but that the principle might possibly enlarge the scope of other paragraphs as well. J

[8] Paragraph (a), as it then stood, referred to "expenses properly chargeable or incurred by the official receiver or the liquidator in preserving, realising or getting in any of the assets of the company". In a liquidation, therefore, on the authority of Lord Hoffmann and the remainder of the House of Lords, rent, where the *Lundy Granite* principle applies, will ordinarily fall within (a) and will be payable mandatorily.

[9] So far as administration is concerned, the relevant rule, as I have said, is r 2.67(1). Two subparagraphs have been highlighted before me, (a) and (f). Rule 2.67(1) contains a list of expenses payable in the following order of priority: (a) refers to "expenses properly incurred
by the administrator in performing his functions in the administration K of the company"; (f) refers to "any necessary disbursements by the administrator in the course of the administration (including any expenses incurred by members of the creditors' committee or their representatives and allowed for by the administrator under r 2.63, but not including any payment of corporation tax in circumstances referred to [in a later subparagraph that I need not consider]". Given the closeness of the opening words of r 4.218(1)(a) as it stood at the time of *Toshoku* (and, indeed, as the rule now stands, although the relevant part is now para 4.218(3)(a)(ii) and a number of other words have been added) to the opening words of r 2.67(1)(a), my inclination would be
to regard rent as falling within r 2.67(1)(a). It seems to me that that L subparagraph, were the matter free from authority, would be apt to refer to all the expenses incurred by the administrator in performing his functions in the administration of the company, making no distinction for this purpose between expenses incurred on his own account, for which he was entitled to be reimbursed, and expenses incurred on behalf of the company.

[10] None the less, that construction did not find favour with David Richards J in *Exeter*. He relied on the closeness of wording between section 19(4) of the Insolvency Act 1986 (the 1986 Act) and the wording of r 2.67(1)(a). The House of Lords previously, in *Centre
Reinsurance International Co* v *Freakley* [2006] UKHL 45; [2006] M 1 WLR 2863, had ruled that section 19(4) applied only to liabilities incurred personally. Section 19 seems to me to be different, however, since the contrast (in cases where that section still applies) is made between "remuneration and expenses" (to which section 19(4) applies) that are due to the administrator, and "sums payable in respect of debts or liabilities incurred" (to which section 19(5) applies) that are due to outside creditors, and are expressly given priority over sums payable to the administrator under section 19(4), as are sums payable under adopted contracts of employment where section 19(6) applies. There is no similar contrast in r 4.218 or r 2.67. Moreover, as Blackburne J observed in *Salmet*, section 19(4) is concerned merely with the ranking

A of remuneration and expenses (as a class) as against the claims of floating charge holders.

[11] It is, however, not necessary for me to reach a final view on the point because if the rent does not fall within (a), it falls, in my judgment, within (f) the disbursement is a necessary one because the application of the *Lundy Granite* principle requires rent to be paid. It is clear from other provisions of the 1986 Act that the administrator has the power to make payments whether or not they are necessary in any ordinary sense of that word. They may, for example, be merely "incidental" to the performance of his functions within para 13 of Schedule 1, without being "necessary" (both words being used in that
B paragraph), and his powers under para 59(1) of Schedule B1 enable him to do (and therefore to incur liability in respect of) anything "necessary or expedient" for the management of the affairs, business and property of the company. It would be surprising if necessary expenses had a measure of priority, but liabilities incurred merely incidentally or through expediency had no priority whatsoever. That suggests either that (a) should be construed more broadly or "necessary" given an extended meaning, or both.

[12] In construing what is "necessary", I was referred to the decision of Briggs J in the recent *Re Lehman Brothers International (Europe) (in administration)* decision [2009] EWHC 2545 (Ch), where Briggs J
C (although his remarks on this point were clearly *obiter*) considered that necessity extended to a payment that ought to be made in fairness and justice to in that case a counterparty that, because of administration, had been or would be prejudiced by Lehman Brothers' retention of securities beyond the date on which they should have been redelivered. He also confirmed that necessity was not confined to a situation in which there had been a specific threat for redelivery. I agree with that observation, which was not, in my judgment, confined to the facts of that particular case. In the present case, there has been no threat by the applicant landlord to bring forfeiture proceedings, although Mr Jourdan said that he would seek permission to do so if that were thought to be
D necessary. In my judgment, it is not, and the application of the *Lundy Granite* principle does not require it.

[13] In those circumstances, whatever the precise extent of the meaning of the word "necessary", it is plainly apt, in my judgment, to extend to a case where the *Lundy Granite* principle applies. Briggs J, in [98], considered the *Lundy Granite* principle to apply to administrations. The difficulty in its application to the *obiter* part of his judgment was that the securities had not been used for the purpose of the administration within the strict confines of the *Lundy Granite* principle: see [109]. That is not a difficulty facing me.

[14] Mr William Trower QC, for the administrators, said that a
E disbursement can be regarded as necessary only if the administrators choose to make it or if the court, founding itself on some proper jurisdictional basis, orders it. He said that the court had no jurisdictional basis in the present case for ordering the payment of rent, there being no unfair harm that would trigger the court's jurisdiction under para 74 of Schedule B1, or no other good and compelling reason for the court to give directions to that effect to the administrators under paras 63 or 68(2) of Schedule B1.

[15] Mr Trower also said that there is a major distinction between r 4.218, as it applies to liquidations, and r 2.67(1), as it applies to administrations, because of the absence in r 2.67(1)(a) of the word
F "chargeable", which, he said, meant chargeable to the estate. For my part, I would naturally read "chargeable" in r 4.218(3)(a)(ii) as chargeable by the liquidators in preserving etc, as the rule goes on to say. I do not consider the absence of that word in r 2.67(1)(a) to have any significance in the present context.

[16] I do not accept the reasons that Mr Trower has advanced for not regarding rent as an administration expense in the present case. The *Lundy Granite* principle seems, in my judgment, clearly to apply, and the court's jurisdiction to order payment derives from the relevant rules that, properly construed in accordance with the *Lundy Granite* principle, compel payment. Any failure on the part of the administrators to recognise this unfairly harms the applicant landlord, and there can be no doubting the appropriateness of the court directing the administrators to act in accordance with the mandatory requirements of the rules, G
properly construed.

[17] I am urged also to construe the rules in accordance with the underlying importance, which I do not dispute for one moment, of the rescue culture on which the 1986 Act is based. Mr Jourdan, however, pointed out that in a number of the older cases in which the *Lundy Granite* principle was developed, liquidators were in fact carrying on business for the purpose of the winding-up in an endeavour to sell the business as a going concern, which is what administrators now do, certainly in the early stages, very often. That did not prevent the *Lundy Granite* principle from emerging and is no reason, in my judgment, for excluding it in the present case. Since, moreover, H
administrators now have the power (with the permission of the court) to make distributions to unsecured creditors under para 65(3) of Schedule B1, administrations may come to resemble liquidations, and it would be surprising were the *Lundy Granite* principle to apply to the one and not the other.

[18] Mr Trower did not of course say that the administrators should be entitled to occupy or use the premises for nothing. What he said is that payments should be tailored to the use that they are making. Mr Jourdan answered that by reference to *Powdrill* v *Watson* [1995] 2 AC 394 and the cases therein referred to. At p450E, Lord Browne-Wilkinson said this in the context of contracts of employment adopted by an J
administrator and section 19(5) of the 1986 Act, as it then stood:

> Although the authorities show that debts incurred before the liquidation do not obtain priority, they indicate that even on the salvage principle all liabilities under a contract incurred after the time of adoption of the contract by a liquidator are entitled to priority.

[19] He then cited *Re S Davis & Co* [1945] Ch 402, a case on bailment, and *Re Levi & Co Ltd* [1919] 1 Ch 416, a landlord and tenant case in which the liquidator that retained a lease for the benefit of the liquidation was held liable to pay as a liquidation expense a dilapidations claim under a covenant to deliver up the premises in good K
repair at the end of the term. He continued, at p450a:

> The salvage principle in liquidation indicates that if a liquidator adopts a contract for the purpose of the more beneficial conduct of a liquidation all such liabilities under such contract after the date of adoption are entitled to priority. This principle is therefore of no assistance in seeking to limit the administrator's liability in this case.

[20] In my judgment, that principle applies here so that, as the rent falling due on the next quarter day is a payment in advance, it is not subject to the Apportionment Act 1870 (see *Ellis* v *Rowbotham* [1900] 1 QB 740) from which it follows, as Mr Jourdan submitted L
and I accept, that the quarter's rent becomes payable in full from that date as one of the costs and expenses of the administration and would not fall to be apportioned should the administrators vacate the premises during that quarter. It follows also from this that the earlier decision of *Shackell & Co* v *Chorlton & Sons* [1985] 1 Ch 378 in the other direction is no longer good law, notwithstanding its application in *Re ABC Coupler & Engineering Co (No 3)* [1970] 1 WLR 702, where the point was conceded. The fuller citation of authority in *Levi*, and its approval in *Powdrill*, establishes that a liquidator electing to hold leasehold premises can do so only on the terms and conditions contained in the lease, and that any liability incurred while the lease M
is being enjoyed or retained for the benefit of the liquidation is payable in full as a liquidation expense. The same principle in my judgment applies in an administration. As Lord Hoffmann recognised in *Toshoku*, in [28], the liability to pay rent is not treated as an expense having priority until (and lasts only so long as) the office holder makes use of or decides to retain the property. Subject to that, any such liability accruing during that period is, in my judgment, to be treated as an expense having the requisite priority.

[21] I mention also that in *Re Linda Marie Ltd (in liquidation)* (1988) 4 BCC 463 rent was accepted to be a necessary disbursement of the liquidator. Since, however, the case proceeded on a concession, it was not a matter of decision and so it remains open to me to

decide the point. I do decide it in the applicant's favour and in so far, therefore, as the rent does not fall within r 2.67(1)(a), it does fall within (f). I should mention that Warner J in *Linda Marie* did not consider that the rent was correctly described as having been incurred in preserving the underlease. It seems to me that that observation is difficult now to square with the observations of the House of Lords in *Toshoku*, which held that rent will ordinarily fall within para (a) of what was then para 4.218(1).

[22] I should also make reference to the decision of the Court of Appeal in *Sunberry Properties Ltd* v *Innovate Logistics Ltd (in administration)* [2008] EWCA Civ 1321; [2009] 1 BCLC 145; [2009] BCC 164. In that case, the Court of Appeal refused permission, after balancing the interests of the applicant landlord against the interests of the administrators, to bring proceedings requiring the administrators to terminate an occupational licence for six months granted by them to a third party in breach of covenants in a lease with a remaining unexpired term of 20 years. That was the main point in the appeal. Under the heading "Consequential matters" starting in [55] and going through to [60], Mummery LJ, after recording that it was accepted that the landlord had no automatic right to be paid the contractual amount during the occupation of the company in administration, limited the amount that the administrators should pay to the sums passing under the licence from the third party. What passed under the licence was a monthly payment equivalent to the rent payable under the lease, whereas under the lease rent was payable quarterly in advance. The Court of Appeal also held that the administrators should pay interest on the passing rent, not at the contractual rate but only to the extent of interest actually received by them.

[23] As a decision of the Court of Appeal, that case might be said to require me to exercise a discretion to consider how much it would be fair for the administrators to pay in this case. If I were required to carry out that exercise, I would conclude that, on the particular facts of this case, it was fair for the administrators to pay the whole of the rent falling due in December because the only evidence of anyone qualified to express an opinion on the matter is that of the applicant's chartered surveyor, demonstrating that there is no realistic possibility of maximising the return from the property so long as the administrators are using part, as issues of data security intrude, and the prospect of redevelopment is adversely affected. The administrators do not contend that the rent payable by them should be calculated at a rate other than that provided for in the leases. Nor do they pray in aid (as was the case in *Sunberry*) an inability to pay. Their point is that they should pay only a proportionate amount of the rent attributable to the floorspace that they occupy, and the applicant landlord is free to have the rest. However, as I have said, that is not, on the evidence, a realistic proposal, or fair to the applicant landlord.

[24] Despite the clear indication of the Court of Appeal in *Sunberry* that the issue of what the administrators should pay should be approached flexibly, that approach was adopted, as I have said, as a result of a concession, and it was accepted on both sides before me that the decision of the Court of Appeal did not strictly bind me to adopt a similar approach. It is not difficult to see why the matter proceeded by way of concession. On the first part of the case, the landlord had argued (see [46] of the judgment) that because it did not have an automatic right to the rent, the balancing exercise should be resolved in its favour by allowing it to bring the proposed proceedings. Consistently with that, the landlord could hardly turn round and say the opposite when considering the terms on which the administrators should be allowed to leave the licence in place. Nor was it in the interests of the administrators to argue that.

[25] The Court of Appeal in *Sunberry* did not consider *Toshoku*, the *Lundy Granite* line of cases or *Powdrill*. Had it done so, it would have needed to address the application of the salvage principle and the rejection in *Toshoku* of the approach of Nicholls LJ in *Re Atlantic Computers Systems plc* [1992] Ch 505, as subsequently applied by him in *Powdrill & Lyle (joint liquidators of Kentish Homes Ltd)* v *Tower Hamlets London Borough Council* [1993] BCLC 1375. In *Powdrill & Lyle*, Nicholls V-C (as he had by then become) considered the issue of whether a post-liquidation liability to community charge on empty flats in the possession not of the liquidator but of an administrative receiver was an expense of the liquidation. He held that it was not, but would rank as such only if the court, as a matter of discretion, directed the liquidators to discharge the obligation out of the assets in their hands. The House of Lords, in agreement in this respect with the Court of Appeal in *Toshoku*, concluded that this case was wrongly decided because the community charge was clearly a post-liquidation liability of the company, and the *Lundy Granite* principle was irrelevant to such liabilities.

[26] The House of Lords in *Toshoku* also disapproved of the proposition that the *Lundy Granite* principle conferred a discretion, Lord Hoffmann observing, at p682, in [39]:

> There is of course no question that section 130(2) of the 1986 Act (the lineal descendant of section 87 of the 1862 Act upon which the *Lundy Granite Co* principle was originally constructed) confers a statutory discretion. But the discretion is as to the remedy which the creditor should be allowed to exercise; whether he should be able to bring proceedings, levy distress or execution or should have to wait for the distribution of the assets in due course of liquidation. The fact that a debt counts as an expense of the liquidation does not necessarily mean that the creditor should be allowed immediately to bring proceedings or levy execution. The order of priorities under rule 4.218(1) may mean that if he is paid at once, the assets to satisfy prior expense claims may be insufficient. So the question of remedy is entirely a matter of discretion. But the discretion does not determine whether a claim is a liquidation expense or not. It is rather the other way round; the claim must be a liquidation expense before the court can have any discretion to grant a remedy which will enable the creditor to obtain payment in priority to other claims.

[27] In other words, the court might have a discretion as to whether to allow forfeiture proceedings to be brought or to allow distress to issue, but there is no discretion to declare something to be or not to be a liquidation expense. The same principle applies, in my judgment, to administrations.

[28] There is one other point. The treatment of rent as a liquidation or administration expense under the *Lundy Granite* principle is not necessarily determinative of the point of time at which the rent should be paid. As Lord Hoffmann recognised in *Toshoku* in the passage that I have just cited, if the sufficiency of the realisable assets is in doubt, the landlord may have to wait and see to what extent the assets will be sufficient to satisfy its claim, even though properly treated as a liquidation or administration expense, since there may be other claims also having priority. There is in that sense no right to immediate payment, although the right for the accruing rent to be treated as an expense having appropriate priority where the liquidator or administrator is making use of the property still remains. None of that is relevant here because the assets are acknowledged to be sufficient to meet the claim of the applicant landlord for rent to be treated as an administration expense, whether under r 2.67(1)(a) or (f), and the administrators are paying administration expenses as they go along, recognising correctly, as is commonly the case, that they have no justification for acting otherwise.

[29] For all those reasons, the application succeeds in respect of the December rent because it is common ground that the premises, even if the administrators now decide to give them up, will not then be vacant, and rent will continue to be payable as an administration expense quarterly in advance under the terms of the two leases so long as the administrators retain or use any part of the premises demised under each such lease for the benefit of the administration. If, of course, they vacate the property demised under one of the leases entirely, the rental liability under that lease, but not under the other, will cease to be payable as an administration expense. If they also vacate the other demised property, the rental liabilities under both leases will cease to be payable as administration expenses.

Application allowed.

Good Harvest Partnership LLP *v* Centaur Services Ltd

CHANCERY DIVISION

[2010] EWHC 330 (Ch)

23 February 2010

NEWEY J

Estates Gazette 10 April 2010

[2010] 14 EG 114

Landlord and tenant — Landlord and Tenant (Covenants) Act 1995 — Guarantee — Lease of commercial premises — Assignment — Landlord consenting to assignment on terms that guarantor of current tenant also guaranteeing obligations of assignee — Landlord seeking to recover unpaid rent from guarantor — Sections 24 and 25 of 1995 Act — Whether guarantee void and unenforceable as frustrating operation of 1995 Act

The defendant guaranteed the obligations of the tenant under a lease of commercial premises let for a term of 10 years at a rent of £206,387 pa, subsequently rising to £245,000 pa. The terms of the lease prohibited assignment without the landlord's consent and specified that the landlord was entitled to require the tenant and its guarantor to enter into an authorised guarantee agreement, in a form permitted by the law, as a condition of giving a licence to assign. An assignment took place in 2004, with the defendant and the current tenant entering into a guarantee agreement by which they guaranteed the performance of the tenant covenants by the assignee from the date of the assignment until the next lawful assignment of the lease. In 2005, the claimant became the registered freeholder of the property and the direct landlord of the assignee.

The claimant brought proceedings against the defendant to recover unpaid rent from December 2008 and March 2009, pursuant to the guarantee agreement, and applied for summary judgment in its favour. The defendant contended that, *inter alia*, the guarantee agreement was void and unenforceable against it by reason of section 25(1) of the Landlord and Tenant (Covenants) Act 1995 since it frustrated the operation of that Act, section 24(2) of which was designed to ensure that the liabilities of a guarantor of the tenant's obligations under a lease came to an end when the lease was assigned.

***Held*: The application was dismissed. The 1995 Act is a significant change in the law on a tenant's liability after assigning the lease. It provides for tenants and others to be released from their obligations under the covenants when a lease is assigned. Section 24 is intended to ensure that any obligations undertaken by a party as guarantor for a tenant should end on the assignment of the lease. If the guarantor has to enter into a further guarantee when the lease is assigned, that guarantee can be said to frustrate the operation of a provision of the Act, within the meaning of section 25(1), in that it would, if valid, impose on the guarantor obligations equivalent to those from which section 24 is designed to secure its release. Had parliament intended a tenant's guarantor to be able to guarantee the obligations of an assignee, it could have been expected to say so explicitly. The provisions enabling tenants to give a guarantee for an assignee in certain circumstances represent an exception to the general prohibition and there is no equivalent provision dealing with guarantors: see sections 16 to 18. Accordingly, the further guarantee into which the defendant had been required to enter as a condition of the licence to assign was invalidated by section 25 so far as it purported to impose liability on the defendant. Section 25 barred the claimant's claim against the defendant.**

The following cases are referred to in this report.

Avonridge Property Co Ltd v *Mashru; sub nom London Diocesan Fund* v *Phithwa*; *Mashru* v *Avonridge Property Co Ltd*; *Avonridge Property Co Ltd* v *London Diocesan Fund*; *London Diocesan Fund* v *Avonridge Property Co Ltd* [2005] UKHL 70; [2005] 1 WLR 3956; [2006] 1 All ER 127; [2006] 1 P&CR 25; [2006] 1 EGLR 15; [2006] 01 EG 100

Wallis Fashion Group Ltd v *CGU Life Assurance Ltd*; *sub nom Wallis Fashion Group Ltd* v *General Accident Life Assurance Ltd* (2001) 81 P&CR 28; [2000] 2 EGLR 49; [2000] 27 EG 145, Ch

This was the hearing of an application by the claimant, Good Harvest Partnership LLP, for summary judgment on a claim against the defendant, Centaur Services Ltd, to recover unpaid rent pursuant to a guarantee of the assignee tenant's obligations.

Romie Tager QC and Philip Kremen (instructed by Brecher) appeared for the claimant; David Holland (instructed by Osborne Clarke, of Bristol) represented the defendant.

Giving judgment, **NEWEY J** said:

[1] This case raises the question of whether the Landlord and Tenant (Covenants) Act 1995 (the Covenants Act) precludes a party that has guaranteed a tenant's obligations under a lease from being required to give a further guarantee in respect of an assignee of the lease.

Basic facts

[2] The basic facts can be stated quite shortly.

[3] By an underlease dated 5 October 2001 (the underlease), Mr David Gladman, Ms Karen Gladman and Mr Jonathan Shepherd, trading together as Gladman Homes, demised to Chiron CS Ltd (Chiron) for a term of 10 years, from 16 July 2001, the premises comprising unit G, Lymedale Business Park, Newcastle-under-Lyme, Staffordshire (the premises). The rent was initially £206,387 pa, but subsequently rose to £245,000 pa. The defendant, Centaur Services Ltd (Centaur), was also a party to the underlease, as guarantor for Chiron.

[4] The underlease contained, among others, the following provisions:

(i) Clause 5.9.3:

By clause 5.9.3, the tenant covenanted: "Not to assign… the whole of the Premises… without the prior consent of the Landlord such consent not to be unreasonably withheld or delayed."

(ii) Clause 5.9.6:

Clause 5.9.6 stated that the landlord was entitled to impose any or all of certain conditions on giving any licence for an assignment of the whole premises. Those conditions included these:

(a) Clause 5.9.6.1:

> upon or before any assignment… the Tenant making the application for licence to assign and its guarantor (if any but not someone who has already given an authorised guarantee agreement) shall enter into an authorised guarantee agreement in a form permitted by Law and agreed between the parties.

(b) Clause 5.9.6.3:

prior to any permitted assignment to procure that the assignee enters into direct covenants with the Landlord to perform and observe all the Tenant's covenants and all other provisions during the residue of the Term so far as permitted by law.

(c) Clause 5.9.6.5:

the Tenant to procure that any… security for the Tenant's obligations under this Lease which the Landlord holds immediately before the assignment is continued or renewed in each case on terms as the Landlord may reasonably require in respect of the Tenant's liability under the authorised guarantee agreement referred to in clause 5.9.6.1.

(d) Clause 5.9.6.6:

the prospective assignee to provide a guarantor or guarantors approved by the Landlord (such approval not to be unreasonably withheld or delayed) where having regard to the financial standing of the prospective assignee it is reasonable for the Landlord to require its obligations under this Lease to be guaranteed.

[5] Chiron assigned the underlease to Total Home Entertainment Distribution Ltd (THED) on 1 September 2004. On the same day, Gladman Homes, Chiron and Centaur entered into a document described as an "authorised guarantee agreement" (the guarantee agreement). This recited that Gladman Homes had agreed to grant Chiron a licence to effect the assignment to THED subject to Chiron and THED entering into a formal licence (as happened) and to Chiron and Centaur entering into the guarantee agreement. The guarantee agreement proceeded to state, in clause 3.1, that Chiron and Centaur each covenanted with Gladman Homes (and their successors) that THED would pay rent, and otherwise perform the lessee covenants, "from the date of the Assignment [to THED] until the next lawful assignment of the Underlease".

[6] On 12 January 2005, the headlease out of which the underlease was carved was surrendered and, on 24 February 2005, the claimant, Good Harvest Partnership LLP (Good Harvest), became the registered freehold proprietor. Good Harvest thus became THED's direct landlord.

[7] The present proceedings, commenced on 9 June 2009, are to recover from Centaur, pursuant to the guarantee agreement, rent due under the underlease on 25 December 2008 and 25 March 2009. The outcome of the claim can be expected to determine Centaur's liability for rent and other liabilities up to the expiry of the term in July 2011.

[8] The application now before me was issued on 1 September 2009. By it, Good Harvest seeks summary judgment in its favour.

[9] Centaur disputes the application on two grounds. First, it contends that the guarantee agreement is void and unenforceable as against it by reason of section 25 of the Covenants Act. Second, it argues that there is a triable issue as to whether the underlease was surrendered or further assigned.

[10] I shall take these points in turn.

Covenants Act point

Covenants Act

[11] As its long title indicates, the Covenants Act makes provision "for persons bound by covenants of a tenancy to be released from such covenants on the assignment of the tenancy". Among other things, the Covenants Act serves to curtail the extent to which an original tenant can be liable to the landlord after it has assigned the lease.

[12] The Covenants Act implements, albeit with significant alterations, recommendations made by the Law Commission in its report *Landlord and Tenant Law: Privity of Contract and Estate* (Law Com no 174, 1988). In *Avonridge Property Co Ltd* v *Mashru* [2005] UKHL 70*, Baroness Hale said, in [39]:

The mischief at which the Commission's recommendations were aimed was the continuation of a liability long after the parties had parted with their interests in the property to which it related.

* Editor's note: Reported at [2006] 1 EGLR 15

[13] The Covenants Act represents a major change in the law. Neuberger J observed in *Wallis Fashion Group Ltd* v *CGU Life Insurance Ltd* (2001) 81 P&CR 28*, in [21], that the Act "represents a sea change in the law relating to a tenant's liability after he assigns the lease".

[14] In keeping with its purpose, the Covenants Act provides for tenants and others to be released from their obligations when a lease is assigned. The relevant provisions include these:

(i) Section 5:

(1) This section applies where a tenant assigns premises demised to him under a tenancy.

(2) If the tenant assigns the whole of the premises demised to him, he

(a) is released from the tenant covenants of the tenancy, and

(b) ceases to be entitled to the benefit of the landlord covenants of the tenancy,

as from the assignment… .

(ii) Section 24:

(1) Any release of a person from a covenant by virtue of this Act does not affect any liability of his arising from a breach of the covenant occurring before the release.

(2) Where

(a) by virtue of this Act a tenant is released from a tenant covenant of a tenancy, and

(b) immediately before the release another person is bound by a covenant of the tenancy imposing any liability or penalty in the event of a failure to comply with that tenant covenant,

then, as from the release of the tenant, that other person is released from the covenant mentioned in paragraph (b) to the same extent as the tenant is released from that tenant covenant… .

[15] The expression "tenant covenant", which appears in both section 5 and section 24, is defined in section 28(1) as follows:

a covenant falling to be complied with by the tenant of premises demised by the tenancy.

A covenant on the part of a tenant to pay rent will thus be a "tenant covenant".

[16] Section 25 of the Act is, as Lord Nicholls noted in *Avonridge*, in [14], "a comprehensive anti-avoidance provision". It states as follows:

(1) Any agreement relating to a tenancy is void to the extent that —

(a) it would apart from this section have effect to exclude, modify or otherwise frustrate the operation of any provision of this Act, or

(b) it provides for

(i) the termination or surrender of the tenancy, or

(ii) the imposition on the tenant of any penalty, disability or liability,

in the event of the operation of any provision of this Act, or

(c) it provides for any of the matters referred to in paragraph (b)(i) or (ii) and does so (whether expressly or otherwise) in connection with, or in consequence of, the operation of any provision of this Act.

(2) To the extent that an agreement relating to a tenancy constitutes a covenant (whether absolute or qualified) against the assignment, or parting with the possession, of the premises demised by the tenancy or any part of them —

(a) the agreement is not void by virtue of subsection (1) by reason only of the fact that as such the covenant prohibits or restricts any such assignment or parting with possession; but —

(b) paragraph (a) above does not otherwise affect the operation of that subsection in relation to the agreement (and in particular does not preclude its application to the agreement to the extent that it purports to regulate the giving of, or the making of any application for, consent to any such assignment or parting with possession).

(3) In accordance with section 16(1) nothing in this section applies to any agreement to the extent that it is an authorised guarantee agreement; but (without prejudice to the generality of subsection (1) above) an agreement is void to the extent that it is one falling within section 16(4)(a) or (b).

(4) This section applies to an agreement relating to a tenancy whether or not the agreement is —

(a) contained in the instrument creating the tenancy; or

(b) made before the creation of the tenancy.

* Editor's note: Also reported at [2000] 2 EGLR 49

[17] In *Avonridge*, Lord Nicholls (with whom Lord Hoffmann and Lord Scott expressed agreement) said, in [18], that section 25 is "to be interpreted generously, so as to ensure the operation of the 1995 Act is not frustrated, either directly or indirectly".

[18] Section 16, to which reference is made in section 25(3), is concerned with "authorised guarantee agreements". It details agreements that tenants can enter into to guarantee the obligations of their assignees up to the point that the lease is assigned again. The section provides as follows:

(1) Where on an assignment a tenant is to any extent released from a tenant covenant of a tenancy by virtue of this Act ("the relevant covenant"), nothing in this Act (and in particular section 25) shall preclude him from entering into an authorised guarantee agreement with respect to the performance of that covenant by the assignee.

(2) For the purposes of this section an agreement is an authorised guarantee agreement if—

(a) under it the tenant guarantees the performance of the relevant covenant to any extent by the assignee; and

(b) it is entered into in the circumstances set out in subsection (3); and

(c) its provisions conform with subsections (4) and (5).

(3) Those circumstances are as follows—

(a) by virtue of a covenant against assignment (whether absolute or qualified) the assignment cannot be effected without the consent of the landlord under the tenancy or some other person;

(b) any such consent is given subject to a condition (lawfully imposed) that the tenant is to enter into an agreement guaranteeing the performance of the covenant by the assignee; and

(c) the agreement is entered into by the tenant in pursuance of that condition.

(4) An agreement is not an authorised guarantee agreement to the extent that it purports—

(a) to impose on the tenant any requirement to guarantee in any way the performance of the relevant covenant by any person other than the assignee; or

(b) to impose on the tenant any liability, restriction or other requirement (of whatever nature) in relation to any time after the assignee is released from that covenant by virtue of this Act.

(5) Subject to subsection (4), an authorised guarantee agreement may—

(a) impose on the tenant any liability as sole or principal debtor in respect of any obligation owed by the assignee under the relevant covenant;

(b) impose on the tenant liabilities as guarantor in respect of the assignee's performance of that covenant which are no more onerous than those to which he would be subject in the event of his being liable as sole or principal debtor in respect of any obligation owed by the assignee under that covenant;

(c) require the tenant, in the event of the tenancy assigned by him being disclaimed, to enter into a new tenancy of the premises comprised in the assignment—

(i) whose term expires not later than the term of the tenancy assigned by the tenant, and

(ii) whose tenant covenants are no more onerous than those of that tenancy;

(d) make provision incidental or supplementary to any provision made by virtue of any of paragraphs (a) to (c).

(6) Where a person ("the former tenant") is to any extent released from a covenant of a tenancy by virtue of section 11(2) as from an assignment and the assignor under the assignment enters into an authorised guarantee agreement with the landlord with respect to the performance of that covenant by the assignee under the assignment—

(a) the landlord may require the former tenant to enter into an agreement under which he guarantees, on terms corresponding to those of that authorised guarantee agreement, the performance of that covenant by the assignee under the assignment; and

(b) if its provisions conform with subsections (4) and (5), any such agreement shall be an authorised guarantee agreement for the purposes of this section; and

(c) in the application of this section in relation to any such agreement—

(i) subsections (2)(b) and (c) and (3) shall be omitted, and

(ii) any reference to the tenant or to the assignee shall be read as a reference to the former tenant or to the assignee under the assignment.

(7) For the purposes of subsection (1) it is immaterial that—

(a) the tenant has already made an authorised guarantee agreement in respect of a previous assignment by him of the tenancy referred to in that subsection, it having been subsequently revested by him following a disclaimer on behalf of the previous assignee, or

(b) the tenancy referred to in that subsection is a new tenancy entered into by the tenant in pursuance of an authorised guarantee agreement;

and in any such case subsections (2) to (5) shall apply accordingly.

(8) It is hereby declared that the rules of law relating to guarantees (and in particular those relating to the release of sureties) are, subject to its terms, applicable in relation to any authorised guarantee agreement as in relation to any other guarantee agreement.

Parties' submissions in outline

[19] Mr David Holland, who appeared for Centaur, argued that the guarantee agreement is avoided by section 25(1) of the Covenants Act in so far as it seeks to impose liabilities on Centaur. Section 24(2) of the Act is designed to ensure that, where a party has guaranteed a tenant's obligations, its liabilities in respect of the lease come to an end when the lease is assigned. To allow such a guarantor to be required to give a guarantee in respect of an assignee's obligations would, accordingly, "exclude, modify or otherwise frustrate the operation" of a provision of the Act, contrary to section 25(1). The only exception to the general rule that tenants and their guarantors should, alike, be relieved of obligations in respect of the lease on assignment is to be found in section 16, which allows tenants to give guarantees for their assignees, but section 16 makes no provision for guarantors to give further guarantees. Mr Holland further suggested that clause 5.9.6.1 of the underlease was a "tenant covenant" within the meaning of the Act, with the result that, by virtue of sections 5 and 24, both tenant and guarantor were released from the provision.

[20] On the other hand, Mr Romie Tager QC, who appeared with Mr Philip Kremen for Good Harvest, submitted that the guarantee agreement did not exclude, modify or otherwise frustrate the operation of the Covenants Act and so was not avoided by section 25. As a result of the guarantee agreement, Centaur undertook new obligations from the date of the assignment to THED; Centaur was, none the less, released from its obligations under the underlease. Had parliament intended to interfere with the ability of guarantors to give guarantees for assignees, it would have made specific provision in respect of guarantors; the fact that it did not do so indicated that it did not consider guarantors to need protection in this respect. There could be no good reason for barring guarantors from giving guarantees for assignees. Where a tenant gives a guarantee for an assignee, it is open to the original guarantor to provide a subguarantee of the tenant's obligations as guarantor; it would be anomalous if the guarantor could not enter into a direct guarantee for the assignee.

[21] Both sides agreed that a trial judge would be in no better position than I am to decide whether section 25 of the Covenants Act is an answer to Good Harvest's claim against Centaur and that I could and should, accordingly, arrive at a final conclusion on the issue notwithstanding that the application before me is for summary judgment.

Conclusion

[22] I have concluded that section 25 of the Covenants Act does bar Good Harvest's claim against Centaur. I can summarise my reasons as follows:

(i) Section 24 of the Act was meant to ensure that any obligations undertaken by a person as guarantor for a tenant should come to an end on the assignment of the lease. Mr Tager suggested that the guarantor's obligations would, in any event, have terminated under the general law relating to guarantees when section 5 of the Act released the tenant from liability. However, I do not think that it matters for present purposes whether Mr Tager is right about this. The important point is that parliament intended section 24 to relieve the guarantor of any liability that it might otherwise have had in respect of the guarantee that it had given;

(ii) If the guarantor is required to enter into a further guarantee when the lease is assigned, it seems to me that the guarantee can, as a matter of language, fairly be said to "frustrate the operation of any provision of [the] Act" (to quote from section 25(1)(a)), in that it would, if valid, impose on the guarantor obligations equivalent to those from which section 24 was designed to secure his release. This conclusion

A is reinforced by the fact that section 25 is, as noted above, "to be interpreted generously";

(iii) Mr Tager argued otherwise on the basis that the guarantor would be undertaking *new* obligations, not reviving old ones. A tenant that gave a guarantee for an assignee could, however, similarly be said to be undertaking new obligations, yet it is abundantly clear that parliament intended a tenant to be able to give no guarantee other than an AGA. As I see it, a premise underlying section 16, which provides for AGAs, is that a tenant could not otherwise give any guarantee: in other words, that section 16 represents an exception to a general prohibition. If (subject to section 16) the Act precludes tenants from giving guarantees for
B assignees, it is difficult to see why guarantors should not likewise be barred from giving such guarantees;

(iv) Had parliament intended a tenant's guarantor to be able to guarantee obligations of an assignee, it could have been expected to say so explicitly, particularly since guarantors are mentioned expressly in the Act more than once: see sections 12(1)(b), 17(3) and 18(3). It has not done so, however;

(v) Section 16 addresses the circumstances in which a tenant can give a guarantee for an assignee, but there is no equivalent provision dealing with guarantors. Nor does section 16 contain any reference to guarantors. There is no indication in the section that an AGA can
C include a guarantee from anyone other than the tenant;

(vi) Were it the case that a tenant's guarantor could be required to give a guarantee for an assignee of the tenant, there would seem to be nothing to limit the guarantor's exposure to the period before that first assignee assigned. Liability under an authorised guarantee agreement (AGA) given by a tenant has to come to an end when the tenant's assignee assigns: see section 16(4). Since, however, section 16 makes no reference to guarantors, there would be no similar restriction on how long such a guarantor's liability could continue. Yet for a landlord to be able to call on a tenant's guarantor to give a guarantee for assignees, other than the first, could drive the proverbial "coach and horse" through
D the legislation. Take a case in which, as must be common, a parent company or a director gives a guarantee for a tenant. If the guarantor could be required to guarantee obligations of assignees other than the particular assignee to which the tenant assigned, it might mean little that the liability of the tenant had to come to an end by the time the assignee assigned on: the parent company or director would remain liable. The effectiveness of the Act could thus be seriously undermined;

(vii) I do not think that it is by any means clear that the Covenants Act permits a guarantor to subguarantee a tenant's obligations under an AGA. Mr Tager suggested that it could be seen from sections 17 and 18 that a guarantor could give such a subguarantee. In this connection,
E Mr Tager drew attention to the fact that section 17(3) envisages a person guaranteeing "the performance by the former tenant of such a covenant as is mentioned in subsection (1)", and he said that section 17(3) would thus relate to a guarantor of a tenant's obligations under an AGA; he made, moreover, a similar point about section 18(3). However, I agree with Mr Holland that sections 17(3) and 18(3) do not obviously refer to guarantees of tenants' obligations under AGAs. So far as section 17(3) is concerned, the words "such a covenant as is mentioned in subsection (1)" appear to relate to "a tenant covenant of the tenancy under which any fixed charge is payable", and section 18(3) speaks of someone guaranteeing the performance "by the former tenant of a tenant
F covenant of the tenancy". In each case, therefore, parliament seems to have in mind a guarantee of the performance by a former tenant of a tenant covenant. Under an AGA, however, a former tenant is not itself under any obligation to perform tenant covenants; its obligation, as section 17(1)(a) recognises, is to guarantee the performance *by its assignee* of tenant covenants. Section 17(3) is thus more obviously applicable to pre-Act tenancies, under which a tenant can remain liable on tenant covenants notwithstanding assignment, than to AGAs given in connection with what the Act terms "new tenancies". In any case, even if a guarantor could subguarantee a tenant's obligations under an AGA, it would not necessarily follow that it should be able to give a direct guarantee for an assignee. In short, I do not think that it helps to try to answer the question of whether a guarantor can give a subguarantee;

(viii) It is fair to say that, if the Covenants Act serves to bar a G
guarantor from giving a guarantee for an assignee, that involves a restriction on freedom of contract. However, the Act is plainly designed to impose restrictions on freedom of contract. The question is how far those restrictions go;

(ix) It is fair to say, too, that, if the Act is to be construed as Mr Holland suggested, it will be capable of operating in ways that look arbitrary. Mr Tager took, for example, the case of a parent company guaranteeing a subsidiary's obligations under a tenancy. If it proved desirable to transfer the lease to another subsidiary, why, asked Mr Tager, should the parent company be unable to act as guarantor? However, it appears to me that the Act could sometimes operate in apparently arbitrary (or H
at least uncommercial) ways even if I accepted Mr Tager's submissions. Suppose, for example, that a company entered into a lease and that the lease was subsequently, as a result of successive group reorganisations, assigned first to one subsidiary and then to another. The Covenants Act would, even on Mr Tager's case, prevent the parent company from giving any guarantee for the second subsidiary however much it wished to and however commercially desirable that was;

(x) In the present case, it is apparent that Centaur was required to enter into a further guarantee when the underlease was assigned to THED. The recitals to the guarantee agreement confirm that Gladman Homes made it a condition of the grant of the necessary licence to J
assign that Centaur should enter into the guarantee agreement;

(xi) Accordingly, the guarantee agreement is invalidated by section 25 of the Act in so far as it purports to impose liability on Centaur.

[23] It follows that, in my judgment, Centaur has a complete defence to Good Harvest's claim and that I should dismiss the proceedings.

Surrender/further assignment point

[24] Centaur's alternative argument was that I should not accede to Good Harvest's application because there is a triable issue as to whether the underlease has been surrendered or further assigned.

[25] Mr Holland argued that Centaur has, at the least, a more K
than fanciful prospect of establishing that the underlease has been surrendered by operation of law or that there has been either an equitable assignment or an assignment by subletting of the underlease. In support of this submission, Mr Holland pointed to evidence to the following effect:

(i) In the autumn of 2006, THED's parent company was acquired by the Woolworths Group;

(ii) In May 2007, Good Harvest was informed that the THED operation was to cease and merge with Entertainment UK Ltd, another wholly owned subsidiary of Woolworths. In the following months, employees of THED were made redundant and the company vacated the premises; L

(iii) In August 2007, THED applied to Good Harvest for consent to assign the underlease to Woolworths plc;

(iv) In February 2008, the entire assets and undertaking of the company were sold, for the most part to Entertainment UK Ltd;

(v) From around the spring of 2008, the premises were being occupied by Wincanton under some form of service agreement for another Woolworths company called "MCR" or "Multi Channel Retail";

(vi) During 2008, several invoices for rent were addressed to "Entertainment Distribution Limited", and that two such invoices, in respect of the rent due on the June and September quarter days, were M
paid by Entertainment UK Ltd rather than THED.

[26] On the other hand, there is evidence before me that Good Harvest has never granted consent to any assignment of the underlease by THED and that neither THED's directors nor the administrators of Woolworths and Entertainment UK Ltd have ever suggested that any assignment or subletting has taken place; in this connection, Mr Tager not only referred to evidence from Good Harvest but pointed out that Centaur has adduced evidence as to what one of its directors has been told by a former director of THED, but it is not said that the latter has spoken of the underlease having been surrendered or assigned by THED. With regard to the invoices addressed to "Entertainment Distribution Limited", these are, as Mr Tager noted, the last three

A words of THED's name ("Total Home Entertainment Distribution Limited"), and the evidence is that the relevant invoices were addressed to THED's registered office. There is evidence, too, that the application for consent to assign the underlease made in 2007 was withdrawn in August 2008 and that the premises are currently being marketed on behalf of the Woolworths Group on the footing that they are "available by way of an assignment or sub-lease of an existing FR&I 10 year lease which commenced in July 2001": in other words, on the basis that the underlease remains in being. Further, Mr Tager made the point that what will have mattered to Good Harvest will have been that the rent was being paid, not which company wrote the cheques.

B [27] As Mr Holland recognised, there is no direct evidence before me that the underlease has been the subject of a surrender or further assignment. Nor, as it seems to me, is there any solid basis for inferring C that the underlease might have been surrendered or assigned on. In all the circumstances, I can see no real prospect of the court accepting at a trial that the underlease had been either surrendered or assigned by THED. It follows that, but for my conclusion on the Covenants Act point, I would have granted Good Harvest summary judgment on its claim.

Conclusion

[28] In the light of the conclusions that I have arrived at on the Covenants Act point, I shall dismiss the proceedings. D

Application dismissed.

LANDLORD AND TENANT
HOUSING ACTS

Earl Cadogan and another v Chehab

ADMINISTRATIVE COURT

[2009] EWHC 3297 (Admin)

26 November 2009

IRWIN J

Estates Gazette 27 March 2010

[2010] 12 EG 98

Landlord and tenant — Housing Act 1988 — Rent assessment — Practice and procedure — Rent assessment committee taking into account supplemental submission from tenant — Submission not copied to landlord — Whether breach of natural justice — Correct approach to valuation of security of tenure

The appellants were the landlords of a flat that was let on a periodic tenancy following the expiry of an earlier lease. In 1998, the respondent took an assignment of the tenancy. In 2001, the rent was increased to £17,000 pa by agreement between the parties. In 2008, the appellants served notice, under section 13(2) of the Housing Act 1988, to increase the rent to £29,120 pa with effect from May 2009. The respondent referred the notice to the rent assessment committee for a determination under section 14 of the 1988 Act.

During the hearing, both parties indicated that they wanted to submit further written representations. After the hearing, such further representations were submitted to the committee, which passed on each party's submissions to the other side. The respondent's surveyor presented the committee with a second supplemental report, which the appellants did not receive. The report included a contention that the rent could not exceed the £25,000 threshold for security of tenure. The committee took that contention into account in reaching its decision, by which it determined the rent at £22,500 pa. The appellants remained unaware of the report until they received the committee's decision.

On appeal, they contended that: (i) natural justice had been breached because the committee's decision had been coloured by controversial and relevant evidence that they had had no opportunity to address; (ii) the committee had erred in making no adjustment for security of tenure on the basis that that factor had advantages and disadvantages for both parties; and (iii) the committee had fettered its discretion by proceeding on the basis of a £25,000 cap or limit to the rent.

***Held*: The appeal was allowed. (1) Evidence that is germane to the decision of the rent assessment committee, and on which it relies in reaching that decision, should not be submitted after the event unless all interested parties have had sight of it. All parties should see all the relevant evidence, and the committee should ensure that this happens. Owing to an error, that had not been the case in the instant proceedings and, accordingly, a breach of natural justice had arisen. If further material is submitted to the committee after the oral proceedings have concluded, the lawyers, surveyors or other agents of the parties should ensure that the material is sent directly to the other side; there is probably a duty to send such material inter partes, rather than merely to the committee. If a rent assessment committee allows a practice whereby the parties do not exchange supplemental submissions directly but instead rely on the committee to pass on material to the other side, the committee will be responsible for ensuring that all relevant material goes to the respective parties. (2) Security of tenure is obviously advantageous to a tenant. The committee had erred in regarding security of tenure as being of no value to the tenant; such an approach was neither appropriate nor rational. (3) A rent assessment committee is obliged to take all relevant factors into account and should not set an artificial limit on the rent it is asked to fix. It must assess security of tenure as being advantageous to the tenant while such security persists. There had been insufficient argument before the committee as to whether, in circumstances where the value of the security of tenure took the appropriate rent above the £25,000 pa threshold and the security of tenure was consequently removed by necessary operation of law, the committee should reassess the rent to reflect the fact that security of tenure no longer existed. Accordingly, the decision of the rent assessment committee was quashed and remitted for rehearing and further argument.**

The following cases are referred to in this report.

Curtis v *Chairman of the London Rent Assessment Committee*; *sub nom Curtis* v *London Rent Assessment Committee* [1999] QB 92; [1998] 3 WLR 1427; [1997] 4 All ER 842; [1998] 1 EGLR 79; [1998] 15 EG 120; (1998) 30 HLR 733, CA, unreported, 27 November 1996, QB

R v *London Rent Assessment Panel, ex parte Cadogan Estates Ltd* [1998] QB 398; [1997] 3 WLR 833; (1998) 76 P&CR 410; [1997] 2 EGLR 134; [1997] 33 EG 88; (1998) 30 HLR 487, QB

This was an appeal by the appellant landlords, Earl Cadogan and Cadogan Estates Ltd, from a decision of a rent assessment committee determining an application under section 14 of the Housing Act 1988 for an assessment of the rent payable by the respondent tenant, Bassam Chehab.

Anthony Radevsky (instructed by Pemberton Greenish) appeared for the appellants; Andrew Dymond (instructed by Denton Wilde Sapte LLP, of Milton Keynes) represented the respondent.

Giving judgment, **IRWIN J** said:

[1] In this case, the appellant landlords appeal to the High Court, under section 11(1) of the Tribunals and Inquiries Act 1992 and pursuant to CPR 52.1(9), against a decision of a rent assessment committee dated 22 June 2009 that determined a rent of £22,500 pa, beginning from 29 May 2009, for flat 5, Bourne House, 189 Sloane Street, London SW1X 9QT, under section 14 of the Housing Act 1988 (the 1988 Act).

[2] The background to the hearing is helpfully summarised by a skeleton argument on behalf of the landlord appellants lodged by Mr Anthony Radevsky, who has been of great assistance in the hearing today.

[3] The respondent to the appeal became a tenant of flat 5 on 6 August 1998, having on that date taken an assignment of a periodic tenancy, which tenancy had arisen in 1993 following the expiry of the lease. The rent had not been increased since 2001. By a decision dated 15 May 2001, a differently constituted rent assessment committee had determined a rent for these premises of £23,600 pa. However, the parties had in fact previously agreed that the rent should be £17,000 pa and the appellants honoured that agreement.

A [4] By a notice under section 13(2) of the 1988 Act, dated 14 April 2008, the appellants sought to increase the rent to the sum of £29,120 pa with effect from 29 May 2009. By an application dated 6 February 2009, the respondent referred the notice to the committee, pursuant to his rights under section 13(4) of the 1988 Act.

[5] Each of the parties put in evidence from a surveyor, a Mr Nicholas Baynes MRICS, on behalf of the appellants, and a Mr Roger Doncom MRICS, for the respondent. Each of those experts acted as the representatives of the parties and gave evidence at the hearing on 12 May 2009.

[6] In the course of the hearing, as I understand the facts and for
B these purposes they are not in dispute both parties indicated that they might wish to submit further written representations, primarily concerning the tenant's repairing obligations, and both surveyors did so. No complaint arises concerning those further written representations because, through the conduit and good offices of the committee, those further representations were passed to the other party.

[7] It appears to be the case that the surveyors, both Mr Baynes and Mr Doncom, did not directly send the further representations to the other side but relied on the committee to pass them on. However, Mr Doncom, surveyor for the tenant, submitted a second supplemental report dated 19 June 2009. He sent that to the committee but did not
C serve that on the appellants and, crucially, the committee did not send it to them either. Accordingly, it is common ground that the appellants had no opportunity to read or comment on the second supplemental submissions from Mr Doncom. The appellants were unaware that such further representations had gone to the committee until receipt of the decision of the committee, which came under cover of a letter dated 26 June. Appended to that decision was a copy of Mr Doncom's second supplemental submission.

[8] It is clear from the nature of its decision that the committee relied on that further evidence from Mr Doncom.

[9] The crucial parts of the further supplemental submission are
D contained in para 3 of Mr Doncom's document, which deal with the effects on the value of a tenancy, of an assured tenancy, in para 1.8, and in para 1.9 they read as follows:

> My client's solicitor has considered Mr Baynes' reliance on Megarry [that was a reference to *Megarry on Assured Tenancies*, (2nd ed)]. In doing so Mr Baynes is now conceding that my client's rent must remain below the £25,000 threshold. Mr Baynes' legal submission is only designed to take account of a benefit which applies to an assured tenancy rather than a tenancy at a rent over £25,000 which has no assured status.

[10] As I have indicated, it is clear that the committee relied on that. In para 6, the committee noted that Mr Doncom had submitted
E written representations, sent after the hearing, and proceeded to set out its computation figures. It is perfectly clear from the content of that paragraph that it is the second supplemental submissions coming from Mr Doncom with which it is dealing.

[11] In para 8, it sets out their conclusions as to rental in part as follows:

> We are required by section 14 of the Act to determine the rent at which the flat might reasonably be expected to be let in the open market by a willing landlord under an assured tenancy on the basis set out in section 14. We have come to the conclusion that the open market rent on that basis would be £22,500 per annum, inclusive of services. We have arrived at that rent by starting from £39,000 as
> F proposed by each of the parties' valuers.

[12] The committee goes on then to set out incremental alterations to that starting point, which lead it to conclude as follows:

> We thus arrive at £22,561.50, which we round to £22,500 per annum. We make no adjustment for security of tenure, which has advantages and disadvantages for both parties.

[13] In appealing this decision, Mr Radevsky, on behalf of the landlord, took four points. First, he said that the evidence lodged with the committee, but not served on his clients, was controversial, was germane to the decision and clearly affected the committee's decision, and that it was a breach of natural justice for the committee to proceed in that way when his client had had no opportunity to address G
the evidence.

[14] Second, he said that it was wrong to treat the argument in the supplemental evidence as setting a £25,000 cap or limit to the rent and that the committee fettered its discretion by reference to such a supposed limit.

[15] Third, he said that the treatment of the security of tenure as a factor that "has advantages and disadvantages for both parties", and which led the committee to make no adjustment for security of tenure, is wrong.

[16] Fourth, although this point was not, in the event, pursued, following an indication from me as to the way my mind was working, H
he said that there was fresh evidence that he should be able to introduce and, if the matter is remitted, can be introduced before the committee on a rehearing.

[17] I have listened with great care to the submissions from Mr Radevsky and I also have in mind the concession by Mr Andrew Dymond, who appeared on behalf of the tenant today, and it seems to me that his first ground is made out. It should never be the case that evidence that is germane to the decision and, indeed, that was relied on by the rent assessment committee, should be submitted after the event without going to all parties that may have an interest in the outcome of the proceedings. It seems to me that this is likely J
to have happened by error on this occasion, certainly not by any deliberate decision on the part of the committee. It may have resulted from the practice that grew up, in at least this case, that supplemental submissions were not exchanged directly between the parties, but rather the surveyors relied on the committee to pass material on.

[18] On any view, if a rent assessment committee permits such a practice, it is taking to itself the responsibility of ensuring that relevant material goes to all the parties. What I would add is that it seems to me very highly desirable that if further material is to be submitted to a committee after the conclusion of the oral proceedings, the lawyers, or surveyors or other agents engaged in the particular hearing must ensure K
that the material goes directly to the other side because reliance on the official support of such committees is not wise. The solution may be different before a hearing, where any deficiency can be put right. However that may be, after the close of a hearing it seems to me that there is almost certainly a duty to send such material *inter partes*, rather than merely to the committee. Perhaps in this instance, Mr Doncom may be forgiven for relying on the committee because that is what had been permitted to take place rather earlier.

[19] It would be perfectly possible to reach the conclusion, on this ground of appeal, from first principles. However, as Mr Radevsky showed me, a parallel situation has arisen previously. The court has L
seen a transcript of a decision by McCullough J in *Curtis* v *London Rent Assessment Committee*, a decision given by the judge on 27 November 1996 with the reference CO/1193/96. The first instance decision is unreported, although the case, for different reasons, came before the Court of Appeal and is reported at [1999] QB 92*.

[20] In similar circumstances, material came to the rent assessment committee that made a difference to the decision of the committee but was not served on the landlord. It was on that basis that McCullough J quashed the decision of that committee and made an order remitting the case for a rehearing.

[21] The reference to the Court of Appeal by the landlord in that M
case was on different grounds, and the appeal from McCullough J was allowed on different grounds, but the report of the Court of Appeal decision makes it clear by implication that the court regarded the decision of McCullough J, when quashing the rent assessment in that case, as being correct.

[22] It must be obvious that all parties should see all relevant evidence and submissions in their case, and the rent assessment committees must take care to ensure that that happens.

* Editor's note: Also reported at *Curtis* v *Chairman of the London Rent Assessment Committee* [1998] 1 EGLR 79

[23] On Mr Radevsky's second point, it does seem to be a fair implication from the way in which the committee expressed its decision that it made no adjustment for security of tenure, which, as it put it, had advantages and disadvantages for both parties.

[24] In my judgment, it is perfectly obvious that security of tenure is an advantage for a tenant. In this instance, the proper course for it to follow would be to assess the value of the rent, including the security of tenure, and then to see whether that caused difficulties of one kind or another. It did not do so. I do not stipulate whether it should adjust for security of tenure after addressing the other factors affecting the value of the rent, or as part and parcel of the same exercise. It seems to me that both approaches might be perfectly proper. Which approach is followed could also be fact-dependent in the individual case. Certainly, to proceed on the basis that security of tenure is of no value to a tenant seems to me not to be either appropriate or, indeed, rational.

[25] The third basis of appeal throws up a problem that was addressed in argument early in this hearing. Given that a rent assessment committee must assess security of tenure as being an advantage to a tenant while it persists, what if the value of that security of tenure takes the appropriate rent above £25,000 pa, which by operation of statue law sets aside or abrogates the security of tenure that has just been valued? Mr Radevsky sought to persuade me that I should rule on that point today, but I decline to do so. It seems to me that that is a difficult point and not entirely met by existing authority.

[26] Mr Radevsky cited to me the decision of *R* v *London Rent Assessment Panel, ex parte Cadogan Estates Ltd* [1998] QB 398*, a decision of Kay J sitting in the Administrative Court. At p403E-F in the judge's decision, he made it clear that, as Mr Radevsky, who appeared in that case, submitted, he was of the view that valuation of an assured tenancy must be approached on the basis of valuing every element within it and every factor that affects the rent level, without imposing any cap or observing any limit pursuant to the statute.

[27] As the judge said, at p405E:

> ... I have come to the conclusion that the applicant's submissions are correct and it is unnecessary to look in detail at the anomalies outlined on behalf of the applicant that would arise if the panel's interpretation was correct.

[28] I agree. It must be the case that all relevant factors should be taken into account by a rent assessment panel and no artificial limit set on the rent on which they fix. The problem that is thrown up, potentially, by a case such as this, is this: if it is the element of security of tenure that brings the rent through the £25,000 limit, that breach of the limit will itself, by necessary operation of law, remove the security of tenure. It is open to argument as to whether, in those circumstances, the committee should again reassess the rent, reflecting the fact that security of tenure no longer exists. The situation is straightforward if, even allowing for security of tenure, the rent does not reach £25,000. It is also straightforward if, without regard to the value of security of tenure, the rent exceeds £25,000. The difficulty arises only if it is the value of the security of tenure itself that carries the appropriate rent across the threshold.

[29] I make no ruling on that. It seems to me that that point is problematic and requires fuller argument on both sides on the proposition before a satisfactory judgment could be reached.

[30] As I have already indicated, Mr Radevsky did not pursue the fresh evidence point for the purposes of today's hearing.

[31] For those reasons, therefore, the decision of the rent assessment committee in this case will be quashed and the matter remitted for rehearing and for reargument.

Appeal allowed.

---

* Editor's note: Also reported at [1997] 2 EGLR 134

A
Idealview Ltd *v* Bello

QUEEN'S BENCH DIVISION

14 October 2009

FLAUX J

Estates Gazette 30 January 2010

[2010] 04 EG 118

Rent review — Delay — Parties acquiring respective interests of tenant and landlord in premises — Previous landlord not carrying B ***out rent review — New landlord seeking review — Arbitrator determining rent due — Whether landlord entitled to possession for non-payment of arrears by tenant — Whether available defence of unreasonable delay in seeking rent review — Whether claim for rent arrears time-barred under section 19 of Limitation Act 1980***

In July 2005, the appellant acquired the residue of a lease of premises. The respondent acquired the freehold reversion in 2006. The 50-year lease ran from May 1969 at a rent of £60 pa for the first 25 years; it provided for a rent review thereafter. If no agreement were reached by the penultimate quarter day before the start of the remaining 25 years, an arbitrator would determine the new rent. Any difference C between that rent and the passing rent would "be added to and be payable with the next instalment of rent due after the said new rent has been determined".

The rent review did not take place. Following its purchase, the respondent raised the issue of a review. It referred the matter to an arbitrator, who, by an award of August 2007, increased the rent from £60 to £1,700 pa. The appellant did not take part in the arbitration. The respondent served a notice under section 166 of the Commonhold and Leasehold Reform Act 2002. It notified the appellant of the rent due from March 1994 to September 2007 on the assumption, in the appellant's favour, that the passing rent had been paid before the D respondent acquired the freehold. The appellant failed to pay, and the respondent brought a claim for possession based on rent arrears. The appellant argued that the claim was time-barred under section 19 of the Limitation Act 1980, and raised further issues of abandonment, estoppel, acquiescence and laches relating to the 13-year delay in seeking a review. Granting a possession order, the county court judge held that section 19 did not bar the claim since the arrears claimed had not fallen due until September 2007, on the next quarter day after the arbitrator's award. The judge also held that: (i) any arguments on delay should have been raised before the arbitrator; (ii) the appellant was bound by the award; and (iii) in any event, there was no evidence to E found an estoppel or similar. The appellant appealed.

***Held*: The appeal was dismissed. The appellant could and should have complained about the delayed review before the arbitrator; he could not now do so by a collateral attack in court. The judge was bound by the arbitrator's determination. Moreover, the lease did not make time of the essence in respect of the rent review, either expressly or impliedly. The lease set the first point at which the landlord could trigger a review, but it did not say that the review had to take place within any particular time. The length of the delay was not a defence to the appellant's claim in the absence of evidence to found an estoppel or similar. The appellant could not point to a representation, when or after it purchased the lease, that the landlord would not seek a rent review. The respondent's** F **claim was not time-barred since it was claiming the passing rent only from the date of its acquisition of the freehold. On the correct construction of the rent review clause, the new rent had not fallen due until September 2007, on the next quarter day following the arbitrator's award.**

The following case is referred to in this report.

Amherst v *James Walker (Goldsmith & Silversmith) Ltd (No 2)* [1983] Ch 305; [1983] 3 WLR 334; [1983] 2 All ER 1067; (1984) 47 P&CR 85; [1983] 2 EGLR 108; (1983) 267 EG 163, CA

This was an appeal by the appellant, Adejola Bello, from a decision G of HH Judge Behar, sitting in Wandsworth County Court, allowing a claim by the respondent, Idealview Ltd, for possession of premises on the ground of rent arrears.

Nicholas Macleod-James (instructed under the direct access scheme) appeared for the appellant; Carl Fain (instructed by Kingsley Napley) represented the respondent.

Giving judgment, **FLAUX J** said:

[1] This is an appeal by the appellant, Mr Adejola Bello, against the judgment of HH Judge Behar at Wandsworth County Court on 23 October 2008, whereby he made an order for possession of premises H at 143 Eardley Road, London SW16, of which the appellant was the tenant, on forfeiture for rent arrears. At a renewed oral application for permission to appeal against that judgment on 9 July 2009 (permission having previously been refused on paper), Blair J gave permission to appeal on one ground alone, as appears in more detail hereafter. The essential facts are not in dispute and can be stated shortly.

[2] By a lease dated 5 May 1969, the then freeholder of the premises let them for a term of 50 years from 25 March 1969. By clause 1 of that lease, the rent payable during the first 25 years of the lease was the sum of £60 pa. Clause 2 was a rent review clause that provided, so far as relevant, as follows: "The Lessee shall pay for such residue." That is to J say, the residue after the first 25 years of the lease:

> hereinbefore mentioned of the said term an annual rent which shall be determined in accordance with the following formula that is to say such rent shall be the rent (but not less than the rent firstly hereinbefore reserved) at which the demised premises first mentioned might reasonably be expected to be let in the open market by a willing landlord by a Lease for a term of years equivalent to such residue as aforesaid without regard for anything built on the land on the same terms and subject to the same incidents in all other respects as this present demise, provided that if no agreement is reached between the parties by the penultimate quarter day prior to the commencement of such residue as aforesaid K as to the rent at which the demised premises might reasonably be expected to be let in the open market on the basis hereinbefore described, then the question shall be referred to the decision of a single arbitrator to be appointed by the President for the time being of the Royal Institution of Chartered Surveyors in accordance with the provisions of the Arbitration Act 1950 or any statutory modification or re-enactment thereof for the time being in force, provided further that until such new rent shall have been determined the rent firstly hereinbefore reserved shall continue to be payable and any difference between that and the said new rent during such period as this last proviso operates shall be added to and be payable with the next instalment of rent due after the said new rent has been determined.

[3] The first 25-year period expired on 24 March 1994. No rent review took place. The appellant apparently purchased the premises at auction on 14 July 2005. At that time, a rent review, under clause 2, had still not taken place. However, as Tomlinson J pointed out in refusing permission to appeal on paper, the terms of sale excluded the seller's liability for arrears of rent above the £60 pa so that the appellant might, had he considered the matter, have realised that there was a potential liability on him as the buyer for any arrears of rent above £60 pa that might result from a rent review.

[4] The respondent company acquired the freehold interest from the original freehold owner on 8 March 2006. In due course, it raised with the appellant the question of a rent review under clause 2. There was no constructive response from the appellant, let alone one suggesting that any such review was too late and, accordingly, the respondent referred the issue of rent for the residue of the term to an arbitrator, Mr Copping Joyce, pursuant to the terms of clause 2 of the lease. The appellant was notified of the arbitration by both the respondent and the arbitrator, but chose, no doubt for reasons of his own, not to participate. On 22 August 2007, the arbitrator issued his award, determining the rent at £1,700 pa, or £425 per quarter, from 25 March 1994. No attempt was made by the appellant to appeal the arbitrator's award or to seek to set it aside before this court.

[5] On 29 September 2007, the rent as determined by the arbitrator fell due under the terms of clause 2 of the lease. On 30 November 2007, the respondent's solicitor wrote to the appellant, enclosing by way of service a notice, under section 166 of the Commonhold and Leasehold Reform Act 2002 (the 2002 Act), of the rent due following the arbitrator's award. The schedule to that notice and the letter itself make clear two things: first, the appellant had not paid any of the passing rent from the date on which the respondent acquired the freehold interest, that is, the figure of £15 per quarter from 25 March 2006; and, second, the respondent assumed, in the appellant's favour, that such rent had been paid before the respondent's acquisition of the freehold. Accordingly, the rent arrears were claimed only for the quarters from 25 March 2006 up to and including the quarter of 24 June 2007 at the original rate of £15 per quarter, and what was claimed thereafter at 29 September 2007 quarter date was, as was explained in the schedule, rent from 25 March 1994 to 29 September 2007 at £425 per quarter less rent at £15 per quarter, assumed to have been paid prior to the respondent's acquisition of the freehold reversion.

[6] The appellant made no reply to this letter. Before the learned judge, he tried to claim that he had not received the letter or the notice, but the judge rejected that contention and held that, on a balance of probabilities, it had been received. In his defence, the appellant ran two substantive defences: first, a point on section 168 of the 2002 Act, which the learned judge held was unsustainable. No permission to appeal that conclusion has been given. The second defence was a defence that any claim for the arrears of rent was time-barred under section 19 of the Limitation Act 1980. That provides as follows:

> No action shall be brought, or distress made, to recover arrears of rent, or damages in respect of arrears of rent, after the expiration of six years from the date on which the arrears became due.

[7] The judge concluded that, pursuant to clause 2 of the lease, the arrears claim did not fall due until 29 September 2007, the quarter day on which the next instalment of rent was due after the determination by the arbitrator. Accordingly, he held that the claim was not time-barred. However, as the judge pointed out, the appellant, who appeared in person before the judge, raised a number of additional arguments, which the judge allowed him to raise, on abandonment, estoppel, acquiescence and laches, all focusing on the delay of 13 years from 1994 to 2007 in seeking a rent review under clause 2 of the lease. The judge held, in [34] of his judgment, that the short answer to those arguments was that the appellant was bound by the determination of the arbitrator, which he had not appealed, so that he could not raise those matters that should have been raised before the arbitrator.

[8] However, the judge went on to deal with the arguments, as he said shortly, and, in dismissing them, he made three findings that are critical for the purposes of this appeal. First, he held, in [36], that clause 2 did not stipulate that time was of the essence and that there was no basis for the implication of a term. Second, he held, in [37], that the appellant was not relying on anything more than the delay and, in those circumstances, he held that the decision of the Court of Appeal in *Amherst* v *James Walker (Goldsmith & Silversmith) Ltd (No 2)* [1983] 3 WLR 334* was a formidable obstacle in the way of the appellant's argument. Third, he held, in [38], that there was no evidence on which to found a discrete defence of estoppel, waiver, acquiescence or abandonment and, in his view, laches did not apply on the facts. Accordingly, the judge held that the defence failed.

[9] The appellant sought to appeal the judge's conclusions on a number of grounds. As I have already said, his application was refused on paper by Tomlinson J in February of this year. He then instructed Mr Nicholas Macleod-James, who appears before me today and who drafted an amended notice of appeal. That was considered by Roderick Evans J on 24 April. He made an order that the application for permission to appeal in respect both of the original grounds and the revised grounds should be listed for oral argument. It was on that basis that the matter came before Blair J, who refused permission to appeal on all grounds but one, and gave permission on that single ground, described as limitation.

[10] Mr Carl Fain, who appeared on behalf of the respondent, as he did before the learned judge, argued that that meant that only the section 19 point was in play before me, but I accept Mr Macleod-James' contention that Blair J must have had in mind the wider issue of delay generally and its effect. Otherwise, I cannot conceive on what basis he could have thought that the limitation defence under section 19 on its own was arguable. Accordingly, I have approached the case on the wider basis.

[11] Having said that, like the learned judge I consider that the complaints about delay in seeking the rent review are all matters that the appellant could and should have raised before the arbitrator. Had he done so, the arbitrator might well have concluded that the review was too late or he might have accepted the arguments now raised by Mr Fain. Either way, those matters were not ventilated when they should have been and, in my judgment, it is not open to the appellant to raise in court, by a collateral attack, such matters. As the judge rightly held, he is bound by the determination of the arbitrator.

[12] That is the short answer to this appeal, but since the other points were fully argued I will deal with them by reference to the points raised in Mr Macleod-James' skeleton argument.

[13] First is the issue of whether time was of the essence under this contract. Mr Macleod-James drew my attention to the point that he said distinguishes this case from the lease considered in *Amherst*, namely that this lease has what he described as a starting point for the review. In other words, had the parties not agreed by the penultimate day of the quarter date prior to the commencement of the residue of the second 25 years, the question, that is, the question of the rent payable, was to be referred to an arbitrator. That penultimate day of the quarter was 24 March 1994. That is no doubt correct, but, in my judgment, all that is doing is setting the first point at which the landlord can trigger a review. There is nothing in clause 2 that says that, thereafter, the review has to take place within any particular time, and I agree with the learned judge that there is nothing in the contract, either expressly or by implication, that made time of the essence.

[14] The second point raised by Mr Macleod-James is the point on which he spent by far and away the most time in his oral argument, which is the point concerning the delay of 13 years. As he submitted, this was longer than any applicable limitation period, be it a six-year period under section 19 or a 12-year period on the basis that the lease is a specialty. He contended that in no previous case had a delay this long been considered and that researches had indicated that the longest delay in any other case was in the region of six years. He recognised that *Amherst* presented an obstacle.

---

* Editor's note Also reported at [1983] 2 EGLR 108

[15] In that case, the relevant delay between the date on which the rent review could first have been sought and the date on which it was in fact sought, was some four-and-a-half years. The Court of Appeal concluded that mere delay in itself in a case where time was not of the essence was nothing to the point and did not provide the tenant with a defence. It is necessary, however, to cite two passages from the judgment that can, it seems to me, legitimately be regarded as qualifying that stark conclusion.

[16] First, from the judgment of Oliver LJ, at p344 of the report. He concluded that there had been no prejudice suffered by the tenant as a result of the delay, and that the delay was not substantial and the failure to seek a review was not in any sense unreasonable. He then went on to say, at p345B:

> Finally, I am encouraged to find that the view that I have formed as to the way in which this type of problem should be approached coincides (save in the one respect of the possibility of abandonment as a separate legal concept) with the views expressed by Slade LJ in the recent decision of the court in *London & Manchester Assurance Co Ltd* v *GA Dunn & Co* (1982) 265 EG 39, 135: "In the absence of binding authority compelling a different conclusion, I am satisfied that delay on its own, even if unreasonable, will not in general disentitle a landlord from invoking a rent review clause in a case where time has not been made of the essence of the contract. There is, I think, no general principle of the law of contract that *mere* delay in the enforcement of a contractual right, or in the performance of a contractual duty, by one party to a contract ('A') will entitle the other party ('B') to regard himself as discharged from the obligation to recognise such right or from the contract as a whole (as the case may be). If in such circumstances A has been guilty of unreasonable delay, then, ordinarily, the prudent and proper course for B to adopt, if he wishes to bring matters to a head, will be to serve a notice on A fixing a reasonable period within which A must exercise his right (if at all) or must perform his part of the contract (as the case may be). In some circumstances, of course, the delay on the part of A may be so gross and inexplicable as to make it so clear that he does not intend to exercise his right or to perform his part of the contract that any such notice is unnecessary. But, ordinarily, it will be necessary for B to serve a notice on A or at least to have some communication with him before he can properly and safely regard himself as being absolved."
>
> It is true that in the circumstances of that case what Slade LJ said was *obiter*, but it forms part of a carefully reasoned analysis which I gratefully adopt.

[17] Then, in the judgment of Lawton LJ, at p347B-C:

> In my judgment there is no justification for reading into the lease an implied term that if the landlord did not serve a rent assessment notice on or before December 25, 1974, he had to do so within a reasonable time thereafter. Such an implied term would not have been necessary to give business efficacy to the lease. The landlord would not have wanted it and the tenants benefited by not having it. Any delay on the landlord's part would mean that the tenants went on paying the original rent until such time as the landlord did serve a notice. If for any reasons of their own, such as a general fall in rental values, they had wanted the landlord to come to a decision about the service of a rent assessment notice they themselves could have served what has come to be known, inaccurately, as a notice "making time of the essence of the contract": see *Stickney* v *Keeble* [1915] AC 386 and *United Scientific Holdings Ltd* v *Burnley Borough Council* [1978] AC 904.

[18] The learned judge continued, at pp347F-348A:

> He would only have been estopped if the tenants could have proved that by his words or conduct he had represented that he did not intend to ask for the payment of a higher rent and in reliance on that representation they had altered their position to their prejudice. In my judgment nothing short of estoppel would have relieved the tenants from their liability to pay a higher rent. The concept of abandonment has been referred to in some of the rent review cases. I do not regard it as a term of art apt to describe a defence to a landlord's claim for a higher rent. If a landlord by his words or conduct leads his tenant reasonably to infer that he did not intend to claim a higher rent he makes a representation to that effect so that the foundation of an estoppel is laid; but the landlord will not be estopped unless the tenant has acted on the representation to his prejudice. A landlord who over a long period makes no attempt to set a rent review procedure in motion may be adjudged to have represented that he did not intend to exercise his rights; but whether he did would be a matter of inference from the circumstances in which the delay had occurred, not from the mere fact of the delay. I can see no reason why mere delay, not amounting to a representation, can be a bar to a landlord in this kind of case claiming a higher rent. He has his contractual rights to a higher rent and the tenant has an obligation to pay it unless he can prove that there is some good reason why he should not. Mere delay would not be a good reason. On the facts of this case, as Mr Rich admitted when opening the appeal, there was no evidence which would have founded an estoppel.

[19] Mr Macleod-James relied on those qualifications to submit that there was here an unreasonable or a gross and inexplicable delay, such as to justify the conclusion that it was too late for the respondent to seek to review the rent. The insuperable difficulty that that argument faces, in my judgment, is the judge's finding of fact in [38] of his judgment that there is no evidence to found a defence of estoppel or similar in the present case. The appellant is bound by that finding and is thrown back on no more than the length of the delay without more. However, as *Amherst* demonstrates, mere delay is not enough.

[20] In that context, I was not helped by hypothetical examples, put forward essentially *in terrorem*, of 99-year leases or even longer leases where there was no review until 50 years after the review was first available or even 100 years after the review was first available. Each case turns on its own facts and it may be that in such a case the facts and circumstances of the case might give rise to an estoppel. It may be that they would not. What matters is that, in the present case the appellant simply cannot point to anything that amounts to a representation that the landlord would not seek a rent review at or after the time at which the appellant purchased the lease. Indeed, as I have already mentioned, the fact that in the conditions of sale the seller undertook responsibility only for the £60 pa does point at least to the possibility that there might very well be a rent review in the future and also, although of course one cannot make too much of this point, the fact that there is unlikely to have been any communication between the landlord and the original tenant from which any representation that there would not be a rent review could be derived. At all events, as the judge rightly pointed out, there was simply no evidence of that kind at all.

[21] The third point relied on by Mr Macleod-James was whether delay after the limitation period limits the extent of recovery. This argument focuses on the fact that the arrears of rent date back to 1994, and it is said that such rent must be time-barred so it cannot be recovered. In my judgment, this argument misses the point that there are two elements to the rent. First, there is the rent at £15 per quarter from 25 March 1994, which was always due irrespective of the rent review. Of course, to the extent that the respondent sought to recover that rent in 2007, the claim would be time-barred for any period prior to 2001 pursuant to section 19 of the 1980 Act. However, as was clarified before Tomlinson J, the respondent is not claiming and was not claiming that portion of the rent, and it assumes that rent had been paid before the appellant became the tenant. Mr Macleod-James essentially complained about that and said that there was no evidence whether the rent had been paid before the appellant became a tenant or not, but it does not seem to me that that matters. All that matters is that the respondent is not claiming that element of the rent until 25 March 2006, when it purchased the freehold reversion, and that period of time after 25 March 2006 is well within the limitation period.

[22] The second element of the rent is the balance of arrears from 25 March 1994, that is, the balance between the £15 per quarter originally due and the £1,700 per quarter assessed by the arbitrator. So far as that is concerned, as the judge rightly held on the correct construction of clause 2, none of that fell due until 29 September 2007, so again there is no question of section 19 being engaged, and, in my judgment, all the attempts of Mr Macleod-James to sidestep that, as I see it, inevitable conclusion, meet the barrier of the clear words of the clause in the lease. That is also the answer to his fourth argument concerning the inability to review by a rent review a statute-barred rent. Whatever the position would be in another case or in a case where time was of the essence, the short answer to this point is that the claim to the arrears of rent is not statute-barred under section 19 because it simply was not due within the meaning of section 19 until it had been ascertained by the arbitrator's award, and then at the first quarter day thereafter, being 29 September 2007.

[23] Finally, Mr Macleod-James sought to make much in his oral submissions of the prejudice suffered by the appellant by reason of liability for rent for a period before he was even the tenant. Much was said about injustice and the effect of the decision on consumer lessees of long leases who could be saddled with such liabilities. Again, I am not over impressed with those sorts of arguments since there are, in my judgment, two clear answers to the suggestion of prejudice in the present case.

[24] First, if at the time of purchasing the lease, the appellant had sought legal advice as to the effect of clauses 1 and 2 and as to the possibility of a rent review, I have little doubt that, if properly advised, he would have been told that there was always a risk of a rent review. There could then have been a discussion with the landlord and it may be that she would have agreed that it was now too late for a review, in which case the respondent could not have pursued the present claim at a later date when it purchased the freehold reversion. Alternatively, if the landlord was not prepared to agree that it was too late for a review, the appellant might have sought and might have obtained some form of indemnity from the seller.

[25] Second, had the appellant participated in the arbitration, all the arguments now put forward so eloquently on his behalf by Mr Macleod-James could have been put before the arbitrator, who might have declined a review or limited it in some way. None of that happened and if, as is submitted, the appellant has suffered prejudice, that is in large part of his own making. I prefer to leave for another occasion issues of the wider ramifications of what might happen on different facts.

[26] In my judgment, this appeal must be dismissed.

Appeal dismissed.

LANDLORD AND TENANT
VIRTUAL ASSIGNMENT

A
Clarence House Ltd *v* National Westminster Bank plc

COURT OF APPEAL

[2009] EWCA Civ 1311

8 DECEMBER 2009

WARD AND JACOB LJJ AND WARREN J

Estates Gazette 27 February 2010

B **[2010] 08 EG 106**

Landlord and tenant — Virtual assignment — Lease of commercial premises — Premises occupied by subtenant — Tenant transferring economic benefits and burdens and management responsibilities to third party — Landlord's consent not obtained — Whether tenant sharing or parting with possession of premises — Whether breach of covenant against alienation without landlord's consent

The respondent was the landlord and the appellant was the tenant under a lease of office premises for a term of 25 years from 1985. The lease contained standard-form covenants against alienation that prohibited the appellant from assigning, underletting, sharing or parting
C with possession of the demised premises or executing any declaration of trust in respect of the property or the lease without the respondent's consent. The appellant did not occupy the premises but had underlet them, with the respondent's consent, to an occupying subtenant for a term that expired three days before the appellant's own lease.

In 2005, the appellant entered into a virtual assignment, whereby it transferred to a third party all the economic benefits and burdens of the lease, including management responsibilities, but without assigning the leasehold interest or changing the actual occupancy of the premises. It also granted a power of attorney to the virtual assignee, enabling it to act in the name of and on behalf of the appellant in respect of the
D property in accordance with the terms of the virtual assignment. The respondent was not informed of those arrangements and its consent was not sought.

When it became aware of the arrangement, the respondent sought a declaration that the appellant had breached the lease by entering into the virtual assignment and damages. The question of liability was determined as a preliminary issue. Finding for the respondent, the judge held that the virtual assignment amounted to a parting with or sharing of possession in breach of the lease because it provided for the assignee to receive the rents and deal with the property in the appellant's place: see [2009] EWHC 77 (Ch); [2009] 2 EGLR 19; [2009] 17 EG 104. The
E appellant appealed.

***Held*: The appeal was allowed. (1) The appellant had not shared or parted with possession of the demised premises or any part thereof by entering into the virtual assignment. The word "possession" should be given its ordinary and strict meaning in a standard alienation clause. The hallmark of the right to possession is the right to exclude all others from the property in question. The context of the alienation clause emphasised the aspect of physical control. At the time of the virtual assignment and at all material times thereafter, the party in possession of the premises was the subtenant under the underlease. Since the appellant was not in possession of the demised premises, it could not part with or share possession by reason of the virtual assignment arrangements.** F **Although, for some purposes, "possession" can include receipt of rents and profits of the land or the right to receive them, the assignee did not receive, or have the right to receive, the rent from the subtenant in the sense of putting the assignee into possession of the demised premises. Under the express terms of the arrangement, it collected the rent as agent of the appellant; the appointment of an agent cannot itself amount to a parting with the right to receive the rents. The fact that the appellant forfeited the rents the moment after receipt and they became money received to the account of the assignee, was a purely contractual arrangement between the appellant and the assignee and did not result in a transfer to** G **the assignee of the right to receive rents and profits in the sense envisaged by section 205 of the Law of Property Act 1925. (2) The virtual assignment arrangements did not breach any other part of the alienation covenants. The relationship between the virtual assignee and the appellant was founded in contract, such that the prohibition against declaring any trust of the property or the lease had not been breached. Nor had there been an assignment in the sense meant by the alienation provisions. "Assignment" is a standard phrase that means a legal assignment unless the context compels a different construction. The context did not so require; the fact that the alienation provisions restricted several specified** H **forms of alienation indicated the parties' intention that only those forms of alienation should fall foul of the restrictions. Further, the prohibition against underletting had not been breached. An underlease is a demise of the property for a term that creates an inferior estate and constitutes the relationship of landlord and tenant between the parties. That was absent from the virtual assignment arrangements, which created no privity of estate.**

The following cases are referred to in this report.

Abbey National plc v *Commissioners of Revenue & Customs (formerly the Commissioners of Customs & Excise)* [2006] EWCA Civ 886; [2006] J
3 EGLR 153; [2006] STC 1961, [2005] EWHC 831 (Ch); [2005] 3 EGLR 73; [2005] 43 EG 190; [2006] STC 852
Akici v *LR Butlin Ltd* [2005] EWCA Civ 1296; [2006] 1 WLR 201; [2006] 2 All ER 872; [2006] 1 P&CR 21; [2006] 1 EGLR 34; [2006] 07 EG 136
Clarence House Ltd v *National Westminster Bank plc* [2009] EWHC 77 (Ch); [2009] 1 WLR 1651; [2009] 3 All ER 175; [2009] 2 All ER (Comm) 273; [2009] 2 P&CR 15; [2009] 2 EGLR 19; [2009] 17 EG 104
Gentle v *Faulkner* [1900] 2 QB 267; 16 TLR 397
Howard v *Miller* [1915] AC 318, PC
Kataria v *Safeland plc* [1998] 1 EGLR 39; [1998] 05 EG 155
Lam Kee Ying Sdn Bhd v *Lam Shes Tong (t/a Lian Joo Co)* [1975] AC 247;
[1974] 3 WLR 784; [1974] 3 All ER 137, PC K
Lyell v *Kennedy (No 4)* (1889) LR 14 App Cas 437, HL
Manchester Airport plc v *Dutton* [2000] 1 QB 133; [1999] 3 WLR 524; [1999] 2 All ER 675; (2000) 79 P&CR 541; [1999] 1 EGLR 147
Pye (JA) (Oxford) Ltd v *Graham* [2002] UKHL 30; [2003] 1 AC 419; [2002] 3 WLR 221; [2002] 3 All ER 865; [2003] 1 P&CR 10
Richards v *Delbridge* (1874) LR 18 Eq 11
Saunders v *Vautier* (1841) Cr & Ph 240; (1841) 4 Beav 115, Ch
Stening v *Abrahams* [1931] 1 Ch 470; [1931] All ER Rep 437; 100 LJ Ch 278; 145 LT 18
Westdeutsche Landesbank Girozentrale v *Islington London Borough Council* [1996] AC 669; [1996] 2 WLR 802; [1996] 2 All ER 961, HL

A This was an appeal by the appellant, National Westminster Bank plc, from a decision of HH Judge Hodge QC, sitting as a judge of the High Court, allowing a claim by the respondent, Clarence House Ltd, for breach of alienation covenants in a commercial lease.

Christopher Nugee QC and Julian Greenhill (instructed by Linklaters LLP) appeared for the appellant; Jonathan Gavaghan (instructed by Brian Drewitt, of Bramhall) represented the respondent.

Giving judgment, **Ward LJ** said:

[1] The issue in this case is whether by entering into a "virtual assignment" of leasehold premises, the tenant of those premises acts
B in breach of the standard-form alienation covenants contained in the lease under which the premises are held. On 23 January 2009, HH Judge Hodge QC, sitting as a judge of the Chancery Division, declared that it did, but only because the tenant either parted with possession or was sharing or was permitting the sharing of possession of the premises: see [2009] EWHC 77 (Ch)*. He ordered an inquiry into damages. Jacob LJ granted permission for this appeal.

[2] For those who do not gambol in this arcane field and have not previously encountered a "virtual assignment", let me endeavour to explain what it is. I cannot do better than repeat the judge's description,
C in [1]:

> A "virtual assignment" is an arrangement under which all the economic benefits and burdens of the relevant lease (including any management responsibilities) are transferred to a third party, but without any actual assignment of the leasehold interest or any change in the actual occupancy of the premises in question. It is typically employed where the relevant lease contains covenants against assigning or parting with the possession of the demised property without the consent of the landlord, and there are concerns either that the landlord may be unwilling to consent to a legal assignment of the lease because of perceived concerns about the financial standing of the assignee, or that the landlord's consent may not be available in advance of the scheduled date for completion of the transaction.

D

It has other uses: it can be employed as a means of removing lease liabilities from the balance sheet; capital can be released to be better invested elsewhere in the assignor's business; property risks can be transferred to a third party; and the assignor's property liabilities can be more closely aligned to business needs because surplus property can be vacated before the term of the lease expires. For these reasons, the clever construct of a virtual assignment was invented and has been deployed by commercial property lawyers in the City. They see the virtual assignment as "an invaluable mechanism which has facilitated a number of vital portfolio transactions for tenants without damaging
E landlords' interests. Reports of its death [a reference to this case] are premature. It is here to stay and the market should be grateful." So wrote Mr Anthony Burnett-Scott, a partner at Ashurst LLP and Mr Nicholas Cheffings, a partner at Lovells LLP, in *Estates Gazette* on 23 May 2009, at p44. The respondent protests that virtual assignments may be "well-known to a handful of large City Law Firms acting for tenants with large property portfolios but they remain a device which is new and (so far) comparatively rarely used and untested between landlord and tenant". This case will therefore decide its fate.

[3] This virtual assignment was made on 10 June 2005 between New Liberty Property Holdings Ltd, a company registered in Gibraltar,
F which I shall call "New Liberty", and National Westminster Bank plc (NatWest), pursuant to the master agreement made the previous day between Royal Bank of Scotland Group plc and New Liberty. The master agreement is a complicated document to which, fortunately, I need not make much reference. It deals with the disposal of a large portfolio of property, both freehold and leasehold. The virtual assignment also covers many properties, some of which are "white leaseholds", where the virtual assignor is no longer in occupation, and "grey leaseholds", where the virtual assignor remains in occupation. The material provisions are these:

> **2. Economic Benefit** G
> The intention of the Virtual Assignment is to pass to the buyer [New Liberty] all of the economic benefits and burdens of the Leases and Underleases in respect of the Properties together with the obligation to manage all dealings with the Landlords and Undertenants as if the Properties had been assigned to the Buyer. Therefore any monies from any Under tenants pursuant to any Underleases, together with all proceeds for the surrender of any Underleases, shall belong to the buyer.
>
> **3. Indemnity**
> … The Buyer will from the Completion Date at all times duly pay to the Landlord in full all Rent on the relevant dates for payment under the Leases and will observe and perform all covenants and obligations contained, and will
> indemnify and keep the Seller indemnified against payment of all Rent and H
> all costs, liabilities and claims arising out of any actual or alleged breach or non-observance of all covenants or obligations contained in [the Leases].
>
> **4. Dealings with Properties**
> … The Seller hereby irrevocably appoints the Buyer to be its agent to act on its behalf and in its name in all dealings connected with the properties. This will include but not be limited to:
>
> 4.1.1 paying all Rent due under the Leases including negotiating any rent reviews under the Leases;…
>
> …
>
> 4.1.3 collecting and getting in all Rent due under the Underleases and negotiating rent reviews;…
>
> … J
>
> 4.1.7 negotiating and completing any surrender;…
>
> 4.1.8 dealing with any renewal of any Underlease…
>
> 4.1.9 negotiating and completing all assignments…
>
> …
>
> 4.1.17 anything else related to the operation of the Leases and Underleases.
>
> 4.2 The Seller will not while this Virtual Assignment is in existence, do or attempt to do any of the things set out in Clause 4.1.
>
> 4.3 Neither the agency appointment in Clause 4.1 nor the Power of Attorney shall entitle the Buyer to do any of the following in the name of the Seller:
>
> 4.3.1 vary the Lease…
>
> 4.3.2 renew or re-grant… the Lease…
>
> 4.3.3 increase the Rent… K
>
> 4.4. To enable the Buyer to perform its functions better, the Seller will provide at the cost of the Buyer:
>
> 4.4.1 confirmation to Landlords and Undertenants that the Buyer is acting on its behalf;…
>
> and will agree themselves not to approach the Landlords or Under tenants…
>
> 4.5 The Buyer will not be entitled to use the name of the Seller in any litigation or other dispute resolution procedure in connection with any of the powers granted by the Seller under this Virtual Assignment without first obtaining the prior written consent of the Seller (which consent shall not be unreasonably withheld or delayed)…
>
> …
>
> **8. Monies Paid and Received Under the Virtual Assignment** L
> All monies paid and received by the Buyer under the Virtual Assignment (whether or not by way of Rent) will be to the account of the Buyer and the Seller will have neither any obligation to pay monies nor any right to receive monies (other than by reason of an Indemnity given by the Buyer in the Virtual Assignment) under or by virtue of the Virtual Assignment.

[4] On the same day, NatWest executed a power of attorney in favour of New Liberty. By clause 2:

> [NatWest] hereby irrevocably appoints [New Liberty] to be its true and lawful attorney to do all acts and things on its behalf and in its name which [New Liberty] may consider properly necessary or advisable to effect or to give
> proper effect to the provisions of the Virtual Assignment… and deliver as M
> its act and deed… [everything] properly necessary or acquired for any of the following purposes [namely the acts similar to those set out in clause 4.1 of the Virtual Assignment itself].

[5] The lease with which we are concerned was a lease made on 3 June 1986 between Sun Life Assurance of Canada, as landlord, and County Bank Ltd, as tenant, relating to a suite of offices being part of the ground and first floors of Clarence House, in Manchester. The term was for 25 years commencing on 25 December 1985 and so expiring on 24 December 2010. The current landlord is Clarence House Ltd, the respondent to this appeal, and the current occupier is William M Mercer Ltd (Mercers) to whom NatWest, the successor to

* Editor's note: Reported at [2009] 2 EGLR 19

the County Bank, underlet the entirety of the premises with the consent of the landlord for a term of years expiring on 21 December 2010, thus retaining a nominal reversion of three days. It is common ground that at no material time has NatWest been in actual occupation of any part of the property.

[6] The alienation covenants in this lease are as follows:

> 30. (1) Not to assign or charge the Demised Premises in any part less than the whole
>
> (2) Not to charge the Demised Premises…
>
> (3) Not to execute any declaration of trust with regard to the Demised Premises or any part thereof or these Presents
>
> (4) Not to share or permit sharing of possession or occupation of the Demised Premises or any part thereof or part with possession or occupation of the same…
>
> (5) Not to underlet any part or parts of the Demised Premises or permit the same to be underlet…
>
> (6)(a) Not to underlet the whole of the Demised Premises nor permit the creation of any derivative underlease…
>
> …
>
> (8) Not to assign the whole of the Demised Premises…
>
> …
>
> (10) Not to assign, transfer, charge or underlet the whole of the Demised Premises or underlet parts thereof pursuant to paragraph 30(5) of this Schedule otherwise than in accordance with and without in each and every such case first complying with the foregoing provisions and subject thereto not without obtaining the previous consent in writing of the Landlord which consent shall not be unreasonably withheld.

[7] The facts giving rise to the litigation can be very shortly stated. Speechly Bircham, which acted for NatWest, wrote to FPD Savills Commercials Ltd, the respondent's managing agent in Manchester, to explain that NatWest had "virtually assigned" its leasehold interest in the premises to New Liberty, and that New Liberty would be paying the rent on behalf of the bank, and also that New Liberty had the bank's authority to deal directly with Savills as agent on all matters relating to the lease. Unfortunately, that letter was never received. It was by chance that the respondent's solicitor learnt from a conversation with a member of the bank's property department that a "virtual assignment" had been entered into six months previously. The respondent's solicitor did not know what a virtual assignment was, and I confess to having sympathy with it. Speechly Bircham refused to produce copies of the documentation and it needed an application for preaction discovery to force the documents from the bank. Not surprisingly, this added to the landlord's suspicions. What was this mysterious arrangement? Why was NatWest refusing to deal with its landlord and insisting instead that it deal with New Liberty or its agent? What was this unforthcoming Gibraltarian company that had been foisted on it, without its knowledge or consent, in place of its approved tenant, a copper-bottomed high street bank? The respondent made plain how unhappy it was dealing with the interloper, especially in the light of the disconcerting fact that, since its involvement, the rent was in arrears. Its concerns could not be assuaged. Its concerns were not without foundation. We have now been informed that, on 28 October, provisional liquidators of New Liberty were appointed.

[8] On 10 April 2008, the respondent issued its claim, seeking a declaration that by entering into the virtual assignment NatWest had acted in breach of the terms of the lease, in that the arrangements constituted a declaration of trust or a sharing of or parting with possession or the creation of a derivative underlease or an underletting or an assignment without consent. It claimed damages and sought an order for an inquiry into damages. By consent, the master directed that there be a trial of the preliminary issue of liability only.

[9] That came before HH Judge Hodge QC, who found for the claimant on 23 January 2009 to the limited extent that I have already indicated.

Judgment under appeal

[10] The judge discussed *Abbey National plc* v *Commissioners of Revenue & Customs (formerly the Commissioners of Customs & Excise)* [2006] EWCA Civ 886; [2006] 3 EGLR 153; [2006] STC 1961 on appeal from the decision of the late Hart J [2005] EWHC 831 (Ch); [2006] STC 852*. The researches of counsel have not revealed any other case in which the courts have had to consider the implications of a virtual assignment. He concluded, however, that although the decisions were a useful aid to understanding the concept of and the commercial motivation for entering into a virtual assignment, he could derive little further assistance from the case because it was concerned with the VAT consequences for the virtual assignor in that case, Abbey National Building Society.

[11] Dealing with the alleged breaches of the alienation covenant, he found, in [13], that "neither in form nor in substance did the virtual assignment amount to an underletting". There was no breach of the covenant against assignment because the covenant covered only a legal assignment and, in any event, it would have to have been completed by registration against the title. There was no declaration of trust involved because "the relationship between the Defendant and New Liberty seems to me to be founded in contract rather than in equity": see [18]

[12] He did, however, find that there was a breach of the restriction against permitting the sharing of possession or occupation. He held:

> 21. … In my judgment, by executing the Virtual Assignment, the Defendant has either parted with possession of the Property to New Liberty or, at least, is sharing, or permitting the sharing of, possession of the Property with New Liberty. … the general definitions section of the Law of Property Act 1925 defines "possession" as including "receipt of rents and profits or the right to receive the same, if any": see section 205(1)(xix). Whilst that definition is provided for the purposes of a particular statute, and may be excluded where the context otherwise requires, in my judgment it encapsulates a concept well known to, and understood by, real property lawyers engaged in the drafting of leasehold documentation. Moreover, the alienation restrictions in the present Lease expressly differentiate between the concepts of "possession" and "occupation" which… should therefore not be conflated. … On the footing that the entire Property is underlet to a third party, the effect of this arrangement is that New Liberty is to deal with the Property as the Defendant, as its head leasehold owner, would otherwise have been expected to deal with it, and the Defendant no longer has the right to do so. In my judgment, that amounts to a parting with or sharing of possession.

[13] NatWest appeals against the declaration granted on that basis and, by its respondent's notice, Clarence House contends that the judge should have accepted as grounds for his declaration that NatWest had acted in breach of the lease by entering into the virtual assignment, and thereby executing a declaration of trust, and/or by assigning the whole of the demised premises to New Liberty without obtaining the previous consent in writing of Clarence House and/or by underletting the whole of the demised premises or permitting the creation of a derivative underlease without consent.

Discussion

[14] It is helpful to begin with an overview of the nature and effect of the virtual assignment. It is an agreement between the tenant and a third party the landlord is not a party and, as in this case, may even be blissfully unaware of these contractual arrangements. That is not to say that the agreement does not have a practical effect on the landlord. The reaction of this landlord shows its concerns. The key feature seems to me to lie in the appointment of New Liberty as NatWest's agent: see clause 4 of the virtual assignment and the power of attorney. In collecting rent from the subtenant, Mercers, and paying it to the landlord, Clarence House, New Liberty acts not in its own right but as agent for NatWest. As such, these actions by the agent do not impinge on the legal relationship that exists and continues to exist between landlord and tenant (the respondent and the appellant). There is a different understanding between NatWest and New Liberty. Although New Liberty must pay rent to the respondent, and does so as agent in discharge of NatWest's debt, the virtual assignment nevertheless imposes the obligation to pay the rent firmly on New Liberty without its having the right of indemnity to recover those payments from NatWest that an agent would ordinarily have against its principal. Likewise,

---

* Editor's note: Also reported at [2005] 3 EGLR 73

A when collecting rent from Mercers, it will call for that rent as NatWest's agent, but, again contrary to an ordinary agency, it keeps the rent and does not account to its principal for it.

[15] The other key provision is the indemnity contained in clause 3. New Liberty must pay the rent and perform all the covenants and obligations contained in the leases and underleases and must indemnify NatWest against any liability in respect of those matters.

[16] Taken together, clauses 4 and 3 operate to achieve the objective set in clause 2, namely to pass all but only the *economic* benefits and burdens of the leases and underleases. What it does not do is alter the underlying relationship between landlord and tenant. The virtual
B assignment does not transfer to New Liberty any proprietary right or interest in the premises whatsoever: NatWest continues to be both tenant under the lease and landlord under the sublease. Nor does the voluntary assignment confer on New Liberty any contractual right in respect of the premises: New Liberty has no entitlement to occupy the premises, the right of occupation being NatWest's.

[17] Thus, as its clever name suggests, we are dealing with what is virtually an assignment almost an assignment, but not quite. Hence, I agree with Mr Christopher Nugee QC, who now appears for the appellant, that:

> C As its name suggests, the VA is not in fact an assignment it merely mimics the economic result of one without changing the legal position vis à vis third parties at all.

[18] In broad terms, this analysis can stand easily alongside this court's analysis of the different virtual assignment made by Abbey National Building Society. There, Abbey transferred all its economic benefits and burdens under its leases to a company called Mapeley Columbus Ltd under a virtual assignment, the terms of which were similar but not always identical to those in NatWest's case. Mr Jonathan Gavaghan, for the respondent, drew attention to and emphasised the caveat that appears in the clause in that case setting out
D the aspirations of the parties, but that does not appear in clause 2 of the virtual assignment before us. There, the relevant clause reads as follows with the italicised words, being those on which Mr Gavaghan relied:

> 4. The intention of the Virtual Assignment is to pass to the Purchaser all of the economic benefits and burdens of the Leases and Underleases in respect of the Properties, together with the obligation to manage all dealings with the Landlords and Undertenants as if the Properties had been assigned to the Purchaser *but without this Virtual Assignment creating, vesting or granting any legal or equitable estate in the Premises to the Purchaser*. To this end, any monies from the Undertenant pursuant to any Underleases, together with all proceeds of any disposal of the Leases or for the surrender of any Underleases, shall belong to the Purchaser.

E

[19] In some cases, Abbey had sublet its leasehold property and was no longer in occupation (the white leaseholds). Mapeley was entitled to receive the rent from the underlessees. In other cases, Abbey remained in occupation (the grey leaseholds) and paid Mapeley "a principal fee" similar to the rent that it would have paid had there been a formal lease back. The Customs & Excise Commissioners concluded that in neither case was there an exempt supply consisting of the "leasing or letting of immovable property" within the meaning of the relevant VAT directive, but rather a standard-rated supply of agency and property management services. The Commissioners contended that the effect of the virtual
F assignment was to give neither a legal nor an equitable interest to Mapeley and, accordingly, Abbey remained the legal and equitable lessee. Since Abbey had conferred no right to occupy on Mapeley, Mapeley could not have made a supply of letting of the property back to Abbey, nor could Mapeley make a supply of "letting" to the subtenants. In the first case, the white leaseholds, the rents due to Abbey under the underleases but paid to the virtual assignee were, as the VAT and Duties Tribunal and Hart J held, exempt from VAT. There was no appeal from this part of the decision. In the other cases, the grey leaseholds, where Abbey remained in occupation, the tribunal held that the assignee had made a standard-rated supply of agency and property letting services to Abbey. Abbey's appeal against that decision was allowed by Hart J, but on this point his decision was reversed by the Court of Appeal.

[20] The Court of Appeal analysed the nature and effect of the G
contractual arrangements between Abbey and Mapeley, both as a matter of English national law and also as a matter of Community law. As to the former, Jonathan Parker LJ held:

> [50] As noted earlier, it is common ground that these contractual arrangements did not involve the transfer of any proprietary interest in the properties in question from Abbey to Mapeley, whether at law or in equity. Further, it is rightly accepted by Mr Goy that they did not, and were not intended to, transfer to Mapeley any contractual right to occupy the properties (for obvious reasons, given the existence of covenants against parting with possession without the consent of the landlord). In consequence, Abbey's status as tenant remained the same after as before the execution of the virtual assignment: it continued H
> to occupy the properties the subject of the virtual assignment in right of its status as tenant under the leases. It follows as night follows day that since Mapeley had itself no proprietary or contractual right to occupy the properties in question, it was not in a position to "lease back" (in the sense of re-transfer) such a right to Abbey. As the Tribunal rightly concluded (see paragraph 52 of its Decision…), the contractual arrangements between Abbey and Mapeley were effective merely to regulate, as between the two of them, Abbey's exercise of its continuing right (*qua tenant*) to occupy the properties.

Paragraph [52] of the tribunal's decision (as quoted by the Court of Appeal in [27]) said:

> 52. The single answer to these two attractive arguments is that [Abbey's] J
> right of occupation, and the right to enjoy the properties by assignment or underletting, derived from the lease with its landlord and not from the virtual assignment. The virtual assignment was a personal collateral agreement between [Abbey] and Mapeley which did not grant rights of occupation but which did regulate how certain rights were to be exercised but only as between [Abbey] and Mapeley. The fact that [Abbey] agreed with Mapeley that [Abbey] would occupy the premises subject to the occupancy regulations made no difference. The regulation of the right to occupy is not the same as the grant of a right to occupy. The right to occupy was given by the lease and the occupancy regulations were merely an agreement as to how the rights of occupation should be exercised as between [Abbey] and Mapeley.

K

[21] As for the application of Community law, Jonathan Parker LJ said:

> 86. In my judgment the Community law authorities which I have examined leave no room for doubt that a right of occupation is an essential and fundamental element of a transaction of leasing or letting for the purposes of Article 13B(b). It follows that since, for reasons given earlier, Mapeley acquired no right of occupation of the properties the subject of the virtual assignment, and hence was never in a position to transfer such a right back to Abbey, the supply made by Mapeley to Abbey under the contractual arrangements in question is not a supply of "*leasing or letting*" within the meaning of Article 13B(b) and hence is not exempt from VAT. In my judgment the Commissioners are correct in characterising the supply by Mapeley as a standard-rated supply of agency and L
> property management services.
>
> …
>
> 89. In the instant case Abbey is, in my judgment, attempting to have it both ways in just the manner described by the ECJ in the last sentence of paragraph 33 of its judgment in *Cantor Fitzgerald* [2001] STC 1453, [2002] QB 546… Any grant by Abbey to Mapeley of a proprietary interest in the properties in question required its landlords' consent, as did any conferring by Abbey on Mapeley of a contractual right of occupation of those properties. Hence, given the absence of such consent, the need to resort to the device of a virtual assignment designed to produce the same commercial results, but without involving Abbey in doing either of those things. Yet Abbey is now contending that for VAT purposes it is to be treated as having done precisely that which the M
> virtual assignment was specifically designed not to do.

[22] Mr Gavaghan argued for the respondent that the added words in the Abbey/Mapeley transfer of the economic benefits and burdens ("without… creating… any legal or equitable estate in the Premises") make that case crucially different from this present case because that reservation is absent from our clause 2, which I have recited in [3] above. Thus, he submitted that *Abbey* can and must be distinguished. I doubt the correctness of that submission. It *may* make a difference to the question of whether the virtual assignment creates a trust that is an issue that arises under the respondent's notice, but Mr Nugee is surely right when he contended that it is not a relevant distinction because it

cannot, I think, be suggested that the virtual assignment did actually transfer any estate in the premises. The respondent also contends that the case is of limited assistance because of the concessions that were made, as recorded in [50] of Jonathan Parker LJ's judgment, concessions that are certainly not being made here by Mr Gavaghan. There, it did not lie in the mouth of the tenant to say that VAT should not be charged because it had transferred a property interest, or a right to occupy without the consent of the landlord that would be to have it both ways, which was the stance that Jonathan Parker LJ criticised in [89]. Even though the concessions played an obvious part in the decision, as I read the judgment of Jonathan Parker LJ he found the concessions to be inevitable, and so I would not regard the authority of that case to have been so undermined that we can disregard it. With respect, I think that it was the right decision. The fact that Abbey was in occupation whereas NatWest is not in occupation makes no difference to that conclusion.

[23] With those introductory observations, I turn to the issues that directly arise on this appeal.

First issue: parting with possession

[24] The judge found and declared that by entering into the virtual assignment NatWest either parted with possession of the property to New Liberty or at least shared or permitted the sharing of possession of the property with New Liberty. Mr Nugee challenged the correctness of that finding. Mr Gavaghan supported it for the reasons given by the judge.

[25] There are, very broadly, two strands to the judge's reasoning that I can précis as follows: first, warning that possession and occupation must not be conflated, he concluded that because of the wide meaning to be given to "possession", the receipt of rents by New Liberty was to put it in possession; and, second, by requiring New Liberty to deal with the property as NatWest would otherwise have done, NatWest was at least parting with or sharing possession.

[26] Mr Gavaghan submitted that it is well settled, *per* Neuberger LJ in *Akici* v *LR Butlin Ltd* [2005] EWCA Civ 1296; [2006] 1 WLR 201*, in [26], that:

> the covenant against parting with possession of the whole or part of the premises… should be given its normal, and technically legally correct, meaning, unless there is any good reason to construe it in some other way.

He emphasised the "technically legally correct" meaning. The technically legally correct meaning that has been accepted both before and since the Law of Property Act 1925 (the 1925 Act) includes possession arising from the receipt of rents and profits or the right to receive them. This was the established meaning applied in *Lyell* v *Kennedy (No 4)* (1889) LR 14 App Cas 437. This is now reflected in the definition of possession to be found in section 205(1)(xix) of the 1925 Act, namely that:

> "Possession" includes receipts of rents and profits or the right to receive the same,…

It is, however, important not to lose sight of the opening words of section 205(1), namely:

> In this Act unless the context otherwise requires, the following expressions have the meanings hereby assigned to them respectively, that is to say…

[27] To establish New Liberty's right to receive the rents, Mr Gavaghan relied on clause 4.1.3 of the virtual assignment under which NatWest appointed New Liberty to collect and get in all rents due under the underlease and clause 4.2 that prevents NatWest doing so itself. Clause 8 makes it plain that the rent belongs to New Liberty ("will be [received] to the account of the buyer"). He submitted that reliance on the agency relationship is misplaced because the true substance of the role taken on by New Liberty is to collect the rents for itself and not for anyone else. In reality, it is New Liberty that can decide whether to sue for any arrears of rent, and it is impossible to envisage that NatWest would ever be able to refuse to give consent to proceedings being brought for the purpose of securing New Liberty's economic benefits.

[28] Mr Nugee's answers are, very broadly, that "possession" in a covenant restraining the parting with or the sharing of possession in the premises is to be given its "normal" meaning of the right to enter and occupy the land to the exclusion of others, and it should not be given its extended meaning as including the right to receive rents and profits of land. Even if that extended meaning applies, New Liberty is not in receipt of the rents nor does it have a personal right to receive them because it merely collects the rents as agent for NatWest and not in its own right. Even if New Liberty had the right to collect rent, it did so as the assignee of a chose in action, not because of any proprietary right or interest in the land and, in particular, not as assignee of the lease.

[29] Having given careful consideration to these respective arguments, I unhesitatingly conclude that I prefer Mr Nugee's analysis. The position appears to me to be as follows.

[30] The starting point is *Akici* because that case, decided by the master in this field, Neuberger LJ, as he then was, dealt with covenants against parting with or sharing possession in this way:

> 23. The difference between possession and occupation is rather technical, and, even to those experienced in property law, often rather elusive and hard to grasp. Nonetheless, it is very well established, and is particularly important, and indeed well known, in the field of landlord and tenant law, especially in relation to the question of whether an agreement creates a tenancy or a licence, and in relation to alienation covenants such as clause 4.18.
>
> 24. While the interpretation of a word or phrase in a document must ultimately depend upon the documentary and factual circumstances in which it was agreed, it is desirable that the courts are as consistent as they properly can be when construing standard phrases in standard contexts. In that connection, a covenant against parting with possession is included in many, quite possibly most, modern commercial leases. Further the courts have consistently given the strict meaning to such covenants as was adopted in unreserved terms by the Privy Council in the *Lam Kee Ying* case [1975] AC 247, and in the five cases therein referred to, at pp255/256.
>
> 25. Accordingly, while one cannot lay down any immutable rule as to how a particular word or expression is to be construed in every document or lease, I consider that any court must be very cautious before construing the word "possession" as extending to occupation which does not amount to possession, especially in a familiarly expressed covenant against parting with possession in a detailed professionally drafted commercial lease, such as that in the present case.
>
> 26. In these circumstances, I consider that it would require a very strong and clear case before a covenant against parting with possession should be construed in any way other than that adopted by the Privy Council in the *Lam Kee Ying* case, particularly in the light of the consistent approach taken in the earlier authorities cited therein. In agreement with Judge Dean QC, therefore, I would hold that the covenant against parting with possession of the whole or part of the premises in the present case should be given its normal, and technically legally correct, meaning, unless there is any good reason to construe it in some other way.
>
> 27. I turn to the covenant against sharing possession. On the face of it, one would expect the word "possession" to have the same meaning each time it appears in clause 4.18, particularly in light of the fact that it is a word which is familiar, especially in the context of leases, to lawyers. The only reason for not giving the word "possession" its normal technical meaning in a covenant against sharing possession appears to be that identified by Sir Douglas Frank QC, namely that possession is, as it were, unitary and cannot be shared.
>
> 28. I do not accept that possession cannot be shared. … as a matter of ordinary language, a lessee who lets another person into possession of the demised premises, so that they are both in possession, can properly be said to "share" possession with that other person. …
>
> 29. … [The covenant] has a real effect, namely to prevent the conversion of a tenancy to a single lessee into what, in practical terms, will amount to a joint tenancy.
>
> 30. It may be said that this conclusion will result in a covenant against sharing possession having relatively little value. The answer to that point may be said to be the same as that given in the *Lam Kee Ying* case [1975] AC 247, 256C, namely that "the words of the covenant must be strictly construed, since if the covenant is broken a forfeiture may result". That approach may well be a little less powerful than it was 30 years ago, on the basis that such canons of construction are now given rather less weight. Nonetheless, the

* Editor's note: Also reported at [2006] 1 EGLR 34

modern approach, namely that such covenants should be given what is, in their documentary, factual and commercial context, their natural and commercially sensible meaning, indicates, to my judgment, the same result. Further, I do not think one should lean in favour of giving a wide meaning to an absolute covenant (ie one which is not subject to a proviso that consent cannot be unreasonably withheld).

[31] The strict meaning given to the covenant in *Lam Kee Ying Sdn Bhd* v *Lam Shes Tong (t/a Lian Joo Co)* [1975] AC 247 was stated, at p256B-C, to be that:

> A covenant which forbids a parting with possession is not broken by a lessee who in law retains the possession even though he allows another to use and occupy the premises.

Their lordships approved the judgment of Farwell J in *Stening* v *Abrahams* [1931] 1 Ch 470, at pp473-474, that a lessee that grants a licence to another to use the demised premises does not commit a breach of the covenant:

> unless his agreement with his licensee wholly ousts him from the legal possession… nothing short of a complete exclusion of the grantor or licensor from the legal possession for all purposes amounts to a parting with possession.

[32] How are we to apply those principles to a consideration of the covenant in this case that is "not to share or permit sharing possession or occupation of the demised premises or any part thereof or part with possession or occupation of the same"? There is no question of sharing or parting with occupation: the case turns on possession. As I see it, the proper approach must be as follows:

(1) We must give possession the same meaning it bears in respect of sharing possession as it bears in respect of parting with possession.

(2) Because this is a common phrase in a standard alienation clause, "possession" must be given its *normal*, albeit also its technically legally correct meaning. The true meaning must take its colour from its context. Although "possession" and "occupation" are different concepts (the difference between the tenant and a licensee, such as the lodger) and although they must not be conflated as Judge Hodge said, nevertheless their juxtaposition is part of the context and serves to emphasise the aspect of physical control that is part of possession. As Lord Browne-Wilkinson pointed out in a different context in *JA Pye (Oxford) Ltd* v *Graham* [2002] UKHL 30; [2003] 1 AC 419, in [40]:

> there are two elements necessary for legal possession: (1) a sufficient degree of physical custody and control ("factual possession"); (2) an intention to exercise such custody and control on one's own behalf and for one's own benefit ("intention to possess").

Moreover, as Chadwick LJ said in *Manchester Airport plc* v *Dutton* [2000] 1 QB 133*, at p142B-C:

> possession is synonymous… with exclusive occupation that is to say occupation (or a right to occupy) to the exclusion of all others, including the owner or other person with superior title.

The hallmark of the right to possession is the right to exclude all others from the property in question. That is the ordinary and normal sense of the word and that is the meaning that it should be given in this covenant.

(3) In the standard covenant against sharing or parting with possession, we should therefore adopt the standard strict meaning and find the breach of the covenant only against: (i) sharing possession if, by entering into this arrangement, NatWest has allowed New Liberty to enjoy joint possession of the demised premises with it; or (ii) parting with possession if, by entering into the virtual assignment, NatWest wholly ousted itself or completely excluded itself from the legal possession of the demised premises for all purposes.

(4) The short answer to the case is that at the time the virtual assignment was entered into, and at all material times thereafter, NatWest was not and has not itself been in possession of the demised premises. The demised premises are the suite of offices on the ground floor and first floor of this building in Clarence Street, Manchester. The occupier and person in possession is Mercers, not NatWest. By virtue of the underlease, NatWest had divested itself of possession and the only party in possession was, is and remains Mercers and Mercers alone. The virtual assignment did not alter that state of affairs. Not having been in possession of the premises, NatWest cannot possibly be said to have parted with possession to New Liberty or to have shared possession by reason of entering into these arrangements.

(5) There is a long line of authorities, now well established, dealing with breaches of the covenant against parting with or sharing possession. For example, *Stening* decided that a licence to erect an advertisement on the wall of premises did not constitute parting with possession of the wall. *Lam Kee Ying* held that the transfer to a newly formed company of the partnership business being conducted on the premises was a parting with possession. In *Akici*, allowing a third party into occupation to carry on business from the premises was a sharing of possession. This stream of cases is consistent with the notion that a leasehold covenant against parting with or sharing possession is concerned with the question of whether the tenant has allowed another into physical occupation with the intention of relinquishing its own exclusive possession of the premises to that other. Consistency demands that this case be approached in that light and, in that light, there has been no breach of the covenant.

[33] It is of course correct, as Mr Nugee conceded, that for some purposes "possession" has a meaning that extends beyond its core meaning and includes receipt of rents and profits of the land or the right to receive them. The crucial question here is whether New Liberty receives the rents or has the right to receive them.

[34] The judge found clause 2 of the virtual assignment to be compelling because it expressed the intention of the parties as being "to pass" to New Liberty "all of the economic benefits" so that "any monies from the undertenants… shall belong to" New Liberty. So, he held that the effect of the arrangement was that New Liberty was to deal with the property as NatWest would otherwise have been expected to do but no longer had the right to do. That, in his judgment, amounted to a parting with or the sharing of possession.

[35] Although that expression of intention informs the agreement, one still has to consider how it was to be achieved. The mechanism is provided by clause 4, which irrevocably appoints New Liberty to be NatWest's agent, to act on its behalf and in its name in all dealings connected with the properties, those dealings including collecting and getting in all rent due under the underlease. Thus, when it collects the rent and when it receives the rent it does so under and by virtue of the express terms of the arrangement as agent of NatWest and in NatWest's name, not in its own name. When New Liberty knocks on the undertenant's door, it may be able to say "you must pay me the rent *on behalf of* your landlord", but it cannot say "you must pay me the rent because you *owe me* the rent". At the moment of receiving the rent, New Liberty held it for NatWest. Thereafter, and perhaps even *eo instanti,* it changes its character by virtue of clause 8, for it then becomes money received to the account of New Liberty. The passage of that scintilla of time remains crucial. The appointment of New Liberty as agent and the grant of the power of attorney to it seem to me to be such integral and essential elements of the scheme that the agency cannot be discounted and swept aside. The appointment of an agent cannot itself amount to a parting with the right to receive the rents, even though NatWest forfeits the rents the moment after receipt. This is a purely contractual arrangement between NatWest and New Liberty and it does not result in a transfer to New Liberty of the right to receive rents and profits in the sense envisaged by the definition in section 205 of the 1925 Act. In my judgment, Mr Nugee was right to test the matter by asking who would be able to sue Mercers if it refused to pay the rent. The answer is obvious. NatWest would have to sue because it is the party to the underlease entitled to receive the rent, and any action brought by New Liberty in its name would be defeated. The fact that NatWest is to all intents and purposes never likely to be able to refuse to give consent to its name being used is nothing to the point.

---

* Editor's note: Also reported at [1991] 1 EGLR 147

[36] The appointment of managing agents to collect rent for NatWest would not put NatWest in breach of the covenant against parting with possession, nor could it be argued that if the managing agent were to keep 10% of the rent that would be a sharing of possession.

[37] Accordingly, I conclude that New Liberty does not, when it collects rent as agent of NatWest, receive that rent itself in the sense of putting it into possession of the demised premises. Nor, prior to collection, does it have the right to receive rent in that sense.

[38] Mr Nugee also made the point that even if the effect of the virtual assignment were to transfer to New Liberty the right to receive the rent, the consequence was not to put New Liberty "in possession" of the premises but merely to pass to it a chose in action separate from the land.

[39] The point here is that a landlord may assign the right to receive the rent without assigning the reversion: *Kataria* v *Safeland plc* [1998] 1 EGLR 39. Such an assignment treats the rent as a chose in action and the assignee acquires no interest in the reversion. The assignee can recover the rent as a debt but cannot forfeit for non-payment of rent or distrain for rent, those rights being an incident of the reversion that does not pass to the assignee: see *Kataria*. I agree with Mr Nugee: even an express assignment of the rent by itself would not amount to sharing or parting with possession of the demised premises.

Conclusion on the first issue

[40] In my judgment, Mr Nugee is correct. NatWest did not, by entering into this virtual assignment share possession or part with possession of the demised premises or any part thereof. Unless, therefore, the judgment can be supported for reasons argued in the respondent's notice, I would allow this appeal.

Respondent's notice: The first question is the virtual assignment a declaration of trust with regard to the property or the lease?

[41] The judge ruled against New Liberty, holding that the relationship between it and NatWest was founded in contract rather than in equity. It was not a bare trust because, under a bare trust, the trustee's sole duties to the beneficiary are not only to allow him to enjoy the trust property but also to obey any direction the beneficiary may give as to how the trust property should be disposed of by putting an end to the trust, (in accordance with the rule in *Saunders* v *Vautier* (1841) Cr & Ph 240, affirming (1841) 4 Beav 115. This is not something available to New Liberty under the terms of the virtual assignment. With some hesitation, the judge held that there was no breach of the prohibition in the lease against declaring any trust of the property or the lease.

[42] Mr Gavaghan submitted that the judge had erred. He submitted that the effect of the voluntary assignment was that NatWest held the bare legal title in the leasehold estate on behalf of New Liberty. All the economic benefits and burdens of the lease passed and all that was left with NatWest was the legal title. The absence of words such as "declaration of trust" were not determinative and the label the parties put on the document or transaction was likewise not determinative of its proper legal characterisation. One must look to the substance, not the form. The judge had been wrong to base his conclusion that no bare trust was created because of a concern about the rule in *Saunders*.

[43] The voluntary assignment does bear some resemblance to a trust in so far as the underlease is vested in NatWest but New Liberty enjoys the economic benefits that flow from it. By itself, that is not sufficient. One must also explore the manner in which the relationship between the parties was created in order to determine whether the arrangements bear the hallmark of a trust that in essence is, as the House of Lords restated in *Westdeutsche Landesbank Girozentrale* v *Islington London Borough Council* [1996] AC 669, the existence of an obligation binding on the conscience of the person vested with the legal ownership. Jessel MR explained the principle thus in *Richards* v *Delbridge* (1874) LR 18 Eq 11, at p14:

> The principle is a very simple one. A man may transfer his property, without valuable consideration, in two ways: he may either do such acts as amount in law to a conveyance or assignment of the property, and thus completely divest himself of the legal ownership, in which case the person who by those acts acquires the property takes it beneficially, or on trust as the case may be; or the legal owner of the property may, by one or other of the modes recognised as amounting to a valid declaration of trust, constitute himself a trustee, and, without an actual transfer of the legal title, may so deal with the property as to deprive himself of its beneficial ownership, and declare that he will hold it from that time forward on trust for the other person. It is true he need not use the words, "I declare myself a trustee," but he must do something which is equivalent to it, and use expressions which have that meaning, for, however anxious the Court may be to carry out a man's intention, it is not at liberty to construe words otherwise than according to their proper meaning.

[44] Here, the proper meaning and cornerstone of the arrangement is agency: New Liberty is appointed the agent of NatWest in respect of the property. New Liberty takes not only the benefits but also assumes the responsibility for managing the property. If it were a trust in the full sense rather than simply a bare trust, the fiduciary obligations would have been owed *to* New Liberty, not *by* New Liberty.

[45] On Mr Gavaghan's analysis, this is a bare trust of the kind described in *Snell's Equity* (31st ed), in para 19-11:

> A bare (or simple) trust is one in which the property is vested in one person on trust for another, the nature of the trust imposing no active duties or powers of management on the trustees. ... In such a case, T's sole duty is to B to enjoy the property and to obey any direction he may give as to how the property should be disposed of.

He noted that New Liberty is given the right to negotiate and complete all assignments of the lease (see clause 4.1.9) and that assignments include the assignment of the property by NatWest to New Liberty. Hence, it is submitted that the rule in *Saunders* applies because the beneficiary is entitled to wind up the trust and require the trustee to assign to it the subject matter of the trust. A bare trustee, or a trustee that is being directed to deal with a lease at the direction of those entitled under the rule in *Saunders*, cannot be compelled to act in a way that would put it in breach of its obligations to the landlord under that lease. There are parallels in that context with the qualified trusteeship that arises on the sale of land, where the vendor is sometimes said to become a trustee for the purchaser as from the date of the contract (with the property correspondingly being at the purchaser's risk). The trusteeship in the case of vendor and purchaser depends on the availability of specific performance: see *Howard* v *Miller* [1915] AC 318, at p326. Specific performance would not be granted of the assignment of a lease in the face of opposition from the landlord if that would result in a breach of covenant and a risk of forfeiture.

[46] Similarly, NatWest could not be compelled to assign the lease to New Liberty when to do so would put it in breach of covenant under the terms of the lease, exposing it to a claim for damages. New Liberty does not have the right to put an end to the alleged "trust", by calling for a transfer of the legal title, because the assignment cannot take place without the consent of the landlord, and even though NatWest is obliged to co-operate in connection with any assignment, NatWest cannot compel Clarence House to agree to the assignment. The point is made clear by reference to clause 3 of the virtual assignment. New Liberty is obliged, among other matters, to observe the covenants in the lease. It could not, without breach of that provision, compel NatWest to assign the lease to it. Accordingly, as between New Liberty and NatWest, there is no present right to call for an assignment and there is therefore no bare trust.

[47] It was with some hesitation that the judge reached the conclusion that there was no breach of the prohibition in the lease against declaring any trust of the property or the lease. I am satisfied that there is no trust. I recognise that virtual assignments are strange new beasts in the forest that one must circle around them suspiciously and cautiously but the moment one gets close and has a good sniff, the overwhelming smell is of contract, not trust. Although the judge would not for a moment have expressed himself in such an inelegant way, so lacking jurisprudential precision, this was the central finding on this point, and he was correct in that conclusion.

Second issue in the respondent's notice: Has there been a breach of the prohibition on assignment?

[48] The lease requires the landlord's consent before the tenant can assign, and the respondent contends that assignment covers both a legal and an equitable assignment and that the voluntary assignment was in effect an equitable assignment of the lease.

[49] Central to this issue is the proper construction of the words "not to assign, transfer, charge or underlet the whole of the demised premises… without obtaining the previous consent in writing of the landlord…". The issue is whether the words should be interpreted as though the word "legally" had been inserted before the word "assign". The general rule is established by *Gentle* v *Faulkner* [1900] 2 QB 267, at pp276-277, where Romer LJ held:

> it seems to me to be clear that a covenant in a lease against assigning the demised premises, in the absence of any context shewing that the covenant is to have an extended meaning, covers only a legal assignment. The covenant against assignment is, therefore, not broken by anything short of a legal assignment. In my opinion such a covenant is not broken by the lessee executing a declaration of trust of the demised premises.

[50] Mr Gavaghan pointed out that the many non-alienation provisions in this lease include the prohibition against entering into a declaration of trust, the underlying commercial reason for which, he submitted, is to ensure that the person with the economic interest in the lease remains the tenant, not some third party that does not have the landlord's consent. Likewise, the prohibition against entering into a transfer of the lease suggests that it is the act of passing the benefit of the lease to someone else, rather than the formal step of registration, that shows the underlying intention. Likewise, the restriction on charging shows the extent of the landlord's control. Hence, the submission is that in the context of the lease as a whole, the commercially sensible meaning of the word "assign" is not restricted simply to legal assignments but to assignments generally.

[51] I do not accept those submissions. As Neuberger LJ made plain in [24] of his judgment in *Akici* (see [30] above), "it is desirable that the courts are as consistent as they properly can be when construing standard phrases in standard contexts". This is a standard phrase in a standard context and it has been settled, since *Gentle,* that "assignment" means a legal assignment unless the context compels a different construction. There is nothing in the context here that requires that wider meaning to be given to the well-known phrase. Clause 30 of the lease dealing with alienation sets out in 10 subclauses precisely what acts must not be done, and it seems to me impossible to conclude that if the draftsman had in mind a bar on equitable assignment he would not have included that. As the judge correctly held, in [16]:

> The fact that the alienation provisions in the lease restrict several, specified forms of alienation leads me to the conclusion that the parties intended that only those forms of alienation should fall foul of the alienation restrictions: the express inclusion of several excludes the implication of other forms of alienation.

[52] In my judgment, the judge was correct for the reasons that he gave. Further, I agree with the submissions advanced by Mr Nugee that there has not been an assignment of the lease to New Liberty because there is no contractual obligation on NatWest to assign the lease to it. New Liberty has, at most, the power to call for an assignment or to effect an assignment as agent of NatWest. Unless and until such an assignment is called for or made, there is no assignment, whether legal or equitable. In any event, any assignment is conditional on the landlord's consent; no consent is forthcoming from the respondent, which has always shown the greatest reluctance to accept New Liberty and there is therefore no possibility of New Liberty obtaining an order for specific performance of the obligation (if there is one) to assign.

Third ground of the respondent's notice: Was there an underletting of the premises?

[53] The respondent submits that although the virtual assignment is not drafted as an underlease, its substantive effect is the same. That submission was rejected by the judge who held, in [13]:

> Neither in form nor in substance did the Virtual Assignment amount to an underletting. Perhaps the most formidable objection to treating it as such is the fact that it neither reserves, nor purports to reserve, any reversion to the Defendant.

[54] In my judgment, the judge was correct. An underlease is the demise of the property for a term that creates a new inferior estate and constitutes the relationship of landlord and tenant between the parties. That is wholly absent from this arrangement. Put very simply, there is no privity of estate.

Conclusion

[55] None of the grounds advanced by the respondent in the respondent's notice succeeds. The judge was correct to decide as he did. His error lay only in his conclusion that NatWest had either parted with possession or at least shared or permitted the sharing of the property with New Liberty. I would therefore allow NatWest's appeal, set the judge's order aside and dismiss the claim.

Jacob LJ said:

[56] I agree.

Warren J said:

[57] I also agree.

Appeal allowed.

LEASEHOLD ENFRANCHISEMENT

McHale and another
A
v Earl Cadogan

COURT OF APPEAL

[2010] EWCA Civ 14

21 January 2010

RIX, RIMER AND PATTEN LJJ

Estates Gazette 10 April 2010

B **[2010] 14 EG 110**

Leasehold enfranchisement — Leasehold Reform, Housing and Urban Development Act 1993 — Collective enfranchisement — Purchase price — Block of flats — Whether positive value to be applied to caretaker's flat — Whether service charge payable by underlessees to include loss of rack-rent from caretaker's flat in circumstances where head lessee obliged by headlease to provide such flat rent-free

The second appellant company was the nominee purchaser for the purpose of a collective enfranchisement application by which qualifying tenants of three flats sought to acquire the freehold of the
C block from the respondent, under section 24 of the Leasehold Reform, Housing and Urban Development Act 1993. The block contained five flats that were let on long underleases and a caretaker's flat. The participating tenants were the first appellant, who held the headlease of the block and was a qualifying tenant of the caretaker's flat, and the underlessees of two other flats.

The headlease referred to the caretaker's flat and contained covenants obliging the first appellant to use his best endeavours to provide a resident caretaker, who was to reside in the caretaker's flat rent-free on a service basis. Lessees of the flats were required under the terms of the underleases to contribute to the first appellant's
D expenses and outgoings in respect of "the costs of employing a housekeeper and… the accommodation (if any) to be provided for such housekeeper" together with "the cost to the Lessor of outgoings for such accommodation (including loss of rack-rent thereon)".

The leasehold valuation tribunal (LVT) determined the enfranchisement price at £770,940. In doing so, it determined the lease value of the caretaker's flat as nil on the ground that the first appellant had no power, under the terms of the headlease or underleases, to charge a rent for that flat. That decision was upheld by the Lands Tribunal, save that it adjusted the enfranchisement price to £780,405 to correct a mathematical error. The appellants appealed in respect of the
E valuation of the caretaker's flat. A further ground of appeal, relating to the LVT's method of calculating marriage value, was adjourned for a separate hearing.

***Held*: The appeal was allowed. Service charge provisions should be construed restrictively, with a policy of not bringing within the general words of such a provision anything that did not clearly belong there. However, in the underleases under consideration, the loss of rack-rent on the caretaker's flat was specifically defined to fall within "the cost of outgoings for such accommodation" and could not be excluded simply because the flat was required to be licensed to the caretaker rent-free. It was precisely because the flat had to be provided for a caretaker, who could not be charged rent, that the value of the flat in terms of rack-rent had been lost.** F
Consequently, the headlease should be valued on the assumption that the first appellant was entitled to be compensated in the service charge for the loss of rack-rent on the caretaker's flat.

The following cases are referred to in this report.

Agavil Investment Co Ltd v *Corner* unreported 3 October 1975
Earl Cadogan v *44/46 Lower Sloane Street Management Co Ltd* LRA/29/2003 and LRA/30/2003 30 July 2004, LT
Earl Cadogan v *27/29 Sloane Gardens Ltd* LRA/9/2005; [2006] 2 EGLR 89; [2006] 24 EG 178, LT
Gilje v *Charlgrove Securities Ltd* [2001] EWCA Civ 1777; [2002] 1 EGLR 41; [2002] 16 EG 182 G
Jollybird Ltd v *Fairzone Ltd* [1990] 2 EGLR 55; [1990] 43 EG 63, CA

This was an appeal by the appellants, Henry McHale and 10 Sloane Gardens Management Co Ltd, from a decision of the Lands Tribunal upholding a decision of the leasehold valuation tribunal on an application for collective enfranchisement of a block of flats under the Leasehold Reform, Housing and Urban Development Act 1993.

The first appellant appeared in person on behalf of the appellants; Anthony Radevsky (instructed by Pemberton Greenish) represented the respondent.

Giving judgment, **RIX LJ** said:

[1] This appeal concerns two questions of law that have arisen out H
of the legislation concerned with the collective enfranchisement of leasehold property. The first raises the issue of whether the leasehold interest, like the freehold interest, should, for the purpose of ascertaining the marriage value that the statute says has to be shared between those interested in each, be valued in what has been called the "no Act world", that is, on the basis that the right to enfranchisement does not exist. That issue has been adjourned to a separate hearing, and this judgment is not concerned with it.

[2] The second issue, with which this judgment is concerned, raises
a question concerning the valuation of a caretaker's flat in the property J
subject to enfranchisement. Should it be valued on the basis that it is dedicated to housing a caretaker rent-free and thus of little value, or should it be valued on the basis that the tenants (underlessees) of the premises collectively nevertheless have to pay a service charge that is designed to provide compensation for that loss of rent? If it is the former, the value of the leasehold interest is lower and the marriage value is correspondingly higher; and vice versa.

[3] The property concerned is 10 Sloane Gardens, London SW1, a building divided into six flats of which the respondent, the Rt Hon Earl Cadogan, is the freeholder (the freeholder) and
the first appellant, Mr Henry McHale is the intermediate landlord K
under a headlease. The second appellant, 10 Sloane Gardens Management Co Ltd, is the nominee purchaser for the purposes of the collective enfranchisement of the property under section 24 of the Leasehold Reform, Housing and Urban Development Act 1993 (the 1993 Act). It will be sufficient if I refer to the two appellants as simply "Mr McHale".

[4] Five of the six flats are underlet on long leases. The sixth is a caretaker's flat in the basement. By an initial notice served on 20 July 2004, three of the tenants, being two of the long underlessees and Mr McHale as tenant of the caretaker's flat under the headlease,

claimed to acquire the freehold of the premises under the 1993 Act. On 15 December 2004, the freeholder applied to the leasehold valuation tribunal (the LVT) to determine the enfranchisement price under Schedule 6 to the 1993 Act. That was before the amendments to the valuation date made by the Commonhold and Leasehold Reform Act 2002 came into effect. The terms of the transfer, other than the price, were agreed on 16 May 2006, which thereby became the valuation date for the purposes of Schedule 6.

[5] The LVT determined the enfranchisement price in the sum of £770,940 on 17 January 2007 in its decision of that date. On 7 February 2007, Mr McHale submitted an application for leave to appeal, which the LVT granted on 23 February 2007. One of the issues raised related to the valuation of the caretaker's flat. On 30 October 2008, the Lands Tribunal (LT) (Mr George Bartlett QC, president, and Mr Paul Francis FRICS) dismissed the appeal, but adjusted the enfranchisement price because of a small mathematical error to £780,405.

[6] The issue for the purpose of the caretaker's flat is whether the headlessee, currently Mr McHale, is entitled to charge the underlessees by way of service charge rent in respect of the caretaker's flat. The LT decided that he was not.

[7] By clause 2(3) of the headlease, the lessee was obliged, at its own expense, to carry out the works for converting the premises into self-contained flats as set out in the first schedule to the lease, which referred to the caretaker's flat in the basement. Clause 2(10) also mentioned the caretaker's flat in the basement as referred to in the first schedule and in drawings incorporated in the lease. Clause 2(11)(c) then contained a covenant on the part of the lessee:

> TO use its best endeavours to provide for the premises throughout the said term a full-time Caretaker… who shall reside in the Caretaker's flat rent-free as a licensee on a service basis…

[8] The underleases of the five other flats reserve a service charge. Thus, clause 2(4) of a sample underlease provides for payment of a proportionate part of the "expenses and outgoings incurred by the Lessor in… the provision of services therein and the other heads of expenditure as set out in the Third Schedule hereto". By para 3 of the third schedule, that service charge includes a proportionate part of:

> The cost of employing a housekeeper or housekeepers and also in respect of the accommodation (if any) to be provided for such housekeeper or housekeepers (a) the cost to the Lessor of outgoings for such accommodation (including loss of rack rent thereon)…

[9] The question therefore became whether the underlessees had to contribute to the cost to Mr McHale of providing the caretaker flat rent-free in the form of "loss of rack rent". Mr McHale, who has represented himself throughout, said that they did, and the freeholder said that they did not. As Mr Anthony Radevsky, on behalf of the freeholder put it to us, there could be no loss of rack rent where Mr McHale was required by his underlease to provide a caretaker's flat for a caretaker to live in rent-free. The LT agreed with the freeholder.

[10] There is some previous discussion of this issue in earlier decisions. In *Earl Cadogan* v *44/46 Lower Sloane Street Management Co Ltd*, a decision of the LT (HH Judge Rich QC and Mr Rose FRICS, 30 July 2004, LRA/29/2003 and LRA 30/2003), the provisions of the headlease and of the underlease were for material purposes identical to those found here: that is to say that the head lessee, by clause 2(12)(c), covenanted that the caretaker "shall reside in the Caretaker's flat rent-free as a licensee on a service basis", and the underlease by para 3 of its third schedule provided for the service charge to include "The cost of employing a housekeeper… and also in respect of the accommodation (if any) within the Building to be provided for such housekeeper… (b) the cost of providing such accommodation (including the loss of rack rent thereon)…". As in the present case, there was a dispute as to whether there had been a "loss of rack rent" in such a situation. But for a separate point of the LT's own taking, not argued before it (as we know because Mr McHale and Mr Radevsky were involved on opposite sides in that case too), Mr McHale would have succeeded in that case, as the following passage from the LT's decision makes it clear:

> 24. In our judgment, if paragraph 3 of the Third Schedule, in providing for the recovery of the cost of housekeeper's accommodation within the building, could only be referring to the caretaker's flat in the basement, the specific inclusion of "the loss of rack rent" within the definition of the "maintenance contribution" would, notwithstanding the provisions for certification of such charges being "incurred", entitle the under-lessor to include the rack rental value of the basement flat, precisely because the language is used in the context of the head-lease which requires it to be let rent-free. We would feel forced to give some meaning to the provision, rather than to reject it as meaningless or mere surplusage or an obvious mistake, and that is the only meaning that it could have.

[11] So far, that was a decision in Mr McHale's favour and if the matter had ended there in that case, Mr McHale would have succeeded there, and, if it were translated into this case and upheld by our decision in this court, would succeed as well here. However, the LT had its own point to the effect that it was possible that the headlease's reference to the caretaker's flat and the underlease's reference to accommodation for a housekeeper were not perhaps speaking of the same thing. The LT ultimately decided the point against Mr McHale, on what was essentially a contra proferentem basis. It said, continuing, in [24] of its decision:

> Once, however, it appears that the words could, at least theoretically, have meaning and effect even if they do not entitle the underlessor to recover a loss which he has not in fact suffered, there is no reason to construe "the cost of providing such accommodation (including loss of rack rent thereon)" as including a loss not incurred or suffered in an amount made irrecoverable by the terms of the head-lease. The underlease should not be construed as entitling the underlessor to recover as part of the maintenance charge a sum in excess of providing the services, unless such construction is unavoidable.

[12] Because that decision was ultimately in Lord Cadogan's favour there could be no appeal by him against the *obiter* observations of the LT previously cited above. In this appeal, however, Mr Radevsky submitted that those observations were wrong.

[13] The issue came up before the LT again in *Earl Cadogan* v *27/29 Sloane Gardens Ltd* (HH Judge Rich QC, 7 April 2006, LRA/9/2005*). The headlease again referred to a caretaker's flat in which the caretaker was to reside rent-free. The underlease, however, provided for a service charge to include "an annual sum equivalent to the market rent of any accommodation provided rent-free by the Lessor". The LT considered *44/46 Lower Sloane Street* and other authorities on service charge provisions and rejected a submission that the headlease prohibited the recovery of a notional rent for the caretaker's flat from the underlessees. The LT said:

> 21… By clause XI(c) the flat is to be provided to the Caretaker rent-free. That does not preclude a provision in any underlease to charge a rent to the underlessee for the flat granted to the underlessee, fixed by reference to the market value of the Caretaker's flat. No more does it, in my judgment, prohibit the requirement of a covenant to pay the same sum under the name of service charge but recoverable as rent. On its proper construction there is nothing in the headlease to affect the question to be decided, save that it is part of the context under which the under-leases must be construed, that the headlessee is bound to provide a Caretaker and to accommodate him rent-free. Mr Munro accepted that such obligations could sensibly be imposed by the freeholders for estate management reasons, namely to avoid the risk of any caretaker obtaining security of tenure, but to maintain the standard of the premises for the benefit not only of the demised premises but also of the surrounding area.
>
> …
>
> 25. I am forced to the conclusion that unless the words of the sub-paragraph are rejected they must be construed as entitling the second respondent to recover a notional rent of the Caretaker's flat as part of the Service Charge. That is the obligation that a reasonable tenant would perceive that he was entering into. There is no ambiguity and no need or right therefore to resort to the *contra proferentem* rule properly so-called…

---

* Editor's note: Reported at [2006] 2 EGLR 89

[14] In the present case, Mr Radevsky did not take the point that appealed to the LT of its own motion in *44/46 Lower Sloane Street* concerned with the theoretical difference between a caretaker's flat and a housekeeper's accommodation. Instead, he submitted that the LT was wrong in its underlying observations concerning the meaning of the identical provisions in the headlease and underlease in that case. He submitted, as he had submitted to the LT there and below, that there could be no loss of rack rent in a situation where the caretaker's flat had to have been licensed rent-free. He also relied on two other authorities concerning service charges, acknowledging that they concerned different terms, but for the propositions that the courts are in general concerned not to recognise anything as being within service charges unless the landlord has actually incurred expenditure, and that clear words are required before an obligation to pay is imposed on tenants in a block of flats.

[15] Thus, he referred the court to *Jollybird Ltd* v *Fairzone Ltd* [1990] 2 EGLR 55, where this court, upholding the decision of Morritt J, refused to extend a clause giving the landlord the costs of central heating to a situation where the landlord could make a profit over and above actual expenditure. The clause concerned was, however, described as "ill-drafted" (at p58M) and the particular issue of construction involved what this court saw as a choice between, on the one hand, an impermissible or absurd construction and, on the other hand, a construction whose only weakness was that its presence was not strictly necessary. I can derive no assistance from that case.

[16] Mr Radevsky also relied on *Gilje* v *Charlgrove Securities Ltd* [2001] EWCA Civ 1777; [2002] 1 EGLR 41, where this court dismissed a landlord's appeal against a decision that it could not recover notional rent for a caretaker's flat, because such rent did not fall within the definition of "monies expended". The leases in that case gave to the landlord a proportionate share of "All monies expended by the lessor in… providing the services… called for…". Those services were scheduled in terms that referred to the caretaker's flat as follows: "Gas, electricity, general and water rates and internal maintenance repairs of the flat occupied by the resident housekeeper or porter." The question in these circumstances was whether the notional rent of the flat was included in "monies expended". This court distinguished an earlier case *Agavil Investment Co Ltd* v *Corner* unreported 3 October 1975, where the notional rent of the caretaker's flat was permitted to be included in the service charge as being within the words "The costs, charges and expenses incurred…". Laws LJ, with whom Mummery and Kennedy LJJ agreed, said that "costs" went beyond "expenses" and could embrace the notional rent as income forgone: para 24. On the basis of that distinction and the ground of *contra proferentem*, it was held that the notional rent was not within "monies expended": para 28. Mummery LJ went on to say, in [33]:

> this underlease does not clearly include the notional rent of the caretaker's flat forgone by the landlord in the various items that the landlord is expressly allowed to recover from the underlessee in the service charge.

[17] I readily acknowledge that it is the policy of the authorities not to bring within the general words of a service charge clause anything that does not clearly belong there. To put the matter another way, service charge provisions have been construed restrictively. In the present case, however, the critical clause expressly included "the cost to the Lessor of outgoings for such accommodation (including loss of rack rent thereon)". Thus, the loss of rack rent on the caretaker's flat was specifically defined to be within "the cost of outgoings for such accommodation". It seems to me to be impossible in such circumstances to say that the loss of rack rent is nevertheless *not* included in such outgoings because the flat is to be licensed to the caretaker rent-free. It is precisely because the flat has to be provided for a caretaker who cannot be charged rent for it that the value of the flat in terms of rack rent has been lost. I agree with the LT in *44/46 Lower Sloane Street* that if this is not the meaning of the provision, it is denuded of all meaning. The LT in the present case was simply content to follow the result in *44/46 Lower Sloane Street* in the belief that that was mandated by the identical language in the present case. It does not seem to have appreciated that that result entirely turned on the housekeeper point, which Mr Radevsky does not pursue (in my opinion rightly so).

[18] As a result, I would allow this appeal on the caretaker flat point. The headlease has to be valued on the basis that Mr McHale is entitled to be compensated in the service charge for the loss of rack-rent on the caretaker flat. For the rest, this appeal is adjourned.

Rimer LJ said:

[19] I agree.

Patten LJ said:

[20] I also agree.

Appeal allowed.

Southern Pacific Mortgage Ltd
A
v Heath

COURT OF APPEAL

[2009] EWCA Civ 1135

5 November 2009

WALLER, DYSON AND LLOYD LJJ

Estates Gazette 13 February 2010

B **[2010] 06 EG 116**

Mortgage — Consumer Credit Act 1974 — Agreement in parts — Mortgage loan made to customer on condition that existing mortgage paid off out of advance — Customer defaulting on payments — Possession orders made — Whether mortgage unenforceable for failure to comply with requirements of section 127(3) of 1974 Act — Whether amount of advance within threshold for application of Act — Whether an agreement in parts within section 18(1)(*a*) such that each part to be considered as separate agreement

In 2002, the respondent's predecessor made a mortgage advance of £28,932 to the appellant, to be secured by a first legal charge on her
C home. It was a condition of the mortgage that the appellant's existing mortgage of £19,000 should be paid off out of the loan.

The appellant fell into arrears with her repayments under the 2002 mortgage and possession orders were made. On appeal against those orders, the appellant contended that the court had no power to enforce the mortgage since it had not been properly executed in accordance with the requirements of section 127(3) of the Consumer Credit Act 1974; although that provision had been repealed in 2007, it continued to apply to mortgages made prior to that date. She submitted that although the mortgage appeared, prima facie, to escape the consumer protection provisions of the Act, since it exceeded the
D then limit of £25,000, it amounted to a "multiple agreement" in two parts, each of which fell below that limit, namely: (i) an advance of £19,000 in the category of restricted-use credit within section 11(1)(*c*), which was to be used to refinance the appellant's existing mortgage; and (ii) the remainder, as unrestricted-use credit within section 11(2). She argued that, pursuant to section 18(1)(*a*) and 18(2), each part was to be regarded as a separate consumer credit agreement regulated by the Act.

Upholding the possession orders and dismissing the appeal, the High Court held that the mortgage was not an agreement in separate parts within section 18(1)(*a*) and 18(2), but was a single agreement
E that fell into more than one category, within section 18(1)(*b*), and was therefore enforceable: see [2009] EWHC 103 (Ch); [2009] 2 All ER 687. The appellant appealed.

***Held*: The appeal was dismissed. Section 18(1) covers two possibilities, in which there are elements of an agreement that fall within two or more disparate categories; the word "category" has the same meaning in section 18(1)(*a*) and 18(1)(*b*). Subparagraph (*a*) deals with the case where the respective elements within the agreement can be seen as constituting separate parts of the agreement; subpara (*b*) deals with the opposite case where they cannot be seen in that way. The two subparagraphs are mutually exclusive. Accordingly, the starting point must be to ascertain whether the agreement, by its terms, contains two or
more parts, in different categories, or whether the agreement F
as a whole, or part of it, falls into two or more categories. In the instant case, the agreement was one whose terms placed the entire agreement in two relevant and disparate categories under the Act, so that section 18(1)(*b*) applied. The terms of the agreement did not place part of it in one category and part in another. It was a single agreement that could not be dissected into parts, and that offered a single credit facility that could be drawn down only as a whole. Accordingly, it was not to be treated as two separate agreements. Since the amount of the credit provided exceeded £25,000, it was
not a regulated agreement.** G

The following cases are referred to in this report.

Burdis v *Livsey* [2002] EWCA Civ 510; [2003] QB 36; [2002] 3 WLR 762
Dimond v *Lovell* [2002] 1 AC 384; [2000] 2 WLR 1121; [2000] 2 All ER 897, HL
London North Securities Ltd v *Meadows* [2005] EWCA Civ 956
McGinn v *Grangewood Securities Ltd* [2002] EWCA Civ 522
National Westminster Bank plc v *Story* [1999] CCLR 70, CA
Southern Pacific Mortgage Ltd v *Heath* [2009] EWHC 103 (Ch); [2009] 2 All ER (Comm) 687

This was an appeal by the appellant, Jayne Heath, from a decision of
HH Judge Purle QC, sitting as a High Court judge, dismissing an appeal H
against possession orders granted to the respondent, Southern Pacific Mortgage Ltd, for mortgage arrears and rejecting a defence based on section 18 of the Consumer Credit Act 1974.

Bradley Say (instructed by Burton & Co LLP, of Lincoln) appeared for the appellant; Malcolm Waters QC and Jonathan Hough (instructed by Glenisters) represented the respondent.

Giving judgment, **LLOYD LJ** said:

Introduction

[1] This appeal raises for decision a question as to the correct
interpretation of section 18 of the Consumer Credit Act 1974 J
(the Act) that has serious consequences for transactions of a commonplace nature, and that has been the subject of strongly divergent academic comment.

[2] The Act seeks to protect consumers by imposing a number of distinct requirements with regard to the documentation of transactions to which it applies. Some of these prescribe the content of the document to be provided to the consumer, for example, that the amount of the credit (a term defined by the Act) must be specified. Others require the consumer to have the opportunity to think about the terms of the agreement, or proposed agreement, and to change his mind. What
requirements apply to any given transaction or proposed transaction K
depends on the nature of the transaction. For that purpose, the Act has a number of definitions of different types of transaction, some of which are or may be cumulative while others are mutually exclusive.

[3] The particular problem with which this case is concerned is a transaction that falls within more than one "category of agreement", at least one of those categories being one mentioned in the Act. Such a transaction would give rise to several concerns, with regard to consumer protection. One is to ensure that all appropriate formalities are observed, having regard to the nature of the transaction. Another is to avoid the possibility of evasion of the statutory regime, by the

A aggregation or combination into one transaction of what are really two or more separate transactions, if, taken separately, they would be subject to the Act's controls whereas as one larger transaction they would escape it, for example, because the amount of the credit exceeds the monetary limits relevant under the Act. One would also expect the legislation to be framed in a way that makes it clear, both for consumers and for providers of credit, what is the regulatory regime and when and how it applies to any given transaction. Otherwise, it would be unfair and inappropriate with regard to one or both parties.

Facts

B [4] The appellant owns a house in Worksop, Nottinghamshire. She had the benefit of a mortgage advance from Halifax, on which some £19,000 was outstanding. She wished to obtain further credit on the security of the house. In January 2002, she obtained an offer from a different lender of a loan of almost £29,000 gross (just over £28,000 net after payment of various fees and costs) to be secured on a first legal mortgage of the house. In order to ensure that the new lender obtained a first legal mortgage, the previous mortgage to Halifax would have to be redeemed, and it was a term of the offer that this should happen. On 7 March 2002, the transaction was completed, the Halifax mortgage was redeemed, and around £9,000 was released to the appellant. Shortly
C thereafter, the benefit of the mortgage was assigned by the lender to the present respondent. Nothing turns on that.

[5] At the time of the transaction, an agreement providing for credit of more than £25,000 was not regulated by the Act. On that footing, the formal requirements of the Act with regard to the content of the documents and with regard to procedures for execution were not followed. On behalf of the appellant, it is now argued that the agreement has to be treated for the purposes of the Act as though it comprised two separate agreements, one relating to the amount that was used to repay the previous mortgage, and the other for the rest. If that were correct, the Act would have applied to each agreement, and, because it had not been complied with, no part of the agreement would be enforceable.

D [6] The appellant fell into arrears under the new mortgage, and the respondent took proceedings for possession. It obtained a first possession order suspended on terms, in 2004, and another, suspended on different terms, in 2006. In 2007, when the appellant faced a warrant for possession, she obtained advice that for the first time suggested the defence that is now raised to the possession claim. The warrant was suspended pending her appeal.

[7] The appeal eventually came on before HH Judge Purle QC in Birmingham, sitting as a High Court judge: see [2009] EWHC 103 (Ch)*. He gave judgment on 29 January 2009, extending time for the appeal but dismissing it. I granted permission for this second appeal.
E The point is important and has not previously arisen for decision in the Court of Appeal.

[8] I say at once that I would dismiss the appeal, largely for the same reasons as were expressed by Judge Purle in his clear and admirable judgment.

Legislation

[9] A consumer credit agreement was defined (at the relevant time) by section 8(2) as a personal credit agreement by which the creditor provides the debtor with credit not exceeding £25,000. It is a personal credit agreement if the debtor is an individual: section 8(1). A consumer
F credit agreement is a regulated agreement unless it is an exempt agreement as provided for in or under section 16.

[10] Sections 10 to 13 set out categories of agreement under the Act. First, in section 10, we meet running-account credit and fixed-sum credit. A facility under a personal credit agreement must be for one or for the other. Section 11 is concerned with restricted-use and unrestricted-use credit. Any particular credit provided under a credit agreement must be one or the other and cannot be both. Section 12 defines debtor-creditor-supplier agreements and section 13 defines debtor-creditor agreements, in a mutually exclusive way.

* Editor's note: Reported at [2009] 2 All ER (Comm) 687

[11] Section 18 is the provision with which this appeal is principally G
concerned. The first four subsections are as follows:

> (1) This section applies to an agreement (a "multiple agreement") if its terms are such as —
>
> (*a*) to place a part of it within one category of agreement mentioned in this Act, and another part of it within a different category of agreement so mentioned, or within a category of agreement not so mentioned, or
>
> (*b*) to place it, or a part of it, within two or more categories of agreement so mentioned.
>
> (2) Where a part of an agreement falls within subsection (1), that part shall be treated for the purposes of this Act as a separate agreement.
>
> (3) Where an agreement falls within subsection (1)(*b*), it shall be treated as H
> an agreement in each of the categories in question, and this Act shall apply to it accordingly.
>
> (4) Where under subsection (2) a part of a multiple agreement is to be treated as a separate agreement, the multiple agreement shall (with any necessary modifications) be construed accordingly; and any sum payable under the multiple agreement, if not apportioned by the parties, shall for the purposes of proceedings in any court relating to the multiple agreement be apportioned by the court as may be requisite.

[12] Because the agreement between the parties in this case was secured by a mortgage of land, the procedural requirements of the Act do not allow the borrower a specified period after execution of the agreement for second thoughts, but rather require that an advance J
copy of the unexecuted agreement be sent to the prospective borrower, containing a notice indicating the borrower's right to withdraw from the prospective agreement: section 58. By virtue of section 61(2), this must be sent at least seven days before the execution copy is sent to the borrower, and the lender must not contact the borrower either during that period or for seven days after the execution copy has been sent (unless within those seven days the agreement is returned executed by the borrower) except in response to a specific request. The substantive provisions of the advance copy must match the execution copy precisely, so that, by the time when the advance copy is prepared, it has to be known exactly what the terms of the eventual agreement will be. K

[13] Under section 61, the agreement is not properly executed unless a document in the prescribed form and containing all the prescribed terms and conforming with the regulations is signed by or for both parties, and section 58 has been complied with as described above. An improperly executed agreement is enforceable against the debtor only on an order of the court. Unless there is a document that contains all the prescribed terms of the agreement signed by the debtor, the court could not make an order for the enforcement of the agreement: section 127(3). This provision has been repealed but not so as to affect its application to previous transactions.

[14] Under the regulations, if the agreement between the present L
parties was a regulated agreement, the document should have set out the amount of the credit to be provided under it, and the rate of interest on the credit.

[15] The Act adopted a novel technique of setting out examples with notes, "for illustrating the use of terminology" used in the Act: section 188, but in the event of inconsistency between an example and some other provision in the Act, the latter prevails: section 188(3). The examples are in Schedule 2. Examples 16 and 18 are said to be of particular relevance; I have set those out in an appendix to this judgment. M

Parties' submissions

[16] The appellant's argument is, in summary, as follows. With regard to the sum that was applied in redemption of the Halifax mortgage, the present agreement was for restricted-use credit, because it was to refinance existing indebtedness of the appellant: see section 11(1)(*c*), while as for the balance of the credit it was for unrestricted-use credit under section 11(2). Therefore, the one transaction contained within it two different types, or categories, of agreement, in terms of the definitions in the Act. Because one part of the agreement fell within one category and the rest within another, section 18(1)(*a*) applied to the respective parts. Each part, therefore, has to be treated as a separate agreement under section 18(2). If the two parts of this agreement are

A treated in that way, one is for restricted-use credit of some £19,000 and the other is for unrestricted-use credit of some £9,000, and both are regulated agreements, but the requirements of the Act have not been complied with as regards either agreement, and neither of them, nor the whole, is accordingly enforceable.

[17] For the appellant, Mr Bradley Say submitted that, if an agreement falls into disparate categories (such as restricted-use and unrestricted-use, or running-account and fixed-sum), section 18(1)(*a*) applies. Section 18(1)(*b*) applies if, and only if, all the categories that are relevant are congruent and compatible, such as unrestricted-use, running-account, and debtor-creditor. The point of section 18(1)(*b*) B is twofold, as can be seen from section 18(2) and (3). If part of the agreement falls within (1)(*b*) (as opposed to (1)(*a*), presumably), then that part is to be treated as a separate agreement. Whether it is part or the whole that falls within (1)(*b*), then section 18(3) provides that the Act is to apply to the agreement on the basis that it falls within each of the relevant categories.

[18] Since every regulated agreement will undoubtedly fall within several compatible categories, on Mr Say's reading section 18(1)(*b*) would apply to every regulated agreement. Even if the agreement does not comprise separate parts for the purposes of section 18(1), there would be several categories, and the effect of section 18(3) is the C somewhat obvious one that all relevant requirements of the Act have to be complied with, whatever the categories may be.

[19] Mr Malcolm Waters QC, for the respondent, submitted that this would leave section 18(1)(*b*) and (3) with no meaningful purpose or function. On his reading, the purpose of section 18(1) is to deal with disparate categories, not with compatible categories, there being no need to make any special provision for categories of agreement that are compatible. He drew attention to the opening words: "This section applies to an agreement... if its terms are such as to place it...". He argued that this does at least two things: it shows that attention has to be given to the effect of the terms of the agreement, and it also indicates D that only in some cases will the section apply, because of the word "if". He submitted that, on Mr Say's reading, all regulated agreements, without exception, would be multiple agreements, whereas the opening of section 18(1) leads the reader to suppose that multiple agreements will be special cases, whether or not frequent. The language of the Act would have been quite different, he suggested, if every regulated agreement was to fall within section 18(1). Moreover, any agreement that falls within section 18(1)(*a*), and therefore falls to be treated as two or more separate agreements for the purpose of the Act, would also fall within section 18(1)(*b*) because each separate part would fall into two or more compatible categories.

E [20] In contrast to Mr Say's reading of section 18(1), Mr Waters submitted that both (*a*) and (*b*) are concerned with disparate categories, and the difference between them depends on whether the agreement, by its terms, does fall to be treated as being in separate parts or not. If it does, it is right to treat the different parts separately, and to apply the Act separately to each. If not, the important point is that, if the Act does apply, all its relevant provisions should apply cumulatively. There should not be scope for an argument that the agreement is, for example, predominantly in one category and therefore only the requirements relevant to that category need to be observed.

[21] One of the functions of section 18(1)(*a*), on that basis, is one of F anti-avoidance, preventing the Act from being rendered inapplicable by combining into one agreement distinct elements that, taken separately, would be or include one or more regulated agreements, but taken together exceed the relevant monetary limit, and so escape the Act altogether. That is consistent with the point that the section applies to "an agreement", rather than to "a regulated agreement"; it does not presuppose that the agreement would, but for the section, fall within the Act.

[22] The other function of the section is to ensure that, if more than one set of requirements is relevant under the Act, both or all are complied with. In this context, the purpose of the respective provisions in paras (*a*) and (*b*) of subsection 18(1) is to identify the agreement in respect of which, for the purposes of the Act, the requirements are to be observed. Financial amounts may be relevant to that, so that it is G necessary to know the details of the agreement with regard to which a particular set of requirements must be complied with. The credit provided under each notionally separate agreement would be different, and would presumably have to be stated separately.

[23] On his reading of the section, Mr Waters submitted that there is nothing in the terms of the present agreement that is such as to place one part of the agreement in one category and another part in another category. On the basis that the agreement provides both restricted-use and unrestricted-use credit, he argued that neither this fact, nor any other feature of the agreement, requires or permits it to be dissected into parts. It was an agreement for the provision of a single advance by H way of fixed sum credit, and the fact that part of that credit was to be used in a specific way, and the rest was not, does not enable or call for the agreement to be treated as being in two separate parts.

[24] A separate point was raised in a respondent's notice, namely that no part of the credit was to be treated as restricted-use credit. In view of my conclusion on the appeal, it is not necessary to decide that point, and I will assume in the appellant's favour that part of the credit was restricted-use credit.

[25] As presented by Mr Say in his submissions, the crux of the debate is between two different approaches to the question of whether section 18(1)(*a*) applies: should one look first at the terms of the J agreement to decide whether it falls into parts, or rather (as Mr Say argued) is it first necessary to consider whether the agreement falls into disparate categories, in which case he said that the paragraph must apply? Although that is an important question, another on which the parties diverge critically is whether section 18(1)(*b*) is concerned with disparate categories or only with compatible categories: is it an antithesis to section 18(1)(*a*) in that it is not concerned with disparate categories, or is it a complement in that it too deals with disparate categories, but in respect of agreements that are not to be separated out into parts?

Decided cases K

[26] The application of section 18 has been touched on, but not decided on the point now at issue, in a number of cases decided at appellate levels. *Dimond* v *Lovell* [2002] 1 AC 384 concerned an agreement under which an individual whose motor car had been damaged in an accident hired a car from a company on terms that the hire charges were not immediately payable by the hirer and were to be recovered out of any damages payable by the other driver's insurance company. It was said that this involved the hire company providing credit to the hirer, so that the agreement was regulated under the Act. The Court of Appeal and the House of Lords accepted this argument. Lord Hoffmann rejected a submission that the agreement was a multiple L agreement, and that the hire element of it was separate and unaffected by the Act. At p396C-D, he said:

> The difficulty I have with this argument is that it seems to sever the provisions that create the debt (hiring the car) from the provisions that allow credit for payment of the debt. Whatever a multiple agreement may be, one cannot divide up a contract in that way. The creation of the debt and the terms on which it is payable must form parts of the same agreement.

[27] Similar arguments were rejected on the same basis in *Burdis* v *Livsey* [2002] EWCA Civ 510*.

[28] Previously, in *National Westminster Bank plc* v *Story* [1999] M CCLR 70, the Court of Appeal had occasion to consider whether section 18 applied to a case where the bank had agreed with two individual customers to lend them £35,000 on three separate facilities, one being an overdraft of £15,000 for one borrower and the others by way of two loan facilities, of £5,000 and £15,000, to the two borrowers jointly. At the relevant time, £15,000 was the financial limit for the credit under a regulated agreement. The borrowers resisted enforcement on the basis first that there were three separate agreements and second that if it was a single agreement it was nevertheless a multiple

---

* Editor's note: Reported at [2003] QB 36

agreement under which each facility was to be treated as provided under a separate agreement, the £5,000 loan facility being provided as restricted-use credit because it was to refinance existing indebtedness of the borrowers to the bank under a previous loan.

[29] The court held that it was a single agreement, and rejected the contention that any part of it provided restricted-use credit. Auld LJ, with whose judgment Lord Woolf MR and Robert Walker LJ agreed, observed, at p77, that:

> The main purpose of section 18 is to prevent frustration of the Act's protection to borrowers by the artificial combination of two or more agreements in one so as to take the total credit negotiated above the limit qualifying for protection.

[30] Because he held that, even if the agreement had been one that fell into parts, all of them were within the same category under the Act, the controversial issues under section 18 did not arise. However, he made these comments about the section at p78:

> Whatever the uncertainties as to the meanings of "part" and "category" of agreement under section 18, they do not require resolution in this case. My inclination, without formally deciding the matter, is that the word "part" in this context includes, but is not restricted to, a facility that is different as to some of its terms from another facility granted under the same agreement or one that can stand on its own as a separate contract or bargain. However, I believe that it would accord with the ordinary and natural use of the word for it to have a broader application so as to include, as here, a separate facility provided with others under an agreement where, even if the facility as a *contractual entitlement* does not stand on its own, the debtor's *use*, or *non-use*, of it does not affect the contractual nature of the agreement as a whole, in particular, his entitlement to *use* those other facilities.

[31] He also expressed an inclination towards the view that "categories" of agreement under the Act means categories mentioned in Part II of the Act, but since this meant that restricted-use and unrestricted-use credit were separate categories, and this was the view most favourable to the borrowers, it was unnecessary for him to decide that point. Nor does that point arise in the present case.

[32] In *London North Securities Ltd* v *Meadows* [2005] EWCA Civ 956, decided by Lord Phillips of Worth Matravers MR, Waller LJ and myself, it was, again, unnecessary to decide any point arising under section 18.

Academic debate

[33] There is disagreement among commentators on the Act as to the correct approach to the provisions of section 18. One of the commentators is Mr Francis Bennion, the draftsman of the Act. We were shown his article "Multiple Agreements under the Consumer Credit Act 1974" [1999] CICC 1. We have also seen the passages from *Goode: Consumer Credit Law and Practice*, and *Guest and Lloyd: Encyclopaedia of Consumer Credit Law*, in which the points are discussed.

[34] Professor Goode uses labels that are not found in the Act: multipart agreement for an agreement in parts, each of which is in a different and disparate category; unitary agreement for one that falls into two or more categories but does not fall into separate parts; mixed agreement for one, part of which is multipart and another part of which is unitary. Although these are not statutory, they are convenient for reference purposes. In his work, it is argued that section 18(1)(*b*) shows that the mere fact that an agreement falls within two or more statutory categories does not of itself make it a multipart agreement within section 18(1)(*a*). The theory advanced is that a multipart agreement is one in which two or more essentially distinct bargains are rolled up into one agreement. By contrast, an agreement providing for a single facility covering different and disparate categories of credit would be a unitary agreement, and therefore within section 18(1)(*b*). It is suggested that, on this basis, examples 16 and 18 in Schedule 2 to the Act are wrong.

[35] Guest and Lloyd, in their notes to section 18, support the view that "two or more categories" in section 18(1)(*b*) must mean disparate categories, since otherwise every agreement would be multiple. They also support the view that the first question to be asked concerns the terms of the agreement, and whether they treat different elements in the transaction differently, with different terms of repayment, different security, different interest rates or the like. If they do, the agreement is likely to be a multipart agreement. Even if they do not, but if there is a substantial degree of disparity between the two elements, having regard to their subject-matter, their legal classification and the operation of the Act, then, they are likely to constitute a multipart agreement. They also argue that, in a case of the present kind, where a loan is advanced, part of which has to be used to repay existing indebtedness secured on the property and the rest is available for use freely, "the agreement is to be treated as an agreement in each of the categories in question and the Act applies to it accordingly, but the two elements do not require to be treated for the purpose of the Act as separate agreements under section 18(2)", unless special factors exist, such as different repayment terms or some other difference of treatment between the two elements in the credit.

[36] A different view is expressed by Mr Bennion in his article. He argues that "category of agreement" includes every type of agreement mentioned in the Act. He takes issue with the reading that "categories" in section 18(1)(*b*) must mean disparate categories, since otherwise every agreement would be a multiple agreement. He sees no problem with his reading, that categories means any kind of category, and no disadvantage in terms of consequences. He suggests that the section has a practical effect only where it needs to and can otherwise be ignored. He expresses the view that "what matters with a multi-agreement document is that it is possible to collect from the document as a whole what amount to the respective terms of two or more separate agreements" to which the Act then has to be applied to see whether each constitutes an agreement falling within a category under the Act.

[37] He disagrees with Professor Goode's view that an agreement cannot fall within both section 18(1)(*a*) and (*b*) and discounts the criticism that, in his view, section 18(1)(*b*) applies to every agreement within the scope of the Act, and achieves nothing in respect of most of them. He sets out a worked example that has some similarities to the present case. He argues that, because it provides both restricted-use and unrestricted-use credit, it is within section 18(1)(*a*) and the several parts must be treated as separate agreements for the purposes of the Act. It is not clear from his discussion how he is able to collect from the hypothetical document governing the postulated agreement as a whole "what amount to the respective terms of two or more separate agreements".

Discussion

[38] Mr Say and Mr Waters sought assistance from the respective commentators, but the debate that we have to resolve is that which arises from the facts of the present case, rather than any wider controversy as to the scope and effect of the section. For example, it is unnecessary to form a view as to the width of the word "category". Having noted the general lines of the academic disagreements, I propose to focus on the points advanced by counsel and on their application to the present case.

[39] Part of the debate before us centred on the practical difficulties that would arise in a case such as this if the two parts of the credit were treated as provided under separate agreements for the purposes of the Act. In order to comply with sections 58 and 61, the advance copy would have to state the amount of the credit that was restricted-use (that is, to be applied in discharge of the existing mortgage) and separately the amount that was unrestricted-use, freely available to the borrower. In the ordinary way, the former would be the amount required by the prior mortgagee in redemption of its mortgage. This would depend on the interest accrued to the date of redemption, and might also be affected by other matters such as the incurring of expenses by the lender in the meantime that it would add to the security. Neither the prospective new lender nor the borrower would be able to specify or anticipate what that amount would be. It would therefore be necessary to get a figure from the existing lender. Of course, lenders are accustomed to being asked for redemption figures in respect of a future date: that is commonplace on the completion of a sale of mortgaged property. However, the existing lender would have to commit itself to accepting

A a stated amount, on a future date far enough ahead to allow for the completion of the procedures under the Act, and the new transaction would have to be completed on that future date. It is by no means clear that such agreement would readily be obtained, and whether the transaction that would in fact be completed on the specified date might be subject to chances and uncertainties against which it would be difficult to provide.

[40] For those reasons, Mr Say's interpretation of the Act might be counter-productive for would-be borrowers, since it might make it unnecessarily difficult for a new lender to be sure that its loan documentation would comply with the Act, and so it might be reluctant
B to make the loan. However, Mr Say's submission was that the Act is clear and that arguments from practical difficulty cannot prevail unless there is an ambiguity that, he argued, cannot be found in the Act. He also contended that any difficulties would not be insuperable, and he showed us *McGinn* v *Grangewood Securities Ltd* [2002] EWCA Civ 522, where an agreement was held unenforceable because of a failure that arose from a similar practical problem.

[41] It seems to me that the argument for the respondent is correct. The starting point is that it is from the terms of the agreement that one must find out whether the agreement is one under which there are two or more parts, in different categories, or whether it, or part of
C it, falls into two or more categories. It is not correct to start from the proposition that more than one disparate category is concerned, and to conclude from this that the agreement must fall into two or more parts. That the starting point is the terms of the agreement is consistent with Mr Bennion's view, as noted in the passage that I have quoted in [36] above. In addition, I agree with Mr Waters that it is significant that it is the agreement that is to be placed in one or more categories, not the credit provided under the agreement.

[42] I also agree with Mr Waters, and with Judge Purle in his judgment under appeal, that categories in section 18(1)(*b*) means disparate categories, just as it does in section 18(1)(*a*). I come to that
D conclusion both because I would expect the same word, used in two different parts of one subsection, to mean the same thing in each, and also because the alternative reading, that it means any categories in section 18(1)(*b*), would mean that there is no point to this paragraph. It seems to me to be logical to read the subsection as covering two possibilities, both being cases in which there are elements of an agreement that fall within two or more disparate categories, at least one of which is a category under the Act. Subparagraph (*a*) deals with the case where the respective elements within the agreement can be seen as constituting separate parts of the agreement; subpara (*b*) deals with the opposite case where they cannot be seen in that way. Moreover, this
E reading of section 18(1)(*b*) is supported also by the opening words "if the terms of the agreement", because on the other reading the paragraph would apply to every agreement, not, as the opening words suggest, only to some.

[43] I accept the submission that the reading for which Mr Say contended would present would-be lenders with serious practical difficulties. They may not be insuperable, but it seems to me that this practical consequence is one that it is legitimate to bear in mind when interpreting the Act. It follows from what I have already said that I do not accept Mr Say's argument that the Act is so clear that this point is not available. A case such as *McGinn* does create real difficulties
F for lenders, but the court in that case made it clear that the problem identified there arose in somewhat exceptional circumstances, and that it would not be too difficult for the lender to avoid it.

[44] I agree that the section is in part aimed against attempts to avoid the application of the Act. Auld LJ said the same in *Story*. The reading of the section that I prefer does not defeat that object.

Statutory examples

[45] Mr Say submitted that Judge Purle had been wrong to hold that examples 16 and 18 in Schedule 2 to the Act were wrong. Mr Waters said that it was not essential to his argument that the examples were wrong, but that, if they were inconsistent with his submissions concerning the primary provisions of the Act, then the inconsistent G
aspects of the examples should be rejected as wrong.

[46] Example 16 in Schedule 2 to the Act is of an agreement relating to a credit card that can be used to obtain cash from the card issuer (unrestricted-use credit) or to pay for goods or services or cash from suppliers or banks that agree to honour the card (restricted-use credit with regard to the goods or services, and unrestricted-use credit as regards cash). On any view of section 18, it is a multiple agreement because it provides these two forms of credit and therefore falls into two disparate categories under the Act. The analysis set out in the Schedule states that the entire agreement falls within several categories, and is therefore to be treated as an agreement in each category under H
section 18(3), which presupposes that section 18(1)(*b*) applies. That seems to me to be correct. What is puzzling is that it goes on to state, in the last sentence, that the restricted-use credit and unrestricted-use credit elements of the agreement are, by section 18(2), to be treated as separate agreements, which assumes that section 18(1)(*a*) applies. I would respectfully disagree with that proposition, as Professor Goode does. It seems to me that it falls within section 18(1)(*b*) and section 18(3), not within section 18(1)(*a*) and section 18(2). In any event, even if section 18(2) were to apply, the credit under the agreement could not be divided between the separate parts, as the entire amount of the credit is available in respect of each aspect of the agreement. On J
that footing, it makes no difference whether the agreement is taken as falling within section 18(1)(*a*) or (*b*). I note also that, in so far as the analysis states that the entire agreement falls into several categories and that section 18(3) therefore applies, it does not support Mr Say's submission as to section 18(1)(*b*), which is that this applies only if all relevant categories are consistent and compatible.

[47] Example 18 is of an agreement for a current account with (implicitly from past conduct) an overdraft facility of an unspecified amount. The statutory analysis states that it is a regulated consumer credit agreement for unrestricted-use running account credit. It also states that, because another part of it, concerned with the operation of the K
account when in credit, is outside the Act, it is a multiple agreement and that section 18(1)(*a*) applies to the two separate parts of the agreement. Again, Professor Goode states that it is not a multipart agreement but a unitary agreement falling into two categories, of which only one is within the Act. Accordingly, he says, section 18(1)(*a*) does not apply and it is not a multiple agreement at all. This is another case in which, even if section 18(1)(*a*) does apply, it does not cause the credit provided to be dissected and attributed in separate parts to different agreements. Nor does the correctness of the analysis make any difference to the way the example would be treated under the Act in practice.

[48] Mr Bennion does not accept as valid the criticisms of the L
analysis in respect of the statutory examples. He suggests that, instead of treating them as erroneous, Professor Goode should have seen that they show his reading of the Act to be wrong, in so far as it treats section 18(1)(*a*) and (*b*) as mutually exclusive. Mr Waters submitted that the examples are not necessarily wrong, for example, because in the case of the credit card different terms (for example, as to interest) may apply to the two different aspects of the facility, and each can be used freely without the other being used at all. He also pointed out that, in the overdraft example, different terms would undoubtedly apply to the account when in credit and to the overdraft facility, and the two aspects of the agreement are in their nature distinct and mutually exclusive in M
the sense that only one can be used at any one time. Mr Say submitted that it was not legitimate to add facts to the statutory examples, and I see some force in that with regard to example 16. In the case of example 18, it seems to me that it must go without saying that different terms apply to a credit balance and to an overdraft.

[49] The significance of the examples in respect of the present debate is not for what they show directly with regard to a case such as the present. They are relevant because the analysis assumes that an agreement can fall within both paras (*a*) and (*b*) of section 18(1). However, even if that assumption were wrong, the application of the Act to the examples would be no different from that stated.

[50] Having reflected on the point, with the benefit of the academic writings as well as counsel's submissions, I have come to the conclusion that the analysis of example 16 in the Act is wrong in one respect, because the better reading of the substantive provisions of the Act is that paras (*a*) and (*b*) of section 18(1) are mutually exclusive. I therefore hold to the view that I have stated above, notwithstanding its being inconsistent with the basis of this aspect of the statutory analysis, and with Mr Bennion's expressed opinion. I disagree with the last sentence of the analysis in respect of example 16. I do not disagree with the treatment of the example for the purposes of the Act that appears from the statutory text, but only with the analysis of how that conclusion is reached. With regard to that aspect of the analysis, it seems to me that it is wrong in the light of the correct interpretation of the primary provisions in section 18. It must give way, as section 188(3) provides, in the case of inconsistency with other provisions in the Act: see [15] above.

[51] With regard to example 18, it seems to me that the statutory analysis may be correct. Professor Goode argues that it is not a multipart agreement but a unitary agreement. However, it seems to me that the agreement may very well fall into parts, the current account aspect being distinct from the overdraft facility. Each of them is available to be used separately, and they will be governed by different terms. On that basis, I agree that section 18(1)(*a*) would apply. Were Professor Goode right to treat the agreement as unitary, rather than multipart, I would agree with him that section 18(1)(*a*) does not apply. Section 18(1)(*b*) alone would be relevant and it would follow that the agreement was not a multiple agreement at all, because section 18(1)(*b*) applies only if both categories are within the Act.

Conclusion

[52] Turning back to the facts of the present case, the appellant had from the respondent the offer of a single facility, which could be drawn down only as a whole. It was to be secured by a first mortgage on the property, which was at the time subject to the existing mortgage in favour of Halifax. It was a term of the loan agreement that any existing mortgage was to be paid off out of the loan. Assuming, in the appellant's favour, that this means that the part of the credit that would be used to redeem the existing mortgage was restricted-use credit, nevertheless I find nothing in the terms of the agreement that permit a conclusion that part of the agreement is to be placed in one category (restricted-use) and part in another (unrestricted-use). It is a single agreement that cannot be dissected into separate parts. To go back to Mr Bennion's words quoted in [36] above, it is not possible to collect from the document as a whole what amount to the respective terms of two or more separate agreements. Nor can it be brought within the words of Auld LJ quoted in [30] above. The entire credit facility had to be drawn together or not at all, as Judge Purle noted in [38] of his judgment.

[53] I consider that the agreement was one whose terms placed the entirety of the agreement in two relevant and disparate categories under the Act, so that section 18(1)(*b*) applied. It was not one whose terms placed part of the agreement in one category and another part in another. The agreement could not be taken apart in that way. It was a unitary agreement, in Professor Goode's language. Accordingly, it is not to be treated as two separate agreements, and because the amount of the credit provided exceeded £25,000 it was not a regulated agreement.

[54] Those are my reasons for concluding that Judge Purle was correct in his decision and that his order should not be set aside or varied. I would dismiss this appeal.

Dyson LJ said:

[55] I agree.

Waller LJ said:

[56] I also agree.

Appendix: Examples 16 and 18 from Schedule 2 to the Act

Example 16

Facts. Under an unsecured agreement, A (Credit), an associate of the A bank, issues to B (an individual) a credit-card for use in obtaining cash on credit from A (credit), to be paid by branches of the A bank (acting as agent of A (credit)), or goods or cash from suppliers or banks that have agreed to honour credit-cards issued by A (credit). The credit limit is £30.

Analysis. This is a credit-token agreement falling within section 14(1)(*a*) and (*b*). It is a regulated consumer credit agreement for running-account credit. Since the credit limit does not exceed £30, the agreement is a small agreement. So far as the agreement relates to goods it is a debtor-creditor-supplier agreement within section 12(*b*), since it provides restricted-use credit under section 11(1)(*b*). So far as it relates to cash, it is a debtor-creditor agreement within section 13(*c*) and the credit it provides is unrestricted-use credit. This is therefore a multiple agreement. In that the entire agreement falls within several of the categories of agreement mentioned in this Act, it is, by section 18(3), to be treated as an agreement in each of those categories. So far as it is a debtor-creditor-supplier agreement providing restricted-use credit it is, by section 18(2), to be treated as a separate agreement; and similarly so far as it is a debtor-creditor agreement providing unrestricted-used credit. (See also example 22.)

Example 18

Facts. F (an individual) has had a current account with the G bank for many years. Although usually in credit, the account has been allowed by the bank to become overdrawn from time to time. The maximum such overdraft has been is about £1,000. No explicit agreement has ever been made about overdraft facilities. Now, with a credit balance of £500, F draws a cheque for £1,300.

Analysis. It might well be held that the agreement with F (express or implied) under which the bank operate his account includes an implied term giving him the right to overdraft facilities up to say £1,000. If so, the agreement is a regulated consumer credit agreement for unrestricted-use, running-account credit. It is a debtor-creditor agreement, and falls within section 74(1)(*b*) if covered by a direction under section 74(3). It is also a multiple agreement, part of which (that is, the part not dealing with the overdraft), as referred to in section 18(1)(*a*), falls within a category of agreement not mentioned in this Act. (Compare example 17.)

Appeal dismissed.

PROCEDURE

Jayasinghe *v* Liyanage

CHANCERY DIVISION

[2010] EWHC 265 (Ch)

18 February 2010

BRIGGS J

Estates Gazette 24 April 2010

[2010] 16 EG 108

Land registration — Practice and procedure — Restrictions — Application to register restriction against property on ground of beneficial interest — Objection to registration — Land Registry adjudicator conducting trial and finding no beneficial interest — Application cancelled — Section 73(3) of Land Registration Act 2002 — Whether adjudicator entitled to conduct trial — Whether obliged to register restriction where arguable claim to relevant interest disclosed pending determination of claim by courts

The appellant applied to register a restriction against the registered title to a property on the ground that she was the sole beneficiary under a resulting trust of the property in her favour. In support of her application, she asserted that she had provided the original funds for the purchase, paid all instalments due under a building society charge of the property, redeemed that charge out of her own funds and borne all the outgoings of the property, as well as receiving all the income from its letting. The respondent opposed the application, asserting that the appellant had no interest in the property and that her version of the history of the matter was a fiction unsupported by the evidence.

The application was determined by a deputy adjudicator to the Land Registry pursuant to section 73(7) of the Land Registration Act 2002. After conducting a two-day trial, including the cross-examination of the appellant, the respondent and a witness, the adjudicator found that the property had been an investment made by the respondent to which the appellant had not contributed financially; he accordingly cancelled the appellant's application.

The appellant appealed to the High Court. She argued that, *inter alia*, the adjudicator had not been entitled to undertake a trial of the issue of whether she had a beneficial interest in the property. Instead, he should have ascertained whether she had an arguable claim to that effect, in which case he should have entered the restriction sought pending a determination of her claim by a competent court.

***Held*: The appeal was dismissed. The procedural code regulating the discharge by the adjudicator of functions conferred by the 2002 Act, as contained in the Adjudicator to Her Majesty's Land Registry (Practice and Procedure) Rules 2003 (SI 2003/2171), is designed to enable the adjudicator to resolve disputes over substantive rights where necessary, rather than merely to conduct a summary process to determine whether an arguable claim exists. Under the statutory scheme, the registrar's task is not confined to deciding whether a person has a sufficient interest in the making of the entry to qualify as an applicant within section 43(3); the registrar may conclude, notwithstanding that an applicant qualifies, that it is unnecessary or undesirable, within section 42(1), to make the requested restriction. Accordingly, where an objection that is not obviously groundless cannot be disposed of by agreement, the matter referred to the adjudicator under section 73(7) is not merely the question of whether the applicant has a relevant right or claim but it involves the additional question of whether the entry of a restriction is necessary or desirable for the purpose of protecting that right or claim. The disposal of the objection is an integral part of the matter referred to the adjudicator. The precise nature of the adjudicator's function on any particular reference under section 73(7) will depend on the precise restriction sought, the nature of the claim or right thereby sought to be protected and the basis of the objection that led to the reference. The adjudicator is given a broad discretion over whether to decide the reference himself or require it to be decided in a competent court: see section 110(1). The former course may properly involve a trial rather than merely a summary review of whether an asserted claim is reasonably arguable. Accordingly, the adjudicator in the instant case had had jurisdiction to conduct a trial of the question of whether the appellant had a beneficial interest in the property under a resulting trust. His decision to take that course had been an appropriate exercise of his discretion.**

The following case is referred to in this report.

Croatia v *Serbia* [2009] EWHC 1559 (Ch); [2010] 2 WLR 555; [2010] 1 P&CR 5

This was an appeal by the appellant, Kusum Jayasinghe, from a decision of Mr Owen Rhys, sitting as deputy adjudicator to HM Land Registry, cancelling an application to register a restriction against the title of a property of which the respondent objector, Don Liyanage was the registered proprietor.

Christopher Maynard (instructed by Lakhani & Co) appeared for the appellant; Nicola Muir (instructed by Seth Lovis & Co) represented the respondent.

Giving judgment, **BRIGGS J** said:

[1] This is an appeal, with permission of Proudman J, from the decision of Mr Owen Rhys, sitting as deputy adjudicator to HM Land Registry, dated 12 May 2008, whereby he directed the cancellation of the application made by the appellant, Ms Kusum Jayasinghe, and dated 25 October 2007 for the registration of a restriction against the registered title to the property known as 37 Reading Road, Hallfield Estate, London W2 6HD (the property).

[2] The appellant applied for a restriction on the ground that she was the sole beneficiary under a resulting trust of the property in her favour, having provided the original funds for its purchase, paid the instalments under a charge of the property in favour of Halifax Building Society, redeemed that charge out of her own money, and borne all outgoings of the property, as well as having received all income accruing from its letting.

[3] Her application was opposed by the respondent, one Mr Don Liyanage, on the ground that the appellant had no interest whatsoever in the property and that (to quote from para 60 of the respondent's statement of case before the adjudicator):

> The Applicant's version of the history of matters is complete fiction unsupported by evidence.

[4] After a two-day trial, which included the cross-examination of the appellant, the respondent and another witness called by the

appellant, in which the opposing parties presented utterly different accounts of the relevant history, the adjudicator concluded, broadly in accordance with the respondent's case, that the applicant's case was indeed a fiction, that her evidence was, save where corroborated, wholly unreliable and implausible and that the property had, from first to last, been an investment made by the respondent, in respect of which, although she had from time to time played a role in its management and received certain benefits in connection with it, she had made no financial contribution of any kind from her own resources.

[5] Two main points are taken on the appeal, by Mr Christopher Maynard on the appellant's behalf. First, he submitted that the adjudicator should not have embarked on a trial of the issue as to whether the appellant had any beneficial interest in the property. He should have ascertained merely whether she had an arguable claim to that effect and directed that her claim should be tested by a competent court, with the restriction for which she had applied remaining in place in the meantime. On that basis, he submitted that she plainly had an arguable case sufficient to withstand any summary determination to the contrary. Second, he submitted that, in any event, the adjudicator's judgment was vitiated by a conclusion that the respondent had funded the repayment of the mortgage, which was irreconcilable with the documents before the court, thereby undermining both the respondent's credibility and the soundness of the adjudicator's general conclusion that the respondent's evidence was, wherever it differed from the appellant's, to be preferred. Furthermore, he submitted that even if the correct conclusion was only that the appellant had paid for the discharge of the mortgage, this would of itself have given her a sufficient beneficial interest in the property (or claim to that effect) to justify the registration of the restriction.

[6] The first ground of appeal found no place in the grounds of appeal appended to the appellant's notice. None the less, it was the basis on which Proudman J considered that permission to appeal ought to be granted. If correct, it would be a matter of major significance to the way in which the Land Registry adjudicators currently perceive that they should carry out their statutory functions.

[7] Mr Maynard's argument may be summarised as follows:

(i) The Chief Land Registrar may enter a restriction in the register for the purpose of protecting a right or claim in respect of a registered estate or charge: see section 42(1)(c) of the Land Registration Act 2002 (the Act).

(ii) An objection that, in substance, denies the existence of a right in respect of a registered estate or charge does not thereby impugn the existence of a claim in relation thereto, unless it demonstrates that the claim is wholly without foundation. Otherwise, all it does is to show that the claim is disputed.

(iii) The adjudicator's function in respect of an objection to an entry of a restriction is simply to ascertain whether the applicant for the restriction has demonstrated an arguable claim within the meaning of section 42(1)(c). If that is demonstrated, the adjudicator has no business to embark on the trial of that claim, a process that should be left to be carried out by a competent court in the usual way while the restriction remains in place.

[8] In support of that submission, Mr Maynard relied primarily on observations of mine in another appeal from the Land Registry adjudicator, namely *Croatia* v *Serbia* [2009] EWHC 1559 (Ch)*, to the effect that a person may have a sufficient interest in the making of an entry to enable him (or it) to apply for the entry of a restriction within the meaning of section 43(1)(c) of the Act on the basis of a claim to an interest in the registered estate, even if that claim is not justiciable in the English courts and is disputed by the registered proprietor. Mr Maynard bolstered his submissions by reference to para 16.1 of the Law Commission Report (Law Comm No 271), entitled *Land Registration for the 21st Century A Conveyancing Revolution*, that led to the passing of the Act, in which it is stated that the newly created office of the adjudicator was designed to offer, for the first time on a wholly independent basis, substantially the same service of dispute resolution formally provided by solicitors to HM Land Registry under the previous statutory regime.

[9] Attractively though those submissions were presented by Mr Maynard, I have been wholly unpersuaded by them. In order to understand my reasons, it is necessary to describe the structure created by the Act with regard both to restrictions and to the resolution by the adjudicator of disputes occasioned by objections to them.

[10] Section 42 of the Act provides that:

> (1) The registrar may enter a restriction in the register if it appears to him that it is necessary or desirable to do so for the purpose of
>
> …
>
> (c) protecting a right or claim in relation to a registered estate or charge.

Section 43 provides that:

> (1) A person may apply to the registrar for the entry of a restriction under section 42(1) if
>
> …
>
> (c) he otherwise has a sufficient interest in the making of the entry.

Section 43(2)(c) states that rules may provide for classes of person to be regarded as being included in subsection (1)(c).

[11] Rule 93 of the Land Registration Rules 2003 (SI 2003/1417) identifies more than 20 classes of person to be regarded as having a sufficient interest in the making of the entry. For present purposes, the relevant class is identified in subrule (b), as follows:

> any person who has a sufficient interest in preventing a contravention of section 6(6) or section 6(8) of the Trusts of Land and Appointment of Trustees Act 1996 and who is applying for a restriction in order to prevent such a contravention.

For a general review of r 93, see paras 47 to 53 of my judgment in *Croatia*.

[12] Section 73(1) of the Act provides that, subject to exceptions that are irrelevant for present purposes, anyone may object to an application to the registrar. By subsection (5):

> Where an objection is made under this section, the registrar
> (a) must give notice of the objection to the applicant, and
> (b) may not determine the application until the objection has been disposed of.

By subsection (6), subsection (5) is disapplied if the objection is one that the registrar is satisfied is groundless. Subsection (7) then provides that:

> If it is not possible to dispose by agreement of an objection to which subsection (5) applies, the registrar must refer the matter to the adjudicator.

[13] Adjudication is dealt with in Part 11 of the Act. Under section 108:

> (1) The adjudicator has the following functions
> (a) determining matters referred to him under section 73(7), …

By subsections (2) and (4), the adjudicator is given the same powers as the High Court to make orders for rectification or the setting-aside of certain classes of documents. Section 109 contains certain general procedural provisions, and makes provision for the establishment of a detailed procedural regime by rules. Under section 110, headed "Functions in relation to disputes":

> (1) In proceedings on a reference under section 73(7), the adjudicator may, instead of deciding a matter himself, direct a party to the proceedings to commence proceedings within a specified time in the court for the purpose of obtaining the court's decision on the matter.

By section 111(1), a person aggrieved by a decision of the adjudicator may appeal to the High Court.

[14] The procedural code regulating the discharge by the adjudicator of functions conferred by the Act is set out in the Adjudicator to Her Majesty's Land Registry (Practice and Procedure) Rules 2003 (SI 2003/2171). Those rules (the Practice and Procedure Rules) expressly incorporate the overriding objective, in a form that, although

* Editor's note: Reported at [2010] 2 WLR 555

A modified to suit the particular functions of the adjudicator, broadly corresponds with that to be found in the Civil Procedure Rules: see para 3. Viewed as a whole, the Practice and Procedure Rules contain a procedural code plainly designed to enable the adjudicator to resolve, where necessary, disputes concerning substantive rights, rather than merely to conduct a summary process designed to ascertain whether there exists an arguable claim. They include, for example, the power to: (i) give detailed directions; (ii) require statements of case; (iii) consolidate proceedings, add or substitute parties and require disclosure and witness statements; (iv) conduct site inspections; and (v) compel the attendance of witnesses.

B [15] The following aspects of the structure that I have described deserve particular emphasis. First, the registrar, under section 42(1) of the Act, is to enter a restriction if it appears to him that it is necessary or desirable to do so for the purposes of protecting a relevant right or claim. Although that process will require him to be satisfied that the applicant is a person entitled to apply within the meaning of section 43(1), and therefore, for present purposes, that the person has a sufficient interest in the making of the entry, that is by no means the whole of the registrar's task. A person may have a sufficient interest to qualify him as an applicant for a restriction, in circumstances where the registrar is none the less not persuaded that it is necessary or desirable C to make the requested restriction for the purpose of protecting the applicant's right or claim.

[16] It follows, in my judgment, that what has to be referred to the adjudicator under section 73(7), where an objection that is not obviously groundless cannot be disposed of by agreement, is not merely the question of whether the applicant has a relevant right or claim, but the additional question of whether the entry of a restriction is necessary or desirable for the purpose of protecting that right or claim. Both of those questions fall within what is described in section 73(7) as "the matter" to be referred to the adjudicator.

[17] It is also apparent from section 73(5) to (7) that determination D of the application for the restriction, where there has been an objection, requires the objection to be "disposed of". The disposal of the objection is therefore an integral part of the matter referred to the adjudicator under section 73(7).

[18] It follows from that analysis that the precise nature of the adjudicator's function on any particular reference under section 73(7) will be significantly affected by an examination of the precise restriction sought, the nature of the claim or right thereby sought to be protected and the basis of the objection that has led to the reference. It is plain from section 110(1) that the adjudicator is given a broad discretion, on a reference under section 73(7), whether to decide "a matter" himself E or to require it to be decided in a competent court, and it is equally plain from the panoply of procedural powers given to the adjudicator under the Practice and Procedure Rules that a decision to decide a matter himself may properly involve a trial, rather than just a summary review directed merely to the question of whether an asserted claim is reasonably arguable.

[19] It follows that, to the extent that this appeal is based on the argument that the adjudicator had no jurisdiction to conduct a trial of the question of whether the appellant had a beneficial interest in the property under a resulting trust, that argument must be rejected. None the less, Mr Maynard submitted that even if the adjudicator had a power F to take that course, his decision to adopt it involved an inappropriate exercise of his discretion. To that question I now turn, bearing in mind that it requires a fact-intensive analysis of "the matter" referred to him under section 73(7).

[20] The appellant's application was for a restriction in form II, namely that:

> No disposition of the registered estate is to be registered without a certificate by the applicant for registration or her conveyancer that written notice of the disposition was given to Kusum Jayasinghe at 25 Brampton Road, Kingsbury, London NW9 9BX.

That restriction was sought to be justified on the grounds that:

> By reason of the applicant providing the original purchase funds, paying the G original charge in favour of Halifax Building Society, and later redeeming that charge, and bearing all costs throughout, and receiving all income from the premises, there exists a resulting trust in her favour.

[21] The respondent's objection, by letter dated 8 November 2007, was, in summary, that he denied that the applicant had provided any money towards the acquisition or maintenance of the property, denied any trust of the property in favour of the applicant, but acknowledged that he had asked her to manage the property for him in his absence and allowed her to keep, by way of gift, any surplus or rent over the cost of discharging mortgage instalments and other outgoings in respect of the property. H

[22] The appellant amplified the grounds of her application in a statement of case dated 1 July 2008, which was prepared by her solicitor and supported by a substantial bundle of documents. After setting out her version of the history, the document continued, in para 20 as follows:

> The Applicant is advised that the Property is held on trust for her. Therefore, the application for a restriction was made on 29 October 2007 (pages 150-152). Without this restriction, the respondent is free to sell the Property and enjoy the sale proceeds. Having contributed not one single penny he would otherwise be allowed to commit a gross fraud. Once the restriction is registered, the Applicant will begin legal proceedings against the respondent and claim J ownership of the Property.

In para 21, it was asserted that even if the adjudicator did not agree that the appellant was the 100% beneficial owner of the property, none the less a restriction should be entered to protect such beneficial interest as the appellant might establish. The statement of case gave notice of an intention to call four witnesses (including the appellant), although acknowledging that the whereabouts of one witness remained unknown.

[23] The respondent's statement of case, dated 25 July 2008, set out his detailed case as to the history, concluded with the assertion (which I have described) that the appellant's version of the history was K a complete fiction and repeated that the basis of his objection was that the appellant had no interest whatsoever in the property. The respondent identified in detail the documents on which he intended to rely, and listed himself as the only witness to be called.

[24] In my judgment, the combined effect of those statements of case was sufficient to demonstrate to the adjudicator that although the appellant contemplated that court proceedings might be necessary for the precise quantification of her interest in the property (if less than full beneficial ownership), the parties were shaping up for a full-scale trial on the merits before the adjudicator, in particular on the issue of whether, as the appellant claimed and the respondent denied, she L had any interest at all in the property sufficient to justify the entry of a restriction.

[25] It is therefore no surprise that, on 19 September 2008, the adjudicator issued written directions for a trial before him. They concluded with a note that if either of the parties wanted further or different directions, they should write to the office of the adjudicator to that effect, setting out any points they wanted to make. No objection to those directions was taken by either of the parties, and the matter was in the event tried by the adjudicator over two days, during which, again without any objection, evidence was tendered by witness statement and examination-in-chief and the witnesses were cross-examined. M

[26] In those circumstances, it seems to me quite impossible for the appellant now to challenge the manner in which the adjudicator dealt with the matter, on the basis that he should merely have limited himself to ascertaining whether the appellant had disclosed an arguable claim for an interest in the property. My reasons follow.

[27] First, the application for a restriction was not based on the mere assertion of a claim but on the assertion of a beneficial interest in the property under a resulting trust. Similarly, the objection contained a wholesale denial of any such beneficial interest. This case is, in that respect, far removed from *Croatia*, in which the restriction was sought to protect a claim to an interest, not justiciable in the English courts,

A pursuant to a treaty agreement between sovereign states regarding the distribution of the assets of the former Socialist Federal Republic of Yugoslavia. Furthermore, the decision that led to the appeal before me concerned only the preliminary issue as to whether the existence of such a non-justiciable claim conferred a sufficient interest on Croatia for the purpose of seeking a restriction, within the meaning of section 43(1)(c) of the Act.

[28] In the present case, by contrast, the matter referred to the adjudicator concerned, not merely the appellant's alleged interest under section 43(1)(c) but the question of whether it was necessary or desirable to enter a restriction to protect the right asserted. It follows B that my conclusion on that preliminary issue, namely that Croatia's claim constituted a sufficient interest under section 43(1)(c), affords no basis for the submission that the adjudicator's task, in a case based on the assertion of a right, is limited purely to ascertaining, on a summary basis, whether the applicant has an arguable claim for the existence of that right.

[29] Second, there was, in my judgment, a real question as to the discretion to be exercised under section 110(1) of the Act, namely whether the adjudicator should decide the matter himself or remit all or part of it to be decided by a competent court. He was faced with a statement of case from the appellant that warned that further proceedings C in court were contemplated, and he could have concluded that the overriding objective might be better served by directing the appellant to issue proceedings in court, so that the restriction question could be determined, along with all other aspects of the dispute between the parties, at a single hearing. Those matters appeared to include issues as to the precise quantification of the appellant's alleged beneficial interest and, by implication, issues as to whether the property should be sold or transferred to the appellant, something that, until then, the appellant had been trying to achieve, albeit by forgery, as she herself admitted.

[30] There were none the less compelling reasons why the adjudicator should have decided not to take that course. The first was D that, on the bizarre history of the matter contended for by the appellant, the registered proprietor of the property was not the respondent at all, but some completely different person who, by sheer coincidence, had substantially the same name, and who, according to the appellant, had disappeared: see paras 2 and 3 of the decision. It was not, therefore, on the appellant's case, a simple two-party dispute easily capable of resolution in court proceedings between the two of them.

[31] Next, the critical issue for the purposes of registration of a restriction was not the quantification of the appellant's beneficial interest (which, in her statement of case, she said that she intended to pursue by court proceedings) but, rather, the question of whether she E had any interest in the property. That was the question squarely raised by the respondent's objection, the disposal of which was the primary purpose of the reference to the adjudicator.

[32] Finally, whatever might otherwise have been the balance of factors for or against deciding the matter himself rather than directing the bringing of court proceedings, the adjudicator was, when he gave directions, faced with two parties plainly determined to have a full-scale contest of the matter on the merits before him. Having given directions for that purpose, the parties subsequently prepared for and conducted a trial of the issues raised by the respondent's objection without the slightest complaint.

F

[33] It follows that, not only is the adjudicator's decision as a matter of discretion to decide the matter himself unchallengeable but it was, in my judgment, plainly correct. For those reasons, the first ground of this appeal fails.

[34] The second ground relates to the adjudicator's conclusion that the respondent, rather than the appellant, had provided the funds for the discharge of the mortgage by which the property had largely been acquired. It is advanced not merely for the purpose of seeking to establish a limited beneficial interest arising from the discharge of the mortgage, but for the wider purpose of seeking to undermine the adjudicator's decision to accept, more or less lock, stock and barrel, the respondent's evidence in preference to that of the appellant.

[35] The adjudicator's decision on this point is to be found in the G final sentence of para 9 of his succinct decision, as follows:

> I also conclude that the repayment of the Halifax Building society mortgage was indirectly funded by the respondent as he states.

Earlier, in para 5, as part of his summary of the respondent's case, the adjudicator said:

> He says that the Halifax mortgage was discharged by the Applicant in 2004, but utilising funds which he had given her.

In para 7, summarising the issues, he said: H

> 7.4 Whose funds were used to redeem the Halifax Building Society mortgage in 2004? Entirely those of the Applicant or, as the Respondent contends, indirectly his?

[36] Reading those passages together, and without any knowledge of what took place at the trial, it might be supposed that the adjudicator's conclusion was that the respondent had made funds available to the appellant, from which she had herself discharged the mortgage. In fact, as was common ground before me, bank statements and other documents deployed at the trial demonstrated that the appellant had discharged the mortgage from a NatWest bank account in her name, on 18 February 2004, from a credit balance on that account largely J attributable to her receipt, on 20 January 2004, of the proceeds of successful litigation by her, rather than from any funds made available to her by the respondent. The amount paid by her to discharge the mortgage was £39,000.

[37] None the less, the documents deployed at trial also showed that the respondent had, on 25 and 30 March 2004, caused to be credited to a bank account of his at HSBC the aggregate sum of £41,150, and that, on 30 March, the appellant had caused an identical sum to be paid out of that account to her sister by using one of a number of cheques presigned in blank by the respondent and completing the relevant details in her own handwriting. K The adjudicator accepted evidence from the respondent that these cheques had been made available by him to the appellant for the purpose of enabling her to make mortgage payments from time to time using his funds whenever the rental income was insufficient for that purpose.

[38] The respondent's evidence at trial, under cross-examination, was that he had told the appellant, at the beginning of 2004, that he intended to transfer money to the UK from abroad into the bank account over which she had effective control by means of the presigned cheques and had asked her to discharge the mortgage from that money. When pressed with the documents demonstrating that the appellant had in fact discharged the mortgage out of moneys of her own, L he said that the £41,000-odd payment from his account that the appellant had arranged to be paid to her sister must have been her way of reimbursing herself for having discharged the mortgage out of her own money rather than his.

[39] Bearing in mind that it appears to have been, or became, common ground at trial that the appellant in fact discharged the mortgage out of her own money, it becomes clear that the adjudicator's reference in para 9 of the decision to the repayment of the mortgage having been "indirectly funded by the Respondent as he states" must have been intended by him to be a reference to the respondent's evidence that his funds had been used to reimburse the appellant in full for her discharge M of the mortgage out of her own money, rather than to a finding that the discharge of the mortgage was made from his own funds. I bear in mind in that context that, for what appear to me to have been good and sufficient reasons, the adjudicator expressly concluded in the decision that the respondent was a reliable and truthful witness with regard to the main issues in the case, and I infer that he accepted the respondent's evidence in cross-examination concerning reimbursement.

[40] Although it is perhaps unfortunate that, without an understanding of what took place at the trial, the adjudicator's terse finding in respect of the repayment of the mortgage might appear to have a different meaning, I am satisfied, for the reasons that I have

given, that by "indirectly funded" he meant to refer to a process of funding by reimbursement rather than by original payment.

[41] Quite separately, that infelicity in the adjudicator's description of his findings on that issue comes nowhere near to undermining his overall conclusion that the appellant's case was, from start to finish, a highly implausible fiction, concocted by someone with no apparent understanding of right and wrong, and wholly insufficient to discharge the burden on her of demonstrating any beneficial interest in the property for the reasons cogently given in paras 8 to 10 of the decision.

[42] In those circumstances, this second ground of appeal also fails. The discharge of a mortgage from her own funds, immediately followed by a reimbursement from the respondent's bank account cannot conceivably give rise to a beneficial interest in the appellant's favour, nor does this aspect of the matter begin to cast doubt on the soundness of the adjudicator's wholesale rejection of her case.

[43] I cannot conclude this judgment without reference to what appears to me to be a disgraceful feature of both parties' conduct in respect of the property. The appellant's case, as summarised in paras 2 and 3 of the decision, was that knowing that an application for a mortgage by her would be refused because she was already a house owner, she arranged with others for the mortgage to be obtained by a stranger, masquerading as the real purchaser, but in fact her mere nominee. The respondent's case, as the adjudicator noted in para 9 of the decision, also involved obtaining a mortgage for the purchase of the property by deceiving Halifax Building Society, since he gave a fictitious address for himself so as to make it appear, contrary to the truth, that he was resident within the UK.

[44] In the event, of course, the adjudicator's rejection of the appellant's case means that she was not, as she asserted that she was, party to a mortgage fraud, but on the other hand it reveals her as advancing a dishonest case, and supporting it by evidence, in a fraudulent attempt to obtain beneficial ownership of the property that in truth belonged to the respondent, her nephew. On any view, her attempt to sell the property for her own benefit involved her forging the respondent's signature on a Land Registry transfer and on a letter to the Land Registry: see para 3 of the decision.

[45] In those circumstances, I shall, after handing down this judgment, invite submissions from the parties as to why the papers in this case should not be referred to the Crown Prosecution Service, with a view to an investigation into whether they disclose serious criminal misconduct on the part of both parties.

Appeal dismissed.

REAL PROPERTY AND CONVEYANCING

A
Davies and others *v* Jones and another

COURT OF APPEAL

[2009] EWCA Civ 1164

9 November 2009

SIR ANDREW MORRITT CVO, SCOTT BAKER LJ AND LEWISON J

Estates Gazette 6 February 2010

B [2010] 05 EG 114

Sale of land — Contract — Benefit and burden — Land in ownership of third parties — Vendor contracting to buy from third parties and assigning rights under contracts to purchaser — Purchaser retaining part of sale price pending completion of works under terms of assigned contract — Purchaser subsequently failing to pay balance — Whether third party entitled to recover retention moneys — Whether purchaser bound to observe vendor's obligations under assigned contract as burden appurtenant to benefit received

By a written contract, the second defendant agreed to purchase certain freehold land and buildings from the first defendant, who did not, C at the relevant time, own them; the greater part of the land was vested in the claimants, and the remainder of it in another party. Accordingly, the first defendant contracted to buy the land from the owners and assigned his right, title and interest in both contracts to the second defendant. The sales were completed in August 2000 by transfers from the claimants and the other owner direct to the second defendant, and the payment by the second defendant of the respective sums due to them. The transfers were duly registered at the Land Registry.

The sale contract for the claimants' land had specified a purchase price of £660,000 but, by a retention clause, provided for the purchaser to carry out certain "Purchaser's Works" and to retain £100,000 of D the purchase price until those works were completed. Thereafter, the purchaser was entitled to keep one-half of the proper costs of those works out of the retention, with any balance to be released to the claimants. The purchaser's works were completed in June 2001, however, the second defendant failed to pay any part of the retention to the claimants. The claimants maintained that the proper cost of the purchaser's works was £30,000, whereas the second defendant claimed to have spent more than £200,000 on them.

In proceedings against the second defendant for the recovery of the retention moneys, a preliminary issue was tried as to whether those moneys were subject to a trust or fiduciary duty requiring the second E defendant to hold or release them on the terms of the retention clause, or, if not, whether the second defendant was bound by some other, and if so what, duty to observe the said terms. The trial judge rejected any finding of a trust or fiduciary duty. However, he held that, in the light of the clear understanding in August 2000 that the benefit of the contract would be given to the second defendant, the latter was bound to observe the terms of the retention clause as a burden appurtenant to the benefit that it took. The second defendant appealed on the latter point.

***Held*: The appeal was allowed. (1) For a party to be bound under the doctrine of benefit and burden, the benefit and burden must be conferred in or by the same transaction. In the case of benefits and burdens in respect of land, it is almost inevitable that the transaction will be effected by one or more deeds or other documents. Further, F the receipt or enjoyment of the benefit must be relevant to the imposition of the burden, in that the former is conditional on, or reciprocal to, the latter. Whether that requirement is satisfied will be a question of construction of the deeds or other documents where the question arises in the case of land, or, in other cases, the terms of the transaction, if not reduced to writing. In each case, it will depend on the express terms of the transaction and any implications to be derived therefrom. Moreover, the party on which the burden is alleged to have been imposed must have had the opportunity of rejecting or disclaiming the benefit. (2) Applying those principles, the second defendant was not bound by the burden of the retention G clause. A "clear understanding" can neither confer benefits nor impose burdens; a transaction having legal effect is required. Since the benefit and burden in the instant case were alleged to arise from land, the relevant transaction had to be effected by a deed or other document. Although the deed of assignment, assigning the first defendant's rights under the contract to the second defendant, was a transaction having legal consequences, the benefit that the second defendant derived from that assignment was not made conditional on it performing and observing the substantial obligations of the first defendant to carry out site works under that contract. Nor did any relevant obligation arise from the completion of the sales in August 2000. Completion of the contract was not dependent on H the second defendant undertaking any of the burdens imposed on the first defendant under the retention clause and no stipulation required the second defendant to perform and observe them. The retention clause was contractually binding on the first defendant and, in the absence of a covenant in the transfer binding the second defendant to perform and observe the vendor's obligations under that clause, those obligations remained the first defendant's alone.**

The following cases are referred to in this report.

Aspden v *Seddon* (1876) LR 1 Ex D 496, CA
Baybut v *Eccle Riggs Country Park Ltd* [2006] WL 3206169 J
Elliston v *Reacher* [1908] 2 Ch 665, CA
Halsall v *Brizell* [1957] Ch 169; [1957] 2 WLR 123; [1957] 1 All ER 371
IDC Group Ltd v *Clark* [1992] 1 EGLR 187; [1992] 08 EG 108
Jenkins v *Young Brothers Transport Ltd* [2006] EWHC 151 (QB); [2006] 1 WLR 3189; [2006] 2 All ER 798
Lidl UK GmbH v *Davies* [2008] EWCA Civ 976
Rhone v *Stephens (executrix)* [1994] 2 AC 310; [1994] Ch 310; [1994] 2 WLR 429; [1994] 2 All ER 65; [1994] 2 EGLR 181; [1994] 37 EG 151, HL
Thamesmead Town Ltd v *Allotey* (2000) 79 P&CR 557; [1998] 3 EGLR 97; [1998] 37 EG 161; (1998) 30 HLR 1052, CA
Tito v *Waddell (No 2)* [1977] Ch 106; [1977] 2 WLR 496; [1977] 3 All ER 129 K
Westhoughton Urban District Council v *Wigan Coal & Iron Co Ltd* [1919] 1 Ch 159, CA

This was an appeal by the second defendant, Lidl UK GmbH, from a decision of HH Judge Jarman QC, sitting as a High Court judge, deciding a preliminary issue in proceedings by the claimants, David Davies and Rachel Davies, for the recovery of moneys retained from the purchase price under a contract for the sale of land.

Jeffrey Littman (instructed by Charles Crookes, of Cardiff) represented the claimants; Jeremy Child (instructed by Clarke Willmott LLP, of Birmingham) appeared for the second defendant.

Giving judgment, **Sir Andrew Morritt CVO** said:

Introduction

[1] By a contract in writing dated 21 June 1999 (the Jones-Lidl contract), the first defendant, Mr Jones, contracted to sell to the second defendant, Lidl, freehold property, comprising land and buildings to the northwest of Perrotts Road, Haverfordwest, for £795,000 (subsequently increased to £898,743). At the time, the freehold of that land was vested for the most part in the claimants, as trustees of a retirement benefit scheme, and to a lesser extent in a Mr and Mrs Thomas. On the same day, Mr Jones contracted to buy from Mr and Mrs Thomas the part vested in them (the Jones-Thomas contract) for £180,000. Four months later, on 20 October 1999, Mr Jones contracted to buy the major part from the claimants (the Jones-trustees contract) for £660,000. By a deed of assignment dated 3 August 2000 (the deed of assignment), Mr Jones assigned to Lidl his right, title and interest in both the Jones-Thomas contract and the Jones-trustees contract. Both were completed on 4 August 2000 by transfers from Mr and Mrs Thomas and from the claimants direct to Lidl and the payment by Lidl to the claimants, Mr and Mrs Thomas and Mr Jones of the respective sums due to them.

[2] Clause 18 of the Jones-trustees contract related to the "Purchaser's Works", which had been defined in clause 14.1 as the ground clearance and site preparation works described in the annexed specification. Clause 18 provided that:

> 18.1 On completion the Purchaser [Jones] shall retain the sum of £100,000 from the purchase monies ("the Retention") payable to the Vendor [the claimants] until the outstanding works shall have been completed by the Purchaser [Jones]
>
> 18.2 The Purchaser [Jones] will at the joint expense of the Vendor and the Purchaser forthwith following the completion date commence and diligently proceed to complete the outstanding works…. and shall complete the same… in any event within the period of three months following the completion date unless prevented from doing so by matters outside his reasonable anticipation and control.
>
> 18.3 Following completion of the outstanding works within the said period the Purchaser shall be entitled to retain one half of the proper costs of so doing from the retention with any balance then being released to the Vendor [the claimants].

[3] Lidl retained the sum of £100,000 from the money due by Mr Jones to the claimants on completion of the Jones-trustees contract. The transfers from Mr and Mrs Thomas and from the claimants to Lidl were duly registered at HM Land Registry on 29 September 2000. The purchaser's works as defined in the Jones-trustees contract were completed by 30 June 2001 at the latest but Lidl failed to pay any part of the retention to the claimants. The claimants contend that the proper cost of those works was £30,000; Lidl claims to have spent in excess of £200,000 in completing them.

[4] These proceedings were instituted by the claimants on 18 October 2005 against both Mr Jones and Lidl seeking the recovery of £100,000 from one or other of them. As against Mr Jones, they were summarily dismissed by Wyn Williams J on 7 January 2008. He also gave the claimants permission to amend their claim against Lidl in some but not all the respects they wished. His decision was upheld by the Court of Appeal on 10 July 2008. The pleadings were duly amended and the claim was fixed for hearing before HH Judge Jarman on 5 and 6 May 2009. An application to adjourn the trial was made to Judge Jarman QC on 27 April 2009 on the basis that necessary expert evidence could not be ready in time. He acceded to that application but, so that the time set aside for the trial should not be wasted, he directed the trial of the following preliminary issue:

> Is the sum of £100,000… subject to a trust and/or fiduciary duty to hold and, as appropriate release, the same on the terms set out in clause 18 of the contract dated 20th October 1999 between the claimants and the first defendant [Jones]?

[5] The trial of that preliminary issue came before Judge Jarman on 5 May 2009. Counsel for the claimants then contended that the preliminary issue was too narrow and should be extended to include the question:

> or is [Lidl] bound by some other, and if so what, duty to observe the said terms.

The extension of the preliminary issue was opposed by counsel for Lidl on the grounds that such a question did not reflect an issue raised by the pleadings, such issue would constitute the addition of a new claim by then barred by the Limitation Act 1980 (the 1980 Act), such new issue did not arise from the same or substantially the same facts as the existing claims and, in any event, was raised too late. The judge was conscious of the fact that the extension would raise a new point and adjourned the hearing to the following day to enable counsel for Lidl to prepare his submissions, but he allowed the extension of the preliminary issue sought by counsel for the claimants.

[6] The judge heard argument on the extended preliminary issue on 6 May 2009 and gave judgment the same day. He determined the original preliminary issue in favour of Lidl but the additional issue that he had allowed the day before in favour of the claimants. In that respect, he concluded that:

> there was a clear understanding that on the 3rd or 4th August, the rights of Mr Jones under his contract with the trustees would be given to Lidl. There was a clear benefit to them of that operation. It gave them the likelihood of completion without going through the intermediary of Mr Jones and being left perhaps with a worthless judgment in damages. There was a clear practical benefit to them. It seems to me that they did have a choice whether to take that benefit or not. That was the basis of the correspondence between the parties' solicitors from July. They had a choice to continue the sub purchaser arrangement. They had no need, from a legal point of view, to deal directly with the trustee's solicitors but that they did. They did so, in my judgment, because they were concerned to have completion sooner rather than later. Having taken the benefit of that arrangement, in respect of the contract in question, in my judgment, they are bound by the burden under it to pay the £100,000 if and in the event it is found the conditions for such payment have been complied with.

The judge declared that:

> [Lidl] is bound by a duty to obseve the said terms [sc those set out in clause 18 of the Jones-trustees Contract] as a burden appurtenant to the benefit it took under the same contract.

[7] Lidl now appeals with the permission of Carnwath LJ. It does so on the grounds that the judge was wrong to have:

(1) allowed the extension of the preliminary issue; and

(2) concluded that Lidl was bound by the terms of clause 18.3 of the Jones-Lidl contract.

I propose to deal first with the second of those issues. However, before I do so, I should set out the facts in somewhat greater detail.

Facts

[8] The land that Lidl acquired had been the site of a garage. Lidl proposed to use it as the site for a supermarket. Thus, it was necessary to remove the garage buildings and fittings and clear the site. The garage premises and other buildings were to be removed by the claimants. Site clearance and preparation works were to be effected by Mr Jones. Provision was made in the Jones-Thomas contract for certain fencing and other works. In the Jones-Lidl contract, clause 14 contained provisions comparable to clause 18 of the Jones-trustees contract. Thus, the vendor, Mr Jones, contracted to commence after the completion date and within three months complete the site preparation work indicated in the definition of the vendor's works. On completion, the purchaser, Lidl, was to be entitled to retain £100,000 from the purchase price until completion of those works to its reasonable satisfaction. If the works were not so completed, the purchaser, Lidl, might carry them out and retain the proper costs of so doing from the retention. Although similar, the provisions of clause 14 of the Jones-Lidl contract were not the same as those of clause 18 of the Jones-trustees contract; under the former, Lidl might reimburse itself for the entirety of the cost of carrying out the works, under the latter, it could retain only one-half of such cost.

[9] On 19 October, namely the day before the conclusion of the Jones-trustees contract, there was an agreement between Mr Jones and

A Lidl (the supplemental agreement) whereby the price payable by the latter was increased to £898,743. In addition, Mr Jones agreed to use his reasonable endeavours to procure that the property was transferred to Lidl direct by way of subsale. As I have already indicated, the Jones-trustees contract was exchanged on 20 October 1999. It provided for a deposit of £30,000 in addition to the retention of the £100,000 on the terms that I have quoted in [2] above. The garage premises and other buildings were excluded from the sale and the claimants were entitled within three months after completion to enter on the land and remove them.

[10] In July 2000, solicitors for all four parties, namely Mr and B Mrs Thomas, the claimants, Mr Jones and Lidl corresponded in respect of completion of the various contracts. On 13 July 2000, the solicitor for Mr Jones sent a draft transfer in respect of the land vested in the claimants to their solicitor together with requisitions on title and a completion statement showing that £630,000 was due to the claimants on completion, namely the price less the deposit already paid. It pointed out that the draft transfer was in favour of Lidl and trusted that the claimants would be prepared to transfer the land direct to Lidl. On 19 July 2000, the solicitor for the claimants confirmed to one for Mr Jones that its clients had no objection to transferring the land direct to Lidl. On 2 August 2000, the solicitor for Lidl indicated to the C solicitor for the claimants that, as agreed with the solicitor for Mr Jones, it would forward the completion moneys due to the claimants direct to their solicitor against the latter's undertaking to forward the transfer and deeds direct to the solicitor for Lidl. On 2 August 2000, the solicitor for Lidl wrote to that for the claimants noting that it had not been able to make the necessary arrangements for completion but had transferred the balance of the completion moneys to it on the basis that the latter would hold them to its order.

[11] The deed of assignment, to which I have referred, was made between Mr Jones and Lidl on 3 August 2000. In addition, following the assignment of the benefit of the Jones-Thomas and Jones-trustees D contracts by Mr Jones to Lidl, it was agreed between them that the Jones-Lidl contract and the supplemental agreement were thereby determined and ceased to have effect. At 3pm on that day, Lidl paid £660,000 to the claimants' solicitor, as it had promised the previous day that it would. The claimants duly executed the relevant transfer to Lidl, which was forwarded to its solicitor under cover of a letter dated 7 August. On 4 August, the claimants' solicitor repaid to the solicitor for Lidl £130,000, being the aggregate of the deposit and the retention. Comparable arrangements were made with the solicitor for Mr and Mrs Thomas, but they are not relevant to anything that we have to decide. Thus, the claimants and Mr and Mrs Thomas received the money E due to them respectively by Mr Jones and the difference between what Mr Jones was liable to pay to them and entitled to receive from Lidl, namely £58,743, was paid or accounted for by Lidl to Mr Jones.

[12] On 25 September 2000, Lidl was registered as the proprietor of the freehold of the land acquired from the claimants and Mr and Mrs Thomas respectively under two different title numbers. Whether Mr Jones ever attempted to execute the works for which both the Jones-trustees contract and the Jones-Lidl contract provided is unclear. They were completed by contractors engaged by Lidl on or before 30 June 2001. Lidl kept the entirety of the £100,000 retention. The claimants deny that they are entitled to retain any part of it. In any event, F they contend, the proper cost of the works was only £30,000.

Proceedings

[13] The claim form was issued on 18 October 2005, one or two days short of six years from the date of the supplemental agreement and the Jones-trustees contract. In their particulars of claim served on 13 February 2006, the claimants relied on the provisions of clause 18 of the Jones-trustees contract. They asserted that Mr Jones had entered into it as agent for Lidl. They claimed against Lidl, alternatively, Mr Jones, £100,000 being the retention sum or that sum less one half of the proper cost of the outstanding works or damages for breach of contract. In its defence served on 12 March 2006, Lidl denied that Mr Jones was its agent. It asserted that it was not a party to the G Jones-trustees contract and was not liable to the claimants thereunder.

[14] In the light of that defence, and a comparable one from Mr Jones, the claimants sought permission to amend so as, among other things, to add claims against Lidl for breach of trust, knowing receipt of trust property, namely the £100,000 retention, and knowing assistance in a breach of trust. By then, more than six years had elapsed since the outstanding works had been completed so that section 21 of the 1980 Act applied and CPR 17.4(2) was engaged. Wyn Williams J gave permission to amend in respect of the trust claims and his decision in that respect was upheld by the Court of Appeal in July 2008, see *Lidl UK GmbH* v *Davies* [2008] EWCA Civ 976. H

[15] The particulars of claim were duly amended and reserved on 7 January 2008 to reflect the trust claims for which permission had been granted. They did not include any claim to the effect that Lidl was bound by the terms of the Jones-trustees contract because it had taken a benefit thereunder. In para 14, it was contended that even if, which was denied, Lidl could lawfully have held the retention moneys, it had unlawfully continued to retain them both in breach of the terms of clause 18.3 and in breach of trust. In paras 17 and 19, the claimants denied that Lidl was bound by the Jones-trustees contract. In paras 25 and 27 of its amended defence, Lidl admitted and averred that it was not liable on the Jones-trustees contract. In para 2 of their reply dated J 4 December 2008, the claimants admitted that Lidl was not party to the Jones-trustees contract and relied on the fact that the claimants were not parties to the deed of assignment.

[16] I have already noted in [4] above that the trust claims raised by the amendments for which Wyn Williams J gave permission were reflected in the original preliminary issue ordered to be tried by HH Judge Jarman QC on 27 April 2009. He concluded on 6 May 2009 that the sum of £100,000 was not subject to a trust or fiduciary duty and so declared. The claimants have not sought to appeal against that declaration. It is in those circumstances that the claim made by the claimants in the extended preliminary issue is the only remaining cause K of action. I will, in due course, consider whether the judge was right to allow the preliminary issue to have been extended in the manner that I have described, but, first, I will consider whether the judge gave the correct answer to it.

Is Lidl bound by the terms of clause 18 of the Jones-trustees contract because it took a benefit under that contract?

[17] Having set out the facts and having dealt with the trust issues Judge Jarman QC dealt with this issue in [20] to [23] of his judgment. In his consideration of it, he referred to the speech of Lord Templeman in *Rhone* v *Stephens (executrix)* [1994] 2 AC 310*, and thereby indirectly to *Halsall* v *Brizell* [1957] Ch 169 and *Tito* v *Waddell (No 2)* [1977] Ch L 106. He also referred to the judgment of HH Judge Pelling QC in *Baybut* v *Eccle Riggs Country Park Ltd* 2 November 2006 [2006] WL 3206169 but not to the judgment of Peter Gibson LJ in *Thamesmead Town Ltd* v *Allotey* (1998) 30 HLR 1052† or Sir Nicolas Browne-Wilkinson V-C in *IDC Group Ltd* v *Clark* [1992] 1 EGLR 187.

[18] I have quoted the judge's conclusion in [7] above. It is criticised by counsel for Lidl on the following grounds:

(1) the "clear understanding" to which the judge referred is an insufficient basis for conferring a benefit or imposing a burden;

(2) the benefit that Lidl did obtain under either the deed of assignment or the transfer of the land was, in each case, unconditional; M

(3) its benefit having been obtained on the execution of the deed of assignment and transfer, thereafter Lidl had no choice whether to accept it;

(4) in any event, the corresponding burden cannot be enforced by action on the contract.

[19] I did not understand counsel for the claimants to place much, if any, emphasis on the "clear understanding" to which the judge referred. Rather, he pointed to the deed of assignment as the source of Lidl's

---

* Editor's note: Also reported at [1994] 2 EGLR 181

† Editor's note: Also reported at [1998] 3 EGLR 97

A rights. Not only did it contain an assignment by Mr Jones to Lidl of the benefit of the Jones-Thomas and Jones-trustees contracts but it provided for the discharge of the Jones-Lidl contract. Accordingly, this case is not a conventional subpurchase. I would accept that submission. Thus, the rights of Lidl were derived successively from the deed of assignment and the transfers executed on completion by the claimants and Mr and Mrs Thomas. The receipt of the benefit of the assignment by the subsequent transfers was conditional on Lidl's performance of the obligations of Mr Jones under the Jones-Thomas and Jones-trustees contracts. Counsel for the claimants then contended that such conditionality was sufficient to impose on Lidl the burden of complying
B with the provisions of clause 18 of the Jones-trustees contract. In addition, he submitted that if the conditions for the imposition of the burden on Lidl existed, it must be open to the claimants to enforce that burden by action, if only by way of a claim for unjust enrichment.

[20] I turn then to the decided cases. The origin, in modern times, for the application of the doctrine of benefit and burden is the judgment of Upjohn J in *Halsall*. That case concerned the enforceability of a deed of covenant entered into by the owners of building plots in a development in Liverpool against the successors in title of the original covenantors. The owners of the plots for the time being enjoyed the benefit of the roads and sewers constructed as part of the development and, on behalf
C of themselves and their successors in title, had covenanted with the trustees to pay part of the cost. Some houseowners, being successors in title to the original covenantors, refused to pay an increased contribution and the trustees and other houseowners applied to the court to determine whether on the true construction of the deed and in the events that had happened the trustees were entitled to require the houseowners to pay the contributions that they claimed.

[21] It was conceded in argument in response to a question from the judge that there was a rule that one who accepts the benefit of a deed must also accept the burden of it: see p180. At p182, Upjohn J observed that it was plain that the defendants could not be sued on the covenants
D contained in the deed because the covenant to pay was positive, aspects of it infringed the rule against perpetuities and because the parties before him were not parties to the original contract. He then considered the foundation of the doctrine of benefit and burden, as formulated in *Coke on Littleton*, at p230b, and applied in *Elliston* v *Reacher* [1908] 2 Ch 665, at p669, and concluded, at pp182-183:

> If the defendants did not desire to take the benefit of this deed, for the reasons I have given, they could not be under any liability to pay the obligations thereunder. But, of course, they do desire to take the benefit of this deed. They have no right to use the sewers which are vested in the plaintiffs, and I cannot see that they have any right, apart from the deed, to use the roads
> E of the park which lead to their particular house, No 22, Salisbury Road. The defendants cannot rely on any way of necessity or on any right by prescription, for the simple reason that when the house was originally sold in 1931 to their predecessor in title he took the house on the terms of the deed of 1851 which contractually bound him to contribute a proper proportion of the expenses of maintaining the roads and sewers, and so forth, as a condition of being entitled to make use of those roads and sewers. Therefore, it seems to me that the defendants here cannot, if they desire to use this house, as they do, take advantage of the trusts concerning the user of the roads contained in the deed and the other benefits created by it without undertaking the obligations thereunder. Upon that principle it seems to me that they are bound by this deed, if they desire to take its benefits.

F

[22] In *Tito*, Sir Robert Megarry V-C considered the principle applied by Upjohn J in *Halsall*, at pp299-303. He concluded, at p302, that the "pure principle" of benefit and burden was distinct from a conditional benefit or the annexation of a burden to property. In *Rhone*, the House of Lords disapproved of the pure principle as enunciated by Sir Robert Megarry but cast no doubt on his comment, at p302, that whether an instrument or transaction has created a conditional benefit or a burden annexed to property depends on the proper construction of the instrument or transaction in question.

[23] In *Rhone*, a successor in title to the original covenantee sought to enforce a covenant to repair a roof against the successor in title to the original covenantor. Lord Templeman, with whom the other four members of the appellate committee agreed, noted that equity G cannot, any more than the law, compel an owner to comply with a positive covenant entered into by his predecessor in title. He then referred to the "pure principle of benefit and burden" enunciated by Sir Robert Megarry V-C and continued, at pp322E-323A:

> I am not prepared to recognise the "pure principle" that any party deriving any benefit from a conveyance must accept any burden in the same conveyance. Sir Robert Megarry V-C relied on the decision of Upjohn J in *Halsall* v *Brizell* [1957] Ch 169. In that case the defendant's predecessor in title had been granted the right to use the estate roads and sewers and had covenanted to pay a due proportion for the maintenance of these facilities. It was held that the defendant could not exercise the rights without paying his costs of ensuring H that they could be exercised. Conditions can be attached to the exercise of a power in express terms or by implication. *Halsall* v *Brizell* was just such a case and I have no difficulty in wholeheartedly agreeing with the decision. It does not follow that any condition can be rendered enforceable by attaching it to a right nor does it follow that every burden imposed by a conveyance may be enforced by depriving the covenantor's successor in title of every benefit which he enjoyed thereunder. The condition must be relevant to the exercise of the right. In *Halsall* v *Brizell* there were reciprocal benefits and burdens enjoyed by the users of the roads and sewers. In the present case clause 2 of the 1960 conveyance imposes reciprocal benefits and burdens of support but clause 3 which imposed an obligation to repair the roof is an independent provision. In *Halsall* v *Brizell* the defendant could, at least in theory, choose J between enjoying the right and paying his proportion of the cost or alternatively giving up the right and saving his money. In the present case the owners of Walford House could not in theory or in practice be deprived of the benefit of the mutual rights of support if they failed to repair the roof.

Thus, Lord Templeman stressed that conditions may be attached to the exercise of a right or power by express terms or by implication, thereby recognising that, as Sir Robert Megarry had considered, the question is one of construction of the relevant instrument or transaction. Even so, as Lord Templeman pointed out, the condition must be relevant or reciprocal to the exercise of the right.

K

[24] The principle was considered again by the Court of Appeal in *Thamesmead*. In that case, the successor in title to the original covenantor refused to pay his share of the costs of repairing and maintaining sewers that he had the right to use as appurtenant to his house. The covenantee in which the relevant housing estate was vested sued for their recovery. The defence was that as the defendant had not been a party to the original covenant he could not be liable for the sums claimed. That defence succeeded at first instance and on appeal. Peter Gibson LJ, with whom Hobhouse and Butler-Sloss LJJ agreed, noted that there was an exception to the rule that the burden of a positive covenant does not run with the land in cases where the covenantor may not take the benefit of a transaction without undertaking the burden imposed by it. L He referred to *Halsall* and *Rhone*. He noted that Lord Templeman had suggested that there were two requirements for that exception to apply, namely: (i) relevance of the burden or its discharge to the exercise of the rights that enable the benefit to be obtained; and (ii) the opportunity to choose whether to exercise the right or, having taken the right, whether to renounce its benefit, as opposed to a choice as to whether to acquire the rights. In respect of those requirements, he concluded, at p1060:

> Similarly, it is not possible to enforce every burden in a conveyance by depriving the covenantor's successors in title of every benefit which he enjoyed under the conveyance. There must be a correlation between the burden and the benefit M which the successor has chosen to take. Lord Templeman plainly rejected the notion that taking a benefit under a conveyance was sufficient to make every burden of the conveyance enforceable. Further, there is no authority to suggest that any benefit obtained by a successor in title, once the property has been transferred to him, to enable the enforcement of a burden under the conveyance is sufficient, even if that benefit was not conferred as of right by the conveyance. In my judgment, it cannot be sufficient that the taking of an incidental benefit should enable the enforcement of a burden against a person who has not himself covenanted to undertake the particular burden. Lord Templeman's reference to rights and power suggests that the successor in title must be able as of right to obtain the relevant benefit. I have already pointed out that not only is there no right conferred on the defendant by the 1988 transfer to use the communal areas but also the plaintiff has no obligation to maintain those areas.

A [Counsel]'s second argument was that the judge erred in holding that the burden of a positive covenant is enforceable only if and to the extent that a successor in title chooses to exercise the corresponding right. That overstates what in fact the judge did say. He did not use the words "to the extent that". [Counsel] submits that once a successor in title enters into the transaction by which he takes title he is liable regardless of whether he has actually chosen to enjoy a benefit. He submits that the judge has confused the acquisition of the right with the exercise of the right. He suggests that any other construction would lead to an impractical result requiring the plaintiff to monitor the exercise of rights by persons living in Thamesmead. For my part, I see considerable force in the common sense of that argument, but in the light of the authorities the argument seems to me impossible. As I have already pointed out, in *Halsall* v *Brizell*, at B 182, Upjohn J was expressing the relevant principle in terms that the successors in title could choose whether or not to take the benefit of the deed. Similarly, in *Rhone* v *Stephens*, Lord Templeman in distinguishing *Halsall* v *Brizell*, expressed himself in terms which indicated that the successors in title had to have a choice whether to exercise the right or, having taken the right, whether to renounce the benefit. Lord Templeman was not expressing himself in terms that the successors in title had to have a choice whether to acquire the rights at all. Accordingly, I must reject this second argument also.

Thus, Peter Gibson LJ also stressed the need for a correlation between the burden and the benefit the successor in title has chosen to take and his ability to choose whether to take the benefit.

C [25] The principle was applied in 2006 in two cases on which counsel for the claimants relied. The first was *Jenkins* v *Young Brothers Transport Ltd* [2006] EWHC 151 (QB); [2006] 1 WLR 3189 and the second *Baybut*. In the former, a solicitor had, in the name of his then firm, entered into a conditional fee agreement with a client. While the litigation proceeded, he moved firms twice. On each occasion, the benefit of the CFA was assigned by the former firm to the subsequent firm. The client was successful and his costs, to be paid by the defendant, were then assessed. The defendant in the action contended that he should not be made to pay the costs due under the CFA after its first assignment because the claimant was not liable to pay them. That contention failed before the costs judge and Rafferty J sitting with D assessors. The latter concluded, at p3198C, that the benefit and burden of the CFA might be assigned because:

> the benefit of being paid was conditional upon and inextricably linked to the meeting by Girlings of its burden of ensuring to the best of its ability that [the claimant] succeeded.

Plainly an inextricable link between benefit and burden would satisfy the tests formulated in all the earlier cases. That is sufficient for present purposes, although I have some doubt whether the relevant benefit and burden were correctly described.

[26] In *Baybut*, the purchaser of a caravan park purported to E terminate the 10-year licences under which the owners of the various caravans occupied their respective pitches. HH Judge Pelling QC concluded, in [59], that the new owner of the caravan park was subject to the burden of the licences not only because they had been novated but also because:

> [he, the purchaser] deliberately chose to take the benefit of the caravan site… and the income stream it represented…. it could not have done so without accepting the burden of the licences entered into as between the claimants and [the former owner].

However, this was said in the context that the sale agreement of the F caravan site had contained a covenant by the purchaser with the vendor to perform and observe the future obligations imposed by the licences. Thus, under the sale agreement, the purchaser took the benefit of the licences conditionally on accepting the burdens thereunder. In addition, the principle that one who takes the benefit of a licence to occupy the land granted to another in the form of an income stream, presumably by receiving periodical payments, will be bound by the burden to permit the licence-holder to occupy his pitch is also well known. Both grounds are in accordance with the principles formulated by Lord Templeman and Peter Gibson LJ.

[27] *Rhone* and *Thamesmead* are binding on us. They establish a number of propositions the application of which are exemplified in the other cases to which I have referred, namely *Halsall*, that part of *Tito* that was not disapproved in *Rhone*, *Jenkins* and *Baybut*. In my view, G those propositions are:

(1) The benefit and burden must be conferred in or by the same transaction. In the case of benefits and burdens in respect of land, it is almost inevitable that the transaction in question will be effected by one or more deeds or other documents.

(2) The receipt or enjoyment of the benefit must be relevant to the imposition of the burden in the sense that the former must be conditional on or reciprocal to the latter. Whether that requirement is satisfied is a question of construction of the deeds or other documents where the question arises in the case of land or the terms of the transaction, if not reduced to writing, in other cases. In each case, it will depend on H the express terms of the transaction and any implications to be derived from them.

(3) The person on whom the burden is alleged to have been imposed must have or have had the opportunity of rejecting or disclaiming the benefit, not merely the right to receive the benefit.

[28] Applying those propositions to the facts of this case, I conclude that the judge reached the wrong conclusion. First, a "clear understanding", such as he referred to in the passage from his judgment that I have quoted in [6] above, is an insufficient base from which to start. An understanding such as he described can neither confer benefits nor impose burdens. What is required is a transaction having legal J effect. In this case, the benefit and burden is alleged to arise from land. Accordingly, the relevant transaction must be effected by a deed or other document.

[29] In his respondent's notice and in argument before us, counsel for the claimants placed principal reliance on the deed of assignment. That deed effected a transaction having legal consequences, namely the assignment of the benefit of the Jones-Thomas and Jones-trustees contracts by the claimants to Lidl. However, the benefit derived by Lidl from that assignment was not thereby made conditional, as it might have been, on Lidl performing and observing the obligations of Mr Jones under those contracts. In the absence of a covenant to K that effect contained in the deed of assignment, the transaction did not purport to impose any burden on Lidl. Both contracts imposed on Mr Jones substantial obligations to carry out various site works. There is nothing in the deed of assignment from which it would be possible to find by implication an undertaking or requirement imposed on Lidl to carry out Mr Jones's obligations. Accordingly, the questions of whether acceptance of the benefit was conditional on performance of the burden and whether Lidl had sufficient opportunity to disclaim or reject the benefit do not arise.

[30] The benefit to Lidl might have been realised by assigning the benefit of the contracts to some third-party purchaser. In that event, L there could have been no question of Lidl undertaking any burden. However, in all probability, the benefit of the deed of assignment would be realised, as it was, by completing the contracts the benefit of which Lidl had thereby acquired. Completion took place on 4 August and some reliance was put on that transaction by counsel for the claimants. However, in my view, that transaction cannot avail the claimants either. Completion of those contracts was not dependent on Lidl undertaking any of the burdens imposed on Mr Jones by clause 18 of the Jones-trustees contract, indeed, they did not arise until after completion. The relevant transfers did not contain any stipulation requiring Lidl to perform and observe the obligations of Mr Jones under the contracts M thereby completed. Accordingly, as with the deed of assignment so with the transfers there is nothing that could be regarded as either expressly or by implication imposing any relevant obligation on Lidl. Again the questions whether receipt of the benefit is sufficiently relevant to or conditional on the imposition of the burden and whether Lidl had sufficient opportunity to disclaim or reject the benefit do not arise.

[31] The fact is that clause 18 of the Jones-trustees contract was contractually binding on Mr Jones. The obligations thereby imposed arose after completion of the contracts. There was no subsequent transaction by which benefits were conferred and burdens imposed capable of attracting the principle relied on. In the absence of a covenant in the transfer binding Lidl to perform and observe Mr Jones' obligations

under clause 18 of the Jones-trustees contract they could not be imposed on Lidl but remained the obligations of Mr Jones alone. Mr Jones was originally a party to these proceedings. Why the claim against him was made only on the basis that he was the agent of Lidl is not for us to speculate. For these reasons, I would allow the appeal and, subject to our conclusion on the other issue, set aside the judge's answer to that part of the preliminary issue and make a declaration in the opposite sense.

[32] In those circumstances, it is unnecessary to consider whether if the burden had been imposed on Lidl it was capable of being enforced by action at the suit of the claimants. However, in the light of counsel's submissions, I wish to add a few words in respect of it. Counsel for Lidl submitted that the burden could be enforced only by withholding the relative benefit, not by action. He relied on the statement of Upjohn J in *Halsall* to the effect that it was plain that the trustees could not sue on the covenant. Counsel also relied on the *dictum* of Sir Nicolas Browne-Wilkinson in *IDC*, at p190G, that:

> The only means of enforcement in a *Halsall* v *Brizell* type of case is to prevent the person from enjoying the rights which they seek to enjoy under the document, save upon the terms that they give effect to the positive obligations imposed on them.

[33] It is not hard to envisage situations in which the failure of the person benefitted to perform the associated obligations could satisfy the conditions for a claim in unjust enrichment. Further, I am unclear how that statement can be reconciled with cases dealing with compensation for letting down the surface of land in performance of a right to do so subject to paying the surface owner compensation. In such a case, it seems that the landowner may sue the mineworker for compensation see, for example, the decision of the Exchequer Division in *Aspden* v *Seddon* (1876) LR 1 Ex D 496 and of the Court of Appeal in *Westhoughton Urban District Council* v *Wigan Coal & Iron Co Ltd* [1919] 1 Ch 159. Further, alternative means of enforcement have been allowed in both *Jenkins* and *Baybut* although this point does not appear to have been raised in either of them. It is not, if, as I understand it to be the case, the other members of the court agree with my conclusion so far, necessary to resolve this issue in this case.

Should the judge have allowed the amendment to the preliminary issue so as to raise the benefit and burden issue?

[34] Given, as I understand it to be the case, the other members of the court agree with my conclusions so far, it is unnecessary to deal with this issue. However, in deference to the argument of counsel, I wish to add a few words in respect of it. In [5] and [12] to [15] above, I have described the state of the proceedings at the time the application was made to the judge on 5 May 2009 to enlarge the preliminary issue. Given that it was common ground that Lidl was not bound by clause 18 of the Jones-trustees contract, it plainly required some amendment to the pleadings to raise an issue capable of being formulated into a preliminary issue that Lidl was bound by clause 18 on the grounds sought to be alleged. Since the transfers had been executed on 4 August 2000 and the works completed by 30 June 2001 any such amendment would have been made outside the period permitted by the 1980 Act for any such new claim. Accordingly, the requisite amendment could not be made unless it satisfied the requirements of CPR 17.4(2).

[35] Although counsel for Lidl told us that he had addressed the judge on both the Act and the rule, neither is referred to in the judge's short judgment given on 5 May. He said:

> In my opinion, neither the particulars of claim nor the rather short response to it in the defence properly clarifies the precise issues between the parties. Although the phrase is not happily worded, in my judgment there is just about enough reference in the amended particulars of claim to the second defendants holding monies in breach of the terms of clause 18.2 of the agreement. This, as everyone accepts, is not a matter which raises any new fact. The question is one of law. I do have sympathy with Mr Child who says that after formulation of preliminary issues last week, he has not come prepared to argue that particular point today. But we have two days for hearing of the preliminary issues, it does seem to me having regard to the overriding objective that all outstanding legal points should be addressed by this court during these two days if that can be done without prejudice to the second defendants. In my judgment by dealing with this point tomorrow that should give a reasonable amount of time for Mr Child to address the legal points now raised. That is the course of action I propose to adopt.

[36] The judge was concerned, and rightly so, to avoid a waste of the court's resources if it was possible to do so. However, in my view, he did not deal adequately with the submissions that counsel told us had been made to him. There were serious points to be considered and determined. First, was an amendment to the particulars of claim needed to justify an enlarged preliminary issue authorised by CPR 17.4(2)? That rule provides that:

> The court may allow an amendment whose effect will be to add or substitute a new claim, but only if the new claim arises out of the same facts or substantially the same facts as a claim in respect of which the party applying for permission has already claimed a remedy in the proceedings.

[37] There can be no doubt that the effect of the requisite amendment would be to add a new claim; nor was it in issue that the period of limitation in respect of that claim had expired. Accordingly, the amendment could not be allowed unless the new claim arose out of the same or substantially the same facts as the existing claim. In the passage that I have quoted, the judge considered that no new fact was involved but no proposed amendment had been formulated and, in his judgment the following day, the judge relied on a "clear understanding" that had not been pleaded and was not relevant to any existing claim.

[38] Second, even if the judge had been satisfied that the reamendment to the particulars of claim necessitated by the proposed extension of the preliminary issue was permitted by CPR 17.4(2), he still had a discretion whether to allow it. The claimants had sought to amend their particulars of claim on three previous occasions. On the most recent, in January 2008, Wyn Williams J had refused to allow them to raise other new claims. The judge did not appear to take account of this relevant consideration at all.

[39] Although the judge was right to try to avoid any waste of the court's resources, I am bound to say that I cannot accept his conclusion. At the very least, he should have insisted on a properly formulated amendment to the particulars of claim. Only then could he have satisfied himself that the amendment to the pleadings, and thereby to the preliminary issue, was permissible under CPR 17.4(2). He was entitled to refuse the application on account of the absence of any properly formulated amendment, the lateness of the application and the fact that earlier similar amendments had been sought and refused.

[40] For these reasons, I consider that the judge's exercise of his discretion was plainly wrong. However, counsel for Lidl, understandably, did not ask for an adjournment so as immediately to raise the matter with the Court of Appeal. The preliminary issue was extended, argued and determined in the manner that I have described. In my view, the proper answer to the preliminary issue so extended is that for which counsel for Lidl contends. In those circumstances, I see no reason to set aside his order extending the preliminary issue if the alternative is to set aside the declaration that he made and substitute the declaration that Lidl seeks.

Summary of conclusions

[41] In my view, the judge was wrong to have:

(1) allowed the amendment to the preliminary issue; and

(2) made the declaration in answer to that part of the amended preliminary issue that he did.

I would set aside the judge's declaration that Lidl is bound by a duty to observe the terms set out in clause 18 of the Jones-trustees contract as a burden appurtenant to the benefit that it took under that contract. Instead, I would grant a declaration that Lidl is not bound to observe those terms.

Scott Baker LJ said:

[42] I agree.

Lewison J said:

[43] I also agree.

Appeal allowed.

Thames Valley Housing Association Ltd and others *v* Elegant Homes (Guernsey) Ltd and others

CHANCERY DIVISION

[2009] EWHC 2647 (Ch)

27 October 2009

MANN J

Estates Gazette 16 January 2010

[2010] 02 EG 86

Sale of land — Solicitor's undertakings — Release of charges — Sale of plots forming part of site over which vendor granting all-moneys charge — Vendor's solicitor undertaking to procure release of all charges on sale — Whether purchasers entitled to summary enforcement of undertakings

The first and second claimant housing associations were associated companies. In 2008, they agreed to purchase, for £675,000, nine plots on a development of 30 houses to be carried out by the first defendant vendor. The latter had acquired the development site and was partly financing the development costs with a loan from the third defendant bank secured by an all-moneys fixed charge over the site. The third defendant envisaged that houses on the development would be sold once completed and that the loan would be repaid in stages from the proceeds of sale of each completed house, although the documents did not provide for a staged release of the charge. However, the nine plots were sold to the first and second claimants bare, with the first defendant agreeing to construct houses on them, under a JCT contract, after the completion of the sale.

In response to requisitions on title, the second defendant undertook to discharge any charge affecting the property. The sale was completed in April 2008. However, the second defendant did not send the purchase moneys to the third defendant or procure the release of the charge. Instead, it paid the purchase price to the first defendant and the third defendant was not informed of the sale.

In December 2008, the claimants brought proceedings for summary enforcement of the undertakings. The second defendant accepted liability under the undertakings but resisted summary judgment. It contended that a full inquiry was required as to the third defendant's entitlement since it was not necessarily entitled to the sum it was demanding, having anticipated that it would receive less than the full amount charged on the sale of each plot. It had indicated that it was willing to release the charge over the nine plots on payment of £1.35m, which represented the sum that it had expected to receive from the sale of those plots, less deductions.

***Held*: Judgment was given for the claimants. The second defendant was ordered to perform the undertakings; it could procure the discharge of the charge by paying the amount required by the third defendant. It had not raised any serious question over the third defendant's strict entitlement such as to justify postponing the enforcement of the undertakings to facilitate an inquiry into the matter. The third defendant's internal documentation, envisaging repayments from the sale of each house, indicated only its own expectations and did not amount to evidence of any agreement to that effect operating between it and the first defendant as its customer. Nor had the parties signed a document to that effect, such as would be required under the Law of Property (Miscellaneous Provisions) Act 1989 for a binding contract as to the terms for release of a charge. In any event, the third defendant's internal documents showed that it was expecting to receive moneys from the sale of each plot as and when built on, rather than plots being sold bare. Further, the amount demanded by the third defendant seemed reasonable. Although it might, as a matter of law, be entitled to require all the moneys owed to it by the first defendant, it was not insisting on that. It was relevant that the sum it required: (i) was less than it had expected to receive, being net of certain payments; (ii) did not exceed the balance of the facility in respect of the development site; (iii) was less than the aggregate of the price paid by the first and second claimants and the amount paid to build the houses on the plots; and (iv) was not alleged to exceed a proper valuation for the nine plots.**

***Per curiam*: Solicitor's undertakings form an important part of the conveyancing mechanism and of the protection for purchasers. A purchaser that accepts such an undertaking should be entitled to rely on it and to know that the solicitor will have to comply with it. It would undermine the sensible practices and procedures of conveyancing if solicitors were entitled to delay compliance with undertakings while they resolved some dispute with the mortgagee, save perhaps in exceptional circumstances.**

The following cases are referred to in this report.

A solicitor, In re [1966] 1 WLR 1604
Angel Solicitors v *Jenkins O'Dowd & Barth* [2009] EWHC 46 (Ch); [2009] 1 WLR 1220; [2009] 1 EGLR 71; [2009] 14 EG 88
Clark v *Lucas Solicitors LLP* [2009] EWHC 1952 (Ch); [2009] 3 EGLR 83; [2009] 46 EG 144
Morgan & Co (L) v *Jenkins O'Dowd & Barth* [2008] EWHC 3411 (Ch)
Udall v *Capri Lighting Ltd* [1988] QB 907; [1987] 3 WLR 465; [1987] 3 All ER 262, CA

This was the hearing of an application by the claimants, Thames Valley Housing Association Ltd, Thames Valley Charitable Housing Association Ltd and Andersons Solicitors, for summary judgment on a claim against the second defendant, Willmett Solicitors, for specific performance of an undertaking to procure the discharge of charges held by the third defendant, Bank of Scotland plc, on a sale of land by the first defendant, Elegant Homes (Guernsey) Ltd.

Adrian Pay (instructed by Andersons Solicitors) appeared for the claimants; John Taylor (instructed by Ahmed & Co) appeared for the first defendant; David Halpern QC (instructed by Barlow Lyde & Gilbert LLP) represented the second defendant; the third defendant did not appear and was not represented.

Giving judgment, **MANN J** said:

Introduction

[1] This is an application made by the first two claimants to enforce an undertaking given by the second defendant (Willmett) in the course of a conveyancing transaction. In these proceedings, those two claimants seek to enforce that undertaking and they have made an application for summary judgment. That is the application that came before me. At the end of the argument, I announced my decision, which was that the undertaking would be enforced and an appropriate order made. Because of other matters that had to be dealt with (including a summary judgment application against the first defendant), it was not convenient to give my reasons at the time and I announced that I would give them later. This judgment contains those reasons.

Facts

[2] The first two claimants (the associations) are associated companies and are both housing associations. The third claimant (Andersons) is their solicitor. The first defendant (Elegant) is a developer and Willmett was its solicitor. For the purposes of carrying out the development referred to below, Elegant borrowed money from the third defendant (the bank). There is a third party (Mr MacPherson) who is a person who is said to be the owner and/or controller of Elegant. Although he figures in some of the events surrounding those with which I have to deal, I will not need to refer to him again.

[3] In 2007, Elegant was minded to acquire a development site in Windsor Road, Bray, in Berkshire. In order to acquire and develop it, it sought finance from the bank, and the bank agreed to provide finance for both purposes. It also had other properties and borrowings were required for that other development as well. It was intended to develop the Bray land by dividing it into 30 units and building a house on each

A of them. An internal bank document in 2007 recorded the "Payment Terms" as being "one bullet repayment on sale of finished units". That particular reference became an important point in the argument of Mr David Halpern QC, who appeared for Willmett, as will appear below. It appears from that provision that the bank at least intended that it would be repaid by a series of repayments, made as and when each of the various units, when completed, was sold in its completed state. The contractual and proprietary documentation, however, did not reflect that. The bank's facility letter for the Bray project was dated 30 April 2007 and provided that the facilities (£6.277m) would be repayable on demand at all times, despite the fact that it was described
B as a "Term Loan". Elegant was to provide security for the loan as set out in the "Security Documents" in schedule 1 to the facility letter. Those documents included:

> 1.1 A first and only debenture from the Borrower.
>
> 1.2 A first and only legal charge over the [Bray] Property.

A debenture was already in place. It was dated 20 May 2005 and was in the familiar form a fixed charge on all freehold and similar properties and a floating charge on a lot of other assets. I do not need to set out its provisions. A fixed charge was given over the Bray property on 1 June 2007; it secured all moneys due from Elegant to the bank. I do
C not need to set out any of the provisions of that document either. The important point to note about both of them is a negative none of them contained any conceivable reference that would support the notion that the bank would be obliged to release any particular part of the Bray property on payment of anything less than the whole of the outstanding debts due from Elegant to the bank. The structure, as it appears from those documents and without prejudging the validity of a submission made by Mr Halpern, was a familiar one. The bank was prepared to lend (and did lend) the costs of buying the land and a percentage of the development costs. It apparently expected to be paid piecemeal by sales of the plots, and no doubt would be highly likely to accept
D repayment in that manner because it was hardly likely that it would be repaid in any other way. The borrower would wish to pay in that manner because it was from sales of individual plots that it would get its money in respect of the development. In other words, that particular repayment mechanism seems, on the face of the contractual documents, to be a matter of commercial practicality rather than legal entitlement, although the documents do contain provisions that forbid sales of any of the properties without the bank's consent. The bank would be highly likely to consent to proper sales because, as I have observed, it would not be repaid in any other way short of full-scale security enforcement.

[4] At the beginning of 2008, Elegant agreed to sell six of its plots
E to the first claimant and three others to the second claimant. For the purposes of this application, there is no need to distinguish between those two sets of plots, and I can treat them as being overall one purchase and the first two claimants as being one purchaser. Those two claimants instructed Andersons to act in the conveyancing; Elegant instructed Willmett. The partner acting at Willmett was a Mr Jonathan Gilbert. The plots sold were not to have houses built on them at the time of sale, contrary to the basis on which the bank understood that it was providing funding. The bank documents demonstrate that it was expecting that the plots would be sold only once houses had been constructed on them. In this case, the arrangement between Elegant and the associations was
F that houses were to be constructed by Elegant for the two associations after the sale of the bare plots, pursuant to a JCT contract.

[5] Obviously, the associations required that their plots be free of any charge and, in order to achieve that, Willmett gave undertakings to procure the discharge of the charges in the usual way. The undertakings were given by the replies to requisitions and by the application of the Law Society's code for completion by post and were to the following effect:

(a) Willmett was asked (in requisition 6.1): "Please list the mortgage or charges secured on the property which you undertake to redeem or discharge to the extent that they relate to the property on or before completion." In response, Willmett stated: "We undertake only in respect of the charge in favour of Bank of Scotland plc dated G
1 June 2007 and registered on 16 January 2008."

(b) Willmett was asked (in requisition 6.2): "Do you undertake to redeem or discharge the mortgages and charges listed in reply to 6.1 on completion and to send us form DS1, DS3, the receipted charge(s) or confirmation that notice of release or discharge in electronic form has been given to the Land Registry as soon as you receive them?" Willmett gave this undertaking by responding "Confirmed".

(c) The following undertaking was given pursuant to the code: "[Willmett] undertake (i) to have the seller's authority to receive the purchase money on completion; and (ii) on completion to have the authority of the proprietor of each mortgage or charge specified under H
paragraph 3 [which included the bank's charge] to receive the sum intended to repay it."

(d) It also gave this undertaking, via the code: "Willmetts undertake… (ii) to redeem or obtain discharge for every mortgage or charge so far as it relates to the property specified under paragraph 3 which has not already been redeemed or discharged."

Willmett accepts that these undertakings were given.

[6] Exchange of contracts and completion took place on the same date, namely 11 April 2008. The contracts and transfers were executed by Mr Gilbert and another member of Willmett's staff pursuant to a power of attorney. On that date, the two associations sent Willmett the J
aggregate purchase price for the two plots, namely £675,000.

[7] Unfortunately, despite its undertakings, Willmett did not procure the discharge of the bank's charge over the nine plots nor did it send the purchase price to the bank. Approximately two weeks after completion, Mr Gilbert procured that the sale proceeds, less around £7,000, were sent to Elegant and paid into a separate account that it had at Barclays Bank. The Bank of Scotland knew nothing about the sale at the time. It did not consent to it.

[8] Over the following months, Andersons chased Willmett for the promised release of the bank's security. By and large, it was fobbed off. Andersons began to rely on the undertaking and, by the end of K
2008, the senior partner of Willmett became involved. I do not need to set out the detail of these contacts. At the beginning of March 2009, Mr Gilbert sent his partners an e-mail, in which he apologised for the mess that he had left them in in respect of various matters and summarily resigned. Since then, he has not been available to Willmett to explain what happened in this transaction and why it happened.

[9] On 24 October 2008, Andersons sent a pre-action protocol letter to Willmett. On 7 November 2008, Willmett, via a solicitor now separately instructed by it, asked for some time before the issue of proceedings. A little time was given and there was more debate in correspondence concerning that. However, nothing came of the delay L
and, on 12 December 2008, the present proceedings were issued, and particulars of claim were served on the same date. The proceedings claim enforcement of Willmett's undertaking and specific performance of Elegant's obligations. On 16 February 2009, the claimants issued an application for summary judgment against Willmett and, on 13 March 2009, issued a similar application against Elegant. I should add that Andersons is a third claimant in these proceedings in case a point were taken as to precisely who the recipient of the undertaking was, but, in the event, nothing turned on that point.

[10] Thus, two summary judgment applications arrived before this court for hearing on 7 October 2009. I heard the application against M
Willmett first.

[11] Since the bank has not received any part of the proceeds of sale, it has refused to provide a discharge of its charge or charges over the land. There have been various dealings between Willmett (or its solicitor) and the bank (and its solicitor) in order to find out what sum the bank requires in order to provide that discharge. The bank's position is that, technically, it has a security over the nine plots (and over around six unsold plots) that secures all the liabilities of Elegant to the bank, which includes the Bray facility but also extends to all other liabilities. The aggregate of those liabilities is several millions of pounds, but the bank has indicated that it would discharge the charge over the nine plots for a sum that is the amount that the bank would have expected

to receive from the nine plots less some deductions; the sum is some £1.35m-odd. That is probably most of the outstanding Bray facility. It does not insist on receiving the entirety of Elegant's outstanding liability. There is no evidence of the value of the nine plots, but there is evidence that the amount paid by the associations to Elegant under the JCT contract was approximately £975,000.

Respective cases of the parties

[12] The case of the associations is very simple. They say that Willmett gave an undertaking and the associations were clearly intended to be the beneficiaries of that undertaking. The undertaking was to secure their position as purchasers and make sure that they did not pay money for land caught by a charge that the solicitor knew about or ought to have known about. The solicitor should have found out about the charge and either procured its discharge or made sure that completion did not take place so that the associations did not lose their money. This is a plain case. The system of conveyancing relies heavily on undertakings by solicitors and their enforceability. So far as Willmett says that judgment should not be given to it until it has been ascertained how much the bank is entitled to ask in order to allow redemption of the charge over the land, the associations say, first, that there is no material dispute and, second, that even if there is some sort of dispute, in the circumstances the associations should not have to wait until it is resolved.

[13] Willmett accepts that it gave the undertaking, and it accepts its effect in that, at the end of the day, it will have to procure the removal of the charge (at least while the bank is asking for sums of the order for which it was asking). However, it said that the bank was not necessarily entitled to ask for the full amount that it was asking for (£1.35m) and there should be an inquiry as to its entitlement in this respect. The reason that it was not necessarily entitled to the full amount was that it plainly anticipated receipt of less than the full amount charged on the whole of the land on the occasion of the sale of each plot, as one could see from the reference to bullet repayments in the bank's internal documents. That was evidence of some form of agreement between the bank and the associations to the effect that the latter would be entitled to redeem on paying less than the full amount owing on the sale of each plot. What was needed was an inquiry as to whether there was some sort of arrangement, and then what that arrangement was. The expression "bullet repayments" was not defined, and the inquiry would also deal with that. Mr Halpern put forward two candidates for what it might mean: see below. Until that inquiry was carried out, it would be wrong to order the solicitor to perform the undertaking and, in the circumstances, summary judgment should not be given. Directions could and should be given for the conduct of such an inquiry. It must be remembered that Willmett does not have a lot of information concerning what happened when the completion took place, since it does not have the assistance of Mr Gilbert. Furthermore, it has even less information concerning the possible circumstances of the alleged agreement between Elegant and the bank. Mr Halpern did not go so far as to say that there could in no circumstances be enforcement of an undertaking such as these when the amount was uncertain, but he said that uncertainty in the amount necessary to procure the discharge of the charges was a circumstance that the court could take into account in its discretion. He accepted that, notwithstanding uncertainty as to the bank's strict entitlement, this court has jurisdiction to enforce the undertaking by ordering the solicitor to procure the discharge even if that would lead to its paying an amount demanded by the bank that might be thought to be less than wholly clear.

Should the undertaking be enforced at this stage?

[14] I turn, therefore, to the determination of those issues. Since there is no dispute concerning the terms of the undertakings or their effect, and since it is accepted that the solicitor could procure the discharge of the charge by paying the amount required by the bank and since the solicitor does not run any defence in the nature of a hardship defence, the real questions become whether Mr Halpern has raised a serious question mark over the entitlement of the bank and, if so, whether that, and the other circumstances, are sufficient to require the associations to wait for enforcement of the charge until an issue as to the bank's strict entitlement has been determined.

[15] Mr Halpern's case on the excessiveness of the bank's demands relies on only two pieces of evidence. First is the internal bank documents to which I have referred. It refers to bullet repayments. He added to this part of a statement by Mr David Squire, a former relationship matter [sic] at the bank, who dealt with Elegant. In para 4 of his statement, he says:

> 4. Quite simply it was standard practice for all BOS advances for construction/development sites to require bullet repayments ie as and when a property on the site was sold, the net proceeds of that sale would be immediately utilised in the direct reduction of the outstanding liability. As can be seen from the documents contained in Exhibit DS1 and annexed to this statement, BOS' standard terms and conditions provide for bullet repayments.

Exhibit DS1 is in fact the internal bank document to which I have already referred. It does not contain standard bank conditions. It refers to "standard conditions precedent to draw down of the land loan" and merely sets out what the bank commonly requires, but it does not amount to standard conditions in documents operating as between the bank and its customer. What it does indicate, not surprisingly, is that the bank would expect to be repaid from those bullet repayments.

[16] That material, which is all that Mr Halpern had, does not amount to evidence of any agreement operating between the bank, on the one hand, and its customer, on the other. Mr Halpern had no other evidence. Nor is there any reason to suppose that, given some time and more investigations, he would be able to obtain any. It is not clear whether his client would have access to Elegant personnel in order to see if they would support an allegation of an agreement, but his client has had the benefit of an early disclosure order against the bank, and I infer that the absence of any further material shows that, at least on the bank's side, there is no evidence of an agreement of the kind that he sought to allege.

[17] Furthermore, Mr Halpern's argument faces more difficulties than that. First, the relationship between the bank and the customer was, as one would expect, actually documented. I have referred to the documentation above. It is completely inconsistent with any agreement of the kind that he alleged. It makes it fundamentally unlikely that there was any binding agreement departing from those. Any agreement would have to be a collateral contract. In law, such a contract could exist, but it would require strong evidence. Not only is there no strong evidence, there is no real evidence at all.

[18] There is also a further technical objection. As Mr Halpern accepted, an agreement as to the terms on which a charge would be released, if it is to amount to a binding contract, must be in writing and signed by the parties pursuant to the provisions of the Law of Property (Miscellaneous Provisions) Act 1989. There is no evidence that such a document exists. Mr Halpern pointed out, correctly, that in some circumstances the full rigour of the statute could be avoided by means of mechanisms such as estoppel and constructive trust. That may be true in different circumstances, but it is even harder to see how that can be the case on the evidence that I have than it is to determine that there was or might have been some form of agreement in the first place.

[19] Mr Halpern also faced considerable difficulty in suggesting how it could now be worked out how much the bank might be entitled to receive. If one looks at the bank's internal document, it indicates that the bank was expecting to receive moneys from the sale of each plot *as and when the plots were completed.* In other words, houses were to be sold, not merely plots. That did not happen in this case. Nor can it now happen. So, in order to ascertain what the bank should now be taken to be obliged to accept, Mr Halpern contrived two methods. The first involved looking at the amount that the bank lent and that could be said to be properly attributable to each plot; that sum would then be reduced by receipts from the sale of plots. The second was to look at the amount that the bank was expecting to receive in respect of the proceeds of sale of each plot; that was then to be reduced by actual receipts from the sale of plots. It is not clear to me how either of those mechanisms is supposed to work; the second is particularly difficult.

A However, the real point is that neither of them has anything much to do with the apparent mechanisms of the transaction in question. One comes back to the documentation. It provides for all moneys to be secured and does not provide in any way for an appropriation of any part of the debt to any part of the land. If the bank was expecting to receive less than the entire sum from time to time (which it plainly was), it was the net proceeds of sale of each plot *as and when built on* and that manifestly has not happened in this case. So, the suggestion that there should be an inquiry to see which of Mr Halpern's methodologies might be applied is an inquiry into something that, in my view, is irrelevant. Mr Halpern sought to support his submissions in this respect by
B showing how the bank advanced moneys from time to time and how, according to him, it could be seen that certain sums could definitely be seen as being lent towards the construction of houses on the nine plots. That, he said, helped to determine how much the bank could properly be said to require for a discharge of the mortgage over the land. I am afraid that material does not support this case.

[20] Thus, Mr Halpern has not succeeded in demonstrating that there is any serious question as to the amount that the bank is entitled, as a matter of law, to require for the discharge of its charge over the nine plots. As a matter of law, it may well be entitled to require all moneys owed to it by Elegant. I do not need to consider the consequences of
C the bank insisting on that because it does not do so. It merely requires a sum that is less than the sum that it expected to receive (that is, net of certain payments) and not exceeding the balance of the Bray facility. Had things gone according to plan, and if there had been an orderly sale of plots with houses on in the manner that the bank expected, the bank would be unlikely to have insisted on a sum calculated in that way. It would have expected (as a matter of commerciality) actual net proceeds. However, things did not go according to that plan. The bank itself was let down because no consent was sought for the sale of these plots, apparently, and it did not expect to have to give consent to a sale of a plot without a house on it, and then still to have to fund the costs of
D building on those plots (which, on the facts, it did since it did not know about the sale of the plots it provided funds for building the houses on them as on the other plots). If further justification of the bank's stance is needed (which it is not), it should be noted that the sum now required by the bank is less than the aggregate of the price paid by the associations for the land and the amount paid to build the houses on them. There was no formal valuation of the nine plots before me, but Willmett does not suggest that the amount demanded by the bank is in excess of a proper valuation of those plots, and the figures just referred to suggest that it is unlikely to be.

[21] There is therefore no evidential case justifying an inquiry as to
E the amount that the bank is strictly entitled to require as the price for releasing the charge over the nine plots. The answer is clear, and the bank's demand seems to me to be reasonable. Accordingly, there is no justification for postponing judgment against Willmetts based on the need to conduct any such inquiry. That means that it is not necessary for me to go into the suggestion, raised by Mr Adrian Pay, that Willmett has had sufficient time to investigate the matter and has failed to keep the associations properly informed of its discussions with the bank. Those suggestions, if properly founded, were to be relied on in opposition to the suggestion that Willmett should have more time to have the position determined. Since there is nothing to have determined, the point does not arise.

F [22] That conclusion also makes it unnecessary to spend much time considering the authorities relied on by Mr Halpern, which, in any event, contain fairly uncontroversial propositions. He took me briefly to *In re a solicitor* [1966] 1 WLR 1604, in which Pennycuick J had to consider the enforcement of conveyancing undertakings given by a solicitor (a Mr Lincoln). He had given undertakings to hold certain leases to the order of the bank, but did not have them. In respect of one of the leases, Pennycuick J said, at p1609F-H:

> Prima facie, it is open to Mr Lincoln to obtain that lease by paying off the first mortgage, in which case he would be entitled to require the first mortgagee to hand over the lease to him and he could then hand over the lease in turn to the bank. There might, of course, be circumstances in which it would be difficult or
> G impossible for him to achieve that result, but there is no evidence before me to show that any such difficulty or impossibility exists, and in the absence of any evidence it seems to me that I ought to proceed on the basis that Mr Lincoln is able to perform this undertaking.

[23] Mr Halpern did not say that it would be impossible for Willmett to achieve compliance with its own undertaking. Obviously, it is not it can procure the discharge by paying the money required by the bank. It has not been suggested that it does not have the money. However, he did rely on the use of the word "difficult" as supporting his proposition that there should be a delay while the amount that the bank is entitled to ask for is determined. He said that that is a similar consideration to
H "difficulty" because one does not know what the bank is entitled to require. However, since I have found that there is no basis for such an inquiry, the "difficulty" does not arise.

[24] Next, he took me to *Udall* v *Capri Lighting Ltd* [1988] QB 907, in which it was pointed out that the jurisdiction to enforce solicitors' undertakings is essentially compensatory and not punitive, although it does have a disciplinary function as well. I accept those propositions. On the facts as I have found them to be, nothing in them prevents the enforcement by me now of the undertaking given by Willmett.

[25] *L Morgan & Co* v *Jenkins O'Dowd & Barth* [2008] EWHC 3411 (Ch) and *Angel Solicitors* v *Jenkins O'Dowd & Barth* [2009]
J EWHC 46 (Ch); [2009] 1 WLR 1220* were relied on as establishing that the court has a degree of flexibility in enforcing undertakings, and that the court might not require a solicitor to comply with undertakings by procuring the discharge of a charge if the amount required was unreasonable and greatly in excess of the market value of the property in question. Remarks to this effect in those cases were *obiter*. It is not necessary for me to consider their correctness (although I would in no way question it) since, on the facts, the point does not arise. In the circumstances, the amount required by the bank cannot be labelled as excessive, and it has not been shown to be in excess of, let alone greatly in excess of, the market value of the nine plots.

K [26] Last, Mr Halpern took me to *Clark* v *Lucas Solicitors LLP* [2009] EWHC 1952 (Ch)†. He pointed out that [60] of that judgment seemed to support his submission that a lack of certainty as to what was legally required in order to satisfy a charge was a relevant factor. I am sure that is right, and it was not challenged by Mr Pay. However, there is no lack of certainty in the present case, on the facts before me. Paragraph [61] of that judgment, which deals with "proportionality", operates against Willmett in this case. It was a case in which a mortgagee was demanding payment of all sums due, which was approximately double the value of the plot in respect of which the undertaking (which was similar to the undertaking in this case) had been given. The deputy
L high court judge (Ms Sarah Asplin QC) said, in [61]:

> Despite the fact that the amount demanded here is approximately double the value of Plot 3, I accept Mr Pay's submissions in this regard. Mr Kenny is entitled to demand the full sum due, and Lucas should be taken to have contemplated that that might well be the case, especially in the light of the fact that their client was developing the Site as a whole. In this regard, I also take into account the extract from the Solicitors' Code of Conduct and the Guide to it, to which I was referred. Such a demand is not therefore something which can be categorised as wholly unreasonable or outside Lucas' contemplation.

That can be said to apply in the present case.

[27] In those circumstances, Mr Halpern has not persuaded me
M that it would be right to postpone any order against Willmett pending the resolution of some issue (which would have been fought between Willmett and the bank) as to what the bank was strictly entitled to receive.

[28] I also bear firmly in mind the position of solicitors' undertakings of this nature in the procedures of conveyancing. They are, as I have already observed, an important part of that mechanism, and a valuable part of the protection for purchasers. A purchaser that accepts such an undertaking should be entitled to rely on it and to know that the

* Editor's note: Also reported at [2009] 1 EGLR 71

† Editor's note: Reported at [2009] 3 EGLR 83

solicitor giving it will have to comply with it. The vendor's solicitor is in a much better position to know what has to be done in order to procure the release of charges than is the purchaser, and the purchaser is entitled to believe and assume that the solicitor will do what is necessary. It would undermine the sensible practices and procedures of conveyancing if solicitors were entitled to delay compliance with their undertakings while they sorted out some dispute with the mortgagee, save perhaps in exceptional circumstances. The solicitor should have taken steps to make sure that there could be no such dispute. If it gives an undertaking without having done so, it cannot sensibly complain if the court requires it to take steps that it would otherwise regard as excessive in order to achieve compliance with the undertaking and to give the purchaser the title that the purchaser was legitimately entitled to expect. Accordingly, even if I had thought that there was some sort of issue of the kind suggested by Mr Halpern, I would almost certainly still have come to the view that the amount demanded by the bank ought to be paid in order to procure the discharge of the charge. The associations are innocent in this matter. It is Willmett that is at fault. One does, of course, have sympathy with the other partners in Willmett, who seem, prime facie, to have been let down by Mr Gilbert, but there is no reason why the associations should be saddled with the consequences of that. In saying that, I do not wish to be seen to be disagreeing with the *obiter dicta* in *L Morgan* and *Angel Solicitors*, which leave open the possibility of modifying enforcement in what might be regarded as extreme cases.

Conclusions

[29] I therefore conclude:

(i) The undertakings given by Willmett were given and broken.

(ii) There is no serious issue as to the entitlement of the bank to require the sum that it has in fact demanded as the price for discharging the charge over the nine plots.

(iii) This court should therefore enforce the undertakings given by Willmett.

(iv) The order for enforcement will be one that requires Willmett to comply with its undertaking and to procure the discharge of the nine plots from the bank's charge.

Judgment given for the claimants.

Weir *v* Area Estates Ltd

CHANCERY DIVISION

18 December 2009

MR ROBIN KNOWLES CBE QC, SITTING AS A DEPUTY HIGH COURT JUDGE

Estates Gazette 13 March 2010

[2010] 10 EG 121

Sale of land — Contract Rescission — Land Registration Act 2002 — Contract for sale of registered freehold land on full title basis under general condition 4.3(a) — Special condition requiring purchaser to accept leasehold interest shown in schedule to freehold title previously determined by surrender — Notice and restriction in respect of bankruptcy of lessee shown against leasehold title in land register and bankruptcy petition and order registered in land charges register — Section 86 of 2002 Act — Whether surrender void by reason of bankruptcy — Whether vendor having notice of bankruptcy — Whether purchaser entitled to rescind on ground that freehold title encumbered by lease

In January 2008, the claimant bid successfully at auction to purchase registered freehold land from the defendant for £400,000. A deposit of £40,000 was paid. The sale contract included the full title guarantee contained in condition 4.3(a) of the general conditions of sale. Further special conditions provided for vacant possession to be given on completion and, by special condition 7, declared the purchaser's acceptance that a certain leasehold interest in the property had determined by operation of law in August 2006 and barred him from requiring further proof or raising any objection or requisition with regard to that interest.

The leasehold interest in question continued to be shown in a schedule to the freehold title, notwithstanding that the lessee had purported to surrender the lease and the defendant had purported to accept that surrender in August 2006. At the time of the surrender, a bankruptcy petition had been pending against the lessee. Both the petition and the bankruptcy order had been registered in the land charges register. Entries had also been made against the leasehold title in the land register, namely a notice of pending bankruptcy proceedings, entered under section 86(2) of the Land Registration Act 2002 prior to the purported surrender, and a later restriction, under section 86(4), in respect of the bankruptcy order.

The sale did not proceed on the contractual completion date. The claimant purported to rescind the contract and brought a claim for the return of his deposit, on the ground that the freehold title continued to be encumbered by the lease, the surrender of which had been ineffective, under section 284 of the Insolvency Act 1986, as a void disposition of property by a bankrupt to a person who had notice of the bankruptcy petition. He argued that the registration of the bankruptcy petition on the land charges register was deemed to give such notice by virtue of section 198 of the Law of Property Act 1925. The defendant argued that section 198 was not relevant since section 86 of the 2002 Act provided a comprehensive code for dealing with the registration of bankruptcies in respect of registered land and, by virtue of subsection (7), did not require a person to whom a registrable disposition was made to make any search under the Land Charges Act 1972. Meanwhile, in May 2008, the defendant secured a disclaimer of the lease from the lessee's trustee in bankruptcy.

***Held*: The claim was allowed. (1) The leasehold title of the lessee's trustee in bankruptcy would become void as against the defendant from the date of the purported surrender only if, at that date, no notice or restriction had been entered in respect of the bankruptcy proceedings and the defendant had had no notice of the bankruptcy petition or adjudication: see section 86(5)(c)(i) of the 2002 Act. Since a notice had been entered on the land register under section 86, the title of the trustee in bankruptcy was not void against the defendant and the purported surrender was ineffective. That position was reached without requiring the defendant to make a search of the Land Charges Register under the 1972 Act. Section 86(7) does not "trump" subsection 86(5)(c)(i). It is concerned solely with whether a search must be made under the 1972 Act and, thus, whether a person is affected by entries on the land charges register, whereas section 86(5)(c)(i) is concerned with the land register, and specifically with entries made under section 86(2) and (4). In the instant case, the entries in the land register were sufficient to determine the question of notice against the defendant. (2) Although special condition 7 contractually bound the claimant to accept that the lease had been determined and disentitled him from requiring any further proof of the determination, it had not been intended as an exception to the full title guarantee basis of the sale under general condition 4.3(a). At the date set for completion, the defendant had not been in a position to convey the unencumbered title that it had contracted to convey and the claimant was accordingly entitled to rescind and to claim the return of his deposit with interest. The defect in title was not merely technical because there was a subsisting leasehold interest, which would continue unless there were either a successful application to the court for ratification of the surrender or a successful request to the lessee's trustee in bankruptcy to disclaim, and the outcome of neither had been a foregone conclusion at the relevant time. In order to save the sale contract, the disclaimer of the lease would have had to occur prior to the contractual completion date.**

The following cases are referred to in this report.

Darvell v *Basildon Development Corporation* (1969) 211 EG 33
Heavsman's and Tweedy's Contract, Re (1893) 69 LT 89
MEPC Ltd v *Christian-Edwards* [1981] AC 205; [1979] 3 WLR 713; [1979] 3 All ER 752, HL

This was the hearing of a claim by the claimant, Graham Weir, against the defendant, Area Estates Ltd, for the return of a deposit paid under a contract for the sale and purchase of land.

Jane Evans-Gordon (instructed by Tarbox Robinson & Partners) appeared for the claimant; Mark Loveday (instructed by Romain Coleman) represented the defendant.

Giving judgment, **Mr Robin Knowles CBE QC** said:

Introduction and facts

[1] The claimant (Mr Graham Weir) seeks summary judgment against the defendant (Area Estates) under CPR 24. Mr Weir's case is that Area Estates has no defence in law, and that that is as apparent now as it would be at trial.

[2] Mr Weir bid successfully at auction for registered freehold land at 22-24 St Andrew Street, Hertford (the property), owned by Area Estates. A contract of sale was entered into on 29 January 2008 (the contract of sale).

[3] A leasehold interest had been granted in respect of the property to a Mr Airey by a lease dated 20 April 2004 for a term of nine years (the lease). This was shown in a schedule to the freehold title.

[4] On 8 August 2006, a bankruptcy petition was presented against Mr Airey. The bankruptcy petition was registered as a pending action in the land charges register and, on 11 August 2006, a notice was entered against the leasehold title under section 86(2) of the Land Registration Act 2002 (the 2002 Act) in respect of that petition.

[5] On 31 August 2006, Mr Airey moved out of the property. He purported to surrender the lease, and Area Estates purported to accept that surrender. For the purpose of this application, the purported surrender is to be assumed to have been for valuable consideration. The bankruptcy petition was, however, still pending and, on 11 October 2006, a bankruptcy order was made against Mr Airey. The bankruptcy order was registered in the land charges register and, on 2 November 2006, a restriction under section 86(4) of the 2002 Act was entered against the leasehold title in respect of that order.

[6] By the time of the auction, more than a year later in January 2008, the freehold title still scheduled the lease, and the leasehold title still showed the notice and the restriction relating to Mr Airey's bankruptcy. The property was knocked down to Mr Weir for £400,000, of which £40,000 was paid by way of deposit.

[7] Although notice to complete was served by Area Estates on 7 March 2008, the sale did not complete. Rather, Mr Weir claimed to be entitled to rescind the contract of sale. He contended that the title was still encumbered by the lease because the surrender was ineffective by reason of the then pending bankruptcy. On 16 May 2008, Area Estates secured a disclaimer of the lease from the trustee in bankruptcy of Mr Weir, but that was too late to save the contract of sale.

[8] In this litigation, Mr Weir sues for the return of his deposit and interest. Area Estates has since sold the property to a third party. It achieved less in this sale and wants to counter-claim for damages against Mr Weir.

Section 284 of the Insolvency Act 1986

[9] Section 284 of the Insolvency Act 1986 (the 1986 Act) provides, so far as material:

> (1) Where a person is adjudged bankrupt, any disposition of property made by that person in the period to which this section applies is void except to the extent that it is or was made with the consent of the court, or is or was subsequently ratified by the court.
>
> …
>
> (3) This section applies to the period beginning with the day of the presentation of the petition for the bankruptcy order and ending with the vesting, under Chapter IV of this Part, of the bankrupt's estate in a trustee.
>
> (4) The preceding provisions of this section do not give a remedy against any person
>
> (*a*) in respect of any property… which he received before the commencement of the bankruptcy in good faith, for value and without notice that the petition had been presented…

[10] Area Estates accepts that a surrender of the lease is a disposition of property for the purposes of section 284(1) of the 1986 Act (and section 86 of the 2002 Act, below). As to the reference in section 284(4)(*a*) to the commencement of bankruptcy, under section 278 of the 1986 Act the bankruptcy of an individual against whom a bankruptcy order has been made commences with the date on which the order is made.

[11] Applying section 284 of the 1986 Act to the facts of the case:

(1) The surrender was within the period to which section 284 applies.

(2) The surrender is void, not having been made with the consent of the court or subsequently ratified by the court.

(3) The above conclusions would not give a remedy against any person "in respect of any property… which he received before [11 October 2006] in good faith, for value and without notice that the petition had been presented".

Section 198 of Law of Property Act 1925

[12] Mr Weir, by Ms Jane Evans-Gordon of counsel, referred to section 198 of the Law of Property Act 1925 (the 1925 Act). This provides, so far as material:

> (1) The registration of any instrument or matter in any register kept under the Land Charges Act 1972 or any local land charges register shall be deemed to constitute actual notice of such instrument or matter, and of the fact of such registration, to all persons and for all purposes connected with the land affected, as from the date of registration or other prescribed date and so long as the registration continues in force.
>
> (2) This section… applies only to instruments and matters required or authorized to be registered in any such register.

[13] If section 198 of the 1925 Act were applicable to the facts of the case, the registration of the bankruptcy petition in the land charges register would be deemed to constitute actual notice of that matter, and of the fact of that registration, to all persons and for all purposes connected with the land affected, as from 11 August 2006.

[14] However, Mr Mark Loveday of counsel, for Area Estates, argued that section 198 of the 1925 Act is not relevant. This is because, he submitted, section 86 of the 2002 Act is "a comprehensive code for dealing with the registration of bankruptcy petitions and bankruptcy orders in relation to registered land", and subsection (7) of section 86 indicates that a person to whom a registrable disposition is made is not required to make any search under the Land Charges Act 1972 (the 1972 Act). In light of this submission, I turn to section 86 of the 2002 Act in order to see whether it produces an outcome that makes a material difference on the facts of the present case.

Section 86 of the 2002 Act

[15] Section 86 of the 2002 Act provides, so far as material:

> (1) In this Act, references to an interest affecting an estate or charge do not include a petition in bankruptcy or bankruptcy order.
>
> (2) As soon as practicable after registration of a petition in bankruptcy as a pending action under the Land Charges Act 1972, the registrar must enter in the register in relation to any registered estate or charge which appears to him to be affected a notice in respect of the pending action.
>
> (3) Unless cancelled by the registrar in such manner as rules may provide, a notice entered under subsection (2) continues in force until
>
> (a) a restriction is entered in the register under subsection (4), or
>
> (b) the trustee in bankruptcy is registered as proprietor.
>
> (4) As soon as practicable after registration of a bankruptcy order under the Land Charges Act 1972, the registrar must, in relation to any registered estate or charge which appears to him to be affected by the order, enter in the register a restriction reflecting the effect of the Insolvency Act 1986.
>
> (5) Where the proprietor of a registered estate or charge is adjudged bankrupt, the title of his trustee in bankruptcy is void as against a person to whom a registrable disposition of the estate or charge is made if
>
> (a) the disposition is made for valuable consideration,
>
> (b) the person to whom the disposition is made acts in good faith, and
>
> (c) at the time of the disposition

A (i) no notice or restriction is entered under this section in relation to the registered estate or charge, and

(ii) the person to whom the disposition is made has no notice of the bankruptcy petition or the adjudication.

(6) Subsection (5) only applies if the relevant registration requirements are met in relation to the disposition, but, when they are met, has effect as from the date of the disposition.

(7) Nothing in this section requires a person to whom a registrable disposition is made to make any search under the Land Charges Act 1972.

[16] Applying section 86 to the facts of the case:

(1) The notice of the bankruptcy petition entered on the Land Register
B in respect of the property continued in force from 11 August 2006 until 2 November 2006, when the restriction was entered after the making of the bankruptcy order.

(2) As from 31 August 2006, the title of Mr Airey's trustee in bankruptcy would have been void as against Area Estates if (the registration requirements were met in respect of the surrender and) at 31 August no notice or restriction was entered under section 86 in respect of the registered estate or charge, and Area Estates had no notice of the bankruptcy petition or the adjudication.

(3) In fact, at 31 August 2006, a notice was entered on the land registry under section 86, with the consequence that the title of
C Mr Airey's trustee in bankruptcy was not void as against Area Estates.

(4) This is the position without requiring Area Estates to make a search under the 1972 Act.

[17] In oral argument, at times, the submissions were distilled into asking whether subsection (7) of section 86 "trumps" subsection (5)(*c*)(i). In my judgment, it is not a question of one provision "trumping" the other. Subsection (7) is addressing the question of whether a search needs to be made under the 1972 Act. Subsection (5)(*c*)(i) concerns instead the land register and specifically what has, pursuant to subsections (2) and (4), been entered in that register by the registrar after there has been registration of a petition in bankruptcy as a pending
D action, and of a bankruptcy order, under the 1972 Act.

[18] This scheme appears to be reflected in the narrative description given by the Law Commission in its 2001 paper (LC 271). Paragraphs 11.40 and 11.41 are in these terms (the italics are in the original):

> 11.40 When a petition in bankruptcy is filed against a debtor, the relevant court official must apply to register the petition as a land charge in the register of pending actions. Where the debtor is the registered proprietor of any land or charge, this can have no direct effect, because registration of a land charge does not affect registered land. However, this registration serves to trigger a procedure for ensuring that an appropriate entry is made on the register of
> E title. The Chief Land Registrar maintains both the land charges register and the register of title. Using the index of proprietor's names, he will attempt to ascertain whether the debtor is the registered proprietor of any land or charge on the register. If he thinks that he is, he must then register a creditor's notice against the title of any land or charge that *appears* to him to be affected. Clause 86(2) of the Bill replicates this procedure. The registrar must enter a notice of the pending action. That entry must remain in the register until either a restriction is entered in the manner explained below in paragraph 11.41, or the trustee in bankruptcy is registered as proprietor.
>
> Procedure in relation to bankruptcy orders
>
> 11.41 A similar procedure applies when a person is adjudicated bankrupt and a bankruptcy order is made against him or her…

F There are further helpful explanations of the scheme in *Ruoff & Roper Law and Practice of Registered Conveyancing*, in paras 13.005.03, 20.021, 34.003, 34.005, 34.011 and 34.012 and in Muir Hunter *Personal Insolvency*, in para 3-700.3.

[19] Mr Loveday further submitted that the presence of subsection (7) of section 86 indicates that constructive notice is not sufficient notice for the purpose of section 284 of the 1986 Act. Again, in my judgment, all that subsection (7) does is to deal in the case of registered land with the question of whether a search needs to be made under the 1972 Act, and thus of whether a person is affected by entries on the land charges register. In the present case, it is the entries on the land register that suffice to answer the question of notice against Area Estates.

Conditions under the contract of sale G

[20] The contract of sale was subject to general conditions, including general condition 4.3(a) in these terms:

> Unless otherwise stated in the special conditions the seller sells with full title guarantee except that:
>
> (a) all matters recorded in registers open to public inspection are to be treated as within the actual knowledge of the buyer…

[21] In addition, the contract of sale was subject to special conditions, including special conditions 1 and 7 in these terms:

> 1. Vacant possession will be given on completion…
>
> … H
>
> 7. The Lease dated 28 April 2004 and referred to on Title Number HD431124 determined by operation of law on 31 August 2006. The Buyer shall accept the position and shall not be entitled to require any further proof of the determination or raise any objection or requisition with regards thereto.

[22] In my judgment:

(1) On its true construction, special condition 7 required Mr Weir to accept as factually accurate the proposition that the lease had determined by operation of law on 31 August 2006.

(2) This requirement was not an exception to the "full title guarantee" basis of the sale, under general condition 4.3(a). Rather,
it was an affirmation of it in this particular. J

(3) Consistently with the requirement, special condition 7 ruled out Mr Weir's being entitled to require any further proof of the determination, or to raise any objection or requisition with regard to it.

(4) Although the land register and the land charges register (both registers under general condition 4.3(a)) revealed the lease and the bankruptcy petition, and did not mention any determination of the lease, the position remained that Mr Weir was contractually required to accept that there had been a determination. It transpired that there had not.

[23] Area Estates urges that during the sale process it "was quite open that the [L]ease was still registered against the freehold title". Indeed,
Mr Weir could see the registration for himself, and it is expressly K
referred to in the first sentence of special condition 7. However, what is involved is not simply an entry on the land register; the full position was that in law and in fact there was a subsisting lease until disclaimed by Mr Airey's trustee in bankruptcy, however easy it might be to persuade the trustee in bankruptcy to disclaim. Special condition 7 said that the lease had been determined by operation of law on 31 August 2006, but in law and fact it had not.

[24] Mr Loveday submitted that Mr Weir was as much fixed with notice of the bankruptcy petition as was Area Estates, and, therefore, as well able to work out the true position. Whether that is correct or
not (I do not see how Mr Weir would know whether the trustee had L
disclaimed, for example, and special condition 7 would stand in the way of his enquiry in that regard) I do not see that it helps Area Estates on the question of whether it was able to convey at completion the unencumbered title it had contracted to convey.

A "technical conveyancing defect"?

[25] Area Estates submits that the defect in title constituted by the subsisting lease is a technical defect insufficient to allow Mr Weir to rescind the contract of sale.

[26] In my judgment, this submission is unsustainable. Either
a successful application to the court seeking ratification of the surrender M
would be required or a successful request to Mr Airey's trustee in bankruptcy to disclaim. The outcome of neither can have been a foregone conclusion at the time. Neither was in Area Estates' sole power, control or grant. Area Estates had no right that would enable it to insist on ratification or disclaimer.

[27] Indeed, it is hard to see why the court would have ratified a surrender by Mr Airey when a bankruptcy petition was pending. As for disclaimer, the outcome of the request would depend on the trustee's assessment at the time whether, in one form or another, there was value to be had for the bankrupt estate from the lease.

[28] Of course, a trustee's assessment will vary from case to case. Even if it is impossible that an assessment by the trustee in the present

A case would have concluded that there was value, the property was still subject to a lease and Area Estates had no right to remove it. I cannot treat the presence of a subsisting lease encumbering freehold property sold with vacant possession as a "technical conveyancing defect".

[29] The present case cannot be characterised, as it was at points in the argument for Area Estates, as equivalent to a failure to remove from the face of the registered title a redundant entry. As at the date for completion, there was a subsisting leasehold interest vested in the trustee in bankruptcy of Mr Airey.

[30] In the course of argument, Ms Evans-Gordon, for Mr Weir, B withdrew, for the purposes of this application for summary judgment, an argument that had the purported surrender been valid Area Estates was still unable to give good title because the entries on the land register did not reflect a surrender of the lease. Mr Loveday urged that the argument that remained, based on an invalid surrender, was just a gloss on the argument that had been withdrawn. It will be clear from what I have said above that, whatever may be the position on the argument that Ms Evans-Gordon withdrew, I regard the combination of a void surrender and an undisclaimed lease as quite different.

[31] Understandably, Mr Loveday, for Area Estates, submitted that one should stand back and look at the facts (assuming these in his favour C for the purpose of the application). More than a year before completion, there had been the purported surrender. The premises were standing empty; no rent had been paid; the first time the trustee was approached to disclaim (in May 2008), he agreed. There is no evidence, urged Mr Loveday, that anyone was going to assert that the lease continued.

[32] These points paint a powerful picture, and perhaps a commercially persuasive one. However, the answer to them lies in remembering that had anyone asserted that the lease continued he or she would have been correct, subject perhaps to arguments about waiver or abandonment. That is not a risk that Mr Weir agreed to take.

[33] The underlying principle was laid down by the House of Lords D in *MEPC Ltd* v *Christian-Edwards* [1981] AC 205 in these terms (*per* Lord Russell of Killowen, at p220C-D, with whom the rest of the House agreed):

> [I]f the facts and circumstances of a case are so compelling to the mind of E the court that the court concludes beyond reasonable doubt that the purchaser will not be at risk of a successful assertion against him of the incumbrance, the court should declare in favour of a good title shown.

[34] There is no prospect in the present case of such a conclusion being appropriate. For all the argument about degree of risk, on which views will differ, Area Estates simply cannot show that as at the date for completion there was no risk of the trustee in bankruptcy asserting successfully that the lease subsisted. The present case is a far cry from that before the House of Lords, where it was concluded that there was no risk of a claim for specific performance being upheld on the basis of a contract some 70 years old that had been abandoned some F 50 years before.

[35] The authorities contain other illustrations of the effect of the presence of the risk of a claim being made that, if successful, would adversely affect title: see, for example, *Darvell* v *Basildon Development Corporation* (1969) 211 EG 33 ("mere conjecture of an attack by what would be idle litigation", *per* Megarry J, at p39) and *Re Heavsman's and Tweedy's Contract* (1893) 69 LT 89: "a claim [that] would now be ridiculous; he would be laughed out of court", *per* Lindley LJ, at p91. The distinction is that, in the present case, the adverse claim was good. The trustee in bankruptcy was a lessee, and Area Estates' own later action in approaching the trustee for a disclaimer is consistent with that. G

Conclusion

[36] In the circumstances, I am satisfied that the claimant must succeed. He was entitled to rescind the contract for sale. Area Estates has no defence to his claim for the return of the deposit and interest. Had Area Estates wanted to avoid these consequences, it should have secured the disclaimer before, on any view, the date for completion under the contract for sale.

[37] I will hear further argument as necessary as to the rate of interest.

Claim allowed. H

Dennis and others *v* Davies

COURT OF APPEAL

[2009] EWCA Civ 1081

22 October 2009

WARD, WILSON AND RIMER LJJ

Estates Gazette 23 January 2010

[2010] 03 EG 104

Restrictive covenants — Nuisance or annoyance — Building — Waterfront properties sold subject to covenants against causing nuisance and annoyance to other owners — Further restrictive covenant requiring approval of management company to any building work — Planning permission obtained for three-storey extension to one of properties — Whether extension a breach of restrictive covenant — Whether consent of management company to proposed works precluding assertion that works constituting annoyance in breach of covenant

The appellant owned a house forming part of a riverside estate of 47 three- and four-storey properties in Caversham, Reading. The respondents owned neighbouring houses on the estate. Each house enjoyed river views, some had mooring rights and most had waterside frontage. The communal areas of the estate and the mooring facilities were held by a management company, which had covenanted to manage and maintain them. Each house was subject to restrictive covenants contained in materially identical transfers between the developer, the management company and the original purchaser of the property. Under the transfers, each owner covenanted not to: (i) erect any building on the plot or any part thereof without the prior approval of the management company (the approval covenant); or (ii) "do or suffer to be done on the Plot or any part thereof anything of whatsoever nature which may be or become a nuisance or annoyance to the owners or occupiers for the time being of the estate" (the nuisance and annoyance covenant).

The appellant commenced work on a three-storey side extension to his property, for which he had obtained planning permission. The respondents claimed that the management company had not approved the extension, which they said would constitute a nuisance or annoyance because it would interfere with their river views, in breach of the restrictive covenants of which they had the benefit. The appellant maintained that approval had been granted and that this precluded the respondents from asserting any breach of the nuisance and annoyance covenant. He submitted that that covenant dealt only with the covenantor's activities on his plot and did not apply to the erection or extension of a building, which was exclusively within the province of the approval covenant; the other owners and occupiers were sufficiently protected by the management company's implied obligation to have regard to their interests when considering an application for approval. The trial judge rejected those submissions, declared that the extension would breach the restrictive covenants and granted an injunction restraining the appellant from carrying out the works. The appellant appealed.

Held: **The appeal was dismissed. The phrase "nuisance or annoyance" in a covenant is directed at conferring a wider protection on the covenantee than would be provided by the remedies available for the tort of nuisance at common law. Although the erection of a house that interferes with a view will not without more amount to a nuisance at common law, that does not require a nuisance and annoyance covenant to be construed as falling short of restraining the erection of potentially annoying buildings. The words of the covenant, on their ordinary and natural meaning, were sufficiently wide as to extend to activities of all natures, including the building of an extension to an existing house that, when built, would constitute an annoyance. Had the parties intended to cut down the meaning of the covenant, they would have done so expressly. Even assuming that the management company was under an implied obligation to take into account the interests of the other owners and occupiers when considering an application for approval, that could not have the implied consequence of reducing the apparently unambiguous scope of the nuisance and annoyance covenant. The giving of an approval was a precondition to any building work, but it did not prevent a plot owner or occupier from asserting and proving that, despite such approval, the proposed building would constitute an actionable annoyance in breach of the nuisance and annoyance covenant. The judge had been entitled to find that a case of "annoyance" was established. Moreover, on the evidence no approval had been granted.**

The following cases are referred to in this report.

C&G Homes Ltd v *Secretary of State for Health* [1991] Ch 365; [1991] 2 WLR 715; [1991] 2 All ER 841; (1990) 62 P&CR 69; [1991] 1 EGLR 188; [1991] 06 EG 170, CA
Dennis v *Davies* [2008] EWHC 2961 (Ch)
Hunter v *Canary Wharf Ltd*; *sub nom Hunter* v *London Docklands Development Corporation* [1997] AC 655; [1997] 2 WLR 684; [1997] 2 All ER 426, HL
Mahon v *Sims* [2005] 3 EGLR 67; [2005] 39 EG 138
Tod-Heatley v *Benham* (1888) 40 ChD 80
Wood v *Cooper* [1894] 3 Ch 671, Ch

This was an appeal by the appellant, Anthony Davies, from a decision of HH Judge Behrens, sitting as a judge of the High Court, granting declaratory and injunctive relief to the respondents, Richard and Margaret Dennis, Daniel and Irene Cameron, Richard Giles, Derek and Ruth Tilsley and Hannah Yentis, in a claim to enforce restrictive covenants.

Tom Weekes (instructed by Pitmans, of Reading) appeared for the appellant; Martin Hutchings (instructed by IBB Solicitors, of Uxbridge) represented the respondents.

Giving judgment, **RIMER LJ** said:

Introduction

[1] This is a defendant's appeal against an order of 21 November 2008 made by HH Judge Behrens, sitting as a High Court judge in the Chancery Division. The judge: (i) declared that certain building works that the then defendant, Mr Anthony Davies, wanted to carry out would, if carried out, breach restrictive covenants whose benefit is enjoyed by each of the respondents (Mr Richard and Mrs Margaret Dennis, Mr Daniel and Mrs Irene Cameron, Mr Richard Giles, Mr Derek and Mrs Ruth Tilsley and Mrs Hannah Yentis); (ii) granted an injunction

restraining Mr Davies from carrying out the works; (iii) ordered him to pay the respondents' costs; and (iv) gave permission to appeal.

[2] The judge had to decide three issues. First, whether, on its true construction, the covenant in para 2 of the relevant schedule applied to Mr Davies's proposed building works. Second, if it did, whether the works would be or become a "nuisance or annoyance" to the respondents within the meaning of that covenant: only if they did would there be a breach. Third, whether Mr Davies had a written approval to carry out the works for the purposes of the covenant in para 1 of the schedule.

[3] The judge decided all three points against Mr Davies. All three were again put in issue by his appellant's notice, although, in the event, the argument focused only on the first and third points. As for the second point, which required a value judgment by the judge based on the evidence, Mr Tom Weekes (who appeared, as below, for Mr Davies) recognised that to persuade this court to substitute a different judgment required the scaling of too high a mountain and he did not press the point. We are thus concerned only to decide the first and third points: construction and approval.

Background facts

[4] I take these gratefully from the judge's judgment; [2008] EWHC 2961 (Ch). The parties live on Heron Island in the River Thames, at Caversham, near Reading. In the mid-1980s, the island was developed by Heron Homes Ltd (Heron) as a residential estate comprising 47 three-storey houses. The judge said of the development:

> 16. … A particular feature of the development was its closeness to the river and the views that each house was afforded of the Thames. The island was described in the advertising brochures at the time as a unique development "lapped by the waters of the Thames on three sides" and, as providing purchasers with a "waterside lifestyle". The open river views were a feature of the development and many houses were also given mooring rights. Nearly every house has a waterside frontage. Views of the River are gained by deliberately designed gaps between the houses and the houses have been carefully orientated in order to take advantage of the river view.

[5] Heron sold the houses by materially identical transfers entered into by itself, its management company, OM Peverel Ltd (Peverel), and the purchasers. It later transferred the communal areas and mooring facilities to Peverel, which covenanted to manage and maintain them, the purchasers covenanting to reimburse Peverel for the cost (the transfers in fact named the management company as OM Ltd, but nothing turns on that).

[6] Each transfer included a series of restrictive and positive covenants by the purchaser, some with just Heron and Peverel, others both with them and with the owners for the time being of other plots on the estate. I will go straight to the relevant provisions of the transfer of 23 Heron Island, which Mr Davies has owned since 1993. The transfer is dated 26 June 1987 and the original transferee was Mr Strange, from whom Mr Davies derived title.

[7] By clause 4(a), Mr Strange covenanted:

> The Purchaser for himself and his successors in title and with the intent to bind the Plot into whosoever hands the same may come hereby covenants with the Vendor and the Management Company and also as a separate covenant with every other person who is now the owner of any part of the Estate for the benefit of the remainder of the Estate including any part thereof for the time being unsold to observe and perform the restrictive and other covenants and stipulations set out in the Third Schedule hereto PROVIDED THAT nothing herein contained shall prevent the Vendor or its successors in title from selling or otherwise disposing of any part or parts of the Estate free from any restriction or stipulation or from waiving compliance with or agreeing to vary any restriction or stipulation now or hereafter affecting any part of the Estate nor be deemed to create a building scheme.

[8] The key covenants in the third schedule are in paras 1 and 2, but I will also set out para 4:

> 1. Not to erect on the Plot or any part thereof any building whether of a permanent or temporary nature except such as shall be in accordance with plans and elevations which shall have been approved in writing by the Management Company and whose proper and reasonable fees shall be paid by the persons submitting such plans and elevations for approval
>
> 2. Not to use the Dwellinghouse for any purpose other than that of a private residence or ancillary thereto and not to carry on from the Plot or any part or parts thereof any trade business or manufacture whatsoever nor to do or suffer to be done on the Plot or any part thereof anything of whatsoever nature which may be or become a nuisance or annoyance to the owners or occupiers for the time being of the Estate or the neighbourhood…
>
> …
>
> 4. (a) Not to allow the wall fence or boundary posts along those boundaries of the Plot indicated by a "T" within the boundary on Plan 1 nor the rear wall or fence to fall into disrepair but to maintain the same in good condition
>
> (b) Not at any time to erect or permit to be erected along any of the boundaries of the Plot any boundary structure of whatsoever nature other than as may exist at the date hereof (and where only boundary marker posts exist along such boundary not to erect any boundary structure of whatsoever nature) without the written consent of the Management Company which shall be entitled to grant or withhold its consent entirely at its discretion
>
> (c) Not to erect construct or plant in front of the front or side building line any wall fence gate or other means of enclosure without the written permission of the Management Company and the Local Planning Authority.

It is not disputed that each respondent is entitled to the benefit of, and to enforce against Mr Davies, the restrictive covenants contained in those paragraphs. Paragraph 4(a), although expressed in negative terms, in fact looks more like a positive covenant; if so, it may not be so enforceable, although I express no final view on that.

[9] The particular problem that has arisen is that, on 19 October 2005, Mr Davies obtained planning permission to build a three-storey side extension to his house. Work started on it in May 2007 but was later halted following complaints from the respondents' solicitors and pending their resolution by these proceedings. Its continuance is now enjoined by the judge's order. Each respondent is a neighbour of Mr Davies and asserts that the proposed extension does not have para 1 approval; and (even if it did) that it will anyway cause them a "nuisance or annoyance" in breach of para 2. Mr Davies's house at no 23 borders the river on the southern side of Heron Island. Mr and Mrs Dennis live at no 16, behind no 23 to the north-west. Mr and Mrs Cameron live at no 17, next door to no 16. Mr Giles lives at no 46, in the central part of the island and approximately to the west of no 23. Mr and Mrs Tilsley live at no 22, on the eastern extremity of the island, roughly north-east of no 23. Mrs Yentis lives next door to them, at no 21.

[10] Mr Davies's proposed extension is intended to comprise a study and utility room on the ground floor, an extension to the sitting room and a balcony on the first floor and an additional bedroom on the second floor. It is 3.2m wide, 6.3m deep, 7.5m high to the eaves and 9.5m to the ridge. The judge summarised the objections of each of the respondents to the extension, which included complaints that it would reduce their river views from parts of their respective houses. During a four-day trial, he had the benefit of the respondents' written and oral evidence, plans, a three-dimensional model, photomontages, expert evidence from two witnesses on planning matters, evidence from a valuation surveyor and a site view. With regard to the evidence directed at proving that the extension would occasion "nuisance or annoyance" to the respondents, in breach of para 2, by interfering with their river views, Mr and Mrs Dennis obviously had the strongest case and Mr and Mrs Tilsley and Mrs Yentis the weakest.

[11] Central to the argument before the judge was whether the covenant in para 2 has any application at all to the erection or extension of a building: Mr Davies's argument was that the regulation of works of that nature was exclusively within the province of para 1. The judge held, however, that, as a matter of construction, the reach of para 2 did extend to restraining the erection of buildings "which may be or become a nuisance or annoyance" to covenantees. Having so held, he had to decide whether Mr Davies's proposed extension would in fact be or become a "nuisance or annoyance" to the respondents. In directing himself as to that, he referred to this court's decision in *Tod-Heatley* v *Benham* (1888) 40 ChD 80. He said:

> 98. In my view therefore the question is whether applying the guidance afforded by the case of *Tod-Heatley* v *Benham* this extension would be or

A become a nuisance or annoyance to the Claimants. Would reasonable people, having regard to the ordinary use of the Claimants' houses for pleasurable enjoyment, be annoyed and aggrieved by the extension? To adopt the words of Lord Justice Lindley, would the extension raise an objection in the minds of reasonable men, and be an annoyance within the meaning of the covenant? Lastly, would the extension reasonably trouble the mind and pleasure, not of a fanciful person or of a skilled person who knows the truth, but of the ordinary sensible English inhabitant of the Claimants' houses?

99. As I have noted the test is an objective one and must be judged by robust and common sense standards.

100. At the end of the day the issue is a relatively narrow one. As I indicated in the course of argument I think that some of the objections of the Claimants B contained in their witness statements would fail that objective test. Thus I agree with Mr Weekes that the objections with regard to the size of the gap, the view on the approach to Number 22, and the visiting of the water front would not reasonably trouble the minds of an ordinary sensible inhabitant of the Claimants' houses.

101. However I have had the benefit of a view from each of the Claimants' houses. I have seen the view of the river admittedly in November. I have seen photomontages of the effect of the extension. I agree with Mr Derbyshire that the loss of view is significant from Numbers 17, 46 and especially from Number 16. In my view the three storey red brick extension would trouble the minds of the ordinary sensible English inhabitant of any of those three houses and in those circumstances it does constitute an annoyance within the meaning C of the covenant.

[12] As I have said, if the judge was right to hold that para 2 is capable of extending to the erection of an allegedly annoying building, we are not asked to review the correctness of the factual conclusion at which he arrived in those paragraphs. I add that, for myself, I would have required much persuading that this court could properly arrive at a view on that matter different from his. With regard to para 2, the only issue argued before us is, thus, whether it is capable of applying to the erection of a building at all. I turn to that.

What is the reach of the covenant in para 2?

D [13] As Mr Weekes emphasised, para 1 is expressly directed at restraining a covenantor from erecting any building on his plot save in accordance with plans previously approved in writing by Peverel, whose reasonable fees must be paid by the covenantor. He submitted that the requirement for the submission of plans for prior approval, linked with the need for business efficacy, had the consequence that two terms had to be implied into para 1, namely that: (i) Peverel could not withhold its consent unreasonably; and (ii) when considering a building application, it must have regard to the interests of the other occupiers of the estate. In support of the first of those assertions, Mr Weekes invoked the decision of Hart J in *Mahon* v *Sims* [2005] 3 EGLR 67. In support E of the second, he referred to Scamell's Land Covenants, 1996, at p239, where the author suggested that "an example of a case of an implied term giving rise to an obligation to act correctly towards the plot owners as a whole (including the owner seeking approval) would be the case of a building scheme". In the present case, clause 4(a) of the standard transfers (quoted above) expressly negatived the creation of a building scheme. We were not referred to any authority supporting the second suggested implied term.

[14] As for para 2, Mr Weekes submitted that it must be construed in the context of a schedule that includes the immediately preceding para 1. He said that, on its true construction in that context, para 2 F did not also extend to the regulation of the erection of buildings on a covenantor's plot. To interpret para 2 as doing so would effectively be to emasculate para 1. It would mean that even if the plans for a proposed building were approved by Peverel under para 1, the building would remain at risk of challenge by one or more covenantees under para 2 on the basis that it may be or become a nuisance or annoyance. If para 2 is construed in a context that includes para 1, he said that it can sensibly be regarded as being concerned only with the covenantor's activities on his plot, not including activities in the nature of the erection of annoying buildings. He said that it would anyway be unsatisfactory if para 2 were interpreted as applying not only to activities on the plot but also to the erection of buildings on it. A covenantor proposing to erect a building will need to obtain both planning permission and an approval under para 1. If para 2 is also capable of applying to his proposal, he will also G need to consider whether the building might prove to be a "nuisance or annoyance" to others. That would or might involve a difficult exercise of judgment. Such matters ought to be exclusively within the province of the planning authority (when considering whether to give planning permission) and of Peverel (when considering whether to give para 1 approval). Mr Weekes supported this submission by pointing out that, quite apart from the colour that para 1 lends to the interpretation of para 2, subparas 4(b) and (c) impose express restraints on the erection of structures on plot boundaries and so assist in the conclusion that para 2 cannot naturally be interpreted as also applying to the erection of buildings or other structures. H

[15] Mr Weekes submitted further that even if para 2 is interpreted in isolation from paras 1 and 4, it still cannot and should not be interpreted as applying to the erection of buildings. It is concerned with the prevention of activities that may be or become a "nuisance or annoyance". That phrase is, he said, borrowed from the law of nuisance at common law, whereas the erection of a building that (without interfering with any easement, such as a right to light or air) interferes with a neighbour's enjoyment of his own property will not amount to a common law nuisance. For that proposition, which was not in dispute, Mr Weekes referred us to the observations of Lord Goff of Chieveley in *Hunter* v *Canary Wharf Ltd* [1997] AC 655, at p685D-G. J

[16] Mr Weekes did not, however, press that submission too far, since he recognised that the relevant part of para 2 ("nor to do etc") is plainly not directed merely at restraining activities that would be or would become a "nuisance" at common law. He accepted that the effect of restraining activities that may be or may become a "nuisance or annoyance" carries a wider import: if it did not, it would add nothing to the covenantees' common law rights. Mr Weekes' moderation in that respect was well judged. Covenants of the nature imposed by para 2 have long been used in conveyances, transfers and leases and it is clear that the expanded phrase "nuisance or annoyance" in such covenants is directed at conferring a wider protection than the remedies available for K the tort of nuisance at common law.

[17] The point is perhaps best illustrated by the decision to which the judge referred, *Tod-Heatley*, and it is worth digressing to that authority and citing from its judgments. The case raised the question of whether the establishment of a hospital for the treatment of various diseases constituted a breach of a covenant in a lease restraining the carrying on of particular trades, including a covenant by the lessee that he would:

> not do or wittingly or willingly cause or suffer to be done any act, matter, or thing in or upon or about the said premises, which shall or may be or grow to the annoyance, nuisance, grievance or damage of the lessor, her heirs and assigns, or the inhabitants of the neighbouring or adjoining houses. L

[18] In this court, Cotton LJ said that the relevant question was not whether a case of nuisance had been made out but whether the covenant had been breached. He rejected the submission that the word "nuisance" gave colour to the entire covenant and required it to be interpreted as though "the thing to be prohibited must be a nuisance or anything that is a legal nuisance": see p92. The relevant phrase included the words "annoyance… grievance or damage". A matter could cause "annoyance" or "grievance" without also causing any pecuniary loss. As to the meaning of those words, he said, at p93:

> Now "annoyance or grievance" are words which have no definite legal meaning. M It has been pressed upon us that we cannot say that it was that which was an annoyance or grievance to reasonable people, because the Judges, in speaking of what would be an annoyance to reasonable people, are only speaking of what they themselves really think would be an annoyance or grievance. That is the difficulty that Judges very often have to deal with; they must not take that to be an annoyance or grievance which would only be so to some sensitive persons. They must decide not upon what their own individual thoughts are, but on what, in their opinions and upon the evidence before them, would be an annoyance or grievance to reasonable, sensible people; and, in my opinion, an act which is an interference with the pleasurable enjoyment of a house is an annoyance or grievance, and within the definition given by V-C Knight-Bruce in *Walter* v *Selfe* 4 De G&Sm 322. It is not sufficient in order to bring the case within the words of the covenant, for the Plaintiffs to shew that a particular man objects

to what is done, but we must be satisfied by argument and by evidence, that reasonable people, having regard to the ordinary use of a house for pleasurable enjoyment, would be annoyed or aggrieved by what is being done.

Lindley LJ made like points. He said, at pp95-96:

> The question which arises is, what is the meaning of the expression "shall or may be or grow to the annoyance, nuisance, or damage" [sic: he omitted "grievance"] of the persons named. Certainly that string of words is introduced in order to give the covenantee a greater protection than he would have had without any such words at all, or if only one of those words were used. … I cannot at all agree with the contention that these words "annoyance or grievance to the inhabitants" mean that which would be according to law a nuisance, or that the covenant is only against such acts as would produce pecuniary damage. … Now what is the meaning of annoyance? The meaning is that which annoys, that which raises objections and unpleasant feelings. Anything which raises an objection in the minds of reasonable men may be an annoyance within the meaning of the covenant.

Bowen LJ said, at pp97-98:

> What is the meaning of the term "annoyance"? It implies more, as it seems to me, than "nuisance." The language of the covenant is, that nothing is to be done, "which shall or may be or grow to the annoyance, nuisance, grievance or damage of the lessor or the inhabitants of the neighbouring or adjoining houses." Now, if "annoyance" meant the same thing as "nuisance" it would not have been put in. It means something different from nuisance. … "Annoyance" is a wider term than nuisance, and if you find a thing which reasonably troubles the mind and pleasure, not of a fanciful person or of a skilled person who knows the truth, but of the ordinary sensible English inhabitant of a house if you find there is anything which disturbs his reasonable peace of mind, that seems to me to be an annoyance, although it may not appear to amount to physical detriment to comfort.

[19] The judge applied that guidance and recognised, as does Mr Weekes, that the para 2 injunction against "nuisance or annoyance" goes wider than would a mere injunction against "nuisance"; and I have quoted the judge's reasons for concluding that a case of "annoyance" was established. However, the prior question before him, and is now before us, was whether para 2 applies to the erection of buildings.

[20] Given that the language of para 2 deliberately extends beyond the restraint of that which may be or become merely a "nuisance", I am not convinced that the fact that the erection of a house that interferes with a view will not without more amount to a nuisance at common law should encourage us to interpret the reach of para 2 as falling short of restraining the erection of potentially annoying buildings. Mr Weekes' submission is, however, that this is nevertheless its correct construction. He said that it cannot be the case that the plot owners must maintain unobscured views across their plots. Is it, he asked, to be said that Mr Davies cannot suffer the growth of trees in the position of the extension that would similarly obscure the views of the river? I understood him to submit that for Mr Davies to suffer his trees so to grow could not give rise to an actionable annoyance under para 2 any more than could his erection of the extension.

[21] Mr Weekes referred us to this court's decision in *C&G Homes Ltd* v *Secretary of State for Health* [1991] Ch 365*. It raised an issue as to whether a health authority's housing of former mental in-patients in two houses on a residential estate resulted in a breach of one or both of two covenants burdening the houses. One covenant, para 20, was in these terms:

> Not to cause or permit or suffer to be done in or upon the property any act or thing which may be or become a nuisance, annoyance, danger or detriment to the transferor or owners or occupiers for the time being of other parts of the estate.

[22] This court, reversing the judge, held there had been no breach of that covenant. The facts of the case were very different from those of this one, so that the decision lends us no direct assistance. However, para 20 has obvious similarities to the relevant part of our para 2 and Mr Weekes drew our attention to two passages in the judgments that he said supported his submission that para 2 is concerned, not with the erection of buildings or other structures on the plots, but only with other activities carried on there. First, Nourse LJ said, at p386C:

> Mr Macdonald's [leading counsel for the appellant] primary submission was that the covenant does not impose any restriction on the persons who may occupy the property. It only restricts the acts or things which the occupants, whoever they be, may do there. Although Ferris J thought that that was too narrow a view, I suspect that it accords with the construction which most conveyancers would put on a covenant in this form, again a very familiar one.

Second, Lord Donaldson of Lymington MR said, at p390C:

> The position in relation to covenant 20 is quite different. Once again I have to look to the object and to the words. It is not directed to the use being made of the property, that being the subject matter of covenant 24(2). It is directed instead at conduct in or upon the premises which causes or may cause nuisance, annoyance, danger or detriment to other owners… or occupiers of other parts of the estate or to the plaintiff. No complaint whatsoever is being made in relation to the conduct of the occupants in or upon the premises. The evidence relied upon as constituting a detriment to the plaintiff relates solely to the use of the premises made by the Secretary of State and amounts to no more than that in a buyers' market a particular purchaser was astute enough to use the general nature of that use as a lever to obtain a small reduction in the purchase price. There has been no breach of covenant 20.

[23] I am not myself convinced that either passage in fact supports the case that para 20 could not have been enlisted in aid of a complaint that the covenantors had erected a building on the burdened land that had caused annoyance to other estate owners. The erection of such a building would, it seems to me, at least be capable of being described as an "act" or "thing" that the occupiers would have done on the land (see Nourse LJ), although I admit to less confidence as to whether Lord Donaldson would have regarded it as embracing whatever he had in mind in referring to "conduct".

[24] Another relevant passage in Nourse LJ's judgment is, however, his observation at p380G, where, in referring to a particular matter to be taken into account in construing the covenants in issue, he said:

> No less to be taken into account is that parties to a conveyancing transaction, having entered into a covenant in a long established and familiar form, must have intended that it should have the effect which earlier authorities have said that it has.

That observation is a pertinent one in the present case because: (i) para 2 is undoubtedly drafted "in a long established and familiar form" (covenants in essentially similar form can be found in the *Encyclopaedia of Forms and Precedents* (2nd ed), 1925, volume XV, at p874 and (5th ed), 2007 reissue, at p226); and (ii) it can also be said to have received a judicial consideration adverse to Mr Weekes' argument. The authority is Romer J's decision in *Wood* v *Cooper* [1894] 3 Ch 671.

[25] The facts of that case were that, by a lease of November 1878, Mr Wood demised a house to Mr Bailey for 99 years. Mr Bailey's covenants included covenants with Mr Wood and his successors (the lessor parties): (a) not, without the licence and consent in writing of the lessor parties being first obtained:

> To erect or build, or cause or permit to be erected or built upon the said piece or parcel of ground hereby appointed and demised, or upon any part thereof, any other building whatsoever save and except a stable and coach-house.

and (b):

> not to do or suffer to be done on the said premises or any part thereof any act, matter, or thing which may be or become an annoyance, nuisance, or disturbance to the neighbourhood or to the tenant of the lessor parties.

[26] I shall refer to these covenants as covenants (a) and (b) respectively. They are essentially equivalent to the covenants in paras 1 and 2 in our case. Mr Wood also granted to Mr Grice a 99-year building lease of adjoining land, on which Mr Grice was to build a house. The house was in fact built by Mr Grice's assignee, Mr Neale, who moved in on 26 February 1894. Mr Bailey had in the meantime assigned his lease to Mr Cooper, the defendant. The inference is that Mr Cooper

* Editor's note: Also reported at [1991] 1 EGLR 188

was visually offended by the proximity of Mr Neale's house and, on 27 February 1894, commenced the erection of a substantial trellis-work screen on his land, fronting Mr Neale's land. Mr Wood objected that this breached the covenants in Mr Cooper's lease, but Mr Cooper completed the work. Mr Wood sued him for an injunction, alleging that: (i) the screen was a "building" erected in breach of covenant (a); and (ii) its erection infringed covenant (b) as being an annoyance, nuisance and disturbance to Mr Neale.

[27] The arguments of counsel, as recorded at p675, show that Mr Neville QC, for Mr Wood, submitted that there was a breach of both covenants. Mr Bigham QC, for Mr Cooper, submitted that the screen was neither a "building" within covenant (a) nor an "annoyance" within covenant (b). Romer J held Mr Cooper to be in breach of both covenants. Although he regarded the case as being on the borderline, he held the screen to be a "building" within covenant (a). He went on to find that its erection also constituted a breach of covenant (b), saying, at p676:

> In the second place, I have no doubt whatever in my own mind, that it is a breach of the covenant that the lessee "will not do or suffer to be done on the premises, any act, matter, or thing which might be or become an annoyance to any tenant of the lessor." To my mind, undoubtedly, what the Defendant has done is an annoyance to Mr Neale, the tenant of the lessor. I think, in the first place, that it does substantially interfere with the access of light to the windows on the ground floor of this building, and that, notwithstanding some parts of the expert evidence; and I feel satisfied beyond that, and irrespective of that, that it causes an annoyance to Mr Neale, the tenant, within the meaning of the words used in the covenant. It falls within the definition of the word "annoyance," in a covenant like this, which was given by the three Lords Justices in the case of *Tod-Heatley* v *Benham* 40 Ch D 80. In the first place, to adopt the language of Lord Justice Cotton, I am satisfied by the evidence before me that reasonable people, having regard to the ordinary use of Mr Neale's house for pleasurable enjoyment, would be annoyed and aggrieved by what has been done by the Defendant. It would be an annoyance or grievance to reasonable, sensible people. It is an act which is an interference with the pleasurable enjoyment of the house. Then, to adopt the words of Lord Justice Lindley, I think it does raise an objection in the minds of reasonable men, and is an annoyance within the meaning of the covenant. Lastly, as pointed out by Lord Justice Bowen, " 'Annoyance' is a wider term than nuisance, and if you find a thing which reasonably troubles the mind and pleasure, not of a fanciful person or of a skilled person who knows the truth, but of the ordinary sensible English inhabitant of a house if you find there is anything which disturbs his reasonable peace of mind, that seems to me to be an annoyance, although it may not appear to amount to physical detriment to comfort."
>
> I have come to the conclusion, therefore, that this is a clear breach of the last-mentioned covenant as well as of the first, and on both grounds I think the Plaintiff is entitled to succeed.

[28] That decision is unhelpful to Mr Weekes, and in coming to his conclusion that the para 2 covenant in the present case was as applicable to catch the erection of annoying buildings as other annoying activities, the judge was influenced by it. The relevance of Wood is that: (i) it concerned alleged breaches of two covenants being the equivalent of those in paras 1 and 2; (ii) the relevant part of the para 2 covenant is in its essentials identical to covenant (b); and (iii) having held that the erection of the trellis infringed covenant (a), Romer J held that its erection also infringed covenant (b).

[29] Mr Weekes sought to meet these difficulties in two ways. First, he submitted that, in approaching the construction of a covenant contained in a transfer dating from 1987, the court should not regard itself as being bound to apply and follow a decision in 1894 on a covenant dating from 1878. Second, he submitted that Mr Bigham anyway did not argue the point that he argued before us. Mr Bigham argued merely that: (i) the screen was not a "building" for the purposes of covenant (a); and (ii) if it was, it was not an "annoyance" for the purposes of covenant (b). What he did not argue was that: (i) it was not a "building"; but (ii) if it was, its erection anyway could not constitute a breach of covenant (b), which was directed only at the restraint of annoying activities of a different type on the burdened land. He invited us to hold that, in assuming covenant (b) to be potentially in point, Romer J accepted a tacit concession that ought not to have been made.

[30] Mr Martin Hutchings, appearing, as below, for the respondents, submitted that the judge's interpretation of para 2 was correct. He pointed out that all the houses on the island were sold by Heron by transfers imposing on the owners a tight scheme of covenants (those in the third schedule), being covenants intended to be directly enforceable by all other houseowners and as reflecting a clear desire by the parties to preserve the original integrity of the scheme of development. He referred in particular to the covenants in para 4, to other covenants of a positive nature requiring the owners to keep their houses and gardens in good order and to a restraint on the cutting down of trees planted by Heron (I have commented on the extent of the enforceability of the positive covenants).

[31] As for para 2, Mr Hutchings submitted that it was in an entirely familiar form. It falls into three distinct parts, respectively restricting: (i) any use of the house other than as a private dwelling-house; (ii) the carrying on from the plot any trade or business; and (iii) the doing or suffering to be done on the plot of "anything of whatsoever nature which may be or become a nuisance or annoyance" to owners or occupiers of the estate or the neighbourhood. He submitted that the quoted words are expressed in terms of the maximum width and that their natural and ordinary meaning obviously embraces the erection of a building that may, by materially interfering with river views, be an "annoyance" to neighbours. In Wood, Romer J apparently had no doubt that covenant (b) restrained the erection of the "annoying" trellis that he had already held to be a "building" for the purposes of covenant (a).

[32] Mr Hutchings submitted further that the natural interpretation of para 2 is not impliedly cut down by the provisions relating to the erection of buildings in para 1. The ordinary sense of the language of para 2 extends to a restraint on the erection of "annoying" structures of one sort or another and the parties are unlikely to have intended to have confined its reach merely to structures that are not a "building" within the meaning of para 1. Although Mr Hutchings agreed with Mr Weekes that Peverel was under an implied obligation not to withhold its para 1 consent unreasonably, he disagreed that it was under a further implied obligation to have regard to the interests of all others on the estate potentially affected by the building proposal, and he submitted that the demands of business efficacy would not require the implication of such a term. He suggested that a fair interpretation of Peverel's responsibilities when considering whether to approve plans for a proposed building was that it would be primarily concerned merely to consider its own interests as the managing agent of the communal areas and facilities and that it would be under no obligation to consider the particular objections of plot owners. The essence of his submission was that there was no sufficient warrant to interpret a para 1 approval as impliedly eliminating a plot owner's right to complain under para 2 that the proposed building would constitute an "annoyance".

Discussion of the reach of para 2

[33] We had the benefit of able arguments on both sides of this point, for which I express my gratitude. Ultimately, the point is a relatively short one. Paragraph 2 contains covenants in a form whose essence has long been familiar to conveyancers and I unhesitatingly accept Mr Hutchings' submission that the ordinary and natural construction of the purchaser's covenant in it "nor to do or suffer to be done on the Plot or any part thereof anything of whatsoever nature which may be or become a nuisance or annoyance" etc is sufficiently wide to be capable of extending to activities of all natures, including building an extension to an existing house that, when built, will be such an "annoyance". Romer J plainly assumed (although did not apparently have a contrary argument) that covenant (b) in Wood extended to restraining the erection of "annoying" buildings and he held Mr Cooper to be in breach of it.

[34] That is not the end of the point, because para 2 must of course be interpreted in the context of the third schedule as a whole, and there remains Mr Weekes' argument that the undoubted width of the language of para 2 must be regarded as implicitly cut down by para 1. The point is that para 1 deals expressly with the giving of approval to erect buildings, without which approval no building may lawfully be erected. It is said that the scheme of the third schedule is therefore that

A para 1 deals exhaustively with whether a proposed building may or may not be built, and if a para 1 approval for its erection is given, there can be no scope for a separate challenge under para 2 to its erection by another plot owner.

[35] The argument can be said to have its attractions and, it seems to me, a like argument could have been advanced in Wood but was not. Quite apart, however, from the fact that Romer J did not expressly address the point, the factual differences between Wood and the present case are such that I would not regard Wood as providing conclusive assistance on the resolution of the point that has been argued before us.

B [36] There was a difference between counsel as to whether, on receiving plans and an application for a para 1 approval, Peverel is impliedly required to have regard to the interests of the other plot owners and occupiers on the island. We were referred to no authority dealing with this and, in the absence of any representation or arguments by Peverel, which is not a party, I would be reluctant to purport to decide the extent of its obligations under para 1. I should, however, say that I found unconvincing Mr Hutchings' submission that Peverel is primarily concerned to consider only its own interests as managing agent of the communal areas and facilities; although assuming (without deciding) that it should direct its considerations more widely, I regard as C uncertain precisely what the scope of those considerations might be.

[37] Whatever their potential reach, I am not, however, persuaded that, on the true construction of paras 1 and 2, the allegedly implied obligations of Peverel in considering a para 1 approval application can have the implied consequence of cutting down the apparently unambiguous scope of para 2. There is, in my judgment, no doubt that the natural sense of the language of the relevant part of para 2 can extend to a restraint on the erection of potentially "annoying" buildings, and I agree with the judge that had it been the parties' intention to cut this down, they would have done so expressly. There is, in my judgment, no good reason why paras 1 and 2 cannot operate alongside each other, D as the judge held they could and did. I can see no sound reason why the giving by Peverel of a para 1 approval should be regarded as impliedly preventing a plot owner or occupier from asserting, and proving if he can, that despite such approval the proposed building will in fact constitute an actionable "annoyance" in breach of para 2.

[38] If the reasonable man were to be consulted as to his view of the relationship between paras 1 and 2, I consider that he would be unlikely to be impressed by the proposition that the ordinary sense of the relevant language of para 2 is impliedly cut down by the existence of para 1. He would recognise that para 1 makes it clear that the obtaining of a Peverel approval is the precondition of carrying out any building work E at all. However, I consider that he would be unlikely to be convinced that the giving of a Peverel approval automatically forecloses the opportunity for a covenantee to object that the approved building will nevertheless constitute an "annoyance" contrary to para 2. Since there is nothing in the language of para 2 to support such a proposition, why, he would ask, should it be thought that it does so?

[39] I would dismiss Mr Davies's appeal on the para 2 ground.

Does Mr Davies have a written approval under para 1?

[40] In the light of my decision on the para 2 issue, this ground of appeal is only of subsidiary interest. That is because, whether or not F Mr Davies does have such an approval, the judge's conclusion on the para 2 issue trumps it. Since, however, we had argument on the point, I will express my views on it.

[41] The story relating to this issue is that Mr Davies submitted an application for approval to Peverel on 10 January 2007. His application form described the building as a "three storey side extension" and included the "plans and planning permission" and Peverel's fee. The individual at Peverel responsible for making the decision was Mr Sheehan, the estate manager. He replied in the third week of January 2007 saying that the proposal had been forwarded to the Heron Island residents' association, which had rejected it, complaining that it would involve a departure from the island's original "open plan" character. Mr Sheehan said that he was not qualified to comment on this and that "in normal circumstances [Peverel] would have no objections G to granting consent to alterations providing the relevant local authority planning has been authorised". He said that he was referring the application to Peverel's solicitor for comment and advice.

[42] Mr Davies e-mailed Mr Sheehan on 8 February, saying that the residents' association had met the previous day and had told him that its only concern was whether the extension encroached on communal land and that the "open plan" objection had not come from it. He explained that he would be engaging a surveyor to ensure that the extension did not so encroach. Mr Sheehan replied on 9 February, saying that he would not personally stand in the way of the extension "and will be glad to grant consent if [the residents' association] has no objection". H

[43] On 20 March, Mr Davies sent Mr Sheehan his surveyor's report showing the extension to be within his plot boundary. Mr Sheehan replied on 2 April saying that there was "stiff opposition" to the proposal and that a committee meeting to discuss it was being held on 4 April. Since the residents' meeting of 7 February, a number of residents had written letters of objection to Peverel. Mr Sheehan said that he was sorry that he did not have a definitive answer on Mr Davies's application at that stage. On 18 April, Mr Davies told Mr Sheehan that he needed a decision by 4 May because his building work was scheduled to start on 7 May; he also asked him to confirm whether he had received his surveyor's report. J Mr Sheehan gave his confirmation on the same day, saying that he had passed the file to Mr Richard Sandler (Peverel's solicitor) "to make a final decision on the consent". On 14 May, Mr Davies e-mailed Mr Sheehan to express his disappointment at the delay and the lack of communication from him. Mr Sheehan replied within 10 minutes to say that Mr Sandler would be advising on the application and to confirm that "no formal decision has been taken in relation to your planning consent from Peverel". He said that he appreciated that the matter had taken some time but "your proposal has raised numerous complaints which we as managers have to consider. There are service charge implications/re-calculations which also need to be considered." K

[44] There was a telephone conversation on the same day between Mr Davies and Mr Sheehan. On the following day, 15 May, Mr Davies e-mailed Mr Sheehan to remind him that he had said in that conversation that a decision "would be made today" (that is, 15 May) and asking whether that was correct. Mr Sheehan replied:

> I have now been advised by my Regional Manager and company solicitor that consent for the application can be given on provision of the following
>
> 1. Professional drawings confirming that the extension is in line with the existing properties. If you have addition [sic] plans in this regard, please can you forward them to me at your earliest convenience to avoid any further delays. Working drawings of the plans/elevations in colour would be helpful. L
>
> 2. Confirmation by an independent RICS surveyor appointed by Peverel OM that your boundary lines are not being altered in relation to the work. I note that you have submitted a surveyor's report but they are unfortunately not RICS. (Please note that the cost of this will be rechargeable to the applicant as per the terms of your transfer Schedule 3.1)
>
> I will be confirming the above by way of hard copy for your records.

(Emphasis supplied.)

[45] This e-mail is central to the argument under this head of the appeal: Mr Davies's case is that it is this that gave him the para 1 approval. As to the phrase "in line", Mr Weekes submitted that it meant M "in keeping", although I understand that there was no agreement on that in the court below. Despite the last line of the e-mail, no hard copy was sent.

[46] Following that e-mail, Mr Davies wrote, on 18 May, to Peverel asking it to arrange for an independent surveyor to review the boundary issue. On 23 May, Mr Sheehan e-mailed Mr Davies to say that he was still awaiting:

> the working drawings from [the] architect. We will need this to ensure the extension is in keeping with the existing properties on the island. These will also need to be inspected by the surveyor who I have now approached and he is happy to complete the survey in order for him to assess measurements etc.

[47] On the same day, Mr Sheehan wrote a circular letter on Peverel notepaper to all the residents. It referred to Mr Davies's application and the fact that Peverel was still awaiting detailed working drawings and plans from him to ensure that the extension was in keeping with the estate and did "not encroach or alter the private estate boundary lines". The letter continued:

> I understand that there has been some opposition to this application and as a result I have sought further advice from our senior management team. They have now advised that legally, we would have no grounds to unreasonably withhold consent for this application providing that the applicant is able to satisfy certain criteria. The crucial clause which dictated the reasoning and rationale behind the decision is outlined in Schedule 3, 1 of the Freehold Transfer Document.
>
> The planning stage of the application which is overseen by the local council should have been the main recourse for objection by those affected by the proposal.
>
> [The letter then set out the covenant in para 1 of the third schedule]
>
> In effect, this clause merely states that the extension must be in keeping with the submitted plans.
>
> We have asked Dunster & Morton Chartered Surveyors… to carry out an inspection of the project as an independent third party to ensure that the proposed extension work does not compromise or alter any of the private estate boundaries. Please note that the cost of this is being covered by the applicant and not your service charge.
>
> Should you wish to make any comments regarding this decision or any legal interpretations of property transfer and covenants, I kindly request that any representations are sent in writing to our Luton office for the attention of our Company Solicitor, Mr Richard Sandler.

[48] On 31 May, Mr Sheehan e-mailed Mr Davies enquiring whether he had had "any success in obtaining those drawings from the architect" and adding that "our surveyor is ready to carry out what is necessary and then we can grant consent providing all is in order. Please note to date this has not been granted." On 4 June, Mr Sheehan sent a chasing e-mail for the architect's drawings. Mr Davies provided them and, on 14 June, Mr Sheehan e-mailed him saying that he had received them and that copies were that day being sent to Mr Walker (of Haslams, chartered surveyor) so that he could carry out an inspection. Mr Walker sent Peverel a four-page report on 22 June that reflected that he had been asked to check the plot's boundary and whether the extension would breach any covenants with regard to service installations.

[49] Mr Sheehan and Mr Davies had a telephone conversation on 26 June, following which Mr Sheehan sent Mr Davies an e-mail repeating that Peverel had not given its approval for the extension and that he must not commence any building work. On 28 June, Mr Gearon, of Peverel, wrote in place of Mr Sheehan to Mr Davies pointing out that Mr Walker's report showed that the foundation piles proposed for the extension were within 1.5m of a storm-water sewer and that, in consequence, Peverel was unable to give its approval to the extension. On 3 July, Mr Davies asserted that, as a result of Peverel's representations to him, "it would now be both unreasonable and unconscionable for you to withhold consent to my extension". On 30 July, Mr Davies's solicitor, Pitmans, put his case higher: it asserted that Mr Sheehan's e-mail of 15 May had given a consent subject to two conditions, both of which had been satisfied.

[50] The issue for the judge under this head was whether Pitmans' contention was correct. I do not understand it to be in issue that the two conditions referred to in the e-mail were duly satisfied. The question is whether the e-mail: (i) amounted to an irrevocable approval subject to the satisfaction of both conditions; or (ii) was no more than an indication that if the conditions were satisfied, Peverel would (without committing itself to do so) issue a formal approval. The judge favoured the latter alternative. He said that the e-mail was:

> 112. … at best a statement of intent that consent will be given in the future if the 2 conditions are satisfied. It does not, however, bind Peverel to grant consent if, on further investigation, other reasonable objections to the granting of consent arise.

[51] The point is a short one. It is of course apparent from the e-mails and correspondence following the e-mail of 15 May that Peverel did not consider that the e-mail had given the required approval. Nor does it appear to have occurred to Mr Davies that it did so: that point was first taken by Pitmans. I doubt, however, if these considerations can be decisive since the answer to the point turns on the construction of the e-mail and subsequent documents cannot ordinarily be invoked in order to assist in its interpretation: those writing them may have been wrong. The judge did not, I understand, place any reliance on the later material, but focused on the e-mail itself.

[52] I consider that he was right to do so and that he also came to the correct conclusion. The key phrase in the e-mail is "consent can be given on provision of the following". That is not the equivalent of a statement that "consent is hereby given conditionally upon the production of the following". It is admittedly close to it, but the phrase used is rather less precise and I consider that the judge was right to regard its language as amounting to no more than a statement as to Peverel's intentions in the future if and when the requested material was provided. Mr Sheehan was in effect saying that if Mr Davies produced that material, Peverel would be able to give him the formal approval that he required. Only, however, when it actually did so would it be giving him that approval. I also agree with the judge that Mr Sheehan's statement did not bind Peverel to give its approval when the further material was provided. His e-mail was, in short, no more than a non-binding statement of intent. The result is that this ground of appeal also fails.

[53] I would dismiss Mr Davies's appeal.

Wilson LJ said:

[54] I agree.

Ward LJ said:

[55] I also agree.

Appeal dismissed.

TOWN AND COUNTRY PLANNING

R (on the application of Clear Channel UK Ltd) *v* Hammersmith and Fulham London Borough Council

COURT OF APPEAL

[2009] EWCA Civ 2142

27 October 2009

SIR ANTHONY MAY PQBD AND WALL AND MOORE-BICK LJJ

Estates Gazette 9 January 2010

[2010] 01 EG 72

Town and country planning — Deemed consent — Town and Country Planning (Control of Advertisements)(England) Regulations 2007 — London Local Authorities Act 1995 — Advertisements on site for many years — Display replaced with digital display showing selection of advertisements for several seconds each — Local authority serving notice under section 11 of 1995 Act requiring removal of display — Class 13 of Schedule 3 to regulations — Whether display having deemed consent — Whether material alteration in use of site — Whether advertisement comprising sequential display — Whether right to revert to previous permitted display

The appellant operated an advertising display site on the wall of a building located near a crossroads in Fulham, London SW6. The site had originally been used to display a static poster hoarding, but the form of the display had changed over the years. In 2008, the appellant replaced the existing, externally illuminated 36m "golden square" display with a new 18m digital, LED-lit display, using the upper part of the area previously occupied by the golden square, and displaying several images at predetermined intervals.

The respondent council served a notice, under section 11 of the London Local Authorities Act 1995, requiring the appellant to remove the new structure. The claimant applied for judicial review of the notice. It claimed that it had deemed consent for the display by virtue of the use of the site for advertising without express consent over the previous 10 years, pursuant to regulation 6 of, and Class 13 of Schedule 3 to, the Town and Country Planning (Control of Advertisements)(England) Regulations 2007.

Dismissing the claim, the judge held that the new display fell within the exceptions to Class 13 because: (i) replacing the golden square with the digital hoarding constituted a material alteration in the manner of use of the site within condition 13(1); and (ii) there was an advertisement comprising a "sequential display" within condition 13(4)(a) that did not meet the conditions thereof: see [2009] EWHC 465 (Admin); [2009] 2 EGLR 81; [2009] 22 EG 120. He further held that the appellant had no right under the regulations to revert to a former use that had enjoyed deemed consent, and the respondents had not been obliged to have regard to any such consent when deciding whether to issue the section 11 notice. The appellant appealed.

***Held*: The appeal was dismissed. (1) The question of whether there has been a material alteration in the manner of use of a site within condition 13(1) is one of fact, to be judged by reference to considerations of amenity and public safety. The evidence before the judge had been sufficient to support his findings and reasons on the matter. (2) The word "advertisement" in the Town and Country Planning Act 1990 can mean both the image and the structure on which it is displayed, and is intended to encompass both unless the context makes clear that only one or the other is intended. In the context of condition 13(4) in Schedule 3 to the 2007 Regulations, it encompasses both the image and the structure. Accordingly, both an animated display and a structure that displays sequential images will fall within condition 13(4)(a). The appellant's displays were sequential, notwithstanding the absence of any rational connection between the advertisements displayed other than that they were displayed for a predetermined period in an unvarying order. Condition 13(4)(a) is intended to apply where there is a change of image, and the expression "sequential displays" is apt to refer to any display in which one image follows another. (3) The 2007 Regulations do not provide a "right to revert" to a previous use, contrary to the position under their predecessors, the Town and Country Planning (Control of Advertisements) Regulations 1992. The different language of condition 13(1) in the 2007 Regulations directs attention to changes in the extent to which the site has been used, and to alterations in the manner of use, during the preceding 10 years. If there has been any such change and it is material, the advertisement on the site falls outside Class 13 and does not enjoy deemed consent. A further change cannot effectively restore the former position. (4) The respondents had not acted unlawfully in exercising their discretion to serve a notice under section 11 of the 1995 Act. Where there has been a material alteration in the manner of use of a site, the only two courses open to the local authority are to tolerate an unlawful use of the site or to require the removal of the advertisement. It is inherent in Class 13 that the site might have been used for advertising for many years and that fact cannot properly influence an authority in favour of tolerating a use that has recently become unlawful. The authority's lack of objection to previous changes of use is likewise irrelevant to the issue.**

The following cases are referred to in this report.

Postermobile plc v *Brent London Borough Council* unreported 11 November 1997

R (on the application of Clear Channel UK Ltd) v *Hammersmith and Fulham London Borough Council* [2009] EWHC 465 (Admin); [2009] 2 EGLR 81; [2009] 22 EG 120; [2009] JPL 1224,

R (on the application of Clear Channel UK Ltd) v *Southwark London Borough Council* [2007] EWCA Civ 1328; [2007] NPC 133

R (on the application of JCDecaux UK Ltd) v *Wandsworth London Borough Council* [2009] EWHC 129 (Admin)

R (on the application of Maiden Outdoor Advertising Ltd) v *Lambeth London Borough Council* [2003] EWHC 1224 (Admin); [2004] JPL 820

This was the hearing of an appeal by the appellant, Clear Channel UK Ltd, for judicial review of a decision of the respondents, Hammersmith and Fulham London Borough Council, to serve a notice under section 11 of the London Local Authorities Act 1995 requiring the appellant to remove an advertising display.

Andrew FraserUrquhart (instructed by Grant Saw Solicitors) appeared for the appellant; Robin Green (instructed by the legal

A department of Hammersmith and Fulham London Borough Council Treasury Solicitor) appeared for the respondents.

Giving the first judgment, **Moore-Bick LJ** said:

[1] This appeal concerns an illuminated digital display advertisement hoarding on the flank wall of a building at the corner of Fulham Palace Road and Winslow Road in south-west London. The building, which occupies 112-114 Fulham Palace Road, fronts that road. The junction between Fulham Palace Road and Winslow Road forms part of a crossroads controlled by traffic lights. The hoarding is attached to the side of the building facing Winslow Road but is angled slightly in the direction of Fulham Palace Road. It is clearly visible from B Winslow Road and also to pedestrians and traffic travelling in a north-westerly direction along Fulham Palace Road.

[2] There has been an advertisement hoarding on this site for many years. For some time prior to 1991, it consisted of a traditional billboard, approximately $18m^2$ in area, of the kind used for paper displays that was sited at or just above ground level. At some time in 1991, a second hoarding, also with an area of approximately $18m^2$, was installed above the billboard. That was illuminated and consisted of what is known as a "trionic" display, that is, a mechanical device consisting of a series of three rotating panels, internally lit, that enable three separate images to be displayed in order, the display changing at C predetermined intervals, usually of a few seconds. In September 1998, the trionic display was replaced with an "ultravision" display, another mechanical device, also internally lit, containing a scrolling device similar to a roller blind, which also enables a series of images to be displayed in order at predetermined intervals, again usually of a few seconds. In 2002, both hoardings were replaced with what is known as a "golden square", a static display in the nature of a traditional billboard, externally lit and approximately $36m^2$ in size. Finally, in April 2008, the golden square was replaced with a digital hoarding of approximately $18m^2$, occupying the upper part of the area previously occupied by the golden square, the space beneath it being left free. The digital hoarding, D which employs LED technology, allows several images to be displayed at predetermined intervals and represents the latest development in equipment used for displaying advertisements. When the digital display was erected, it was angled slightly in the direction of Fulham Palace Road to render it more visible from that direction.

[3] Since 6 April 2007, the display of advertisements has been controlled by the Town and Country Planning (Control of Advertisements)(England) Regulations 2007 (the 2007 Regulations), made under the Town and Country Planning Act 1990 (the 1990 Act). They provide, so far as is material for present purposes, as follows:

E **Deemed consent for the display of advertisements**

6.(1) Subject to regulations 7 and 8… consent is granted for the display of an advertisement of any class specified in Part 1 of Schedule 3, subject to

(a) the standard conditions; and

(b) in the case of any class other than Class 12, the conditions and limitations specified in that Part in relation to that class.

…

SCHEDULE 3

Class 13 Advertisements on sites used for preceding ten years for display of advertisements without express consent

Description

13. An advertisement displayed on a site that has been used continually F for the preceding ten years for the display of advertisements without express consent.

Conditions and Limitations

13.(1) An advertisement does not fall within this description if, during the relevant 10-year period, there has been either a material increase in the extent to which the site has been used for the display of advertisements or a material alteration in the manner in which it has been so used.

…

(4) An advertisement that

(a) comprises sequential displays; or

(b) otherwise includes moving parts or features; or

(c) features intermittent lighting in a manner designed to give the appearance of movement,

is not permitted unless

(i) it is displayed on 6th April 2007 and falls within the description specified in any of sub-paragraphs (a) to (c); or G

(ii) it is first displayed after that date, and the advertisement most recently displayed fell within any such description.

[4] On 3 April 2008, the respondent council, as the local planning authority, wrote to the appellant noting that the digital display had been erected adjacent to the flank wall of 112-114 Fulham Palace Road and requiring its removal on the ground that it had not received express consent and did not benefit from deemed consent under the 2007 Regulations because it was materially different from the hoarding that had previously been displayed on that site. The appellant declined to comply with the council's demand on the ground that the hoarding H was no larger than the illuminated hoarding it had replaced and that the site had the benefit of deemed consent under the 2007 Regulations. On 23 May 2008, the council served on the appellant a notice, under section 11 of the London Local Authorities Act 1995 (the 1995 Act), requiring the appellant to remove the hoarding within 21 days. The parties were unable to resolve their differences in correspondence and, accordingly, on 12 June 2008, the appellant started proceedings for judicial review seeking to have the council's decision quashed.

[5] The grounds on which the appellant sought judicial review were that the hoarding enjoyed deemed consent because the site had been used continuously for advertising since 1991 and there had been no J material increase in the extent or alteration in the manner in which it had been used. There was, it contended, no material distinction between changing displays manually (as in the case of traditional billboards), mechanically (as in the case of trionic or ultravision displays) or digitally. Moreover, the appellant contended that in deciding to serve a notice under section 11 of the 1995 Act, the council had failed to have regard to two material considerations, namely that the site had been used for advertising for a long time and that, if given a reasonable opportunity to do so, the appellant could revert to a former use of the site for which there would be deemed consent and would thus not be deprived of its use altogether (the so-called "right to revert").This latter K argument was based on the decision of Collins J as to the effect of the previous regulations in *R (on the application of Maiden Outdoor Advertising Ltd)* v *Lambeth London Borough Council* [2003] EWHC 1224 (Admin)*, as explained by this court in *R (on the application of Clear Channel UK Ltd)* v *Southwark London Borough Council* [2007] EWCA Civ 1328†.

[6] The appellant's claim for judicial review was heard by Irwin J. He found that the replacement, in April 2008, of the golden square by a digital hoarding represented a material alteration in the manner of use of the site; he also held that the digital hoarding was an advertisement that comprised a sequential display. As to the exercise by the council L of their discretion, the judge held that, whatever may have been the position under the previous regulations, under the 2007 Regulations once deemed consent had been lost as a result of a material alteration in the manner in which the site was used it could not be recovered by reverting to a former use and that the council had not therefore failed to take that possibility into account. He therefore dismissed the claim and it is against his decision ([2009] EWHC 465 (Admin)‡) that the present appeal is brought.

[7] Mr Andrew Fraser-Urquhart, for the appellant, accepted that it was for the judge to decide as a matter of fact whether there had been a material alteration in the manner in which the site had been used M within the 10 years immediately preceding the date of the notice, but he submitted that the judge had not in fact asked himself the correct question and that, in any event, the evidence before him was insufficient to support his finding that there had been any such material alteration. He also submitted that the judge was wrong to hold that the hoarding was an advertisement that comprised sequential displays and wrong again to hold that there was no "right to revert". Finally, he submitted

* Editor's note: Reported at [2004] JPL 820

† Editor's note: Reported at [2007] NPC 133

‡ Editor's note: Reported at [2009] 2 EGLR 81

that the council had failed to take into account a number of relevant factors when deciding whether to serve a notice under section 11 of the 1995 Act requiring the removal of the hoarding, among which was their power to proceed by way of a prosecution.

[8] Four issues arise for determination on this appeal, namely whether:

(i) the judge's approach to the issue of material alteration was correct and there was sufficient evidence to support his finding;

(ii) the digital hoarding was an advertisement that comprised sequential displays;

(iii) under the 2007 Regulations, there exists a "right to revert"; and

(iv) the council's decision to serve a notice under section 11 of the 1995 Act was unlawful because it failed to take relevant matters into account.

(i) Material alteration

[9] It was common ground that whether there had been a material alteration in the use of the site was to be judged by reference to considerations of amenity and public safety, those being the criteria by reference to which the council's powers under the 2007 Regulations are to be exercised: regulation 3(1). It was also common ground that whether there had been any such alteration was a matter of fact for the judge to decide.

[10] The judge was provided with, among other things, colour photocopies of the previous golden square display together with six brief video recordings of the digital display, four made during the day and two after dark. He found, in [20] of his judgment, that the alteration of the structure was material in that it was capable of affecting amenity in the area. He said:

> 20. I do find that the alteration of this structure is material so that it is capable of affecting amenity in the area. I do so on two mutually consistent and supportive bases. Firstly, the size, bulk and position, and the "nature of the object", if I can use that term, make this structure markedly more obtrusive in the area, reasonably capable of representing a material change in the amenity value to the immediate vicinity. Secondly, the nature of the illumination and display itself the changing light sometimes meaning that an 18-square-metre screen is extremely bright, casting light on to other buildings, and sometimes suddenly dark, or partly dark, represents a strikingly new illumination. This is capable of affecting amenity in the way that a fixed light from a shopfront, or the fixed light on to the previous hoarding, it seems to me, does not. The position of the illuminated screen is deliberately very high and it casts light high up on to the surrounding area, where residential property may be likely to be in question.

[11] In the following paragraph, he said:

> 21. In relation to both of these conclusions as to the materiality of the alteration, I have considered the question by testing it in this way: if permission for express consent was applied for, it seems to me clear that perfectly rational people could lodge objections to the effect on amenity of this structure. We are not in the position of an application for express consent and I am not intending to pre-judge any such application, if that is what ever transpires. But it seems to me that it is plain that a rational person with an interest in the amenity of the area could say this affects amenity to a considerable degree.

[12] Mr Fraser-Urquhart submitted that [21] betrays the true reasoning that lies behind the judge's finding in [20] and shows that, instead of deciding whether, as a matter of fact, there had been a material alteration in the manner in which the site had been used, he had considered whether a reasonable person could think that the amenity of the area had been affected. I am unable to accept that. I think that it is clear that the judge asked himself the correct question and answered it in [20], giving his reasons for finding as a fact that there had been a material alteration in the manner of use. All he did in [21] was to test his conclusion by reference to the response of a rational observer. If he had concluded that no rational observer could sensibly think that the alteration affected the amenity of the area, that would no doubt have caused him to reconsider his finding. As it was, it tended to confirm him in the conclusion that he had already reached. In my view, this criticism of the judge's decision is without foundation.

[13] Mr Fraser-Urquhart's next submission was that the quality of the video recordings and the circumstances under which they had been made did not provide a fair impression of what is observed by the naked eye, and did not therefore provide a satisfactory basis for finding that there had been a material alteration in the manner of use of the site. I have had the benefit of viewing the video recordings as well as seeing the colour photocopies of the golden square that were before the judge. Colour photocopies are less informative than photographic prints because they do not have the same sharpness of focus or depth of colour, but for many purposes they are an adequate substitute. A video recording made by an amateur is likely to be less valuable in some respects than one made by a professional, but again may provide evidence that is sufficiently reliable for the particular purposes for which it is produced. It is true, as Mr Fraser-Urquhart pointed out, that, in this case, the judge had no scientific evidence to assist him, such as measurements of the degree and intensity of illumination, but where amenity and public safety are concerned the effect on the viewer is likely to be more important than bald measurements of that kind. The question is not whether the judge had available to him all the evidence that might conceivably have been provided, but whether the evidence that he did have was sufficient to support the finding he made. In my view it was and the reasons that he gave for his decision demonstrate that he had in mind factors of a kind that the evidence enabled him to assess satisfactorily.

[14] Finally Mr Fraser-Urquhart submitted that in reaching his decision on this issue the judge failed to take into account the fact that the council apparently did not regard any of the previous alterations in the use of the site as material. In my view, however, this is a false point. It was for the judge to decide whether the change from a golden square to the digital display was material. That was a matter of fact on which he had to make his own finding. What the council thought about the materiality of the change was not relevant; nor was it relevant what they had thought about any of the previous changes.

[15] For these reasons, I consider that the judge's finding that there had been a material alteration in the manner of use of the site is unassailable.

(ii) Sequential displays

[16] The appellant's case was that since the images displayed on the hoarding are unrelated to each other and each is entirely static (there being no animation or movement of any kind in the image), there is no "advertisement that comprises sequential displays" within the meaning of condition 13(4). The judge held that there was a sequential display because the advertisements followed each other in the same order for the same length of time, although he also expressed the view that even if they had been displayed at random they would have constituted a sequential display on the assumption that they would be displayed one after the other, each for the same total period of time.

[17] In my view, the answer to this question depends to a large extent on the meaning of the word "advertisement". "Advertisement" is defined in section 336(1) of the 1990 Act as follows:

> any word, letter, model, sign, placard, board, notice, device or representation… in the nature of, and employed wholly or partly for the purposes of, advertisement… and (without prejudice to the previous provisions of this definition) includes any hoarding or similar structure used… for the display of advertisements…

It is thus capable of meaning both the image itself and the structure on which it is displayed. This is potentially of some importance since none of the individual advertisements contains any sequential display, in the sense of changes to its appearance, although the structure on which they are displayed does.

[18] Mr Fraser-Urquhart submitted that the meaning of the word depends in each case on its context: sometimes it means the image and sometimes the structure. I quite accept that the context in which the word is found is important and may make it clear that only some of its many meanings is intended. However, I think that Mr Robin Green was right in submitting that the definition is worded in a way that suggests that it was intended to encompass both the image and the structure,

A unless the context makes it clear that one or other was not intended. In the present case, subpara (b), which refers, among other things, to "moving parts", suggests that it was intended to apply to structures because images do not have moving parts, although they may have moving features and sequential displays. Mr Fraser-Urquhart submitted that it follows from the use of the word "otherwise" in para (b) that in order to fall within para (a) an advertisement must in some other way include moving parts or features and so must exclude static images such as those displayed by the digital hoarding in this case. That might be a possible interpretation of condition (4) if the word "advertisement" meant simply the image displayed, but the difficulty that I have with
B the submission is that, in view of the way in which the condition is structured, the word cannot sensibly have been intended to bear a different range of meanings in respect of different paragraphs. In my view, it was intended in this context to encompass both the image itself and the structure on which it is displayed. Accordingly, an animated display falls within para (a), as does a structure that displays sequential images; and such a structure also falls within para (b) if it has moving parts or features. Although I would agree that the expression "comprises sequential displays" is slightly awkward to describe the action of a hoarding that displays sequential images, I do not consider it to be so inapposite as to indicate that para (a) was not intended to bear the
C meaning that I have indicated.

[19] Are the displays in this case sequential? Mr Fraser-Urquhart submitted that they are not because a series of images cannot properly be described as sequential unless they are displayed in a rational order, so that each is in some way related to that which has gone before and that which comes after. In this case, it is said, there is no rational connection of any kind between the advertisements other than the fact that they are displayed for a predetermined period in a predetermined and unvarying order.

[20] Once again, I think that the context in which the expression is used is important. Sometimes, the words "sequence" and "sequential"
D are used to denote a series of events arranged in a rational order, in which each constituent is related in some way to that which has gone before and that which comes after, but, in my view, they can equally well be used in respect of a series of events that from some exterior perspective are, or may be considered to be, linked in some way. I agree with the judge that if the equipment had been set to display the images in a random order, and even for random periods of time, they could as a whole properly be described as a sequential display merely because, from the perspective of the viewer, one follows another. I think that in the context of para (4), which covers advertisements with moving parts or features and intermittent lighting designed to give the appearance
E of movement, para (a) is intended to apply where there is a change of image and that the expression "sequential displays" is apt to refer to any display in which one image follows another. In my view, therefore, the judge was right to hold that the case fell within para 4(a).

(iii) "Right to revert"

[21] In order to understand this issue, it is necessary to set out the relevant parts of the Town and Country Planning (Control of Advertisement) Regulations 1992 (the 1992 Regulations), the predecessors of the 2007 Regulations. Regulation 6 of the 1992 Regulations was in substantially the same terms as regulation 6 of the 2007 Regulations, but Class 13 and its conditions and limitations
F were somewhat different. They provided as follows:

> Description.
>
> 13 An advertisement displayed on a site which was used for the display of advertisements without express consent on 1st April 1974 and has been so used continually since that date.
>
> Conditions and Limitations
>
> 13(1) No substantial increase in the extent, or substantial alteration in the manner, of the use of the site for the display of advertisements on 1st April 1974 is permitted; …

[22] In *Maiden*, Collins J held that in cases where consent was deemed to have been given to the use of a site for the display of advertisements but such consent had been lost as the result of a material alteration in the manner of use of the site, the advertiser could revert G
to the previous lawful use and thereby recover the deemed consent, hence the description "right to revert". In those circumstances, he held that it was unlawful for the council to serve a notice under section 11 of the 1995 Act without considering whether to give the advertiser an opportunity to revert to its previous use.

[23] In *Clear Channel* v *Southwark*, this court, of which I was a member, agreed in substance with the view expressed by Collins J as to the effect of the 1992 Regulations, although it favoured a rather different analysis. Sir John Chadwick, with whom the other members of the court agreed, said: H

> 33. If I may say so, I think that Mr Justice Collins was right, in the *Maiden* case, to accept and adopt that reasoning in relation to the construction and effect of class 13 in schedule 3 to the 1992 Regulations. But I do not think it apt to describe the removal of offending hoardings and their replacement by others as having the effect that the deemed consent was "revived"; or as giving rise to a "right to revert". The true analysis is that the question whether the display of an advertisement (or hoarding) on a site is authorised by deemed consent under regulation 6(1) has to be answered by reference to the facts as they are at the time. At paragraph 25 of its skeleton argument, filed for the purpose of this appeal, the appellant asserts that "The issue of deemed consent is to be taken as a 'snapshot'; it gives legality to a particular display at a particular time". In my view that is a correct statement of the position. J

[24] Mr Fraser-Urquhart submitted that there was no reason to think that when making the 2007 Regulations the minister had intended to alter the previous position, and that the effect of doing so was so serious for advertisers that they should be construed in the same way as the 1992 Regulations unless it was impossible to do so.

[25] Two features of Class 13 and condition 13(1) of the 1992 Regulations enabled the court to hold that they should be construed as requiring the question of whether there was deemed consent to the display of an advertisement on a particular site to be answered by reference to the facts as they were at the time the question fell to be K
answered (the so-called "snapshot" approach). The first was that the description of the class referred only to continuity of use of the site for advertising; the second was the use in condition 13(1) of the expression "No substantial increase… is permitted", which directed attention to the existing use and required one to determine whether or not it was permitted. The language of condition 13(1) of the 2007 Regulations is quite different. It directs attention to changes in the extent to which the site has been used and to alterations in the manner in which it has been used during the preceding 10 years. If there has been any such change or alteration, and if it was material, the advertisement currently on the site falls outside Class 13 and does not enjoy deemed consent. I can L
see no basis on which it could be argued that the position under the 2007 Regulations equates to that under the 1992 Regulations, nor do I see how it can plausibly be argued that a further change can effectively restore the former position. The choice of language is itself, in my view, sufficient evidence, if evidence were needed, of an intention to depart from the previous position. I am quite satisfied, therefore, that there is no right to revert under the 2007 Regulations of the kind that was held to exist under the 1992 Regulations. The judge rejected this argument on these grounds and, in my view, he was right to do so.

(iv) Discretion M

[26] The conclusion that I have reached on the existence of a right to revert largely undermines the appellant's challenge to the exercise by the council of their decision to issue a notice under section 11 of the 1995 Act. Mr Fraser-Urquhart submitted that their decision was flawed because in exercising their discretion to take that course they had failed to take into account a number of relevant matters. Apart from the right to revert, these were said to be the fact that: (a) the site had been used for advertising for many years; (b) the council had not objected to previous changes in the use of the site as being material; and (c) when the digital display was erected, it was not apparent that, if it amounted to a material alteration in use, it could not be removed and the previous deemed consent revived.

[27] In my view, there is no substance in any of these complaints. In the light of the 2007 Regulations, there were only two courses open to the council if there had been a material alteration in the manner of use of the site, namely: to tolerate an unlawful use of the site or to obtain the removal of the advertisement. They could not serve a notice of discontinuance under regulation 8, since that applies only to advertisements for which there is deemed consent. In my view, none of the three matters to which the appellant referred was relevant to the council's decision. It is inherent in Class 13 that the site may have been in use for advertising for many years; it is difficult to see how that could properly influence the council in favour of tolerating a use that had recently become unlawful. The fact that they had not objected to previous changes of use is likewise irrelevant; it was not suggested that by reason of their nature or the circumstances in which they had occurred the council's failure to act had created a legitimate expectation on the part of the appellant that the erection of a digital display would not be regarded as material. The replacement of the 1992 Regulations by the 2007 Regulations was a matter of public record and was known to the appellant, which was able to obtain legal advice concerning their effect.

[28] Finally, Mr Fraser-Urquhart sought to contend that the council should have given consideration to pursuing the matter by way of prosecution rather than by the issue of a notice under section 11 of the 1995 Act. I am not sure that I understand why it is said that that course should have been considered, since the display would have remained unlawful and the only purpose of bringing a prosecution would have been to have it removed. I am not at present persuaded, in so far as this was suggested, that the opportunity of challenging a prosecution as being an abuse of process (as was held possible, albeit in rather different circumstances, in *Postermobile plc* v *Brent London Borough Council* unreported 11 November 1997) would have been more favourable to the appellant than the opportunity of challenging the council's decision by way of a claim for judicial review; but even if it were, I doubt whether that would be a factor that the council ought to have taken into account when deciding what course to take. It is unnecessary to decide either of those questions, however, because this particular complaint did not form any part of the grounds on which the appellant sought judicial review and it was not addressed by the council in evidence or argument or by the judge in his judgment. It is far too late for it to be raised now.

[29] For all these reasons, I am satisfied that in exercising their discretion to serve a notice under section 11 of the 1995 Act, the council did not act unlawfully.

[30] I would therefore dismiss the appeal.

Giving the second judgment, **Wall LJ** said:

[31] I agree that this appeal should be dismissed for the reasons given by Moore-Bick LJ, whose judgment I have had the advantage of reading in draft. I add a short judgment of my own in order to make two points, which I regard as emphasis for the decision that we have reached, rather than as the expression of any original thought.

[32] Speaking for myself, I have no difficulty with the phrase "An advertisement that (a) contains sequential displays" in para 4(a) of Class 13 in Schedule 2 to the 2007 Regulations. In my judgment, the noun "sequence" and the adjective "sequential" are ordinary words carrying the meaning of one person or object following another. This is the primary meaning given to the words in the second edition of the *Oxford English Dictionary* and, in my judgment, the words are apt to describe the advertisements to which the local authority take objection. As I understand it, a number of still advertisements follow each other for an allotted period of time: this sequence of advertisements is then repeated.

[33] Mr Fraser-Urquhart submitted, as I understood him, that for there to be a "sequential display" there had to be a connection between each item, so that each advertisement related to its successor or predecessor in some way. He thus seized with unfeigned enthusiasm on the example of traffic lights given by the judge in [24] of the judgment. He agreed with the judge that traffic lights were indeed a "consequential sequence" because, he argued, contrary to the judge's view, there was indeed a connection between the content of the colours in a sequence of traffic lights.

[34] I do not think that [24] of the judgment indicates any error on the part of the judge, and I am wholly unable to give the words "sequential displays" the limited and restricted construction for which Mr Fraser-Urquhart argued. The judge, in my view, was entirely correct in concluding, as he did in [26] of his judgment, with the following words:

> … Here these seven advertisements follow each other in the same sequence, for the same length of time, and simply rotate, while the display has the content unchanged. I therefore find that this is indeed a sequential display within Class 13, paragraph 4(a), and is thus not permitted…

[35] My second point relates to the decision of Blake J in *R (on the application of JCDecaux UK Ltd)* v *Wandsworth London Borough Council* [2009] EWHC 129 (Admin). I would like to express my specific agreement with what the judge said in [54] of his judgment concerning the 2007 Regulations and the need for co-operation and consultation between the owners of advertising sites and local authorities.

Conclusions

[36] It seems that, in the light of the wording of the new regulation 13, there can be no return to the status quo as of right, and there would have to be an application for planning permission that is clearly not for this court to determine. As things stand, therefore, the school stands to lose the licence fee and the appellant to lose the use of the site that it had been using until 2007, where its hoarding has long been displayed. It seems therefore that those in the position of the appellant that are proposing to make amendments to the site need to be very careful and think through the consequences of any changes before making them, and that might involve consultation with the relevant local authority.

[37] As I understand the matter, if an application for express consent is refused by a local authority, the owner of the advertising hoarding can appeal. That seems to me to be the proper and democratic process, rather than risk a breach of Class 13, which, as here, has lost the appellant its previous use of the site.

[38] As I indicated earlier, these observations are, essentially, supplemental to the judgment of Moore-Bick LJ, with which I am in complete agreement.

Sir Anthony May PQBD said:

[39] I agree that this appeal should be dismissed for the reasons given by Moore Bick LJ.

Appeal dismissed.

R (on the application of Fidler) *v* Secretary of State for Communities and Local Government and another

ADMINISTRATIVE COURT

3 February 2010

SIR THAYNE FORBES, SITTING AS A HIGH COURT JUDGE

Estates Gazette 17 April 2010

[2010] 15 EG 96

Town and country planning — Enforcement notice — Building operations — Limitation — Town and Country Planning Act 1990 — Erection of new dwelling without planning permission — Works concealed under straw bales — Bales removed after four years — Local planning authority issuing enforcement notices — Section 171B of 1990 Act — Whether limitation period for enforcement action expiring — Whether erection and removal of straw bales forming part of relevant building operations

In 2002, the appellant built a new dwelling on his land without seeking planning permission from the second respondent local planing authority. He surrounded the building with straw bales, covered with a tarpaulin, in order to hide the construction process. His aim was to keep the building hidden until the four-year time limit for taking enforcement action against building operations, under section 171B of the Town and Country Planning Act 1990, had expired.

In 2006, the appellant removed the straw bales and tarpaulin. The second respondents issued enforcement notices requiring the demolition of the dwelling and the removal of all materials. A planning inspector appointed by the first respondent Secretary of State dismissed the appellant's appeal against the notice. The inspector held that the time limit for enforcement had not expired because the relevant building operations included the erection and removal of the straw bales and, accordingly, these had not been substantially completed, within the meaning of section 171B, until the straw bales were removed in 2006. The appellant appealed under section 289 of the 1990 Act.

***Held*: The appeal was dismissed. A number of ancillary activities can take place on a construction site that, if considered in isolation, would not be a building operation. However, these activities can none the less form part of the contemplated and intended building operations when considered as a whole. Whether such an activity forms part of the overall building operations will be a matter of fact and degree. The term "operations", in the context of the definition of "development" in section 55(1) of the 1990 Act, is capable of covering a wide range of activities relating to the actual work of building. Although the appellant's erection and removal of the straw bales was not itself, when considered in isolation, a building operation, it formed part of building operations when the totality of the works, as originally contemplated and intended, was considered. The appellant had always intended to remove the straw bales, and he could not enjoy a reasonable level of residential amenity, consistent with normal expectations of what a dwelling-house should provide, until he did so. The inspector's findings made it clear that the erection and removal of the bales constituted an integral and essential part of building operations that were intended to deceive the local planning authority and to achieve, by deception, lawful status for a dwelling built in breach of planning control. Substantial completion of those building operations had not taken place until the straw bales had been removed.**

The following cases are referred to in this report.

Dilworth v *Commissioner of Stamps* [1899] AC 99
Sage v *Secretary of State for the Environment, Transport and the Regions*; *sub nom Sage* v *Maidstone Borough Council* [2003] UKHL 22; [2003] 1 WLR 983; [2003] 2 All ER 689; [2003] 2 P&CR 26; [2003] 1 PLR 121; [2003] JPL 1299

This was an appeal by the appellant, Robert Fidler, from a decision of an inspector appointed by the first respondent, the Secretary of State for Communities and Local Government, upholding enforcement notices issued by the second respondents, Reigate and Banstead Borough Council, in respect of building operations.

Stephen Hockman QC (instructed by Wright Hassall LLP, of Leamington Spa) appeared for the appellant; Paul Brown QC (instructed by the Treasury Solicitor) appeared for the first respondent; Rupert Warren (instructed by the legal department of Reigate and Banstead Borough Council) represented the second respondents.

Giving judgment, **SIR THAYNE FORBES** said:

Introduction

[1] This is an appeal brought under section 289 of the Town and Country Planning Act (the 1990 Act), permission having been granted by Wyn Williams J on 26 January 2009. In these proceedings, the appellant (Mr Fidler) challenges the decisions of an inspector (the inspector), duly appointed by the first respondent (the Secretary of State), as set out in his decision letter dated 7 May 2008 (the decision letter), in respect of three out of the 12 appeals under section 174 of the 1990 Act that the inspector determined on that date (that is, appeals 1, 2 and 3), whereby he dismissed, *inter alia*, the claimant's appeal against enforcement notices issued by the second respondents on 16 February 2007 in respect of: (i) a new dwelling, together with; (ii) its associated conservatory; and (iii) the patio that Mr Fidler had constructed on his land at Honeycrock Farm, Axes Lane, Salfords, Surrey RH1 5QL (the land). The second respondents are the local planning authority (the LPA) for the area in which the land is situated.

[2] It is common ground that the central aspect, indeed the key to the outcome, of these proceedings is the appeal against the inspector's decision in respect of appeal 1 (that is, the section 174 appeal against the enforcement notice relating to the new dwelling, hereafter "Appeal 1"). On behalf of Mr Fidler, Mr Stephen Hockman QC very properly accepted that the challenge to the inspector's decisions in respect of appeals 2 and 3 (that is, the section 174 appeals with regard to the enforcement notices directed at the new dwelling's associated conservatory and patio) could not succeed unless the appeal against the inspector's decision in appeal 1 were to be successful. For that reason, the submissions of the parties to these proceedings were almost entirely directed to challenging (in the case of Mr Fidler) and defending (in the case of the Secretary of State and the LPA) the inspector's decision in respect of appeal 1. The reason why this was clearly the appropriate approach to adopt will become readily apparent as this judgment proceeds.

[3] The breach of planning control alleged in the enforcement notice with which appeal 1 was concerned (the enforcement notice) is expressed as follows (see para 3 of the enforcement notice): "Without planning permission, the erection of a dwellinghouse in the approximate position on the attached plan." The reasons for issuing the enforcement notice include the following (see para 4): "It appears to the Council that the above breach of planning control has occurred within the last four years. The building in question was substantially completed less than four years ago… ." The requirements of the notice are as follows (para 5): "(1) Demolish the dwellinghouse and remove all the resultant materials. (2) Reinstate the land to its former condition." The time for compliance is stated to be: "Twelve months from the date this notice takes effect."

[4] So far as material, section 174 of the 1990 Act provides as follows:

> 174 Appeal against enforcement notice
> (1) A person having an interest in the land to which an enforcement notice relates… may appeal to the Secretary of State against the notice, whether or not a copy of it has been served on him.
> (2) An appeal may be brought on any of the following grounds —

(a) that, in respect of any breach of planning control which may be constituted by the matters stated in the notice, planning permission ought to be granted…

…

(d) that, at the date when the notice was issued, no enforcement action could be taken in respect of any breach of planning control which may be constituted by those matters.

…

(f) that the steps required by the notice to be taken exceed what is necessary to remedy any breach of planning control which may be constituted by those matters…

[5] Before the inspector, Mr Fidler relied on each of grounds (a), (d) and (f) in support of his section 174 appeal in appeal 1 and, in due course, the inspector dismissed the appeal on all three grounds. However, the section 289 proceedings before me are concerned only with the inspector's decision in respect of ground (d), which was itself founded on the provisions of section 171B of the 1990 Act (often referred to as "the four-year rule"), which, so far as material, provides as follows:

171B — (1) Where there has been a breach of planning control consisting in the carrying out without planning permission of building, engineering, mining or other operations in, on, over or under the land, no enforcement action may be taken after the end of the period of four years beginning with the date on which the operations were substantially completed.

[6] In brief, the basis of Mr Fidler's appeal on ground (d) was that the new dwellinghouse had been substantially completed by around June 2002, although it was still concealed within a shield of straw bales, the top of which was covered with a tarpaulin, that he had deliberately erected in order to conceal the construction of the new dwelling and to take advantage of the four-year rule. The new dwelling was revealed when the straw bales and tarpaulin were eventually removed in July 2006, by which time the four-year period from substantial completion of the new dwelling had expired. Accordingly, it was Mr Fidler's case, on ground (d), that although he readily accepted that the building operations involving the construction of the new dwelling constituted development for which he had not been granted planning permission, the time limit for taking enforcement action had expired long before the LPA came to issue the enforcement notice on 16 February 2007.

[7] Stated broadly, the inspector dismissed Mr Fidler's appeal on ground (d) on the basis that the overall building operations relating to the construction of the new dwelling included the erection and the removal of the straw bales and tarpaulin that had been deliberately put in place to conceal the construction and the existence of the new dwelling in order to take advantage of the four-year rule. The inspector then went on to decide that, when considered as a whole, the building operations were not substantially completed until the removal of the straw bales in July 2006. The inspector therefore concluded that the four-year time limit for taking enforcement action had not expired by February 2007 and accordingly dismissed the appeal on ground (d).

Statutory framework

[8] In addition to the material terms of section 171B and section 174, quoted respectively in [4] and [5] above, the 1990 Act also contains the following relevant provisions:

55 — (1) Subject to the following provisions of this section, in this Act, except where the context otherwise requires, "development," means the carrying out of building, engineering, mining or other operations, in, on, over or under land, or the making of any material change in the use of any buildings or other land.

(1A) For the purposes of this Act "building operations" includes —

(a) demolition of buildings;

(b) rebuilding;

(c) structural alterations of or additions to buildings; and

(d) other operations normally undertaken by a person carrying on business as a builder.

(2) The following operations or uses of land shall not be taken for the purposes of this Act to involve development of the land —

(a) the carrying out for the maintenance, improvement or other alteration of any building of works which —

(i) affect only the interior of the building, or

(ii) do not materially affect the external appearance of the building,

…

…

57 — (1) Subject to the following provisions of this section, planning permission is required for the carrying out of any development of land.

…

171A — (1) For the purposes of this Act —

(a) carrying out development without the required planning permission…

…

constitutes a breach of planning control.

(2) For the purposes of this Act —

(a) the issue of an enforcement notice…constitutes taking enforcement action.

Facts

[9] In his decision letter, the inspector gave the following succinct account of the main relevant factual circumstances as part of his findings of fact relating to the new dwelling and its construction:

148. The new dwelling is sited on what was the north-western corner of the western yard. It is constructed of stone and brickwork with some oak framing and a second-hand clay tiled roof. Two redundant grain silos have been used to form the casing for the stone/brickwork which creates two castellated features at the north-eastern and north-western corners. There is a stain-glass lantern feature over a central hall/gallery area. The house comprises kitchen, living room, study, shower room/WC and separate WC on the ground floor and 4 bedrooms on the first floor, and another room which the appellant said was still to be fitted out as a bathroom. The windows are uPVC double glazed units set within stained timber sub-frames. To the south of the house is a gravelled forecourt whilst to the north is the new patio and conservatory at the north-western corner.

149. Mr Fidler made it quite clear that the construction of this house was undertaken in a clandestine fashion, using a shield of straw bales around it and tarpaulins or plastic sheeting over the top in order to hide its presence during construction. He stated that he knew he had to deceive the Council of its existence until a period of 4 years from substantial completion and occupation had occurred as they would not grant planning permission for its construction.

…

150. As an observation, it is evident from the documentary evidence provided that he house was in existence in some form at the time Mr Morden made his inspection in May 2002. The photographs from that period do show the presence of a large stack of straw bales covered in blue plastic sheeting — which it is accepted were actually straw walls with a void inside in which the house was constructed. …

…

156. … [Mr Fidler] produced a substantial number of bills and invoices concerning the purchase of building materials. There are also letters and statements from tradesmen who worked on the construction of the house and from friends who visited during the building period and those living at Honeycrock Farm. Taken together, along with the evidence of Mr Fidler and his family, I consider this is sufficient to show that, on the balance of probability, the house as a building was constructed by June 2002 and had been lived in for some time prior to that date. …

…

162. … In this case, the weight of evidence before me indicates that at the date the notice was served (16/02/07) the dwelling had been built and in occupation for over 4 years (ie by 16/02/03). Some internal decoration may have been outstanding and there is still no bathroom upstairs but a shower room was said by Mr Fidler to have existed by that time, and this was not contested by the Council. …

163. The main point at issue is whether the presence or otherwise of the straw bales encasing the dwelling are of relevance in terms of the consideration of the matter of substantial completion. There is no dispute that they were not part of the structure of the dwelling and would not have required the skills of a builder for erection or demolition/removal. Nevertheless, they were without doubt put there for a purpose and that was to conceal the dwelling whilst under construction and until it was considered that the legal argument on the 4 year rule would succeed. The walls of straw were not placed there by chance but were fundamentally related to the construction of the dwelling. I accept that the act of concealment does not in itself provide a legitimate basis for the Council to succeed as hiding something does not take away lawful rights that may accrue due to the passage of time. However, the ultimate intention as to whether the

walls of straw were to remain or be removed is, in my view, material having regard to the authority in *Sage*.

> 164. Mr Fidler was questioned by the Council's advocate on whether it was his intention to live behind a wall of straw bales with no outlook other than at a wall of straw and very limited amounts of natural light. He said he could have gone on living that way if need be but I consider this answer to be disingenuous. From his own evidence and submissions it was always his intention to remove the bales once he thought that lawfulness had been secured. It is therefore quite obvious he never intended to continue to live within a straw stack and until the straw was removed he could not enjoy a reasonable level of residential amenity, consistent with normal expectations of what a dwelling house should provide. It might be argued that people choose to live in caves or enclosures with little or no light or outlook. That may be so, but that was not Mr Fidler's intention. He built a house in a traditional form with large numbers of windows in the walls. If he had intended to look out on straw bales 3m away then it begs the question as to why one would go to the trouble of inserting windows at all. The presence of these windows demonstrates his intentions for outlook, not least the tall picture window in the northern elevation at both floor levels which lights the central hall/gallery area.

[10] In paras 169 to 171 of his decision letter, the inspector went on to express his conclusions with regard to the section 174(2)(d) ground of appeal in appeal 1 in the following terms:

> 169. … it was never Mr Fidler's intention to build a house which remained encased within walls of straw covered in sheeting. It was always his intention to remove the straw walls thus revealing his edifice once he thought that sufficient time had passed for the lawfulness of the construction to be secured. The day-to-day existence within the dwelling when encased by straw was seen as a temporary situation which would be endured for as long as it took to secure lawfulness. It was not a normal living environment (limited, if any, natural light; no outlook; poor ventilation) or one which was intended as a final outcome. Rather it was a situation that would be tolerated for the time being.
>
> 170. As a matter of fact and degree, I therefore find that the straw bales were part of the totality of the operations and it was necessary for them to be removed before the point of substantial completion was reached. The matter of substance is not that the bales hid the dwelling (although they certainly did) but that they formed part of the totality of the operations in the holistic sense accorded by the authority in *Sage*. The situation that existed prior to the removal of the straw bales fell short of what the appellant contemplated or intended to carry out and his intentions were not realised until removal had occurred. As this did not happen until July 2006, substantial completion did not occur until that time and this is well within the 4 year period from the date of the service of the notice.
>
> 171. I appreciate that this is a most unusual case and I am not aware of any clear authority that I can draw on which is directly comparable. *Sage* is a valuable authority on the matter of legal principles but differs in terms of the actual facts. I have interpreted and applied these principles to the best of my judgement. Having regard to the circumstances of the case, I consider that it is right to find against the appellant because the case of lawfulness is not made out on the balance of probability. Unless this test is met the appeal should not be allowed. Accordingly, having regard to the wording of s171B(1), the appeal on ground (d) fails.

[11] *Sage* v *Secretary of State for the Environment, Transport and the Regions* [2003] UKHL 22; [2003] WLR 983. As can be seen from the terms of para 171 of his decision letter, the inspector sought to interpret and apply the legal principles stated in *Sage*, while acknowledging that the facts of that case were very different. In *Sage*, the local planning authority had served an enforcement notice alleging that the erection of a dwellinghouse constituted a breach of planning control. No building work had actually been carried out on the structure during the four years preceding the service of the notice. However, the dwelling itself was not fit for habitation because the ground floor consisted of rubble, there were no service fittings or a staircase, the interior walls were not plastered and the windows were unglazed.

[12] The landowner appealed against the enforcement notice on the ground that, *inter alia*, the building operations were substantially completed more than four years before the local planning authority had taken any enforcement action (that is, relying on sections 171B and 174(2)(d) of the 1990 Act). The inspector dismissed this ground of appeal on the basis that the time limit of four years did not begin to run until the building operation involving the construction of the dwelling was substantially completed and that, as a question of fact and degree, the dwelling in question was still in the course of construction and was not "substantially completed".

[13] The landowner appealed to the High Court on the ground that, for the purpose of determining the starting point of the four-year period under section 174B of the 1990 Act, "what have to be substantially completed are those operations and works which amount to a breach of planning control and that operations and works which do not amount to development because they fall within section 55(2)(a) are not to be taken into account" (see [5]); since all the work remaining to be done on the dwelling in question was either internal work or work that did not materially affect the external appearance of the building, it was, pursuant to section 55(2)(a), work that did not amount to the development of land for which planning permission was required and that the building operation falling within the compass of section 174B(1) must therefore have been completed more than four years before enforcement action was taken by the local planning authority. The judge allowed the appeal on those grounds and the Court of Appeal upheld that decision.

[14] In allowing the local planning authority's appeal, the House of Lords rejected the argument summarised in the previous paragraph in favour of a holistic approach. Thus, in [24], Lord Hobhouse said:

> 24. The same holistic approach is implicit in decisions on what an enforcement notice relating to a single operation may require. Where a lesser operation might have been carried out without permission or where an operation was started outside the four-year period but not substantially completed outside that period, the notice may nevertheless require the removal of all the works including ancillary works.

Agreeing with Lord Hobhouse, Lord Hope put the matter in this way in [6] and [7]:

> 6. … it makes better sense of the legislation as a whole to adopt the holistic approach which my noble and learned friend has described. What this means, in short, is that regard should be had to the totality of the operations which the person originally contemplated and intended to carry out. That will be an easy task if the developer has applied for and obtained planning permission. It will be less easy where, as here, planning permission was not applied for at all. In such a case evidence as to what was intended may have to be gathered from various sources, having regard especially to the building□s physical features and design.
>
> 7. If it is shown that all the developer intended to do was to erect a folly, such as a building which looks from a distance like a complete building… but was always meant to be incomplete, then one must take the building when he has finished with it as it stands. It would be wrong to treat it as having a character which the person who erected it never intended it to have. But if it is shown that he has stopped short of what he contemplated and intended when he began the development, the building as it stands can properly be treated as an uncompleted building against which the four-year period has not yet begun to run.

Parties' submissions

[15] On behalf of Mr Fidler, Mr Hockman QC stressed that the breach of planning control alleged in the enforcement notice was "erection of a dwelling house". He also emphasised that the construction (that is, the erection) of the dwellinghouse in question was complete by the beginning of June 2002, having been occupied since October 2001.

[16] Mr Hockman submitted that the inspector had fallen into error in deciding that the erection and removal of the straw bales formed part of the overall building operations relating to the erection/construction of the new dwellinghouse. It was Mr Hockman's submission that: (i) "building operations" in section 55(1A) of the 1990 Act is exhaustively defined in subparas (a) to (d) of that subsection; (ii) the erection and removal of the straw bales was not covered by any of those subparagraphs and; (iii) therefore, those activities were neither building operations in their own right nor did they form any part of the building operations involved in the erection/construction of the new dwelling. Mr Hockman also submitted that nothing in the speeches of their lordships in *Sage* was authority for the proposition that you can take into account an activity that is not a building operation as defined (that is, the erection and the removal of the straw bales not forming part of the building), as forming part of the actual building operation itself (that is, the construction of the dwellinghouse in question).

[17] In support of his submission that the expression "building operations" is exhaustively defined in section 55(1A), Mr Hockman referred to *Dilworth* v *Commissioner of Stamps* [1899] AC 99, at p106, and contended that the context of the 1990 Act shows that the word "includes" in section 55(1A), which precedes the activities identified in subparas (a) to (d), is not merely used for the purpose of adding to the natural significance of the expression "building operations" but should be construed as equivalent to "means and includes", thus affording an exhaustive explanation of the meaning of that expression for the purposes of the Act.

[18] Mr Hockman submitted that the inspector should have directed himself by reference to the statutory definition of "building operations", but that he had singularly failed to do so. Mr Hockman referred to the second sentence of para 163 of the decision letter (see above) and submitted that the inspector's findings of fact in that sentence showed that the straw bales formed no part of the actual structure of the dwellinghouse and that this clearly placed the erection and removal of the bales outside the exhaustive definition of "building operations" in section 55(1A) because such an activity would not normally be carried out by a builder (as the inspector acknowledged). Mr Hockman suggested that the approach adopted by the inspector, in which he had failed to apply the discipline of the statutory definition of "building operations" to the material facts, necessarily led to uncertainty and arbitrariness, both of which undesirable outcomes are avoided if the statutory definition is strictly applied to the factual circumstances of the case.

[19] In the alternative, Mr Hockman submitted that if, contrary to his primary submission, the erection and removal of the straw bales did form part of the relevant building operations, the inspector had erred in law in concluding that those operations were not substantially completed until the removal of the bales. In support of that submission, Mr Hockman relied, in particular, on the following matters:(i) the terms of the enforcement notice (see [3] above);(ii) the terms of para 156 of the decision letter (see [9] above);(iii) the fact that the dwellinghouse had been built and occupied for more than four years before enforcement action was taken (see para 162 of the decision letter); and (iv) the fact that the straw bales did not form any part of the structure of the actual dwellinghouse: see para 163 of the decision letter.

[20] On behalf of the Secretary of State, Mr Paul Brown QC (supported by Mr Rupert Warren on behalf of the LPA) submitted (correctly, in my view) that Mr Fidler's case was entirely dependent on the submission that, as a matter of law, the erection and removal of the straw bales could not amount to *or form any part of* building operations within the meaning of the Act. For the reasons developed below, Mr Brown and Mr Warren both argued that this submission was wrong.

[21] Both Mr Brown and Mr Warren readily accepted that the erection and removal of the straw bales was not itself a building operation when considered in isolation. However, they submitted that such an activity could nevertheless, as a matter of fact, be found to form part of building operations when the totality of the operations, as originally contemplated and intended, is considered (applying the *Sage* holistic approach). I agree with that submission.

[22] As it seems to me, the question of whether section 55(1A)(a)-(d) is an inclusive or exhaustive definition of "building operations" is not critical to the outcome of these proceedings, although I incline to the view that Mr Brown and Mr Warren are right in their submission that the expression is not exhaustively defined in and thus limited to those matters set out in subparas (a)-(d). In my view, there is nothing in the context of the 1990 Act that would justify construing "includes" in that subsection as meaning "means and includes".

[23] More to the point, and as Mr Brown observed, on analysis it is clear that the essence of Mr Hockman's crucial submission, that the activity of erecting and removing the straw bales did not and could not form part of the overall building operations (in this case, the clandestine construction of a dwelling in breach of planning control that would have the benefit of the four-year rule), is that the activity in question had itself to be a building operation within the meaning of the Act. As Mr Brown submitted, there is nothing in the Act or in the authorities to justify such a conclusion that is, in my view, plainly wrong. I agree with Mr Brown that there can be a number of ancillary activities on a construction site that, if considered in isolation, would not be a building operation within the meaning of the Act (for example, the provision of temporary canteen facilities) but that could nevertheless form part of the contemplated and intended building operations when considered as a whole (in line with *Sage*). In each case, it is a matter of fact and degree as to whether such an activity does form part of the overall building operations. In this context, it is significant, in my view, that the definition of development in section 55(1) of the 1990 Act includes, *inter alia*, the term "operations" (a term that is capable of covering a wide range of activities relating to the actual work of building) as opposed to, say, "works" (a term that is obviously more restricted in meaning when linked to the expression "building").

[24] I therefore agree with both Mr Brown and Mr Warren that the inspector was quite right to consider the evidence and make appropriate findings of fact with regard to the totality of the building operations that Mr Fidler originally contemplated and intended to carry out. In this particular case, crucial to the inspector's later findings were his initial conclusions of fact that Mr Fidler knew perfectly well that he would not be granted planning permission for the new dwelling, that his intention was to construct it in a clandestine fashion, using a shield of straw bales to conceal it during construction and in order to deceive the LPA of its existence until he believed that he was in a position to take advantage of the four-year rule: see para 149 of the decision letter.

[25] It is true that the inspector went on to make findings that the new dwelling was constructed and occupied by June 2002 (paras 156 and 162 of the decision letter) and that the straw bales did not form part of the actual structure of the dwelling: para 163 of the decision letter. However, as it seems to me, the inspector's findings of fact that are crucial to the outcome of these proceedings appear in the following sentences of paras 163, 164 and 169 of his decision letter, which, although already quoted, bear repetition:

> 163. … [the bales] were without doubt put there for a purpose and that was to conceal the dwelling whilst under construction and until it was considered that the legal argument on the 4 year rule would succeed. The walls of straw were not placed there by chance but were fundamentally related to the construction of the dwelling. …
>
> 164. … From [Mr Fidler's] own evidence and submissions it was always his intention to remove the bales once he thought that lawfulness had been secured. It is therefore quite obvious he never intended to continue to live within a straw stack and until the straw was removed he could not enjoy a reasonable level of residential amenity, consistent with normal expectations of what a dwelling house should provide. …
>
> …
>
> 169. … it was never Mr Fidler's intention to build a house which remained encased within walls of straw covered in sheeting. It was always his intention to remove the straw walls thus revealing his edifice once he thought that sufficient time had passed for the lawfulness of the construction to be secured.

[26] The inspector's foregoing findings of fact were clearly ones that he was entitled to reach on the evidence and, as it seems to me, were findings that fully justified his critical conclusion that the erection and removal of the straw bales formed part of the totality of the building operations in question (see para 170 of the decision letter), that is, that this particular activity formed part of the totality of the building operations that Mr Fidler originally contemplated and intended to carry out: see para 6 of the speech of Lord Hope in *Sage*. As Mr Brown observed, in the light of his findings of fact, the inspector was fully entitled to find (as he in effect did) that there was such a close and intimate connection between the erection/removal of the straw bales and the construction of the dwelling as to lead to the conclusion that the former was a necessary part of the overall building operations relating to the latter. In my view, the inspector's findings of fact make it abundantly clear that the erection/removal of the straw bales was an integral, indeed an essential ("fundamentally related"), part of building operations that were intended to deceive the LPA and to achieve, by deception, lawful status for a dwelling built in breach of planning control. In my view, the approach adopted by the inspector in this case

cannot be faulted and I have no hesitation in rejecting Mr Hockman's submissions to the contrary effect.

[27] So far as concerns Mr Hockman's alternative submission (see [18] above), I am satisfied that there is nothing in this particular argument. In the light of his findings of fact, it was a matter of judgment on the part of the inspector as to whether substantial completion of the operations did not occur until the removal of the straw bales. Nothing that Mr Hockman submitted about this aspect of the matter persuades me that the inspector's judgment is even arguably wrong. In fact, as it seems to me, and in all the circumstances of this case, the inspector was plainly right to reach the conclusion that he did.

Conclusion

[28] For all the foregoing reasons, I have come to the firm conclusion that this appeal must be and is hereby dismissed.

Appeal dismissed.

Welwyn Hatfield Borough Council *v* Secretary of State for Communities and Local Government and another

COURT OF APPEAL

[2010] EWCA Civ 26

29 January 2010

PILL, MUMMERY AND RICHARDS LJJ

Estates Gazette 3 April 2010

[2010] 13 EG 84

Town and country planning — Lawful use certificate — Residential use — Enforcement action — Limitation periods — Town and Country Planning Act 1990 — Planning permission granted for erection of hay barn for agricultural storage — Building fitted out and used as dwelling — Section 171B(1) and (2) of 1990 Act — Whether construction of building and its residential use immune to enforcement action after four years — Whether owner entitled to lawful use certificate

In December 2001, the respondent council granted planning permission to the second appellant to erect a hay barn to be used for agricultural storage. The second appellant had deceived the respondents in his planning application; although the building looked like a barn from the outside, it was fitted out internally as a three-bedroom dwelling, with all utilities, and the second appellant had always intended to use it as such.

In August 2006, he applied to the respondents for a certificate of lawfulness of existing use, on the ground that the four-year time limit for taking enforcement action in respect of a change of use to a single dwellinghouse, under section 171B(2) of the Town and Country Planning Act 1990, had expired. He claimed to have moved into the building in August 2002 and to have lived there continuously thereafter, although he had sought to avoid the residential use from coming to the respondents' attention.

The respondents refused the certificate on the ground that the building was not a dwelling, since its external appearance lacked the features of such, and that the applicable time limit was therefore not the four-year period applying to dwellings but the 10-year period under section 171B(3) of the 1990 Act. The second appellant's appeal to the first appellant was allowed by an inspector, but that decision was subsequently reversed by a High Court judge, who held that no "change of use" had taken place to which section 171B(2) could apply because the building had not been used other than as a dwelling.

The appellants appealed. In addition to section 171B(2), they relied on the four-year limit under section 171B(1) for enforcement against the carrying out of building operations without planning permission, arguing that the initial construction of the building had breached planning control as being the construction of a dwelling rather than the hay barn for which permission had been granted.

***Held*: The appeal was allowed.**

(1) The second appellant was entitled to the lawful use certificate for which he had applied. The relevant planning permission permitted the building to be used for agricultural storage only. Its use as a single dwellinghouse was therefore to be regarded, for the purpose of section 171B(2), as a change of use in breach of planning control. The application of section 171B(2) did not depend on establishing actual use for agricultural storage for a period prior to the residential use. Accordingly, the four-year time limit applied in respect of the residential use. The courts should not adopt a strained construction of the section in reaction to a deliberate deceit practised by an applicant or out of concern for the difficulties that such conduct created for local planning authorities in enforcing planning control. (2) The physical and design features of the second appellant's building were those of a dwelling, not a hay barn, and it was properly classified as such for planning purposes. Therefore, the building operation carried out by the second appellant did not, internally and externally, fully accord with the planning permission for a hay barn and the building had been constructed in breach of planning control. The respondents could no longer take enforcement action in respect of the construction of the building because the four-year time limit under section 171B(1) had expired.

***Per curiam*: An issue remained as to whether an immunity from enforcement, in respect of the construction of a building under section 171B(1) of the 1990 Act, also conferred immunity on its use for the purpose for which it was designed such as to entitle the owner to a lawful use certificate in respect of the residential use of the building.**

The following cases are referred to in this report.

Arun District Council v *First Secretary of State*; *sub nom R (on the application of Arun District Council)* v *First Secretary of State*; *First Secretary of State* v *Arun District Council* [2006] EWCA Civ 1172; [2007] 1 WLR 523; [2007] 1 P&CR 10; [2007] JPL 237

Murfitt v *Secretary of State for the Environment* (1980) 40 P&CR 254; [1980] JPL 598, QB

R (on the application of Welwyn Hatfield Council) v *Secretary of State for Communities and Local Government* [2009] EWHC 966 (Admin)

R v *Westminster City Council, ex parte Monahan* [1990] 1 QB 87; [1989] 3 WLR 408; [1989] 2 All ER 74; (1989) 58 P&CR 92; [1989] 1 PLR 36; [1989] JPL 107, CA

Sage v *Secretary of State for the Environment, Transport and the Regions*; *sub nom Sage* v *Maidstone Borough Council* [2003] UKHL 22; [2003] 1 WLR 983; [2003] 2 All ER 689; [2003] 2 P&CR 26; [2003] 1 PLR 121; [2003] JPL 1299

Westminster City Council v *Great Portland Estates plc*; *sub nom Great Portland Estates plc* v *Westminster City Council* [1985] AC 661; [1984] 3 WLR 1035; (1984) 50 P&CR 34; [1985] JPL 108, HL

This was an appeal by the appellants, the Secretary of State for Communities and Local Government and Alan Beesley, from a decision of Collins J, sitting in the Administrative Court, allowing a claim by the respondents, Welwyn Hatfield Council, to quash the grant by the first appellant to the second appellant of a certificate of lawfulness of existing use in respect of the residential use of a building.

Sarah-Jane Davies (instructed by the Treasury Solicitor) appeared for the first appellant; Alexander Booth (instructed by Perrins, of Harpenden) appeared for the second appellant; Wayne Beglan (instructed by the legal department of Welwyn Hatfield Borough Council) represented the respondents.

Giving the first judgment, **Richards LJ** said:

[1] This case concerns the time limits under section 171B of the Town and Country Planning Act 1990 (the 1990 Act) for enforcement action in respect of breaches of planning control.

[2] In December 2001, Mr Alan Beesley was granted planning permission by Welwyn Hatfield Council for the erection of a hay barn (the development proposed in the planning application and shown on the accompanying plan). The permission was subject to a condition that "the building hereby permitted shall be used only for the storage of hay, straw or other agricultural products and shall not be used for any commercial or non-agricultural storage purposes".

[3] Mr Beesley has admitted that he deliberately deceived the council when he applied for planning permission, in that he filled in the application on the basis that the building was to be a hay barn but he had always intended to build and reside in it as a dwelling. The building was constructed between January and July 2002 with a steel frame and the external appearance of a barn, but fitted out internally as a dwelling. The external walls have chipboard with fibreglass insulation and are finished internally with fibreglass. The internal walls are made of timber and plasterboard. There are a garage/store, entrance hall, study, lounge, living room, kitchen, WC, storeroom, gym and three bedrooms, including two with en-suite bathrooms. The building does not have conventional windows, but roof lights and ceiling openings allow natural light to penetrate into all but two rooms. The property is connected to mains electricity, water and drainage and has heating, a waste disposal unit and a domestic telephone line.

[4] Mr Beesley's case was that he and his wife moved into the property on 9 August 2002 and lived there continuously thereafter. He was evidently careful, however, to avoid the residential use of the building from coming to the attention of the council. Building regulations approval was not sought and there was no record of him on the electoral register.

[5] On 15 August 2006, he applied, under section 191 of the 1990 Act, for a certificate of lawfulness of existing use on the basis that the time for enforcement action against the use of the building as a dwelling had expired.

[6] The time for enforcement action is governed by sections 171A and 171B of the 1990 Act, which provide in material part as follows:

> 171A(1) For the purposes of this Act
>
> (a) carrying out development without the required planning permission; or
>
> (b) failing to comply with any condition or limitation subject to which planning permission has been granted,
>
> constitutes a breach of planning control.
>
> ...
>
> 171B(1) Where there has been a breach of planning control consisting in the carrying out without planning permission of building, engineering, mining or other operations in, on, over or under land, no enforcement action may be taken after the end of the period of four years beginning with the date on which the operations were substantially completed.
>
> (2) Where there has been a breach of planning control consisting in the change of use of any building to use as a single dwellinghouse, no enforcement action may be taken after the end of the period of four years beginning with the date of the breach.
>
> (3) In the case of any other breach of planning control, no enforcement action may be taken after the end of the period of ten years beginning with the date of the breach.

I should also mention section 55(1), which provides that "except where the context otherwise requires, 'development' means 'the carrying out of building, engineering, mining or other operations in, on, over or under land, or the making of any material change in the use of any buildings or other land'".

[7] The focus of attention in the council's consideration of Mr Beesley's application for a certificate of lawful use was section 171B(2). The council reasoned that only use as a single dwellinghouse benefited from the four-year (as opposed to 10-year) time limit for enforcement; the building was not a dwellinghouse, since its external appearance did not have the features of a dwelling; and the four-year limit therefore did not apply. On that basis, the council refused a certificate.

[8] Mr Beesley appealed, under section 195 of the 1990 Act, against the council's refusal. By a decision dated 8 July 2008, an inspector appointed by the Secretary of State allowed the appeal. He said that the breach of planning control concerned the use of the building as a single dwellinghouse and that it was clear from section 171B(2) that in all such cases the relevant time limit was four years. He stated the issue as being whether there had been a breach of planning control by the use of the building as a single dwellinghouse and, if so, whether that breach had continued without material interruption throughout a period exceeding four years before the date of the application for a certificate. He considered it unnecessary to determine whether the building *was* a dwellinghouse, and he expressly avoided any determination in respect of its status as a dwellinghouse. He found, however, that it was *capable of being used* as a dwellinghouse and that its actual use as a single dwellinghouse had commenced more than four years before the date of the application and had continued without material interruption up to the date of that application. On that basis, he granted a certificate that, on 15 August 2006, the use of the building as a dwellinghouse had become lawful by reason of immunity from enforcement action.

[9] The council applied to the High Court, under section 288 of the 1990 Act, challenging the inspector's decision. The case was heard by Collins J: [2009] EWHC 966 (Admin). The council contended, first, that the inspector's finding that the building had been used as a single dwellinghouse for four years was flawed on a variety of grounds. Those arguments were rejected by the judge and are not pursued before us.

[10] However, the council also took a new point, namely that there could be no "change of use" to use as a single dwellinghouse because the building had no use prior to its occupation for residential use, and thus section 171B(2) could not apply. Collins J accepted that contention. He held, in summary, that the construction of the building was lawful and not in breach of planning control, and, since there was never any intention to use it otherwise than as a dwellinghouse and it had never been used for any purpose other than a dwellinghouse, there had been no change of use ("it was constructed for that use and has ever since been used in such a way"). Accordingly, the four-year time limit did not apply.

[11] Although the judge arrived at that conclusion as a matter of construction of section 171B and after consideration of the cases cited to him, he had serious concerns about the deceit practised by Mr Beesley and was understandably relieved to arrive at a conclusion that would prevent Mr Beesley from profiting from that deceit. He said, in [36]:

> I am bound to say that seems to me a conclusion which accords with not only the merits of this case but prevents a particular sort of fraud being perpetrated because otherwise Mr Beesley would have got away with a plot to breach the planning laws and to obtain a dwelling house in breach of the Green Belt policies and a development that he would never have been able to achieve if he had gone about things in an honest fashion.

He also raised the question of whether a criminal offence had been committed in obtaining a planning permission by deception.

[12] Both the Secretary of State and Mr Beesley have appealed against the judge's order, in each case with permission granted by Elias LJ. The arguments presented by Ms Sarah-Jane Davies, for the Secretary of State, and by Mr Alexander Booth, for Mr Beesley, are broadly to the same effect, albeit with differences of emphasis, and I can treat them together. There are two main aspects to the case advanced by them:

(i) First, it is said that, contrary to the judge's finding, the building was built in breach of planning control. The planning permission was for the erection of a hay barn, whereas what was erected was not a hay barn but a dwellinghouse: the building had not been built in accordance with the planning permission. The case therefore falls squarely within section 171B(1) and the four-year time limit for enforcement action applies.

(ii) Second, and irrespective of whether the construction of the building was in breach of planning control, there was a change of use of the building to use as a single dwellinghouse: the use permitted by the planning permission was use as a hay barn, but Mr Beesley used the building as a dwellinghouse. The judge was wrong to find that because the building had never been used, in fact, for any other purpose than as a dwellinghouse, there had been no change of use. Thus, the case falls within section 171B(2) and the four-year time limit for enforcement action again applies.

[13] Mr Wayne Beglan, for the council, met that case in this way. He submitted that: (i) section 171B(1) is not relevant, since it is concerned with operational development, whereas the certificate sought relates to *use* of the building; (ii) the judge was right to find that there had been no change of use of the building and that section 171B(2) did not therefore apply; but (iii) it would be open to the council to enforce against the use of the *building* on the ground that it constitutes a material change of use of the *land*, which is subject to a 10-year time limit under section 171B(3). The council's argument based on material change of use of the *land* is a new one, not taken before the inspector or, at least in so clear a form, before Collins J.

[14] Before I consider those submissions, I should refer to the policy behind section 171B. In that connection, we had cited to us *Arun District Council* v *First Secretary of State* [2006] EWCA Civ 1172; [2007] 1 WLR 523. The facts of that case were that planning permission had been granted for an extension to a dwellinghouse to provide accommodation for a dependant relative. A condition was attached that were the extension to be vacated, it was to be used only for purposes incidental to the main dwellinghouse. In breach of that condition, however, the extension was used as a separate dwelling. The court held that the expression "breach of planning control" in section 171B(2) applied both to development without planning permission and to the failure to comply with a condition subject to which planning permission was granted, and that the four-year time limit therefore applied to a change of use (whether or not material) to use as a single dwellinghouse in breach of condition. In reaching that conclusion, the court relied heavily on the legislative history.

[15] Auld LJ, giving the leading judgment, commented that: "The rationale for the different periods of limitation according to the nature of the breach of planning control was to impose a shorter period, four years, where it was considered that a longer period could cause serious loss and/or hardship in the event of enforcement proceedings long after the event, including in the case of homeowners the loss of their homes, and ten years for any other breach of planning control": see [5]. In [18], he quoted the following passage from the February 1989 report by Mr Robert Carnwath QC (as he then was), *Enforcing Planning Control*, as summarising the effect of the law as it stood at that date:

> In four specific cases, there is immunity four years after the breach (all operations; breaches of conditions "relating to the carrying out of operations"; changes of use to single dwellings; and breaches of a condition prohibiting such a change)… The logic behind these exclusions is not entirely clear. Special protection was no doubt thought desirable for people's homes. In the case of operations, the governing considerations presumably were the relative ease of detection, the potential costs involved in reinstating the land, and the need to provide certainty for potential purchasers.

The report recommended that the four-year immunity be retained for all those categories save for breaches of conditions relating to operational development. As Auld LJ explained in [21], the recommendation was implemented by the Planning and Compensation Act 1991, which introduced sections 171A and 171B into the amended 1990 Act, effecting in simpler form (and save for the omission of breaches of conditions relating to operational development) the relevant provisions of the predecessor legislation. A similar analysis of the history is to be found in the judgment of Carnwath LJ (as he had by then become) in [38] to [48].

[16] It is clear from the legislative history as examined in both judgments that a deliberate and informed decision was taken to retain a four-year limitation period for the categories of case now to be found in section 171B(1) and (2), and that one of the policy reasons was the desirability of affording special protection for people's homes.

[17] In agreeing in *Arun* that any enforcement action had to be taken within the four-year time limit in section 171B(2), Sedley LJ said, in [36]:

> I can entirely understand the local planning authority's sense of frustration about this. Their planning department is not a police station, and the discovery that a person such as Mrs Brown has not to put too fine a point on it cheated on a conditional grant of permission, to the detriment of her neighbours and of planning control, may well be a matter of time and of chance. The ordinary ten-year period might well have been thought reasonable for such cases, but in circumstances which Carnwath LJ's judgment illuminates it is not what Parliament decided to provide.

[18] It seems to me that a similarly restrained approach is called for in the present case. The court should not be tempted to adopt a strained construction of the section in reaction to the deliberate deceit practised by Mr Beesley or out of concern for the difficulties that such conduct creates for local planning authorities in enforcing planning control. The outcome should be the same, as though, for example, there had been a genuine change of mind in the course of construction of a building for which planning permission had been obtained in good faith. The question is whether the situation, viewed objectively, is one for which the statute has provided a four-year time limit or a 10-year time limit. If it is considered that there should be a different outcome in a case of dishonesty or deliberate concealment, it is for parliament to amend the legislation accordingly.

[19] That brings me back to the ways in which the case for the appellants was put. The first argument, as I have said, is that what was erected was not a hay barn but a dwellinghouse and that this was not in accordance with the planning permission: in terms of section 171B(1), the erection of something other than a hay barn constituted the carrying out of building operations without planning permission, so that the four-year time limit applied. The inspector did not think it necessary to make a finding on this issue.

[20] In support of their argument, the appellants referred the court to *Sage* v *Secretary of State for the Environment, Transport and the Regions* [2003] UKHL 22; [2003] 2 All ER 689, a case under section 171B(1). The local planning authority in that case had served an enforcement notice on the ground that the claimant was in breach of planning control in partially erecting a dwellinghouse. The claimant contended that the notice had been served outside the four-year time limit on the basis that "the date on which the operations were substantially completed" within the meaning of the subsection was the date after which the building work remaining to be done would no longer itself involve a breach of planning control because, if taken on its own, it would not require planning permission. The House of Lords rejected that contention, holding that a holistic approach should be adopted and that regard should be had to the totality of the operations that the person originally contemplated and intended to carry out (in that case, the creation of a dwellinghouse). The particular issue was different from that arising in the present case, but certain aspects of the analysis are pertinent.

[21] Lord Hobhouse of Woodborough, giving the leading judgment, said that there could be no doubt that the underlying purpose of section 171B(1) was "to introduce a single easily-applied limitation period for operations": [10]. In respect of one of the issues before the inspector, as to whether the building in question was an agricultural building that had not in the circumstances required planning permission, Lord Hobhouse observed:

> 14. The inspector rightly did not investigate the intentions of Mr Sage at various stages in the history nor the uses he had made of the structure from time to time. The character and purpose of a structure falls to be assessed by examining its physical and design features. The relevance of the assessment is to determine whether or not the building operation is one requiring planning permission. The actual use made of the building does not alter the answer to be given. Keeping a pig in the sitting-room or hens in the kitchen does not turn a dwelling house into an agricultural building even if the humans move out.

> Permission for a change of use may have to be applied for but that would be a separate question.

He agreed with the inspector's conclusion that having regard to its layout and appearance, the building was not an agricultural building but was best described as a dwellinghouse in the course of construction.

[22] Later in his speech, when considering the central issue in the case, Lord Hobhouse stated:

> 23. When an application for planning consent is made for permission for a single operation, it is made in respect of the whole of the building operation. There are two reasons for this. The first is the practical one that an application for permission partially to erect a building would, save in exceptional circumstances, fail. The second is that the concept of final permission requires a fully detailed building of a certain character, not a structure which is incomplete. This is one of the differences between an outline permission and a final permission… As counsel for Mr Sage accepted, if a building operation is not carried out, both externally and internally, fully in accordance with the permission, the *whole* operation is unlawful. She contrasted that with a case where the building has been completed but is then altered or improved. This demonstrates the fallacy in Mr Sage's case. He comes into the first category not the second.

[23] Lord Scott of Foscote, although in general agreement with Lord Hobhouse, went further and questioned whether the building in the case was in fact an uncompleted dwellinghouse. He said that the classification of a building for planning purposes can change if the use to which the building is put changes; and the classification of a building in course of construction can also be changed by use, or by intentions for future use, of the uncompleted building inconsistent with its original classification. He gave the following example:

> 39. For example, under the Town and Country Planning General Development Order 1988…, planning permission is in general not necessary for the erection of a building which is reasonably necessary for the purposes of agriculture. A farmer who commenced the construction of such a building would not, by doing so, be in breach of planning control. But if, before the building operations were complete, his intentions changed and he began to install a bathroom and other features indicative of a dwelling, the operations would be in breach of planning control.

[24] The passages in *Sage* that I have quoted have a direct bearing on the present case. Looked at as a whole, the physical and design features of what was built by Mr Beesley were those of a dwellinghouse, not a hay barn. Its character and purpose, and its proper classification for planning purposes, were those of a dwellinghouse. Yet the planning permission was for the erection of a hay barn. If one asks whether the building operation carried out was, both externally and internally, fully in accordance with the planning permission, the answer is clearly "no": what was created internally was not on any sensible view a hay barn. The internal fitting-out of the building, with the rooms and features of a dwellinghouse, meant that it had been built in breach of planning control.

[25] It follows that the construction of the building fell within section 171B(1) and that the four-year time limit under that subsection applied. Since, on the inspector's findings, the building had been completed more than four years before Mr Beesley applied for a certificate, it was too late for the council to take enforcement action against the operational development constituted by the construction of the building.

[26] The certificate applied for by Mr Beesley related, however, to the use of the building, not to its construction. For the council, Mr Beglan submitted that section 171B(1) is simply irrelevant to the question of use. For the Secretary of State, Ms Davies countered with the contention that if the construction of a building has become immune from enforcement by virtue of section 171B(1), use of the building for the purpose for which it was designed must also be lawful: otherwise, the legislation would make no sense. I see some attraction in that contention, but I am concerned that the relationship between operational development and use was not fully explored before us and I prefer not to express a conclusion on the issue because, for reasons to which I now turn, I take the view that section 171B(2) provides immunity for the use of the building in this case and no decision is therefore needed on the outstanding issue under subsection (1).

[27] The appellants submit that the case falls within section 171B(2) because there was a breach of planning control consisting in a change of use of the building to use as a single dwellinghouse. The planning permission was for the erection of a hay barn and was conditional on the building being used for the storage of hay, straw and other agricultural products. Section 75(2) and (3) of the 1990 Act provides that where planning permission is granted for the erection of a building, the grant of permission may specify the purposes for which the building may be used and, if no purpose is specified, the permission shall be construed as including permission to use the building for the purpose for which it is designed. In this case, whether one looks at what was implicit in the grant of permission for a hay barn or at the express terms of the condition as to use, it is plain that the permitted use of the building was for agricultural storage. Mr Beesley, however, did not use the building for agricultural storage. He used it as a dwellinghouse. That was a different use from the permitted use.

[28] Up to that point there is no difficulty in the argument. Collins J found, however, that since there was never any intention to use the building otherwise than as a dwellinghouse, and since it had never been used in fact for any other purpose than a dwellinghouse, there had been no change of use of the building. The council relies strongly on that reasoning. Mr Beglan submitted that the *building* had only one use, namely that of a dwellinghouse from the time of its completion, and there could not therefore have been a change of use of the building. Use of the building for residential purposes did constitute a material change of use of the *land* (or very probably did so the inspector made no finding of fact as to the prior use of the land); but section 171B(2) is deliberately limited to change of use of a *building*, and if there were a material change of use of the land but not of the building the situation fell within section 171B(3) and was subject to the 10-year time limit for enforcement provided for by that subsection.

[29] I do not accept the council's submissions on this issue, and I respectfully take a different view on the issue from that taken by Collins J below. The building in question was permitted to be used only for agricultural storage. In those circumstances, as it seems to me, its use as a single dwellinghouse is properly to be regarded, for the purposes of section 171B(2), as constituting a change of use in breach of planning control. I do not think that the application of the subsection can depend on establishing actual use for agricultural storage for a period, however short, prior to the residential use: use of the building for a purpose other than the permitted use is sufficient. An alternative way of looking at it, but with the same result, is that in the short period between completion of the building and its residential occupation the building had no use, and the change from no use to use as a single dwellinghouse constituted a change of use to which the subsection applied. A finding that the situation falls within section 171B(2) and is subject to the four-year time limit provided for by that subsection accords with the statutory policy of affording special protection for people's homes. It also avoids an artificiality, as I consider it to be, in the distinction that the council draw in this context between use of the building and use of the land.

[30] If, as I have found, the situation falls within section 171B(2), the council's reliance on section 171B(3) must fail. The plain legislative intention is that once the four-year time limit is found to apply, it displaces the 10-year time limit even if the situation could be analysed by another route as one to which the longer time limit also applied. I did not understand Mr Beglan to argue otherwise.

[31] I want, however, to say a little more about the case advanced by the council under section 171B(3). I have explained that the case depended on the proposition that use of the building as a dwelling did not constitute a change of use of the building but did constitute a material change of use of the land. Mr Beglan submitted that this provided a basis on which the council could enforce against the use of the building, which would be sufficient to disentitle Mr Beesley to the certificate for which he applied. However, Mr Beglan's argument went further than that. He submitted, in reliance on *Murfitt* v *Secretary*

A *of State for the Environment* (1980) 40 P&CR 254, that an enforcement notice based on a material change of use of the land could require not only the cessation of the residential *use* of the building but also the *removal* of the building itself, even though the construction of the building was immune from direct enforcement action by virtue of section 171B(1).

[32] I am very doubtful about that elaboration of the council's argument. *Murfitt* was a very different case, in which there had been a material change of use of land to use for the parking of heavy goods vehicles in connection with a haulage business, and the enforcement notice required the cessation of that use and the removal of hardcore B that had been placed on the site for the purpose of parking the vehicles. In rejecting a submission that the placing of the hardcore was operational development immune from enforcement action by reason of the four-year time limit, the Court of Appeal plainly accepted that the hardcore was so integral to the use of the site for the parking of vehicles that it could not be considered separately from the use, or that it was properly to be regarded as ancillary to the use being enforced against. I do not think that similar reasoning can be applied to the building in question here, and I would be reluctant, in any event, to accept that an enforcement notice directed against use of the land could properly require the removal of a building that enjoys immunity from C enforcement by virtue of section 171B(1). However, it is unnecessary for me to say anything more on the point, both because of my finding that the council's basic case under section 171B(3) must fail and because Mr Beglan made it clear that the council would want to enforce against the residential use of the building even if they could not secure the removal of the building itself.

[33] In conclusion, for the reasons that I have given, I take the view that, in this case, there was a breach of planning control consisting in the change of use of the building to use as a single dwellinghouse within section 171B(2). On that basis, a four-year time limit applied and Mr Beesley was entitled to the certificate of lawfulness of existing use D for which he had applied.

[34] I would therefore allow this appeal and restore the inspector's decision.

[35] If my analysis is correct, the lesson for local planning authorities is clear. When checking whether a building has been built in accordance with planning permission and is being used in accordance with the permitted use, they need to look carefully at the *inside* of the building and not just at the exterior. External appearances can be highly misleading, as this case shows, and authorities need to be alert to the possibility of deception. The legislation in its existing form is open to abuse. Whether it should be amended so as to prevent dishonest E advantage being taken of the shorter time limit under section 171B(1) and (2) is, as I have said, a matter for parliament.

[36] I have had the benefit of reading in draft the judgment of Mummery LJ. For my part, I doubt whether, under the legislation as it stands, the false representations made in Mr Beesley's application for *planning permission*, reprehensible though they were, could have the effect of disentitling him to a certificate of lawfulness of existing use. Anyone can apply for a certificate under section 191 and, if a certificate is granted, it affects the status of the land and can be relied on by third parties. It seems to me that the question of whether the statutory conditions for the grant of a certificate are fulfilled F must be given the same answer whoever is making the application: entitlement depends on objectively ascertainable facts concerning the building/land and its use rather than on the personal position or conduct of the particular applicant. It is far from obvious how dishonesty by Mr Beesley in his application for planning permission could be relied on to defeat a third-party application for a certificate; and if a third party would be entitled to a certificate, I cannot see why Mr Beesley should not also be entitled to one. It may also be relevant that the legislation makes specific provision in respect of dishonesty in an application for the *certificate* itself: section 193(7) confers a power to revoke a certificate if, on the application for a certificate, a statement was made or document used that was false in a material particular or material information was withheld; and section 194 makes it a criminal offence to provide false or misleading information, etc, for the purpose G of procuring a decision on an application for a certificate. There is, however, no suggestion that Mr Beesley's dishonesty extended to his application for the certificate or his evidence in support. I mention those points in reaction to the issues canvassed by Mummery LJ. I agree with him, however, that since they were not raised or explored in argument at any stage of this litigation, it would not be right to decide this appeal by reference to them.

Giving the second judgment, **Mummery LJ** said:

[37] I agree with Richards LJ that, on the basis of the legal arguments advanced in this court, the appeal by the Secretary of State H should be allowed.

[38] It is a surprising outcome that decent law-abiding citizens will find incomprehensible: a public authority, deceived into granting planning permission by a dishonest planning application, can be required by law to issue an official certificate to the culprit consolidating the fruits of the fraud. If a public authority behaved in a deceitful way, their planning decisions would rightly be set aside by the court as an unlawful abuse of power. If, however, the public authority have been deceived, it seems that they can be required, at the end of the requisite four-year period, to provide the person who deceived them with a certificate of lawfulness. J

[39] In his applications for planning permission, Mr Beesley made the following representations to the council: he proposed to build a new hay barn on land at Cooper's Lane Road, Northaw; that was the purpose for which the land and the buildings were to be used; the proposal did not involve any change of use of land; and provision for sewage disposal was "N/A" (not applicable) to the hay barn. The council granted planning permission for the proposed building on the express condition that it should be used only for the storage of hay, straw or other agricultural products. Since the site was in the metropolitan green belt, only agricultural use was appropriate.

[40] The representations were not true. They were intended to K deceive and did deceive the council into granting planning permission for a new hay barn on the site. Unknown to the council, Mr Beesley always intended to construct on the site a building for use as a single dwellinghouse. Externally and consistently with the dishonest scheme, the building was disguised to look like a hay barn. Internally, it was built and fully fitted-out as a single dwellinghouse connected to main drainage. Building regulations approval was not obtained. Domestic rates were not paid. Mr Beesley never intended to implement, and never in fact implemented, the planning permission granted by the council, since the building was not a hay barn within the meaning of the permission granted. He never in fact used the building as a hay barn. He L only ever used it as a single dwellinghouse.

[41] After four years of living in his pretend hay barn with its concealed living quarters and without storing any hay, straw or other agricultural products in it, Mr Beesley applied to the council that he had duped for a certificate of lawfulness for an existing use, that is, unlawful occupation as a single dwellinghouse for four years. The basis of the application was that that use was a breach of planning control dating from 9 August 2002, when he moved in, and that the four-year time limit for enforcement barred legal action by the council against him.

[42] The council's refusal of his application does not surprise me. Richards LJ has set out the history of the subsequent proceedings M before the inspector and Collins J, the relevant statutory provisions and case law and his conclusions. In the matter of the construction of section 171A(1) and section 171B(1), (2) and (3), I am driven to the same conclusions as Richards LJ. There has been a breach of planning control in failing to comply with the condition or limitation subject to which planning permission had been granted; that breach consisted in the change of use of the building from permitted use as a hay barn to use as a single dwellinghouse; no enforcement action could be taken by the time the true situation was disclosed because more than four years had passed from the date of the breach of planning control.

[43] I am puzzled by the total absence of argument from the council (or the Secretary of State) concerning the effect of Mr Beesley's

reprehensible conduct in obtaining planning permission by deception and in failing to implement it. Does such conduct not have any effect on his right to assert, as against the council deceived by him, that the time limits in section 171B(1) and (2) bar the council's rights of enforcement of his breaches of planning control?

[44] We have not been referred to anything in the planning legislation that expressly precludes Mr Beesley from relying on the time limits by reason of his false representations, so I assume that there is no such provision. There are, however, areas of fundamental public policy, general principles of law and justice and established legal doctrines inhabiting the vast legal world outside the Planning Acts. In appropriate cases, the policies, principles and doctrines in the larger legal world can have an effect on the interpretation and operation of legislation, unless parliament has expressly or impliedly excluded their application.

[45] I recognise that, while aiming to protect the interests of potential purchasers of property and to eliminate the hardship of enforcement proceedings long after the event, the limitation provisions in question confer benefits on people who are in breach of planning control. However, it is very difficult to believe that parliament could have intended that the certificate procedure in section 171 should be available to someone who has dishonestly undermined the legislation by obtaining a planning permission that would not have been granted had the council been told the truth.

[46] It is, in general, legitimate for the courts to interpret and apply legislation, if possible, in harmony with fundamental aspects of public policy and principles of law and justice. Take just a few examples. As a matter of public policy, the courts have not allowed statutory provisions to be used to accomplish or cover fraud or, as it is put in the cases going back 300 years, as "an engine of fraud". As to the effects of illegality, the courts will not in general enforce rights with the object of deceiving public authorities in their functions, such as the collection of tax. There is a general principle that man should not be entitled to rely on illegal conduct to make a claim or be allowed to profit from his own wrong; there is the maxim that "fraud unravels everything"; and there is the doctrine, which has been applied to interests in and uses of land, that a person who has made representations on which another has acted may, as against the person to whom the representations were made, be bound by his representations and estopped from asserting the contrary.

[47] None of those matters has been raised or explored in argument at any stage in this litigation. It would not be right to decide this appeal on the basis of arguments that have not been made or to express views on them that are unnecessary for the disposal of this appeal. The arguments may be raised in a future case. The points may be relevant if and when amendments to the legislation are considered.

Giving the third judgment, **Pill LJ** said:

[48] I agree that the appeal should be allowed for the reasons given by Richards LJ. I agree with his penetrating analysis of the legislative scheme. Immunity for the use of the building is provided by section 171B(2) of the 1990 Act, as amended.

[49] I, too, see force in the contention of Ms Davies, recited by Richards LJ in [26]. However, a distinction is maintained in section 171B between building and other operations, on the one hand, and a change of use, on the other. That is also maintained in the interpretation section in the 1990 Act section 336(1) in which it is stated that "use" in respect of land does not include the use of the land for the carrying out of any building or other operations on it. I would leave over the argument that although enforcement proceedings cannot be taken against the dwellinghouse as a building because of section 171B(1), its use as a dwellinghouse can be prevented. I agree with Richards LJ that it is not necessary to decide the point.

[50] In his judgment, Mummery LJ refers powerfully to "fundamental aspects of public policy and principles of law and justice" that inhabit "the vast legal world outside the Planning Acts". Respectfully, I acknowledge the force of his strictures. However, although I am not comfortable in the role of an apologist for those Acts, a role to which I do not aspire, the Acts provide a scheme for the control of the development of land by building and other operations and by changes of use. Under the scheme, decisions as to the use being made of land and buildings need to be taken objectively on the facts and, for the purposes of section 171B, I see no place for an analysis of the morality of a particular occupier.

[51] In *Westminster City Council* v *Great Portland Estates plc* [1985] AC 661, Lord Scarman, stated, at pp669G-670A:

> In the course of his judgment, with which the other members of the court agreed, Lord Parker CJ said, at p491, [*East Barnet Urban District Council* v *British Transport Commission* [1962] 2 QB 484] that when considering whether there has been a change of use "what is really to be considered is the character of the use of the land, not the particular purpose of a particular occupier". These words have rightly been recognised as extending beyond the issue of change of use: they are accepted as a statement of general principle in the planning law.

Lord Scarman did accept, at p670E-G, a limitation on that statement:

> Personal circumstances of an occupier, personal hardship, the difficulties of businesses which are of value to the character of a community are not to be ignored in the administration of planning control. It would be inhuman pedantry to exclude from the control of our environment the human factor. The human factor is always present, of course, indirectly as the background to the consideration of the character of land use. It can, however, and sometimes should be given direct effect as an exceptional or special circumstances. But such circumstances, when they arise, fall to be considered not as a general rule but as exceptions to a general rule to be met in special cases. If a planning authority is to give effect to them, a specific case has to be made and the planning authority must give reasons for accepting it.

A broad view of what are "material considerations" may sometimes be permissible when planning decisions are taken: for example, *R* v *Westminster City Council, ex parte Monahan* [1990] 1 QB 87. However, Lord Scarman does not appear to me to have in mind in that paragraph a power in local planning authorities to override section 171B for misbehaviour.

[52] Richards LJ has considered, in [15] and [16], the background to section 171B. In his report in 1989, Mr Robert Carnwath QC (as he then was), when reviewing the existing law, stated that "special protection was no doubt thought desirable for people's homes". The "need to provide certainty for potential purchasers" was also recognised. These factors provide the background to section 171B and to the requirement to construe the section in the manner stated by Richards LJ. If exceptions to prevent someone benefiting from blatant deceptions such as those practised by Mr Beesley in this case are to be devised, it would best be done through the statutory scheme, in my view. On the present scheme, Mr Beesley is entitled to a certificate of lawful use.

[53] I add a further comment. The inspector did not expressly find that the building was a dwellinghouse, confining himself to a finding that "the building is capable of being used as a dwelling house". That was sufficient to justify the conclusion he reached but, in my view, on the facts found by him and applying the principle stated by Lord Hobhouse in *Sage* v *Secretary of State for the Environment, Transport and the Regions* [2003] UKHL 22; [2003] 1 WLR 983, in [15], the building plainly was a dwellinghouse. "The character and purpose of a structure falls to be assessed by examining its physical and design features": cited by Richards LJ, in [21].

[54] Part VII of the 1990 Act confers extensive powers on local planning authorities to take enforcement action in respect of breaches of planning control. The ruse practised by the applicant in this case was such that the breach may not have been readily detectable. It was practised not in obtaining the certificate of lawful use but in the duplicity and design ingenuity deployed at an earlier stage. I am conscious of the many demands on the resources of local planning authorities, but there are ways in which they can protect their position and that of those they represent. Some inspection of buildings, internal as well as external, for which planning permission has been granted would appear to be appropriate on or before the completion of the building. A power to require information is provided (section 171C) and rights of entry: sections 196A and 196B.

A [55] Moreover, Mr Beesley was able to connect what purported to be a barn with mains electricity, water and drainage and a domestic telephone line. Although the question has not been investigated in detail, on the face of it it appears surprising that the local planning authority B did not become aware of these connections with mains services.

Appeal allowed.

LANDS CHAMBER
RESTRICTIVE COVENANTS

Re Hutchinson and another's appeal

UPPER TRIBUNAL: LANDS CHAMBER

[2009] UKUT 182 (LC)

21 September 2009

Mr George Bartlett QC, President

Estates Gazette 30 January 2010

[2010] 04 EG 114

Restrictive covenant — Discharge or modification — Objections — Section 84 of Law of Property Act 1925 — Application to discharge or modify restrictive covenant to allow second dwelling on plot — Owners of nearby land objecting — Whether entitled to be admitted as objectors — Whether having such right as original covenantee — Whether entitled as owners of land to which benefit of covenant annexed

In 2007, the applicants applied, under section 84(1) of the Law of Property Act 1925, to discharge or modify a restrictive covenant that affected their property. They wanted to construct a further dwelling in the curtilage of the property, which they had acquired in 1998. The covenant was contained in a 1991 transfer of the property by the then owner, G, to the applicants' precedessor in title. The transfer had been lost but the Land Registry entry described the land transferred as forming "part of the land comprised in a Deed of Gift dated…1975" in G's favour and stated that the covenant was to benefit "the retained land or any part thereof". The deed of gift had also been lost.

Subsequent to the 1991 transfer, G had transferred a nearby area of woodland and agricultural land (the orange land) and a house and buildings (the green land) to the trustees of a family settlement, of which he was the life tenant, although he continued to occupy the land. He repurchased the green land from the trustees in 2004.

G and the trustees sought to be added as objectors to the applicants' application as the owners of land to which the benefit of the covenant was annexed by the 1991 transfer. G further claimed that he was entitled to object: (i) as the original covenantee entitled to enforce the covenant or to maintain a claim for nominal damages against the original covenantor for its breach; and (ii) under section 78(1) of the 1925 Act, as the occupier of the trustees' land or as a person deriving title under them by reason of his beneficial interest as tenant for life.

***Decision*: The objectors were not admitted. (1) G was no longer entitled to enforce the covenant, in the capacity of original covenantee, against a successor in title of the original covenantor. Under the rules relating to the transmission of the benefit and burden of a restrictive covenant, he had lost the status of original covenantee, and surrendered the power to enforce as such, when he disposed of the orange and green land. That power could not subsequently be revived on reacquisition of the green land; G was in no better position than his immediate predecessor in title or a purchaser who had not previously owned the land. Moreover, there was insufficient evidence that the green land formed part of the land for the benefit of which the covenant had been imposed or that it was capable of being benefited by the covenant. (2) Neither G nor the trustees could enforce the covenant as holders of an interest in land to which the benefit had been annexed. The benefit of a covenant is annexed only to such land as is both identified in the conveyance by express words or necessary implication as intended to be benefited and is described in such terms as enabled it to be easily ascertained from other evidence. The words "the retained land" in the 1991 transfer did not sufficiently identify the land to be benefited as the orange and green land. Further, since the Land Registry entry described the land conveyed by reference to that included in a deed of gift, which had been lost, it was impossible to establish what land was included, nor was it implicit that the "retained land" was the whole of that land or, if a part of it, which part. There was nothing in respect of either the orange nor the green land that would enable a purchaser of the application land readily to ascertain what constituted the "retained land". Accordingly, the covenant was not annexed to any land and neither G nor the trustees could enforce it by virtue of their interests in the orange or green land. (3) G's ability to maintain a claim for nominal damages against the original covenantors did not entitle him to be admitted as an objector. The power of the Lands Tribunal to discharge or modify a covenant, which extends to "any freehold land affected by a restriction", does not apply to a covenant that is enforceable in contract only, since in that situation no freehold land would be affected in a way that was relevant to the right of the covenantee. Therefore, where an application under section 84 is made, only a party appearing to have the benefit of the covenant by virtue of an interest in the application land is entitled to be admitted as an objector; a party with a purely contractual right cannot object.**

The following cases are referred to in this report.

Crest Nicholson Residential (South) Ltd v *McAllister* [2004] EWCA Civ 410; [2004] 1 WLR 2409; [2004] 2 All ER 991; [2004] 2 P&CR 26; [2004] 2 EGLR 79; [2004] 24 EG 150

London County Council v *Allen* [1914] 3 KB 642

Millbourn v *Lyons* [1914] 2 Ch 231, CA

Newton Abbot Co-operative Society Ltd v *Williamson & Treadgold Ltd* [1952] Ch 286; (1952) 1 TLR 284; [1952] 1 All ER 279

This was a hearing to determine the admissibility of objections by the objectors, Joseph Goddard, Mark Dunkley, Michael Goddard and Gregory Whiting, to an application by the applicants, Ian and Lorraine Hutchinson, to discharge or modify a restrictive covenant under section 84 of the Law of Property Act 1925.

Martin Hutchings (instructed by IBB Solicitors, of Uxbridge) appeared for the applicants; Daniel Gatty (instructed by Richard Wilson & Co, of Goring-on-Thames) represented the objectors.

Giving his decision on the admission of objectors, **Mr George Barlett QC, president**, said:

[1] The substantive application in this case relates to an area of land forming part of the curtilage of the applicants' house, 1 Captain's Gorse, Upper Basildon, Reading. It seeks the discharge or modification of a restrictive covenant contained in a transfer of title of the subject land dated 17 May 1991. The transfer was from Mr Joseph Nigel Goddard to Mr Andrew Paul Greener and Mrs Wendy Greener, the applicants' predecessors in title. Mr Goddard objected to the application, claiming to be the owner of land to which the benefit of the restriction attaches.

A The applicants have challenged his entitlement to object. There is also an application by Mr Mark Kenneth Dunkley, Mr Michael Holland Goddard and Mr Gregory John Whiting (referred to as the trustees), who are the owners of land that is occupied by Mr Goddard and that he claims to have owned at the date of the transfer. The trustees seek to be added as objectors, and the applicants oppose this. Whether Mr Goddard or the trustees or all of them should be admitted to object to the application is the subject of this decision following a hearing at which submissions were advanced by counsel for each of the parties.

[2] The application was made in July 2007. It named Mr Goddard as the person entitled to the benefit of the covenant and, in due course,
B notice of the application was served on him. On 29 April 2008, he filed an objection claiming that he was the original covenantee and also that he owned land at Rushdown Farm adjacent to the application land. Some months later, the applicants instructed a solicitor that, on 26 March 2009, raised with Mr Goddard's solicitor the question of his entitlement to object to the application. It sought from the solicitor a copy of the transfer of 17 May 1991.

[3] Because of the tribunal's concern about the slow progress of the proceedings, a pre-trial review (PTR) was fixed before a member, Mr Peter R Francis FRICS, on 7 May 2009. In his order following the PTR, Mr Francis ordered disclosure of the
C parties' respective title deeds by 13 May 2009 and a stay of the proceedings in the event that the applicants, before 20 May 2009, made application to the High Court under section 84(2) of the Law of Property Act 1925 and gave other procedural directions. The substantive hearing was fixed for 16 September 2009. No application to the High Court was made. The objector's title documents, which consisted of Land Registry entries only, were produced on 20 May 2009, and the applicants' solicitor informed him that it was not satisfied that he was entitled to object. On 6 August 2009, his solicitor applied to have the trustees added as objectors, and the applicants objected to this. As a result, the hearing on 16 September 2009 was changed to
D one for the purpose of determining the admission for objectors under section 84(3A).

[4] The case for the applicants is that neither Mr Goddard nor the trustees are entitled to the benefit of the covenant, and that, in any event, the trustees should not be permitted to object in view of the delay in their request to do so.

[5] The transfer of 17 May 1991 cannot be found. Reliance is therefore placed on the Land Registry entry. This identifies the property as "Land adjoining No 1 Captains Gorse Upper Basildon", the consideration as £12,500, Mr Goddard as the transferor and Mr and Mrs Greener as the transferees. The extent of the land
E transferred is described by reference to a plan and is said to be "part of the land comprised in a Deed of Gift dated the 1st day of December 1975 and made between Joan Winifred Goddard (1) and the said Joseph Nigel Goddard (2)". The entry sets out the restrictive covenant as follows:

> The Transferees hereby jointly and severally covenant with the Transferor that neither they nor their Successors in Title shall do or permit or suffer to be done on the property or any part thereof any act or thing which shall be or may become a nuisance or annoyance to the owner or occupier for the time being of the retained land or any part thereof and in addition the Transferees covenant that at all times the land shall be used for agricultural purposes or as garden
> F area and no building or structure for residential or commercial purposes shall be constructed thereon provided however that nothing herein shall prevent the Transferees and their successors in Title from erecting on the property a shed greenhouse or workshop.

[6] The applicants' house, 1 Captain's Gorse, is one of nine houses in a close development. It is one of the two plots that are on the eastern side of the development. The application land abuts it to the east. Immediately to the east of the application land is the southerly tip of an area of land (the red land) owned by Mr Anthony Pattinson, consisting of a house, Rushdown House, set in extensive wooded grounds. On the north side of the red land are the house and other buildings of Rushdown Farm that are owned and occupied by Mr Goddard. The land on which these stand (the green land) is around 300yd from the application land at its nearest point. Wrapping round G
the red and green land on the west, north and east is an area of land (the orange land), perhaps 50 acres or so in extent, consisting of woodland and agricultural land. This land is owned by the trustees and is occupied by Mr Goddard. The application land is separated from the orange land by a plot of similar size to it that adjoins 2 Captain's Gorse. To the west of this plot and, like it, abutting the orange land are the curtilages of four of the houses in Captain's Gorse and the curtilage of a further house. This part of the orange land, for a substantial distance northwards, consists of woodland.

[7] A document entitled "Response by Joseph Nigel Goddard" was
filed by his solicitor in July 2009. It is in the form of a statement by H
him, although he has not signed it. It is said to be made in response to the applicants' claim that he should be debarred from objecting. In the statement, Mr Goddard says that he owned the orange land and the green land at the time of the transfer in 1991 and that he has remained the owner of the green land. The orange land, he says, was transferred in 1998 pursuant to a power of sale by Coutts & Co, banker, to Mr Michael Holland Goddard and Mrs Joan William Goddard, who were acting as trustees of a settlement of which he is the life tenant. He produced the 1998 transfer and the settlement, which is dated 5 October 1951. In a statement dated 15 September 2009 signed
by him (but containing no declaration), he says that his earlier J
statement, that he had remained the owner of the green land since 1991, was wrong. In fact, owing to his financial difficulties in the 1990s, his family bought the land, which had been mortgaged, and he was finally able to buy it back in 2004. The applicants pointed out that Mr Goddard's two statements contained no declaration and that he had not signed the first one, but they did not seek to cross-examine him.

[8] The application seeks discharge or (as now amended) modification of the covenant on grounds (*a*), (*aa*) and (*c*) in section 84(1). In support of ground (*aa*), it is stated, among other things, that, on 17 August 1977,
12 October 1983 and 18 July 1984, Mr Goddard was refused planning K
permission on applications each of which sought permission for the erection of a house on the strip of land comprised by the application land and the plot to the north.

[9] For the would-be objectors, Mr Daniel Gatty advanced four reasons why Mr Goddard is entitled to the benefit of the covenant. First, it is said, Mr Goddard, as original covenantee, still has the benefit of the restriction even if he is unable to enforce it against the covenantors' successors in title. This would entitle him to nominal damages against the covenantors in the event of a breach. Second, it is said, Mr Goddard is the freehold
owner of part of the land, the green land, for the benefit of L
which the restriction was imposed, the "retained land" in the 1991 transfer being the orange land and the green land, and on this account he has the benefit of the covenant. The third basis that is advanced is that Mr Goddard is the occupier of the orange land, now owned by the trustees, and since, it is said, the covenant related to that land, he has the benefit of it under section 78(1) of the 1925 Act. The fourth basis advanced is that the trustees are Mr Goddard's successors in title in respect of the orange land, but since he has a beneficial interest in it as tenant for life he is a person deriving title under successors in title to the original covenantee and the covenant
is deemed to be made with him under section 78(1). As far as the M
trustees are concerned, the contention is simply that they are the successors in title to Mr Goddard in respect of the orange land, and the benefit of the covenant was annexed to that land by the 1991 transfer.

[10] Leaving aside, for the moment, Mr Goddard's claim that he is entitled to object because he has a contractual right to obtain nominal damages against the original covenantors in the event of a breach of covenant, it is to be noted that his claim to be entitled to the benefit of the covenant is made on two alternative bases: that he is the original covenantee and that he is not the original covenantee. The latter basis applies in any event to the trustees. On the latter basis, he and the trustees must show that the benefit of the covenant was annexed to land

A of Mr Goddard by the 1991 transfer: see *Crest Nicholson Residential (South) Ltd* v *McAllister* [2004] EWCA Civ 140; [2004] 1 WLR 2409*. On the former basis, it is sufficient for him to show that the covenant was taken to protect land retained by him at the time of the transfer (see *Millbourn* v *Lyons* [1914] 2 Ch 231; *London County Council* v *Allen* [1914] 3 KB 642) and that, as a matter of evidence, it does benefit that land (see, for example, *Newton Abbot Co-operative Society Ltd* v *Williamson & Treadgold Ltd* [1952] Ch 286). I consider first, therefore, whether Mr Goddard is properly to be treated as the original covenantee for the purposes of these rules.

[11] In my judgment, Mr Goddard cannot claim to be the original
B covenantee for the purpose of enforcing the covenant against a successor in title to the covenantor. He disposed of both the orange land (in 1998) and the green land (some time in the 1990s), and although he bought back the green land in 2004, it is difficult to see why he should be in any different a position from a purchaser who had not previously owned the land. For the purpose of the rules relating to the transmission of the benefit and burden of a restrictive covenant, Mr Goddard ceased to be the original covenantee when he disposed of the land, and I can see no reason in principle why, by purchasing the green land in 2004, he should be in any better position as to the enforcement of the covenant than his immediate predecessors in title. If, during the period
C from 1998, when the applicants bought the land, and 2004 there was no one who could enforce the covenant against them, it seems to me inconceivable that the power to enforce, surrendered by Mr Goddard when he sold the green land, could subsequently be revived. *London County Council*, on which Mr Gatty sought to rely, provides no support for his contention. It did not address the situation of a covenantee who has divested himself of the land said to be benefited only to become owner of part of it at some later date.

[12] The effect of this conclusion is that neither Mr Goddard (under section 78(1) or otherwise) nor the trustees can enforce the covenant unless there has been annexation. The benefit of a covenant is annexed
D only to such land as is identified in the conveyance by express words or necessary implication as intended to be benefited, being described in such terms as enable it to be easily ascertained from other evidence: see *Crest Nicholson, per* Chadwick LJ, in [31] to [34]. The justification for the requirement that the land should be readily ascertainable was stated by the lord justice as follows, in [34], at p2422 G-H:

> It is obviously desirable that a purchaser of land burdened with a restrictive covenant should be able not only to ascertain, by inspection of the entries on the relevant register, that the land is so burdened, but also to ascertain the land for which the benefit of the covenant was taken — so that he can identify who can enforce the covenant. That latter object is achieved if the land which is intended
> E to be benefited is defined in the instrument so as to be easily ascertainable. To require a purchaser of land burdened with a restrictive covenant, but where the land for the benefit of which the covenant was taken is not described in the instrument, to make inquiries as to what (if any) land the original covenantee retained at the time of the conveyance and what (if any) of that retained land the covenant did, or might have, "touched and concerned" would be oppressive. It must be kept in mind that (as in the present case) the time at which the enforceability of the covenant becomes an issue may be long after the date of the instrument by which it was imposed.

[13] Mr Gatty said that the words "the retained land" sufficiently identify the land to be benefited as the orange land and the green land.
F I do not accept this. The entry refers to the land conveyed as being part of the land comprised in the 1975 deed of gift. The deed of gift has been lost, so that it is impossible to establish what land was comprised in it, nor is it implicit that "the retained land" was the entirety of this land or part of it and, if part of it, what part. Mr Gatty said that the 1998 conveyance to the trustees of the orange land shows that Mr Goddard owned the freehold of that land in 1998, and Mr Goddard's statement that he owned the land in 1991 is sufficient to establish that he did. However, in respect of neither the orange nor the green land is there anything in the entry that would enable a purchaser of the application land readily to ascertain what "the retained land" was. Accordingly, the G
covenant was not annexed to any land and, since it was not, neither Mr Goddard nor the trustees are able to enforce it by virtue of their ownership of, respectively, the green land and the orange land, nor can Mr Goddard rely on section 78(1) to enforce it.

[14] I have said that Mr Goddard cannot claim to be the original covenantee for the purpose of enforcing the covenant against a successor in title of the covenantor. Even if he were, he would have no entitlement unless the covenant were imposed for the benefit of the green land and he could show that it is capable of being benefited by it. Although I am inclined to think that the first part of the covenant (against nuisance or annoyance) can be said, through a combination H
of the 1998 conveyance and Mr Goddard's statements, to have been imposed for the benefit of the orange land, there is insufficient evidence, in my view, to show that the green land formed part of the retained land. I also accept Mr Martin Hutchings' submission that the second part of the covenant (restricting the use of the application land and the erection of a building or structures on it) was not, on a proper construction, intended to benefit the retained land. (This last point would not, however, defeat the objection since the application is made in respect of the totality of the covenant.) In any event, however, there is nothing before me to establish that the green land is capable of being benefited by the restrictions. The application land is a small parcel of J
land adjacent to a succession of plots in residential use and separated from the green land by around 300yd of wooded land. No evidence has been called on Mr Goddard's behalf to suggest how the green land might benefit from the restrictions. The use and building restriction cannot, it seems to me, on any realistic basis, be said to benefit the green land, and, taking a similarly realistic view, the same would go for the restriction relating to nuisance and annoyance. For this additional reason, also, Mr Goddard has no entitlement as the original covenantee and the owner of the green land.

[15] This leaves the claim that Mr Goddard should be admitted to object to the application as a covenantee who could maintain a claim for K
nominal damages against the original covenantors, who sold the subject land in 1998. I am happy to be able to conclude that he is not entitled to be admitted on this basis. The power of the tribunal to discharge or modify a covenant applies in respect of "any freehold land affected by any restriction". That power does not extend to a covenant that is enforceable in contract only because there would be no freehold land affected in any way that was relevant to the right of the covenantee. It follows, in my judgment, that where an application under section 84 is made only a person appearing to have the benefit of the covenant by virtue of an interest in the application land is entitled to be admitted as an objector. L

[16] Mr Gatty, however, pointed to section 84(5), which provides:

> Any order made under this section shall be binding on all persons, whether ascertained or of full age or capacity or not, then entitled or thereafter capable of becoming entitled to the benefit of any restriction, which is thereby discharged, modified, or dealt with, and whether such persons are parties to the proceedings or have been served with notice or not.

Mr Gatty said that, under this provision, Mr Goddard would lose his entitlement to nominal damages for breach of covenant and so should be admitted to object in order to defend that entitlement.

[17] It is, of course, the case that an objection from a person in such M
a position would fail because ground (*c*) (probably) or ground (*aa*) (certainly) would apply, but this is not determinative of the issue. Under section 84(5), one possibility is that a person with a purely contractual right loses his entitlement if the restriction is discharged. The other possibility is that he retains it despite the discharge. It does not seem to me to be necessary to decide which of these two possibilities is correct. In either event, it does not seem to me that the scheme of section 84 is such that a person with a purely contractual right should be treated as being entitled to object.

[18] Neither Mr Goddard nor the trustees are admitted to object to the application. The parties are now invited to make submissions on costs, and a letter dealing with this accompanies this decision.

* Editor's note: Also reported at [2004] 2 EGLR 79

A *Representations on costs*

[19] I have now received representations on costs. The applicants ask for their costs of the proceedings on the admission of objectors against Mr Joseph Goddard, Mr Dunkley, Mr Michael Goddard and Mr Whiting, and there is no resistance to this. I can see no reason why such costs should not be awarded. Those persons must accordingly pay the applicants' costs of the proceedings, such costs if not agreed to be the subject of detailed assessment by the registrar on the standard basis. B

Objectors not admitted.

SUPPLEMENT
CASES NOT REPORTED IN "ESTATES GAZETTE"

ABUSE OF PROCESS

Land Securities plc and others
A *v* Fladgate Fielder (a firm)

COURT OF APPEAL

[2009] EWCA Civ 1402

18 December 2009

Mummery, Moore-Bick and Etherton LJJ

Tort — Abuse of process — Judicial review — Alleged collateral or improper purpose in seeking judicial review of planning permission
B ***for development — Developer bringing claim for damages — Whether tort of abuse of process existing — Whether collateral or improper purpose to bring judicial review claim with aim of obtaining financial advantage to compensate for adverse effect of development on neighbouring property***

The respondent firm of solicitors occupied offices facing a site, near Oxford Street, London W1 (site A), that the appellants wanted to develop by demolishing the existing buildings and replacing them with a 10-storey structure. Planning permission was granted for the development, which was expected to take three years. The respondent was concerned that this would adversely affect its plans to relocate to
C larger premises by damaging the marketability and value of the residue of its lease, which was due to expire in 2013.

The respondent considered that it might have recourse to judicial review in the light of the local planning authority's resolution to waive the requirement of 30% affordable housing by "crediting" the appellants with over-provision on another site (site B). It formulated objections to the planning application and held discussions with the appellants to resolve the situation, to no avail.

The respondent applied for judicial review of the planning permission for site B on the ground that the credit resolution was unlawful. The appellants withdrew their reliance on the affordable housing credit
D and, in a revised planning application for site A, proposed to make a payment into the local planning authority's affordable housing fund in place of on-site affordable housing. That application was granted. The respondent withdrew its application for judicial review in respect of site B but obtained permission for judicial review of site A.

The appellants claimed damages of £17m against the respondent on the ground that its objections and review claims were an abuse of process, being motivated by the collateral and improper purpose of procuring a financial benefit. The respondent applied to strike out the claim, repudiating the existence of a tort of abuse of process. The High Court held that such a tort did exist and could be applied
E to judicial review proceedings, accordingly refusing to strike out the claim for failure to disclose a cause of action. However, the court granted summary judgment to the respondent on the ground that no abuse had been established on the facts: [2009] EWHC 577 (Ch). The appellants appealed.

***Held*: The appeal was dismissed. (1) It is possible to bring a claim for the tort of abuse of process. In cases where such abuse is alleged, it is irrelevant whether or not there was reasonable and probable cause for the proceedings, in whose favour they ended or whether they have in fact ended. However, the tort has been invoked successfully only in cases involving a blatant abuse of the process of arrest and imprisonment to achieve a collateral advantage by compulsion.** F
Even if a broader principle did apply and the tort could be committed outside circumstances of compulsion by arrest, imprisonment or other forms of duress, there is no basis for extending it beyond the particular heads of damage that must exist to invoke the tort of malicious prosecution. In civil cases, these are confined to the three well-established heads of damage to: (i) a person's fair fame and credit; (ii) the person, such as in cases of imprisonment; or (iii) property, in the sense of being put to charges and expenses in defending litigation. They do not extend to all cases of economic loss. Given that such severe limits are imposed in malicious prosecution cases, where it is necessary to show an absence of reasonable and probable cause and that the G
proceedings ended in favour of the person maliciously prosecuted, it would make no sense to permit wider recovery for abuse of process, which can apply even where the alleged abuser has a good cause of action. (2) There is no authority for the proposition that the tort of abuse of process can apply to judicial review proceedings for which the court has given unconditional permission despite the raising of objections as to abuse, and where the damage relied on is economic loss outside the limited categories of damage for which the tort of malicious prosecution can be invoked. Even if a wider formulation of the tort of abuse of process were to be adopted, such a tort could not be committed by a party that instituted proceedings with a genuine interest in, and an intention to secure, H
their successful outcome, even if that party's motives were mixed and it hoped also to achieve an objective not within the scope of the proceedings. That is often the situation in judicial review proceedings concerning planning matters, and there is a public interest in bringing judicial scrutiny and remedies to bear on improper acts and decisions of public bodies. A party alleging abuse of process must be able to show that the predominant purpose in bringing proceedings was not to obtain the remedy that the law offers but to achieve some other object that lies outside the range of remedies granted by the law. J
The respondent's interest in respect of its property relocation was insufficiently collateral to the judicial review proceedings as to render those proceedings abusive.

The following cases are referred to in this report.

Bradford Corporation v *Pickles* [1895] AC 587, HL
Broxton v *McClelland (No 1)* [1995] EMLR 485
Dowling v *Colonial Mutual Life Assurance Society Ltd* [1915] HCA 56
Equitable Life Assurance Society v *Ernst & Young* [2003] EWCA Civ 1114; [2003] 2 BCLC 603
Gilding v *Eyre* (1861) 10 CB(NS) 592; (1861) 142 ER 584
Goldsmith v *Sperrings Ltd*; *Goldsmith* v *Various Distributors* [1977] 1 WLR 478; [1977] 2 All ER 566, CA K
Grainger v *Hill* (1838) 4 Bing NC 212, QB
Gregory v *Portsmouth City Council* [2000] 1 AC 419; [2000] 2 WLR 306; [2000] 1 All ER 560
Hanrahan v *Ainsworth* (1990) 22 NSWLR 73
Land Securities plc v *Fladgate Fielder (a firm)* [2009] EWHC 577 (Ch); [2009] 13 EG 143 (CS)
Majory (a debtor), Re; sub nom Debtor (No 757 of 1954), R [1955] Ch 600; [1955] 2 WLR 1035; [1955] 2 All ER 65, CA
Metall und Rohstoff AG v *Donaldson Lufkin & Jenrette Inc* [1990] 1 QB 391; [1989] 3 WLR 563; [1989] 3 All ER 14, CA
Parton v *Hill* (1864) 10 LT 414
QIW Retailers Ltd v *Felview Pty Ltd* [1989] 2 QdR 245

A *Quartz Hill Consolidated Gold Mining Co* v *Eyre* (1883) LR 11 QBD 674, CA

R (on the application of Feakins) v *Secretary of State for the Environment, Food and Rural Affairs* [2003] EWCA Civ 1546; [2004] 1 WLR 1761

R (on the application of Mount Cook Land Ltd) v *Westminster City Council, sub nom Mount Cook Land Ltd* v *Westminster City Council* [2003] EWCA Civ 1346; [2004] 2 P&CR 22; [2004] 1 PLR 29; [2004] JPL 470, CA

Savill v *Roberts* (1698) 12 Mod Rep 208

Speed Seal Products Ltd v *Paddington* [1985] 1 WLR 1327; [1986] 1 All ER 91, CA

Three Rivers District Council v *Bank of England (No 3)* [2001] UKHL 16; [2003] 2 AC 1; [2001] 2 All ER 513

Varawa v *Howard Smith Co Ltd* [1911] HCA 46

B *Williams* v *Spautz* [1992] 174 CLR 509

This was an appeal by the appellants, Land Securities plc, LS Wilton Plaza Ltd and LS Park House Ltd, from a decision of Mr Bernard Livesey QC, sitting as a deputy High Court judge, giving summary judgment to the respondent, Fladgate Fielder, on a claim against it for damages for abuse of process in the bringing of judicial review proceedings.

Christopher Nugee QC and Jonathan Evans (instructed by Linklaters LLP) appeared for the appellants; Alan Steinfeld QC and Tom Leech (instructed by Barlow Lyde & Gilbert LLP) represented C the respondent.

Giving the first judgment, **Etherton LJ** said:

Introduction

[1] This is an appeal from an order, dated 25 March 2009, of Mr Bernard Livesey QC, sitting as a deputy judge of the Chancery Division, by which he gave summary judgment for the respondent against the appellants under CPR 24.2 and dismissed the action.

[2] The appellants allege that the respondent is liable for substantial damages in tort for threatening and bringing proceedings against Westminster City Council (Westminster) for judicial review of the D grant of planning permission for development of a site in Victoria, London SW1 (the Wilton Plaza development). The essence of the claim is that the respondent's predominant purpose for threatening and bringing those proceedings was not to obtain relief against Westminster by quashing the permission, but to put pressure on the appellants to assist the respondent to relocate its offices. The issue before the deputy judge, and on this appeal, is whether the appellants' claim is sufficiently arguable to be allowed to proceed to trial.

Facts

[3] The appellants (LS) are part of a group of property development E companies. It is not necessary, for the purpose of this appeal, to distinguish between them.

[4] The respondent is a firm of solicitors, practising from offices at 25 North Row, just south of Oxford Street, London. At the material time, the respondent occupied the offices under a lease dated 5 April 1988 for a term of 25 years from 25 March 1988, and so expiring on 24 March 2013, at an annual rent of £820,000. By the end of 2005, the respondent's management committee had decided to make plans for the future expansion of the firm into larger and newer premises with a longer term. In January 2006, it decided to instruct property agents to arrange the disposal of the residue of their existing lease and to find F those other premises.

[5] Opposite the respondent's offices, on the other side of the road, which was a narrow one, was a 1960s retail and residential block with a prominent tower, called Park House. On 31 January 2006, LS applied for planning permission to develop Park House by the erection of a modern retail, office and residential block with two basements and 10 floors above ground level. The work of demolition and reconstruction was expected to take in the region of three years, following which fitting-out was to take place.

[6] The respondent received notice of that application in early February 2006. It was concerned that the proposed development would adversely affect the marketability and value of the residue of its lease and frustrate its plans for relocation.

[7] A planning partner of the respondent, Mr Mark Harnett, inspected the planning file in respect of Park House at Westminster's offices. He G discovered that, in addition to the application for the development of Park House, LS had, on 13 August 2005, made an application for the Wilton Plaza development. He considered that there was a connection between the two developments concerning affordable housing. In an e-mail to the respondent's managing partners, he explained that a certain percentage (usually approximately 30%) of affordable housing must be provided on-site in connection with any residential development. LS was proposing, however, that no affordable housing should be provided on-site at Park House. It was intending to "over-provide" affordable housing on the Wilton Plaza development and to secure H Westminster's agreement that the "over-provision" would be used to offset the failure to provide affordable housing on other schemes of LS, such as Park House (the housing credit policy). Mr Harnett said in the e-mail that he would like to discuss the possibility of taking judicial review proceedings in respect of the planning application for the Wilton Plaza development. He continued:

> I believe that we should give serious consideration to attacking this for the following reasons:
>
> 1. The idea of an affordable housing credit is novel and I believe that there may be reasonable grounds for challenging this in the High Court;
>
> 2. The concept is important in connection with the proposed development J of Park House. If affordable housing will have to be provided on site at Park House the profitability of the scheme will be significantly affected.
>
> 3. The idea of the affordable housing credit is strategically important for a developer such as Land Securities who would seek to use it in connection with other schemes also.
>
> 4. It is an early shot across the bows of Land Securities and would give a firm signal of our intentions.

[8] Mr Harnett attached to the e-mail a note on strategy (the strategy note), in which he said that he thought that LS would ultimately obtain a consent, but that a planning objection and judicial review, with an attack on the housing credit policy, would cause potential problems K with the development of Park House, including delay and making the development less valuable because of the need to accommodate more low-value elements, such as social housing and community facilities. The strategy note included the statements: "Need persuade LS that… we can cause problems they can avoid" and "the AH credit scheme key to Park House development and strategically important for LS generally — it is novel and believe prospect could be declared unlawful — prospects moderate" and "Note Don't collaborate with the other objectors — makes deal more difficult".

[9] On 23 February 2006, Westminster's planning committee resolved to grant permission for the Wilton Plaza development. The L resolution included a paragraph specifically agreeing the housing credit policy (the credit resolution). The broad effect intended to be achieved by the credit resolution is reflected in the following passage of the planning officer's report:

> A high proportion of affordable housing is proposed, above normal policy requirements. The applicant, Land Securities, is requesting that the increase in both market units (over and above the floorspace in the previous permission on this site) and affordable units proposed, form a "residential credit" under a "land banking strategy" to be used against other mixed use/commercial/housing schemes by Land Securities in the future (subject to planning), where it is not viable/feasible to provide the full requirement of market housing and/or M affordable housing sought under mixed use policies on site. The principle of a "credit" has been accepted by the Council on previous occasions.

[10] The respondent took advice from Mr Christopher Lockhart-Mummery QC. He advised in writing on 13 March 2006 that he could see no basis on which the credit resolution could possibly be lawful, and set out the grounds for an application for judicial review. He summarised the position in para 12 of his opinion, as follows:

> For the above reasons, it is my opinion that the formulation of the land banking strategy is unlawful. The essential grounds for any application for judicial review would, in brief, be (1) that it is unlawful to formulate such a planning strategy which is not a general land use strategy but one designed to

benefit exclusively one individual, and (2) that in any event, the substance of the strategy, and the process by which it is evolving, constitutes an unlawful departure from established processes for the making of land use policy.

[11] On 31 March 2006, LS applied to Westminster for permission to develop Park House, intending to make use of the credit resolution.

[12] At the end of March 2006, the respondent approached LS for a meeting. Before the meeting, the respondent sent LS a copy of Mr Lockhart-Mummery's opinion. A meeting took place on 5 April 2006 between Messrs Allen Cohen, Mark Harnett and John Goreing, of the respondent, and Mr Mike Hussey, of LS. What was said at the meeting forms a critical part of LS's allegations.

[13] Witness statements of those who attended the meeting show similarities to a significant degree, but also some differences of emphasis. Mr Cohen said in his witness statement:

> I also said that the firm was anticipating three years of "hell" while the development was under construction. Mr Hussey asked me what we wanted. I remember him saying "Do you want money?" to which I replied: "Absolutely not".
>
> I suggested that one way out would be for Land Securities to take an assignment of our lease. I said that this would enable us to "slip away" with Land Securities' assistance. Otherwise an alternative would be to delay the development of Park House until the end of our lease in 2013. … Mr Goreing asked Mr Hussey if Land Securities had any stock in W1 (which would have enabled us to swap our existing building for other premises).

[14] Mr Hussey said in his witness statement, however, that he thought that the respondent was looking for a financial payment and a windfall. He said:

> My view is that the financial co-operation they sought, whilst unidentified, was intended by them to go beyond assistance in their plans to relocate, although it was presented in part in that context. … I believe that when they discovered that Land Securities were applying for planning permission to redevelop Park House in February 2006, Fladgate Fielder saw an opportunity to use the possibility of judicial review (and other tactics as in Mr Harnett's note) as a weapon to force us to the negotiating table, with a view to assisting them to implement their plans by financial payment. In spite of Mr Cohen's denials, I do think they were looking for a windfall.

[15] The respondent communicated with Mr Hussey, either directly or through LS's solicitor, on various dates in April and May 2006, holding out the hope that the respondent would not have to issue proceedings and that there could be a mutually acceptable solution.

[16] On 19 May 2006, the respondent issued an application against Westminster, settled by leading and junior counsel, for permission to apply for judicial review of the planning consent for the Wilton Plaza development on the ground that the credit resolution was unlawful. LS was joined to the proceedings as an interested party. In its statement of facts and grounds for judicial review, the respondent explained its standing to bring the proceedings and, in particular, its proximity to Park House, LS's application for planning permission in respect of that site and the likely reliance of LS on the credit resolution in connection with that development.

[17] LS filed summary grounds of opposition, in which it submitted that permission to apply for judicial review should be refused because: (i) the respondent did not have sufficient interest in the matter to which the application related; and (ii) the claim had no realistic prospect of success. The issue of the standing of the respondent to bring the proceedings was addressed in paras 4 to 12 of the summary grounds. It was there stated that: the respondent's premises were not located near the Wilton Plaza development; the respondent could not demonstrate that it had any interest in the Wilton Plaza development; the respondent's only interest would be in respect of a future planning permission for Park House; at the meeting in April 2006 between representatives of LS and the respondent, Mr Cohen, of the respondent, stated that the proposed development of Park House looked fine and this was simply a business transaction in that the respondent wanted to move from its current location opposite Park House and saw this as an opportunity to negotiate some financial settlement with the interested party; that was the respondent's true motivation.

[18] The respondent responded to those allegations in a letter to the court dated 23 June 2006, in which it set out its interest in bringing the proceedings owing to the proposed development of Park House. It said that, at the April 2006 meeting, Mr Cohen had made it expressly clear that the respondent was not looking for a cash settlement but simply looking to ensure that its business did not suffer. It denied that the bringing of the claim to protect its business interests was a collateral purpose amounting to an abuse of process.

[19] On around 5 July 2006, Messrs Cohen, Harnett and Goreing circulated to all their partners a memorandum headed "Proposed redevelopment of Park House" (the July memo). The July memo explained the background to the judicial review proceedings. It expressed views on the consequences of the failure or success of those proceedings. In that respect, the July memo said, as follows:

> nevertheless there clearly is a risk that the court may refuse to hear our application on the basis that our principal concern is with another development, namely Park House. However, if the court refuses to grant "standing" for this reason this may pave the way for a further application for a reconsideration of the issue when the Park House scheme falls to be considered by Westminster Council.
>
> …
>
> If the application is successful, then Land Securities will have to reconsider how it intends to proceed. To some extent, the options available to it will depend upon the action taken by the court. It may for example decide to quash the council's declaration of the affordable housing credit but refuse to quash the permission itself. Alternatively, it may of course take the opposite action or quash both decisions. In any event, a successful action would require Land Securities to substantially rethink its proposals for Park House which would result in a considerable delay to the implementation of its scheme. This has already been partially achieved to the extent that Westminster City Council has put on hold its consideration of the planning application for the redevelopment of Park House until the outcome of this challenge is known. We have put in a holding objection to the grant of planning permission for Park House and will put in further detailed objections in due course.
>
> There are a number of reasons why delaying the development is the best feasible option available to the firm, all of which relate to the achievement of our long term ambitions.
>
> We have identified during the business plan process we can only thrive as a business with continued growth. We require increased critical mass and spread of expertise in order to continue to attract the high value or complex transactions we are targeting.
>
> We are approaching capacity in our building and have identified the need to expand our premises in order to accommodate our plans.
>
> …
>
> Relocation… would allow us to maintain the efficiencies we enjoy here but with increased facilities and project the positive image of a thriving firm required to attract those who will contribute to our development.
>
> Unfortunately these plans are now untenable due to the Park House development. These premises have effectively become un-assignable given their size, the length of the lease and the effect of the development.
>
> We have explored our options and concluded that leaving these premises vacant or paying compensation to withdraw from our obligations are both currently too expensive.
>
> Each month that passes towards the expiry of our lease reduces these potential obligations by some £100,000.
>
> It follows therefore that the longer we can delay the development the greater the options open to us albeit with a short term requirement for overflow space.

[20] Also on 5 July 2006, Newman J granted unconditional permission for the judicial review proceedings under CPR 54.5.

[21] LS then revised its application for planning permission for the development of Park House by proposing a financial contribution towards Westminster's affordable housing fund rather than the provision of off-site affordable housing on the Wilton Plaza development. That would make the credit resolution irrelevant to the application in respect of Park House. LS proposed to Westminster that the financial contribution should be £5.7m. On learning of the proposal, the respondent suggested to Westminster that if a financial contribution were to be paid it should be a considerably higher figure, namely some £11.4m.

[22] Westminster's planning committee considered the revised Park House application in September 2006. Planning permission was subsequently formally granted on 10 November 2006.

[23] In January 2007, the respondent agreed to withdraw its judicial review application in respect of the Wilton Plaza development, and a draft consent order to that effect was lodged on 13 February 2007.

[24] In the meantime, by letter dated 9 February 2007 addressed to each of the respondent's partners personally, LS's solicitor indicated a claim against each of them for losses in the region of £17m on the ground that the partners who had conducted the meeting with Mr Hussey on 5 April 2006 had "demonstrated the illegitimate motivation of the Judicial Review — driven by financial motives — not any concern as to the Wilton Plaza planning permission".

[25] In early February 2007, the respondent issued an application for permission to claim judicial review of the Park House consent. It was granted permission to proceed on 29 March 2007 by Sir Michael Harrison. Following a further application for planning permission for Park House and the granting of permission pursuant to that application in May 2008, the respondent claimed judicial review of that second consent in June 2008. Those applications for judicial review of Park House were heard by Collins J on 27 February 2009. He dismissed the claim in respect of the second Park House consent. He held that there had been a defect of procedure in respect of the earlier consent, but declined to quash the permission.

Claim

[26] Paragraphs 5 to 8 of the amended particulars of claim contain an overview of the claim as follows:

> 5. The Claimants claim substantial damages for the tort of abuse of civil process.
>
> 6. The Defendant threatened to issue and then issued and pursued two applications for judicial review of planning permissions obtained by the Second and Third Claimants in respect of Wilton Plaza and Park House respectively. In so doing, the Defendant sought to pressurise the Claimants into making financial contributions and assisting it to relocate its business from its premises opposite Park House. The Defendant was not motivated by concern about the lawfulness of the planning permissions and its purpose was not to prevent the developments from taking place, but rather to force the Claimants to assist it to move. That objective was beyond the scope of the judicial review proceedings and amounted to an improper and collateral advantage. That conduct by the Defendant amounts to an abuse of civil process and is tortious.
>
> 7. The acts of the Defendant relied on as constituting the tortious conduct are set out in more detail below, but by way of summary:
>
> 7.1 The Defendant's dominant purpose in threatening and then issuing and pursuing the judicial review proceedings was not to prevent either development, but to bring pressure to bear on the Claimants to force them to assist the Defendant to relocate its business.
>
> 7.2 The Defendant had no interest in preventing the development at Wilton Plaza, which was some distance away from its premises and would not affect the Defendant at all, but rather objected to that development as a means of bringing pressure on the claimants by impeding the Second Claimant's proposed development of Wilton Plaza and the third Claimant's proposed development of Park House.
>
> 7.3 In relation to the proposed development at Park House, the Defendant stated in terms that the development itself "looked fine" and that it "wished [the Claimants] well with it".
>
> 7.4 Although the Defendant expressed concern at the potential of the development at Park House to disrupt and adversely affect its business, it took no steps to discuss the development project with the Claimants in an attempt to minimise that disruption, as would be normal in such circumstances.
>
> 7.5 The Defendant has never contended that either redevelopment would result in an actionable nuisance in respect of noise and disturbance to the Defendant.
>
> 7.6 The Defendant sought by means of threatening and then issuing and pursuing the judicial review proceedings to obtain protection from alleged future disturbance of its use and enjoyment of its premises in circumstances where no such protection was available in the law of nuisance or otherwise and sought thereby to exert pressure on the Claimants to facilitate a relocation from its premises.
>
> 8. The Claimants have suffered substantial loss as a result of the delay to both developments caused by the Defendant's actions and seek to recover such loss as damages.

[27] LS claims damages caused by delay in carrying out the two developments as a result of the judicial review proceedings. The loss is estimated in the region of £17m.

[28] In its amended defence, the respondent alleges that the credit resolution had a direct effect on the grant of planning permission for the development of Park House, and its primary and dominant purpose in issuing and pursuing the two judicial review applications was to avoid the damage that the development of Park House would have on its business.

Respondent's applications

[29] The respondent issued an application notice on 31 July 2008 that, as subsequently amended, applied for an order striking out the claim pursuant to CPR 3.4(2)(a) and/or (b); alternatively, for "reverse" summary judgment against the appellants pursuant to CPR 24.2. The grounds were that: (a) the tort of abuse of process does not exist; (b) if the tort does exist, it does not apply to an application for judicial review; and (c) if the tort does exist and applies to judicial review claims, it applies only if the application for judicial review fails and the defendant had no reasonable or probable cause for bringing that application; (d) LS should have opposed the applications for judicial review or applied to strike them out as an abuse of process; and (e) on the evidence, LS has no real prospect of succeeding on its claims because the respondent's primary or dominant purpose was as set out in the defence, namely to avoid the damage that the redevelopment of Park House would have on its business, and was not unlawful.

[30] Witness statements in support of the application notice were made by Mr Cohen, Mr Harnett and Mr Goreing, of the respondent. Mr Hussey made a witness statement on behalf of LS.

Judgment of the deputy judge

[31] The deputy judge dismissed the application to strike out the claim, but he granted the respondent summary judgment pursuant to CPR 24.2.

[32] He concluded that there were virtually no issues of fact that needed to be determined before the issue of law at the heart of the applications could be considered. The factual basis on which he proceeded to consider the merits of the applications was summarised in [40] of his judgment ([2009] EWHC 577 (Ch)) as follows:

> What appears plain is that Fladgate saw that it had a valuable negotiating position vis a vis the claimants and sought to use it to assist to get relocated and that it intended that the assistance should be at some material cost to the claimants and of financial benefit to themselves: it is fair to assume for the purposes of this application that it was this purpose which was driving their actions. It may also have had the additional intention, in the event that it did not reach satisfactory terms with the claimants, to oppose the application to delay it, but I do not have this in mind for the purposes of considering the applications.

[33] The deputy judge summarised the principal authorities relied on by both sides. He concluded that there is no general tort of maliciously instituting civil proceedings, but that there is a tort of abuse of process, which was first formulated in *Grainger* v *Hill* (1838) 4 Bing NC 212, and for which it is not necessary to prove malice or want of reasonable and probable cause or that the proceedings have been terminated, let alone in favour of the plaintiff. He accepted LS's case that the tort can apply to proceedings for judicial review. In view of the different ways in which the elements of the tort have been described in the authorities, the deputy judge did not consider it appropriate for him to define the ingredients and limits of the tort. For those reasons, he refused to strike out the claim on the ground that it did not disclose a cause of action. He said that the heart of the case was whether the respondent, in bringing the judicial review proceedings, was acting for an improper or collateral advantage. He concluded that there was no realistic prospect of LS succeeding on that issue. His conclusion on the point was stated in [84] and [85] of his judgment, as follows:

> 84. The factor which in my judgment breaks down the claimants' argument is the fact that there is an obvious connection between the Credit Resolution passed in relation to the Wilton Plaza application and the use intended to be made of it in relation to the third claimant's application to redevelop Park

A House. That it was a real connection is evident from the fact that the claimants did not seek to defend the Judicial Review proceedings to the end but instead severed the connection by offering to make payment of some £5.769 million in lieu of affordable housing at Park House to the Westminster Affordable Housing Fund and amended both planning applications accordingly.

85. The matters in the preceding paragraphs do to my mind greatly affect what amounts to a collateral purpose for the purposes of the application of the tort to judicial review proceedings. If, as I am satisfied is the case, a property owner is entitled to use judicial review proceedings for the protection of his interest in and enjoyment of his property, why should he not compare the diminution in the value of his interest and enjoyment of his own property, which is likely to attend the development, with the countervailing (and perhaps very B substantial) development gain which the developer will enjoy from the same development? Why should it be a collateral purpose if the property owner seeks to use any negotiating position he may have to restore the value he believes he will lose by the development by getting it from the person who caused it who happens also to be the only person who might be prepared to pay it?

[34] Finally, the deputy judge held that the respondent had no basis for striking out the proceedings on the ground that LS should have opposed the judicial review proceedings or applied to have them struck out because they were an abuse of process. He reached that conclusion on the basis that the procedure in judicial review proceedings does not give a proper opportunity for the determination of contested questions C of abuse of process.

Appeal

[35] LS appeals on the grounds that the deputy judge had wrongly: (i) characterised and identified the respondent's predominant purpose in bringing the judicial review proceedings; (ii) formulated the legal test for "collateral purpose"; (iii) applied the "collateral purpose" test; and (iv) applied the principles for granting summary judgment. LS says that the respondent's predominant purpose should have been assumed to be simply that which the judge himself had set out in [40] of his judgment, and that assumed purpose was outside the scope of the judicial review proceedings and was not reasonably related to the D grievance of which the respondent complained in the judicial review proceedings; the respondent's actions therefore were, or arguably were, tortious on the basis of *Grainger*. LS says that, in any event, this is not an appropriate case for summary judgment because the law is both difficult and uncertain, and there are material facts in dispute on which the court cannot properly reach a conclusion until after disclosure and a full trial.

[36] The respondent has served a respondents' notice to uphold the judgment on the various points and arguments on which the deputy judge found against it. It includes the respondent's contention that there is no general tort of abuse of process, and any such tort as E exists is limited in ways that make it inapplicable on the facts of the case; in particular, that it is limited to ancillary process within existing proceedings and it cannot apply to judicial review proceedings.

Law

[37] LS's starting point is that *Grainger* established the existence of a tort of abuse of process.

[38] The facts in that case were that the plaintiff was the master and owner of a ship. In September 1836, he was loaned £80 by the defendants, repayable on 28 September 1837. The loan was secured by a mortgage of a ship. In the meantime, the plaintiff retained the ship's register so F as to continue to use the ship for his profit. In November 1836, the defendants became concerned as to the sufficiency of their security and decided to obtain possession of the register. Having threatened to arrest the plaintiff unless he repaid the money, they commenced an action of *assumpsit*, made an affidavit of debt and procured a writ of *capias*, indorsed for bail in the sum of £95 17s 6d. That was a writ directing the sheriff to arrest the defendant to ensure his attendance in court. They sent two sheriff's officers with the writ to the plaintiff, who was lying ill in bed from the effects of a wound. The officers told the plaintiff that they had not come to take him but to get the ship's register, but that if he failed to deliver the register or to find bail, they would either have to take him or leave one of the officers with him. He was unable to procure bail. After being imprisoned for 12 hours, and "being much alarmed", he gave up the register. He subsequently came to an arrangement with G the defendants, was discharged from arrest, paid the costs and repaid the loan. No further steps were taken in the action of *assumpsit*. The plaintiff brought proceedings against the defendants for the loss caused generally by his inability to carry on business owing to the defendants' conduct, as well as a specific claim in trover in respect of the ship's register. A verdict having been given for the plaintiff at the trial, the defendants applied to enter a nonsuit on various grounds, including that the action could not be maintained unless the writ of *capias* had been issued without reasonable and probable cause, which had not been averred, and which could never be established without showing that the action of *assumpsit* had determined against the defendants. H

[39] The Court of Exchequer Chamber (Tindal CJ, Park, Vaughan and Bosanquet JJ) dismissed the nonsuit. Tindal CJ held that it was not necessary to prove the termination of the *assumpsit* proceedings and the absence of reasonable and probable cause for those proceedings because the action was not one for malicious arrest or malicious prosecution, for which those would have been necessary elements. He said, at p221, that the action was, rather, "for abusing the process of the law, by applying it to extort property from the Plaintiff". The directions given to the plaintiff to compel him to give up the register were not part of the duty imposed by the writ of *capias*. The complaint of the plaintiff was "that the process of the law has been abused, to J effect an object not within the scope of the process". Park J, at p222, said that the case was not one for a malicious arrest, but "a case *primae impressionis*, in which the Defendants are charged with having abused the process of the law, in order to obtain property to which they had no colour of title". He said that, as to the count in trover, the compulsion under which the register was detained was tantamount to a conversion. Vaughan J said, at p223, that what the plaintiff complained of was "an abuse of the process of law, for the purpose of extorting property to which the Defendants had no claim", and that, as to the claim in trover in respect of the register, the "taking was as much a forcible taking as if a pistol had been presented to the Plaintiff's head". Bosanquet J said, K at p224, that the action was not one of malicious arrest or prosecution or for maliciously doing that which the law allows to be done. Rather, "the process was enforced for an ulterior purpose; to obtain property by duress to which the Defendants had no right". The action was "not for maliciously putting process in force, but for maliciously abusing the process of the Court". He said that "the register was illegally obtained by duress, under an abuse of the process of this Court". Tindal CJ and Park J expressly acknowledged that there was no prior precedent for the cause of action of abuse of process on which the plaintiff succeeded.

[40] A similar conclusion was reached by the Court of Common Pleas in *Gilding* v *Eyre* (1861) 10 CB(NS) 592; (1861) 142 ER 584. L In that case, following judgment against the plaintiff for a debt and substantial repayment of it by him, the defendant issued a writ of execution for the full amount of the debt, in consequence of which the plaintiff was arrested by the sheriff's officers. The plaintiff brought proceedings to recover the amount that he had overpaid on the basis of malicious arrest. The court (Williams, Byles and Keating JJ) rejected the defendant's contention that the claim was not sustainable because the original proceedings had not terminated in the plaintiff's favour. The single judgment of the court distinguished cases of malicious prosecution, emphasising that the proceedings had, subject to execution, actually terminated and that the defendant "by means of a M regular writ of execution extorted money which he knew had already been paid and was no longer due on the judgment": p604. *Grainger* was not mentioned.

[41] If *Gilding* is properly to be regarded as an example of a *Grainger* case of tortious abuse of process, it is the last reported case in which such a claim has succeeded in this jurisdiction. If not, *Grainger* itself is the first and last such case. Accordingly, the last reported successful action in this jurisdiction for the tort abuse of process was either around 140 or 170 years ago.

[42] Both *Grainger* and *Gilding* were cases that concerned the abuse of process within existing proceedings. In each case, the process was issued without judicial permission or intervention. The process

was intended for a specific purpose in support of the proceedings, but the defendants used the process for an improper purpose outside its scope. In both cases, the improper use to which the process was applied involved compulsion amounting to duress, namely arrest and imprisonment. The difference between the facts in the two cases and the matters on which the claim is based in the present proceedings is striking. In the present case, the claim concerns the bringing of proceedings for public law remedies by way of judicial review, for which the permission of the court was necessary and was obtained. In the case of the judicial review proceedings in respect of the Wilton Plaza development, the standing of the respondent to bring those proceedings, its motives for doing so and the prospects for the success of the proceedings were expressly raised by LS in opposing the grant of permission, but the judge nevertheless gave unconditional permission. It is not now suggested that the respondent did not have standing to bring the judicial review proceedings or that it had no prospect of success. How then is LS able to argue, on the basis of *Grainger*, that the present proceedings have a sufficient prospect of success to be permitted to proceed to trial? It claims to be entitled to do so on the basis of statements in subsequent cases in this jurisdiction as to the existence and ambit of the tort of abuse of process and also in the context of applications to stay or strike out proceedings on the ground of abuse of process. It also relies on Australian cases in which the tort has been successfully invoked. In my judgment, none of those statements or authorities provides an arguable basis for the claim in the present case or, on the statements of case and the evidence, any reason to permit this action to proceed to trial. I turn to consider the further cases on which LS relies.

[43] Mr Christopher Nugee QC, for LS, referred to *Parton* v *Hill* (1864) 10 LT 414. I do not find this of any assistance in clarifying the extent of the *Grainger* tort. It concerned an allegation of the malicious issue of an attachment of a debt owing from a third person to the plaintiff, without reasonable or probable cause. It was therefore similar to an action for malicious prosecution. It was not a *Grainger* abuse of process case at all. There is a brief and uninformative reference in the judgment of Blackburn J, at p415, to "the doctrine laid down by Tindal CJ in *Grainger*", but it is unclear from the very short and rather confusing report whether or not he was there simply repeating a comment of counsel for the unsuccessful plaintiff.

[44] *Re Majory (a debtor)* [1955] Ch 600 was a case in which the debtor disputed a bankruptcy petition and opposed the making of a receiving order on the ground that the creditor had attempted, in connection with the proceedings, to extort £8 15s from the debtor in excess of the sums lawfully due under a court judgment. It was contended that this was "extortion" in bankruptcy law, that is to say, the oppressive use or threat of proceedings in order to obtain some collateral advantage such that the court would not permit the creditor to make use of the process that he had abused. In the course of delivering the judgment of the Court of Appeal, Evershed MR made the following general observation, at pp623-624:

> (3) The so-called "rule" in bankruptcy is, in truth, no more than an application of a more general rule that court proceedings may not be used or threatened for the purpose of obtaining for the person so using or threatening them some collateral advantage to himself, and not for the purpose for which such proceedings are properly designed and exist; and a party so using or threatening proceedings will be liable to be held guilty of abusing the process of the court and therefore disqualified from invoking the powers of the court by proceedings he has abused.

[45] Evershed MR did not refer to *Grainger*. His observation was expressly limited to abuse that would disqualify a party from invoking court proceedings. It said nothing about a cause of action in tort for abuse of process.

[46] In *Goldsmith* v *Sperrings Ltd* [1977] 1 WLR 478, the plaintiff commenced proceedings for damages for libel and an injunction against the publishers, the editors and the main distributors of *Private Eye*. In addition, he issued writs against a large number of other wholesale and retail distributors of the paper for the same relief. Some of the distributors applied for an order that the actions against them should be stayed or dismissed as an abuse of process of the court on the ground that the plaintiff's purpose in pursuing the actions against them was not to protect his reputation but for the collateral purpose of destroying the paper by cutting off its retail outlets. The case was not therefore one in which damages were claimed for the tort of abuse of process. The judgments of the Court of Appeal contain, however, the most detailed analysis of the "collateral advantage" principle in the context of abuse of process.

[47] All three judges, Lord Denning MR, Scarman LJ and Bridge LJ, cited both *Grainger* and *Majory*, but they all differed in their analysis. Lord Denning MR dissented in the result. Scarman LJ said (at p498F) that: "If Sir James Goldsmith's purpose in initiating or pursuing his actions against the secondary distributors be to destroy 'Private Eye', namely, to use his wealth so as to suppress it, he is abusing the process of the court." Such a wide statement of principle goes beyond anything in *Grainger* or, indeed, *Majory,* is inconsistent with subsequent authority and was not supported by Mr Nugee. It appears to merge motive and purpose.

[48] Bridge LJ said, at p503D, that, for the purpose of Lord Evershed's general rule in *Majory*, the phrase "collateral advantage" cannot embrace every advantage sought or obtained by a litigant that is beyond the court's power to grant him. He said, at p503F-H, that one can certainly go so far as to say that when a litigant sues to redress a grievance, no object that he may seek to obtain can be condemned as a collateral advantage if it is reasonably related to the provision of some form of redress for that grievance. On the other hand, if it can be shown that a litigant is pursuing an ulterior purpose unrelated to the subject matter of the litigation and that, but for the ulterior purpose, he would not commence proceedings at all, that is an abuse of process. Bridge LJ said that those two cases were plain, but acknowledged that there is a difficult area in between. He postulated the situation of a litigant with a genuine cause of action, which he would wish to pursue in any event, who could be shown also to have an ulterior purpose in view as a desired by-product of the litigation. Bridge LJ very much doubted whether such a litigant could be debarred from proceeding. The deputy judge sought to apply Bridge LJ's analysis to the facts of the present case, but Mr Nugee submitted that he had failed to apply it correctly.

[49] Mr Nugee next referred us to *Speed Seal Products Ltd* v *Paddington* [1985] 1 WLR 1327. The question in that case was whether the defendant should be permitted to amend to add a counter-claim asserting that the action was brought in bad faith for the ulterior motive of damaging the defendant's business, and not for the protection of any legitimate interest of the plaintiff's. Fox LJ, with whom the other two members of the Court of Appeal agreed, decided that the decision in *Grainger* provided a basis for an arguable case that there had been an actionable abuse of the process of the court. He said, at p1335H, that he expressed no view as to the strength of the defendant's case.

[50] In *Metall und Rohstoff AG* v *Donaldson Lufkin & Jenrette Inc* [1990] 1 QB 391, the plaintiffs claimed, among many other heads of relief, damages for abuse of process. The defendant applied to have the order for service of the writ out of jurisdiction set aside on the ground, among other things, that the claim for abuse of process was misconceived. The substance of the complaint against the defendant, in respect of this head, was that it had abused the process of the court by adducing false evidence and submitting a false case for the primary purpose of defeating claims to the return of the plaintiff's metal and other assets and to prevent the grant of an injunction restraining the defendant from dealing with such assets in the meantime. Slade LJ, giving the judgment of the Court of Appeal, said that those matters did not expose the defendant to an action for damages in tort under the *Grainger* principle. Having acknowledged, at p469B, that the Court of Appeal in *Speed Seal Products* had established that "it is at least well arguable that there exists a tort of abuse of the process of the court of a nature established by the decision… in *Grainger*", said that a party alleging such abuse had to show that the predominant purpose of the other party in using the legal process had been one other than that for

A which it was designed. Holding that the requirements of the tort were not satisfied on the facts of the case, he said, at p470D-F:

> Relief in tort under the principle of *Grainger* v *Hill* is not, in our judgment, available against a party who, however dishonestly, presents a false case for the purpose of advancing or sustaining his claim or defence in civil proceedings. This may well cause hardship to an injured party who cannot be sufficiently compensated by an appropriate order for costs. However, if there is a gap in the law it rests on sound considerations of public policy, as does the rule of law which gives immunity to witnesses against civil actions based on the falsity of evidence given in judicial proceedings. If the position were otherwise, honest litigants might be deterred from pursuing honest claims or defences and honest
> B witnesses might be deterred from giving evidence…

[51] Those comments have a particular resonance in proceedings for judicial review to challenge the unlawful conduct of public bodies.

[52] In *Broxton* v *McClelland (No 1)* 30 January 1995*, the defendant issued various applications to strike out the claim, including a claim of abuse of process. The action was being financially maintained by a third party. The defendant contended that the maintainer's purpose was to oppress and ultimately bankrupt the defendant, and for that reason the action should be struck out as an abuse of process. Simon Brown LJ, with whom the other two members of the Court of Appeal agreed, referred to *Grainger*, *Goldsmith* and *Speed Seal Products*, and said
C that the proceedings should not be struck out since the plaintiff was not seeking to achieve a collateral advantage beyond the scope of the action. He said that the motive for bringing proceedings is irrelevant, and that a plaintiff is entitled to seek the defendant's financial ruin if that would be the consequence of properly prosecuting a legitimate claim. That was a clear rejection of the approach of Scarman LJ in *Goldsmith*. Simon Brown LJ appears to have preferred and followed the analysis of Bridge LJ in *Goldsmith*.

[53] Mr Alan Steinfeld QC, for the respondent, laid emphasis on two further English authorities relevant to the tort of malicious prosecution. In *Quartz Hill Consolidated Gold Mining Co* v *Eyre* (1883) 11 QBD
D 674, the Court of Appeal held that there was a cause of action for falsely and maliciously presenting a winding-up petition. The Court (Brett MR and Bowen LJ) considered that there was no general cause of action for maliciously bringing civil proceedings without reasonable and probable cause. An action for malicious prosecution could, however, be brought in specific cases. Both Brett MR and Bowen LJ endorsed the three heads of damage capable of giving rise to a claim for malicious prosecution set out by Holt CJ in *Savill* v *Roberts* (1698) 12 Mod Rep 208. Brett MR set out the principles as follows, at p683:

> When we look back to the decisions of the judges of earlier times… we find it laid down by Holt CJ in *Savill* v *Roberts* that there are three heads of damage
> E which will support an action for malicious prosecution. There is damage to a man's person, as when he is taken into custody, whether that be, as in former times, upon mesne process or upon final process, or whether it be upon a criminal charge. To take away a man's liberty is damage, of which the law will take notice. Secondly, to cause a man to put to expense is damage, of which the law will take notice. But Holt CJ adds a third head of damage, and that is where a man's fair fame and credit are injured. This is also a head of damage of which the law will take notice. Under the old law as to bankruptcy it was held that where a man was falsely and maliciously and without reasonable or probable cause made a bankrupt, two kinds of legal injury were inflicted upon him: first, in order to get rid of the bankruptcy, he was obliged to incur expense, and that was an injury; secondly, it was held that to allege of a trader that he
> F was insolvent and liable to be made a bankrupt, was injury to his fair fame and credit, of which the law would take notice. Therefore under the old system of bankruptcy a trader had a good cause of action, if he was made a bankrupt falsely and maliciously and without reasonable or probable cause.

[54] Mr Nugee submitted that the reference to "expense" in that description of the second head of damage is to general economic loss of the kind claimed in the present proceedings. That is not correct. In *Savill*, the Court of Common Pleas held that the plaintiff, Roberts, was entitled to recover £11 damages in proceedings for malicious prosecution, the defendant having maliciously caused Roberts to be indicted for causing a riot and Roberts having been acquitted. The £11 G was the cost incurred by Roberts in successfully defending himself on the indictment. Holt CJ, giving the judgment of the court, explained the right to recover "charges and expenses" as damages for malicious prosecution very briefly in the following way, at p209:

> Thirdly, that a man put to answer an indictment is put to charges is notorious; and if so, it is an injury to his property; and if this injury be occasioned by a malicious prosecution, it is reason and justice he should have an action to repair him the injury…

[55] Holt CJ appears there to have limited damages for purely H pecuniary loss in an action for malicious prosecution to the costs of defending the maliciously prosecuted proceedings. There is nothing whatever in *Savill* to indicate that general economic loss is recoverable in such an action.

[56] Indeed, Mr Nugee's gloss on *Savill* would be contrary to the whole approach in *Gregory* v *Portsmouth City Council* [2000] 1 AC 419. In that case, the House of Lords refused to extend the tort of malicious prosecution for civil proceedings beyond the heads of damage recognised in *Quartz Hill*. Lord Steyn, with whom the other members of the judicial committee agreed, said, at pp426F-427A:

J

> The inquiry must proceed from the premise of the law as it stands. The tort of malicious prosecution is narrowly defined. Telling lies about a defendant is not by itself tortious: *Hargreaves* v *Bretherton* [1959] 1 QB 45. A moment's reflection will show what welter of undesirable relitigation would be permitted by any different rule. … Damage is a necessary ingredient of the tort. This element of the tort was explained in a dictum of Holt CJ in *Savill* v *Roberts* (1698) 12 Mod Rep 208. Holt CJ defined the interest protected by the tort as follows:
>
> "there are three sorts of damages, any one of which is sufficient to support this action. First, damage to [the plaintiff's] fame, if the matter whereof he be accused be scandalous. Secondly, to his person, whereby he is imprisoned. Thirdly, to his property, whereby he is put to charges and expenses." K

[57] Lord Steyn explained his refusal to contemplate the extension of the tort of malicious prosecution to civil proceedings generally as follows, at pp432G and 433A:

> for essentially practical reasons I am not persuaded that the general extension of the tort to civil proceedings has been shown to be necessary if one takes into account the protection afforded by other related torts… Instead of embarking on a radical extension of the tort of malicious prosecution I would rely on the capacity of our tort law for pragmatic growth in response to true necessities demonstrated by experience.

L

[58] Mr Nugee referred to, and relied on, various Australian cases. In particular, he cited *Varawa* v *Howard Smith Co Ltd* [1911] HCA 46, *Dowling* v *Colonial Mutual Life Assurance Society Ltd* [1915] HCA 56, *QIW Retailers Ltd* v *Felview Pty Ltd* [1989] 2 Qd R 245, *Hanrahan* v *Ainsworth* (1990) 22 NSWLR 73, and *Williams* v *Spautz* [1992] 174 CLR 509. *Varawa* was a case of alleged abuse of process by malicious arrest in proceedings for breach of contract, in which the court held that the evidence did not support the case for damages for abuse of process. *Dowling* was an application to dismiss a petition for sequestration (in insolvency), and so was not a claim for damages in tort. The court held that, on the facts, there was no abuse of process since the petitioning M creditor wanted to use the process for the very object for which it was designed by law, namely sequestration. In *QIW Retailers*, the claim was for abuse of process in seeking an order for the winding-up of the plaintiff, which had been dismissed as an abuse of process. In upholding the claim for abuse of process, the judge proceeded expressly by way of analogy with the *Quartz Hill* categories of damage. In *Hanrahan*, the parties had conceded that the tort of abuse of process was part of the common law of Australia, and the only question was whether the adjudication of the claim of abuse of process should have been left to the jury. *Williams* concerned an application to stay proceedings for criminal conspiracy to defame and to injure on the ground of abuse of process, and so was not a claim for damages in tort.

---

* Editor's note: Reported at [1995] EMLR 485

LS's submissions

[59] On the basis of the authorities, LS formulated the relevant law as follows:

> It is a misuse of legal proceedings for a person to use or threaten them for the predominant purpose of obtaining some collateral advantage to himself beyond the proper scope of the proceedings. Such a misuse (traditionally labelled "abuse of process") is both a ground for staying or striking out the proceedings (if the relevant purpose can be established to the necessary standard on a stay/strike out application); and is a wrongful act which, if it causes damage, is actionable as a tort on the *Grainger* v *Hill* principle.

[60] Mr Nugee accepted that not all the statements of principle in the cases can be easily reconciled. He submitted that the authorities do not make any clear distinction between the jurisdiction to stay and dismiss cases and the tort of abuse of process. He said that, in the context of both, the question is not one of motive but of purpose. If the claimant wants to obtain relief within the proper scope of the proceedings, the reason why is irrelevant. On the other hand, if the predominant purpose is to achieve an objective not within the proper scope of the proceedings, it is irrelevant whether the proceedings have been successful or even brought to a conclusion and whether the claimant had reasonable grounds to believe that they would be successful.

[61] Mr Nugee submitted that the respondent's predominant purpose in bringing the judicial review proceedings can be established only at trial. He said that the witness statements show a difference of recollection and tone, and in some cases outright contradiction, in respect of the critical meeting on 5 April 2006. He submitted that the evidence as a whole discloses a well-arguable case that the dominant purpose of the respondent was not to quash the planning permissions but to put pressure on LS to assist it to relocate its business and to threaten LS with the prospect of financial loss. He submitted that there was no relevant difference between that and simply asking for a cash payment. He emphasised that the respondent was already actively seeking to relocate before it became aware of LS's application for planning permission for the development of Park House. He particularly stressed the significance of LS's evidence that, at the meeting of 5 April 2006, the respondent's representatives had said that they wished LS well with the development of Park House and that their opposition was "purely a business transaction", and Mr Harnett's statements in his strategy note that he expected that LS would ultimately be successful in obtaining planning permission and that there should be no collaboration with other objectors. He pointed out that the respondent was sending to LS, rather than to Westminster, various documents from time to time in connection with the judicial review proceedings, with suggestions that there should be a meeting. He further emphasised that there had never been any suggestion by the respondent that LS was or would be likely to be in breach of any private law obligations in respect of the development of Park House, and, in particular, that LS would be likely to carry out the development in such a manner as to cause actionable nuisance. He submitted that the respondent had no interest in a successful outcome to the judicial review proceedings. It was the threat of the proceedings, he said, that enabled the respondent to have a negotiating position. Success in the proceedings would not have advanced its objectives. In short, LS says that it has an arguable case that the respondent was, through the threat of the judicial review proceedings, seeking to obtain assistance with relocation, which was not something within the scope of the proceedings. That factual case, LS contends, is sufficiently strong to be permitted to go to trial, which would give it the advantage of full disclosure of documents and the opportunity for cross-examination.

[62] Mr Nugee submitted that it is irrelevant that the purpose of the respondent in challenging the planning consent for the Wilton Plaza development in the judicial review proceedings was raised expressly by LS in opposing the grant of permission for those proceedings, and that, notwithstanding LS's objections, Newman J granted unconditional permission. Mr Nugee emphasised that the standing to bring judicial review proceedings has been steadily relaxed. He also submitted that, bearing in mind the normal practice of the Administrative Court, Newman J was not in a position to hear extensive argument and evidence, including oral evidence, on the matter of collateral advantage, which was put in issue by the denial of the respondent that the real reason it was applying for judicial review was to obtain money from LS. He submitted that had the respondent admitted that the purpose of the judicial review proceedings was merely to obtain money from LS, or if the court had considered all the relevant evidence and the arguments, Newman J could and should have refused permission for the judicial review proceedings.

[63] Mr Nugee made the general commercial point that if the respondent's conduct was permissible, it would be possible for others, without any genuine interest in a land development, to bring judicial review proceedings for the purpose of extracting money. He said that LS, in particular, was exposed to such claims because it carries out large and complex developments in central London and, as the present case shows, any delay in the implementation of planning permission can cause significant financial damage.

[64] Finally, Mr Nugee emphasised that the deputy judge found against LS on the summary judgment application on the single ground that the respondent's purpose in bringing the judicial review proceedings was not sufficiently collateral. He submitted that, in reaching that conclusion, the deputy judge had not adhered to his assumptions of fact in [40] of his judgment, and, moreover, the deputy judge had selected one of the many different judicial formulations of the ingredients of the tort, namely that of Bridge LJ in *Goldsmith*, but he had not even applied that test correctly.

[65] In summary, Mr Nugee submitted that the present case raises difficult, novel and arguable points of law that, in the light of the arguable factual case advanced by LS, should be finally determined only after trial: *Three Rivers District Council* v *Bank of England (No 3)* [2001] UKHL 16; [2003] 2 AC 1*, in [95] ff (Lord Hope) and *Equitable Life Assurance Society* v *Ernst & Young* [2003] EWCA Civ 1114.

Discussion and conclusion

[66] Notwithstanding Mr Nugee's eloquent and attractive submissions, I do not consider that the deputy judge was wrong to grant summary judgment against LS.

[67] The authorities provide no basis for extending a tort of abuse of process to the respondent's proceedings for judicial review. The following points appear clearly from the authorities. First, there is no general tort of malicious prosecution of civil cases. On policy grounds, the tort of malicious prosecution in respect of civil cases is confined to the three well-established heads of damage recognised in *Quartz Hill* and *Gregory*. Second, essential ingredients of a claim for malicious prosecution are the absence of reasonable and probable cause and that the proceedings have ended in favour of the person maliciously prosecuted. Third, *Grainger* has never been overruled. It is authority for a tort of abuse of process. The only other case within this jurisdiction in which the tort has arguably been successfully invoked is *Gilding*, more than 140 years ago. Both cases concerned a blatant misuse of a particular process, namely arrest and execution, within existing proceedings. In both cases, the abuse of that process involved compulsion by arrest and imprisonment to achieve a collateral advantage. Fourth, in cases of abuse of process, it is irrelevant whether there was reasonable or probable cause for the proceedings or in whose favour they ended or whether they have ended at all. Fifth, statements in the English authorities describing a broader application of the test of abuse of process than the critical factual elements of *Grainger* and *Gilding* were all *obiter*. In particular, *Majory,* which has been cited in support of a wider formulation, was not a claim in tort but of opposition to a receiving order; it was in any event one of the *Quartz Hill* special circumstances and the opposition to the receiving order failed on the facts. There was no reference to *Grainger*, and Evershed MR referred only to the sanction of prohibiting the abuser from invoking the power of the court by the proceedings he had abused. Sixth, as to the broader statements of principle, there is no clearly accepted approach for

* Editor's note: Reported at [2003] 2 BCLC 603

A identifying what is sufficiently collateral to establish the tort of abuse of process. The analysis that appears to have received most support is that of Bridge LJ in *Goldsmith*.

[68] It is not necessary on this appeal, as it was not necessary for the deputy judge, to define the precise limits of the tort of abuse of process. I consider, however, that even if the tort can be committed outside circumstances of compulsion by arrest, imprisonment or other forms of duress, there is no reasonably arguable basis for extending the tort beyond the other particular heads of damage that must exist for the invocation of the tort of malicious prosecution. A different conclusion would not only go beyond the factual context of *Grainger*
B and *Gilding* but would be inconsistent with the refusal of the House of Lords in *Gregory* to extend the tort of malicious prosecution to all civil proceedings. It makes no sense severely to limit the cause of action for malicious prosecution, an essential ingredient of which is that the proceedings had been brought without reasonable or probable cause, to three particular heads of damage, but to extend to all cases of economic loss a tort of abuse of process that can apply even where the alleged "abuser" had a good cause of action. The dangers of parallel litigation and — echoing the concerns of Slade LJ in *Metall* — deterring the pursuit of honest claims are obvious. The wider descriptions of the tort of abuse of process in cases prior to *Gregory* must be reappraised in the
C light of the decision of the House of Lords in that case and the policy considerations underlying it.

[69] I do not consider that, having regard to their particular facts, the Australian cases assist the analysis further.

[70] Those points apply with particular force to proceedings for judicial review. We were not referred to any case in any jurisdiction in which it has been held or suggested that the tort of abuse of process can apply to judicial review proceedings for which the court has given permission, let alone when the damage relied on is general economic loss outside the limited categories of damage for which the tort of malicious prosecution may be invoked. There is a public interest in
D bringing judicial scrutiny and remedies to bear on improper acts and decisions of public bodies. The permission stage of judicial review is intended to weed out claims without sufficient prospects of success. In *R (on the application of Feakins)* v *Secretary of State for the Environment, Food and Rural Affairs* [2003] EWCA Civ 1546; [2004] 1 WLR 1761, Dyson LJ, with whom the other members of the Court of Appeal agreed, specifically addressed the question of abuse of process in the context of judicial review proceedings. He said:

> 23. In my judgment, if a claimant has no sufficient private interest to support a claim to standing, then he should not be accorded standing *merely* because he raises an issue in which there is, objectively speaking, a public interest.
> E As Sedley J said in *R* v *Somerset County Council, Ex p Dixon* [1997] JPL 1030, when considering the issue of standing, the court had to ensure that the claimant was not prompted by an ill motive, and was not a mere busybody or a trouble-maker. Thus, if a claimant seeks to challenge a decision in which he has no private law interest, it is difficult to conceive of circumstances in which the court will accord him standing, even where there is a public interest in testing the lawfulness of the decision, if the claimant is acting out of ill-will or for some other improper purpose. It is an abuse of process to permit a claimant to bring a claim in such circumstances. If the real reason why a claimant wishes to challenge a decision in which, objectively, there is a public interest is not that he has a genuine concern about the decision, but some other reason, then that is material to the question whether he should be accorded standing.

F

[71] The effect of LS's submissions in the present case is that even though an allegation of abuse is expressly raised as an objection to permission for judicial review and the judge nevertheless grants unconditional permission, as did Newman J in the judicial review proceedings concerning the Wilton Plaza development, it would nevertheless be possible for a party in the position of LS immediately to institute proceedings for abuse of process. It could do so since the tort does not depend on the absence of reasonable and probable cause in bringing the "abusive" proceedings or their failure or, indeed, their termination. Further, the logic of LS's submissions is that even if the respondent had succeeded in the judicial review proceedings for which Newman J had given permission, the economic loss caused to LS from its successful outcome would be recoverable from the respondent in G
tort for abuse of process. There is nothing whatever in the authorities to justify such a remarkable conclusion.

[72] Accordingly, both for those reasons and because the damages claimed by LS are general economic loss outside the *Quartz Hill* categories of damage, LS's claim has no real prospect of success.

[73] Furthermore, even if a wider formulation of the tort of abuse of process was adopted than I consider to be justified on the authorities, whether the wider formulation be that of Bridge LJ in *Goldsmith* or otherwise, the deputy judge was fully entitled to conclude that the interest of the respondent in respect of its property relocation was insufficiently collateral to the judicial review proceedings as to render H
those proceedings abusive. What, in my judgment, emerges clearly from the authorities is that the tort is not committed by a party that institutes proceedings with a genuine interest in, and an intention to secure, their successful outcome, even if the claimant's motives are mixed and it hopes that it may also achieve an objective not itself within the scope of the proceedings. That is often the situation in judicial review proceedings concerning planning matters, as was recognised by the Court of Appeal in *R (on the application of Mount Cook Land Ltd)* v *Westminster City Council* [2003] EWCA Civ 1346; [2004] 2 P&CR 22*. In that case, Auld LJ, with whom the other two members of the Court of Appeal agreed, said in [46]: J

> judicial review applications by would-be developers or objectors to development in planning cases are, by their very nature, driven primarily by commercial or private motive rather than a high-minded concern for the public weal.

[74] LS itself has alleged in para 7.6 of the amended particulars of claim that the intention of the respondent, in issuing and then pursuing the judicial review proceedings, was to obtain protection from alleged future disturbance of the use and enjoyment of its premises. Further, it is not alleged in the amended points of claim or the reply that the judicial review proceedings were instituted without any intention of bringing them to a conclusion. On the contrary, it is perfectly clear from the K
July memo sent to all the respondent's partners that the respondent did indeed address the possibility and consequences of succeeding in the judicial review proceedings, recognising that success in striking down the credit resolution would be likely to have a direct effect on LS's proposals for the development of Park House to the advantage of the respondent. Whatever may have been said at the 5 April 2006 meeting, it has not been suggested, and it would be fanciful to suggest, that the July memo was not a genuine document intended to be taken at face value by the respondent.

Conclusion

[75] For all those reasons, I would dismiss this appeal. L

Giving the second judgment, **Moore-Bick LJ** said:

[76] I agree with the judgment of Etherton LJ and add a few words of my own in view of the importance of the issues raised in this appeal.

[77] Although the concept of abuse of process is well known to the law, both in civil and criminal proceedings, it has rarely been treated as giving rise to a cause of action. More commonly, it is relied on as a ground for terminating proceedings that constitute an abuse and, as such, is a flexible concept for preventing injustice. The circumstances in which the court will regard conduct as amounting to an abuse of process are not narrowly defined, nor should they be. Although certain types M
of abuse are well recognised, it is necessary for the courts to have the power to control their own proceedings and to prevent abuse, whatever guise it may take.

[78] It has been recognised for more than 150 years, however, that in some circumstances abuse of process is actionable as a tort at common law. The existence of such a cause of action can be traced back to *Grainger* v *Hill* (1838) 4 Bing NC 212, in which the defendants had obtained a writ of *capias*, not for the purposes of ensuring the plaintiff's attendance at trial, but in order to extort from him under the pressure

* Editor's note: Also reported at [2004] 1 PLR 29

of arrest the register of the vessel on which the defendants' loan was secured. At the time the writ was issued, the loan had not fallen due and no cause of action had arisen to recover it. The defendants had no interest in obtaining the arrest of the plaintiff; they simply wanted to put pressure on him to hand over the document. Possession of the register was essential to enable the plaintiff to trade the vessel and it had been left in his hands so that he should be in a position to do so. The plaintiff sought to recover the losses that he had suffered as a result of being unable to trade.

[79] The defendants argued that the plaintiff had not alleged, and could not show, that the action had been terminated in his favour (as would be necessary to sustain a claim for malicious prosecution), but Tindal CJ held, at p221, that to be irrelevant since:

> this is an action for abusing the process of the law, by applying it to extort property from the Plaintiff, and not an action for a malicious arrest or malicious prosecution,… In the case of a malicious arrest, the sheriff at least is instructed to pursue the exigency of the writ: here the directions given, to compel the Plaintiff to yield up the register, were no part of the duty enjoined by the writ. If the course pursued by the Defendants is such that there is no precedent of a similar transaction, the Plaintiff's remedy is by an action on the case, applicable to such new and special circumstances; and his complaint being that the process of the law has been abused, to effect an object not within the scope of the process, it is immaterial whether the suit which that process commenced has been determined or not, or whether or not it was founded on reasonable and probable cause.

[80] Vaughan J said, at p223:

> It is an action for abusing the process of law, by employing it to extort property to which the Defendants had no right: that is of itself a sufficient cause of action…

[81] As Etherton LJ has observed, the only other case in which a successful claim has been made in England to recover damages for abuse of process is *Gilding* v *Eyre* (1861) 10 CB(NS) 592, another case in which the defendant obtained the arrest of the plaintiff, this time under a writ of *capias ad satisfaciendum*, a process that issued without the need for judicial intervention. In that case also, the defendant's purpose was to extort property from the plaintiff (in that case, a sum of money in excess of that which remained due under a judgment) under pressure of arrest. *Grainger* does not appear to have been cited, but the court had no difficulty in holding that the pleading disclosed a good cause of action.

[82] *Re Majory (a debtor)* [1955] Ch 600 concerned abuse of a rather different kind. A debtor admitted his liability to the creditor and offered to pay the debt by instalments. The creditor issued a judgment summons for the debt and costs in the sum of £12 5s, having warned the debtor that he would do so. Having entered judgment, the creditor demanded payment of the debt in two substantial instalments, with £21 costs. He then issued bankruptcy proceedings that the debtor opposed on the ground that they amounted to an attempt to extort from him a greater sum than was due. The case is of interest mainly because of the discussion of the meaning of "extortion" in the context of bankruptcy, but it is clear that the court was not concerned with a claim of the kind considered in *Grainger*. Rather, the question was whether the creditor should be prevented from pursuing the proceedings because they involved an abuse of the process. I agree, therefore, that it does not shed any light on the question that we have to decide in this case.

[83] The other English cases on which the appellants rely likewise deal with abuses of a kind that might cause the court to stay or set aside the proceedings. In *Goldsmith* v *Sperrings Ltd* [1977] 1 WLR 478, the plaintiff brought an action for libel against the publishers, editor and main distributors of *Private Eye*, claiming damages and an injunction. He also issued 111 writs against individual wholesale and retail distributors of the magazine, seeking similar relief. Many of the distributors applied for an order that the actions be stayed or dismissed on the ground that they were an abuse of process, being brought for the collateral purpose of destroying the magazine by cutting off its commercial outlets.

[84] Lord Denning MR expressed himself in characteristically broad terms. At p489C-E, he said:

> In a civilised society, legal process is the machinery for keeping and doing justice. It can be used properly or it can be abused. It is used properly when it is invoked for the vindication of men's rights or the enforcement of just claims. It is abused when it is diverted from its true course so as to serve extortion or oppression: or to exert pressure so as to achieve an improper end. When it is so abused, it is a tort, a wrong known to the law. The judges can and will intervene to stop it. They will stay the legal process, if they can, before any harm is done. If they cannot stop it in time, and harm is done, they will give damages against the wrongdoer.

[85] However, Mr Nugee relied on this case principally for passages in the judgments of Scarman and Bridge LJJ. Scarman LJ said, at pp498H-499A:

> In the instant proceedings the defendants have to show that the plaintiff has an ulterior motive, seeks a collateral advantage for himself beyond what the law offers, is reaching out "to effect an object not within the scope of the process": *Grainger* v *Hill* (1838) 4 Bing (NC) 212, 221 *per* Tindal CJ. In a phrase, the plaintiff's purpose has to be shown to be not that which the law by granting a remedy offers to fulfil, but one which the law does not recognise as a legitimate use of the remedy sought: see *In re Majory* [1955] Ch 600, 623.

[86] When Scarman LJ referred to "a legitimate use of the remedy sought", I think that he must have intended to refer to the process by which the remedy afforded by the law is obtained, because he said, at pp499H-500C:

> Men go to law to redress a grievance. They may not know or understand the limits of the remedies provided by law — though no one suggests that Sir James Goldsmith's advisers could be said to suffer from ignorance of the law. But, equally, a man, while pursuing the remedies offered by law, may negotiate to secure, by arrangement with the parties sued, terms more favourable than, or different from, what he would get in the absence of agreement. Such a negotiation, undertaken by properly advised parties, each of whom may have a legitimate interest in avoiding litigation and may be prepared to concede more than the law requires of them to achieve that end, does not necessarily mean that the plaintiff by his litigation is reaching out to secure a collateral advantage. In the context of libel, he may reasonably see in settlement a more effective way of protecting his reputation than by action; and, whether he pursues his litigation to judgment, or settles it, he may in either case be seeking no more than the way he thinks best in the circumstances to protect his reputation. Since that is the object of the law of libel (see *Gatley on Libel and Slander*, 7th ed (1974), Chap 1, p1, para 1) it would, in my judgment, need strong evidence that the plaintiff was in fact seeking something beyond the protection and vindication of his reputation before the court could stay his action as an abuse of process.

[87] That certainly appears to have been the view of Bridge LJ, who said, at p503D-H:

> For the purpose of Lord Evershed's general rule [sc the rule in *Re Majory*], what is meant by a "collateral advantage"? The phrase manifestly cannot embrace every advantage sought or obtained by a litigant which it is beyond the court's power to grant him. Actions are settled quite properly every day on terms which a court could not itself impose upon an unwilling defendant. An apology in libel, an agreement to adhere to a contract of which the court could not order specific performance, an agreement after obstruction of an existing right of way to grant an alternative right of way over the defendant's land — these are a few obvious examples of such proper settlements. In my judgment, one can certainly go so far as to say that when a litigant sues to redress a grievance no object which he may seek to obtain can be condemned as a collateral advantage if it is reasonably related to the provision of some form of redress for that grievance. On the other hand, if it can be shown that a litigant is pursuing an ulterior purpose unrelated to the subject matter of the litigation and that, but for his ulterior purpose, he would not have commenced proceedings at all, that is an abuse of process. These two cases are plain; but there is, I think, a difficult area in between. What if a litigant with a genuine cause of action, which he would wish to pursue in any event, can be shown also to have an ulterior purpose in view as a desired byproduct of the litigation? Can he on that ground be debarred from proceeding? I very much doubt it. But on the view I take of the facts in this case the question does not arise and it is neither necessary nor desirable to try to lay down a precise criterion in the abstract.

[88] In *Speed Seal Products Ltd* v *Paddington* [1985] 1 WLR 1327, the plaintiff sought an injunction to restrain the use by the defendant

of allegedly confidential information. The defendant sought to make a counter-claim for damages for abuse of process, alleging that the proceedings had been brought for the purpose of deterring potential customers from dealing with them rather than for the protection of any legitimate interests. They relied on *Grainger*. Fox LJ, with whom the other members of the court agreed, held that the decision in that case provided the basis for an arguable case that there had been an actionable abuse of the process so as to justify giving leave to amend. However, I do not regard this decision as shedding much light on the question we have to decide. The court was required to do no more than determine whether the claim was sufficiently arguable to go to trial. The threshold for those purposes is not very high and there is no discussion in the judgments of issues of the kind that arise in this case.

[89] In *Broxton* v *McClelland* 30 January 1995, Simon Brown LJ identified as the essential element in abuse of process the misuse of the court's process to achieve something not available in the course of (or, I would say, by means of) properly conducted proceedings. I respectfully agree. It seems to me that whether the question is one of staying or of striking out the proceedings themselves or of the existence of a cause of action, the claimant must be able to establish that the defendant's predominant purpose in bringing the proceedings is not to obtain the remedy that the law offers (disregarding for this purpose the use it may seek to make of that remedy once it has obtained it) but to achieve some other object that lies outside the range of remedies that the law grants. At the level of this principle, I see no difficulty in assimilating the decisions on abuse of process as a tort with the decisions concerning staying or striking out the proceedings.

[90] Abuse of process in the sense described above and malicious prosecution are at one level related since both involve the misuse of the court's process, but at another level they clearly differ, because a claim for malicious prosecution depends on establishing that the claim was not well founded (the claimant must show that the proceedings were determined in its favour and that the defendant had no reasonable and probable cause for bringing them), whereas abuse of process is not concerned with the merits of the claim but with the purpose for which the proceedings are brought. The distinction was well drawn by Isaacs J in *Dowling* v *Colonial Mutual Life Assurance Society Ltd* [1915] HCA 56 in these terms:

> In English law there has long been recognized a form of wrong by malicious use of process — such as by malicious arrest. But in order to maintain an action for malicious *use* of the process there must have been a termination of the suit in the plaintiff's favour. If, however, there has been an *abuse* of the process, as distinguished from the *use* of it, it is unnecessary to show any such termination of the suit. If the object sought to be effected by the process is within the lawful scope of the process, it is a *use* of the process within the meaning of the law, though it may be malicious, or even fraudulent, and in the circumstances the fraud may be an answer; if, however, the object sought to be effected by means of the process is outside the lawful scope of the process, and is fraudulent, then — both circumstances concurring — it is a case of *abuse* of that process, and the Court will neither enforce nor allow it to afford any protection, and will interpose, if necessary, to prevent its process being made the instrument of abuse. *Grainger* v *Hill* 4 Bing (NC) 242 laid down the distinction.

[91] In *QIW Retailers Ltd* v *Felview Pty Ltd* [1989] 2 QdR 245, the first defendant sought a winding-up order against the plaintiff and the second defendant arranged for an article referring to the proceedings to be published in a trade magazine. The primary purpose of doing so was, as the court found, to force the plaintiff's board to enter into negotiations for the control of the company to be ceded to the first defendant. The issue and publicising of a petition for a winding-up order was held to be sufficient to support a claim for damages by analogy with the cases on malicious prosecution.

[92] Finally, it is necessary to refer to *Hanrahan* v *Ainsworth* (1990) 22 NSWLR 73. In that case, the claimant, a police officer, brought proceedings against the defendant alleging that he had started an action for defamation against him for the purposes of intimidation and in order to coerce him into abandoning a criminal investigation. The court held that there was sufficient evidence of abuse to go to the jury and that if abuse were established the claimant was entitled to claim damages in respect of it, and was not confined to the remedy of having the proceedings struck out.

[93] If abuse of process consists in making use of proceedings in order to achieve some object other than one of the remedies that it lies in the court's power to grant, it might seem obvious that proceedings for judicial review could be abused as much as proceedings of any other kind. In principle, that may be so, but it is much less likely to be the case in practice. There are at least two reasons for that. The first is that it is necessary to obtain permission to proceed with a claim for judicial review, and that requires the court to consider whether the claim has a real prospect of success and, more generally, whether the applicant should be allowed to pursue the claim. If the defendant or an interested party considers that the proceedings constitute an abuse, that can be drawn to the attention of the court, which will consider the position before permission is given. The remarks of Dyson LJ in *R (on the application of Feakins)* v *Secretary of State for the Environment, Food and Rural Affairs* [2003] EWCA Civ 1546; [2004] 1 WLR 1761, to which Etherton LJ has referred, make that clear.

[94] The second reason lies in the public law nature of the proceedings themselves, the essential nature of which is to ensure that a public body complies with the law. That does not mean that the claimant will not be seeking to serve a private interest of its own; in very many cases, it will, and will be expecting to further that interest as a direct or indirect result of obtaining the relief that it seeks. Whatever may be the claimant's private purpose in commencing and continuing the proceedings, however, the public has an interest in ensuring that breaches of the law by public bodies are identified and, where appropriate, corrected. It is difficult, therefore, to contemplate a case in which the Crown, in whose name the proceedings are brought, does not have a proper interest in obtaining whatever remedy the court may see fit to grant.

[95] Although rarely applied, I do not think that the principle in *Grainger* has ever been seriously doubted. It is treated as being good law in the leading textbooks on tort and has been applied in Australia, as the cases to which I have referred show. I am therefore unable to accept the broad submission made by Mr Steinfeld that the issue and pursuit of civil proceedings can never amount to a tort, even where the predominant purpose of the claimant is to obtain a "collateral benefit", by which I understood it to mean some object other than the relief available from the court or some other advantage that depends on obtaining such relief. However, it does not follow that any conduct of that kind will be actionable as an abuse of process. Indeed, for reasons to which I shall come in a moment, I am satisfied that the tort has a very limited application under modern conditions.

[96] In *Gregory* v *Portsmouth City Council* [2000] 1 AC 419, the claimant, a local councillor, brought an action for malicious prosecution against the council as a result of his having been subjected to internal disciplinary proceedings that were eventually resolved in his favour. The claim was struck out as disclosing no reasonable cause of action. Lord Steyn, with whom the other members of the appellate committee agreed, said, at p427B:

> In English law the tort of malicious prosecution is not at present generally available in respect of civil proceedings. It has only been admitted in a civil context in a few special cases of abuse of legal process. Sometimes these cases are described as constituting a separate tort of abuse, but in my view *Fleming, The Law of Torts,* 9th ed. (1998), p687 is correct in observing that they "resemble the parent action too much to warrant separate treatment."

[97] It is not clear from that passage whether Lord Steyn was intending to refer only to the malicious institution of unfounded civil proceedings or to the malicious abuse of process more generally. *Clerk & Lindsell on Torts* (19th ed), in para 16-05, treats his remarks as extending to the latter, but *Grainger* was not cited in the House and the various examples of the tort to which Lord Steyn refers all involve acting maliciously and without reasonable and probable cause. In fact, in the same paragraph of *Fleming* the author draws a distinction between those torts that are said to resemble the parent action too much to warrant separate treatment and those cases in which a legal process, not itself devoid of foundation, has been perverted for some

predominant extraneous purpose. *Grainger* is cited as the leading example. *Gregory* is not, therefore, direct authority for the scope of the tort of abuse of process, but, even so, it has profound implications for any development of the tort of the kind that would be required to enable the appellants to succeed in this case.

[98] The origins of the tort as an action on the case (see *Grainger*) mean that damage is of the essence of the claim and it is therefore relevant to enquire whether an action will lie in respect of damage of any particular kind. In both *Grainger* and *Gilding*, the damage consisted of an interference with the plaintiff's liberty. No English case appears to have gone beyond that, but I think it is arguable that damage of any kind that would support a claim for malicious prosecution of civil proceedings (which is a different form of abuse of process) would suffice. If so, the tort would extend to cases of injury to person, damage to property and damage to reputation ("fair fame and credit"): see *Quartz Hill Consolidated Gold Mining Co* v *Eyre* (1883) LR 11 QBD 674. To hold that it extends to damage of other kinds, however, would involve a significant departure from the position hitherto recognised. Although the early authorities recognise that damage to property includes pecuniary loss, they do not support the conclusion that it extends to consequential economic loss rather than the immediate loss of property obtained by coercion and the cost of defending the proceedings. Before a general right to recover costs was created, the expense incurred by a defendant in contesting proceedings was damage of a kind that would support a claim for malicious prosecution (see *Savill* v *Roberts* (1698) 12 Mod Rep 208), but that is no longer the case since, as Bowen LJ pointed out in *Quartz Hill*, an award of costs is deemed to provide a complete indemnity in respect of the reasonable costs of the action.

[99] As Mr Nugee has demonstrated, there are arguments in favour of recognising other kinds of damage, including consequential economic loss, at least in cases where the predominant purpose of instituting proceedings has been to inflict such loss as a means of coercion, but there is no clear authority that supports such a development of the law. As Etherton LJ has noted, most of the Australian cases were concerned with other questions. There is some discussion in the judgment of Macrossan J in *QIW Retailers* of the nature of the damage that will support an action for abuse of process, in particular whether it is limited to those kinds of damage that support an action for malicious prosecution, but it was unnecessary for the judge to reach a firm conclusion on the question because he was satisfied that the nature of the proceedings were such that the case was covered by the principles set out in *Quartz Hill* and earlier cases, which he applied by way of analogy. Only *Hanrahan* might be said to support the proposition that the tort has a wider scope. It is difficult to identify the nature of the damage in respect of which the claimant in that case sought to recover, but all the members of the court appear to have accepted that damage of any kind would suffice to support a claim. The only English case that lends any support to the proposition that a claim for abuse of process may lie in respect of economic loss is *Speed Seal Products*. The precise nature of the loss that the defendant sought to recover by its counter-claim in that case is not clearly identified in the report, but since it was complaining that the action was being used as a weapon to persuade potential customers not to deal with it, I think it can safely be inferred that it was seeking to recover economic loss in one form or another resulting indirectly from the institution of the proceedings. The case appears to have been put primarily as one of malicious prosecution and, alternatively (and perhaps as something of an afterthought), as one of abuse of process, relying on *Grainger*. In his judgment, Fox LJ discussed the essential elements of the tort but without referring to the nature of the damage that would support it. The decision in *Gregory* still lay well in the future. For that reason and for the reasons mentioned earlier, I do not think that the decision provides any support for the claimant's case.

[100] It may be that the development of the tort of abuse of process has taken a different path in Australia from that which it has taken in England. In my view, however, there are strong policy reasons for limiting to a necessary minimum the range of circumstances in which the prosecution of well-founded civil proceedings will give rise to a cause of action. In general, people should be free to take action to vindicate their rights without facing the threat of collateral proceedings. In *Gregory*, their lordships did not think it necessary or appropriate to extend the tort of malicious prosecution to civil proceedings generally or (it would seem) to any kinds of damage other than those that had been recognised in *Quartz Hill* and the earlier cases. That was partly because they considered that there were other remedies capable of providing adequate protection to a person against whom civil proceedings had been taken maliciously and without reasonable and probable cause. In my view, similar considerations apply in the case of abuse of process. There is an effective remedy available in the exercise of the court's inherent power to stay the proceedings and to award costs against the claimant on an indemnity basis. I am unable to accept Mr Nugee's submission that this case proves the contrary. The fact is that the appellants could have taken more vigorous steps to persuade the court that the claim for judicial review was being used as an aid to extortion and, therefore, as nothing more than a device to cause them harm, and if the court had been persuaded that the proceedings were improper relief would have been granted.

[101] I am not persuaded that there are sufficient grounds for extending the scope of the tort of abuse of process in order to recognise damage of a kind that has not hitherto been regarded as supporting a cause of action. The result is that the abolition of arrest in support of civil proceedings and the introduction of a power to award costs when it is in the interests of justice to do so means that the tort of abuse of process now has a much reduced role. I should not wish to exclude the possibility that a claim could be made for personal injury in the form of psychiatric harm resulting from abusive proceedings, but such cases are likely to be very rare. Although the tort may still live on in the context of bankruptcy and winding-up proceedings, even there the likelihood is that its role will be limited. In those circumstances, Mr Steinfeld's broad submission, to which I referred earlier, is not wholly wide of the mark.

[102] Mr Nugee submitted that the matter should be allowed to go to trial in order for the issues of law to be determined on the basis of the judge's eventual findings of fact. In many cases that would be a powerful argument, since a clear understanding of the facts is essential to enable the court to reach a sound conclusion on the law. In the present case, however, I do not think that it would be right to take that course. The nature of the loss that the appellants seek to recover is not in doubt; nor is it in doubt that it is not the kind of loss that has hitherto been accepted as supporting a claim for abuse of process. The critical question on which the claim turns is whether the scope of the tort should be extended to economic loss of the kind in issue and that can be answered without the need for a trial.

[103] For the reasons that I have given, I do not think it should and I therefore agree that the appeal should be dismissed.

Giving the third judgment, **Mummery LJ** said:

[104] I agree with both judgments. Etherton LJ's judgment on the arguments and the authorities is comprehensive. This judgment is confined to a few brief general points on the excellent arguments from each side about the current size of the tort of abuse of legal process.

[105] First, on the breadth of the tort, I see that *Oxford Principles of English Law: English Private Law* (2nd ed) Andrew Burrows QC, treats "harassment by legal process" as an instance of an intentional wrong. The discussion covers both malicious prosecution and, "on a somewhat lower plane", the tort of abuse of legal process. The statement that the tort "is committed when litigation is instituted in pursuit of an ulterior motive" is qualified:

> 17.368 … the exact shape of this tort remains uncertain, and even its existence has been viewed with scepticism. Nevertheless its existence is warranted by *Grainger* v *Hill*, where the defendant was made liable for using arrest for debt as a means to force the claimant to give up the registration documents of his vessel, without which he could not sail.

[106] The definition proposed by Mr Nugee is very broad: instituting legal proceedings for the predominantly collateral or ulterior purpose of obtaining a benefit for which the proceedings were not designed. Its

A potential scope was demonstrated by Mr Nugee's assertion that the action lies even in cases where: (a) the defendant to the tort claim has been successful in the allegedly abusive legal process; and (b) the only compensation sought, as here, is for pure economic loss. Liability stems from the presence of an improper ulterior purpose in the legal process. Unlike malicious prosecution, liability does not depend either on the outcome of the earlier litigation or on the special nature of the damage inflicted — damage to reputation, to person or to property, and including the pecuniary costs and expenses of defending malicious criminal or civil proceedings.

[107] Second, recovery for pure economic loss. There is no **B** generalised tort in English law of intentionally inflicting economic loss on another person. No case was cited of recovery for pure economic loss intentionally inflicted by malicious criminal or civil proceedings or by abuse of legal process.

[108] Third, the relevance of ulterior purpose. The institution and pursuit of legal proceedings is an exercise of the fundamental right of access to the courts. As a general rule, the exercise of a legal right is not an unlawful abuse of that right merely by reason of a predominant improper or ulterior purpose: *Bradford Corporation* v *Pickles* [1895] AC 587.

[109] Fourth, the importance of procedural law. In legal proceedings, **C** the interests of the parties and of the public are protected by procedural law, not just by substantive law. The right of access to the courts does not confer an unlimited licence to litigate. The courts have a range of wide discretionary powers to control the conduct of the parties, to safeguard the public interest and to protect the courts themselves from misuse of legal process, for example, powers to stay or strike out proceedings, to impose a range of conditions on the conduct of the litigation and to order payment of legal costs and expenses. There is no pressing need to supplement procedural law by expansive substantive tortious liability in order to protect litigants in civil proceedings from malicious or abusive claims.

D [110] Fifth, the settled shape of an action for malicious prosecution of civil proceedings. An action for malicious prosecution of civil proceedings lies only if: (i) the abusive legal proceedings have failed; (ii) they were brought without reasonable cause and maliciously; and (iii) there has been special damage to reputation, person or property, and including the pecuniary costs and expenses of contesting the malicious proceedings. Since the House of Lords in *Gregory* v *Portsmouth City Council* [2000] 1 AC 419 recently declined to expand the tort of malicious prosecution to cover civil proceedings generally, it does not seem to me to be consistent with current legal policy to mould a tort of abuse of legal process into a shape as large as Mr Nugee contended.

E [111] Sixth, the special character of judicial review proceedings. I doubt whether it is sensible to lump together and make broad generalisations about all kinds of legal process, treating their abuse **F** as giving rise to the same policy or practical considerations. In this case, the process was judicial review challenging the lawfulness of an exercise of planning powers by a public authority. Access to a court for that purpose is an important right of every citizen affected by the decision or act under challenge. It would be unfortunate if a person could be pressured into desisting from such action by the threat of third-party civil liability for substantial damages, even if the judicial review application succeeded.

[112] The special judicial review procedure under CPR 54 governing claims of abuse of power has robust inbuilt protections from abuse of process for those directly affected by the claim. Unlike the case of an **G** ordinary action, the prior permission of the court is required to proceed with a claim for judicial review. There is a short limitation period. There is no automatic right to disclosure of documents or to test evidence by cross-examination. Interested parties can be joined or can apply to be joined in the application at any stage. As in this case, the interested parties in judicial review proceedings are usually beneficiaries of the exercise of power the lawfulness of which is challenged. A person receives the benefit of a planning decision. The lawfulness of the permission is liable to be judicially reviewed. The recipient of the planning permission naturally wants to support the validity of the decision under challenge. It is essential that such a person has the right **H** to participate and be heard in the protection of his interests. It is also desirable that the applicant for judicial review should not be deterred from pursuing an application, once permission has been obtained for an arguable case that is considered fit for further consideration, for fear that, win or lose, there may be exposure to a liability to compensate the interested party for pure economic loss suffered in consequence of the claim.

[113] Finally, on the authorities, the wrong complained of in the seminal case of *Grainger* is very far removed from the claim in the present case. I do not doubt the correctness of that early decision on its facts. I only doubt the width of the legal principle that eminent **J** writers, such as Professor Prosser, and distinguished judges, such as Lord Denning, have managed to extract from it. That case was one of blatant and unjustifiable misuse of a coercive and intimidating process violating the personal freedom of an individual for an improper and ulterior purpose — arrest and extortion of valuable property to which the plaintiff in the first action was not entitled. It is nothing like a claim for pure economic loss alleged to have been suffered as the result of an application for judicial review challenging the validity of a public authority's grant of planning permission. I hesitate over placing too much reliance on it, 170 years later, as a good guide to the shape of a tort for abuse of process by an application for judicial review. **K**

Appeal dismissed.

LEASEHOLD ENFRANCHISEMENT

The Wellcome Trust Ltd *v* Baulackey

CENTRAL LONDON COUNTY COURT

6 April 2009

MR RECORDER MURRAY ROSEN QC

Leasehold enfranchisement — Leasehold Reform, Housing and Urban Development Act 1993 — Lease extension — Lessee serving section 42 notice on freeholder claiming lease extension — Notice served more than two years after lessee completing purchase of flat but less than two years after registration of title — Section 39(2) of 1993 Act — Whether qualifying tenant for required period of two years prior to service of notice — Whether section 39(2) concerned only with period during which lessee having registered legal title

The defendant contracted to purchase a leasehold interest in a flat. The contract stipulated that the vendor would serve a notice on the claimant freeholder, under section 42 of Chapter II of the Leasehold Reform, Housing and Urban Development Act 1993, claiming a new lease of the flat. This was done in July 2007 and the sale was completed shortly thereafter. By a counternotice, the claimant argued that the vendor was not entitled to a new lease because she had not, at the relevant time, been a qualifying tenant for the necessary period of two years under section 39(2) of the 1993 Act. It relied on the fact that although the vendor had purchased the flat in June 2005, more than two years before the service of her section 42 notice, she had not been registered as the lessee until September 2005.

The claimant applied to the court for declaratory relief. It contended that the 1993 Act was concerned only with the legal title rather than with the equitable rights acquired on exchange of contracts and that section 101(2) of the Act, which provided that an agreement for a lease or tenancy qualified as a tenancy for the purposes of Part I of the Act, did not support the contrary view but merely set out a specific exception to the general rule in a situation where there was no risk of a conflict with the section 5(3) requirement that no more than one person could be regarded as a qualifying tenant of a particular flat at any one time.

***Held*: The claim was allowed. The system set out in Chapter II of the 1993 Act depends on there being one qualifying tenant at any one time. Accordingly, subject to specific exceptions, only the legal owners on both sides, namely the landlord and the tenant, can be involved in the process and have the necessary rights and obligations. The requirement for the tenant to have been a qualifying tenant for two years before the relevant date is intended to be a simple test and requires registration throughout that period. To hold otherwise would damage the overall structure of the Act and create a risk of there being more than one qualifying tenant. The provision in section 101(2) for agreements for leases does not derogate from the general principle that a tenant becomes a tenant only on acquiring the legal ownership of the lease; a distinction should be made between the vertical transactions contemplated by section 101(2) and the horizontal transaction involved in a sale of the leasehold interest by the lessee. The declaration sought by the claimant was granted.**

The following cases are referred in this report.

Biles v *Caesar* [1957] 1 WLR 156; [1957] 1 All ER 151, CA
Brown & Root Technology Ltd v *Sun Alliance & London Assurance Co Ltd* [2001] Ch 733; (1997) 75 P&CR 223; [1997] 1 EGLR 39; [1997] 18 EG 123, CA
Gentle v *Faulkner* [1900] 2 QB 267; 16 TLR 397
Pearson v *Alyo* [1990] 1 EGLR 114; [1990] 25 EG 69, CA
Roberts v *Church Commissioners for England* [1972] 1 QB 278; [1971] 3 WLR 566; [1971] 3 All ER 703, CA
Scribes West Ltd v *Relsa Anstalt (No 3)*; *sub nom Scribes West Ltd* v *Anstalt (No 3)* [2004] EWCA Civ 1744; [2005] 1 WLR 1847; [2005] 2 All ER 690; [2005] 2 P&CR 3; [2005] 1 EGLR 22; [2005] 09 EG 190
Smith v *Express Dairy Co* [1954] JPL 45
Walsh v *Lonsdale* (1882) LR 21 Ch D 9, CA

This was the hearing of a claim by the claimant freeholder, The Wellcome Trust Ltd, against the defendant, Nazli Baulackey, for declaratory relief in respect of a leasehold enfranchisement claim by a tenant under the Leasehold Reform, Housing and Urban Development Act 1993.

Anthony Radevsky (instructed by CMS Cameron McKenna LLP) appeared for the claimant; Andrew Skelly (instructed by Percy Short & Cuthbert) represented the defendant.

Giving judgment, **MR RECORDER MURRAY ROSEN QC** said:

Introduction

[1] This case concerns the validity of a claim to a new lease of flat B, 3 Egerton Gardens, London SW3, which is made by Ms Michelle McLaughlan. Ms McLaughlan was at the time a lessee of the flat and served a notice under section 42 of the Leasehold Reform, Housing and Urban Development Act 1993 (the 1993 Act). The claimant, The Wellcome Trust Ltd, is the freeholder of the flat and the defendant is now the lessee, having succeeded Ms McLaughlan. I will refer to the defendant simply as Ms Baulackey.

[2] The Wellcome Trust issued a claim form under CPR 8 on 15 November 2007, seeking an order, under section 46 of the 1993 Act, declaring that as at the relevant date on which Ms McLaughlan served her notice, namely on 5 July 2007, she had no right to a new lease, essentially because she was not a qualified tenant for the purposes of the 1993 Act. The Wellcome Trust served a counternotice essentially to that effect on 26 September 2007.

Brief facts

[3] The relevant facts are set out in witness statements from Mr Charnock, on behalf of the claimant, and Ms Baulackey, the defendant. The facts are not in dispute and there has been no oral evidence and no cross-examination before me.

[4] The timetable, which is all that is needed, essentially, for present purposes, is as follows. Ms McLaughlan was registered as the lessee of the flat on 8 September 2005, having completed her purchase of it from an unknown predecessor more than three months earlier, on 3 June 2005. She contracted to sell the leasehold interest to Ms Baulackey by exchange on 27 June 2007. The sale and purchase was completed quickly on 6 July 2007. Just before completion, Ms McLaughlan served the notice now in issue under section 42. She had contracted to do that and the notice was in a satisfactory form as far as Ms Baulackey was concerned. The registration of Ms Baulackey's leasehold acquisition was effected on 16 August 2007.

A [5] The crucial point as far as that chronology is concerned is that the notice now in issue, served by Ms McLaughlan on 5 July 2007, was more than two years after she had purchased the leasehold on 3 June 2005 but less than two years before her leasehold interest was registered on 8 September 2005. I ought to mention that the lease in question was a lease dated 25 June 1986, for 32 years less 10 days from 25 September 1982 and so expires in September 2014.

1993 Act

[6] The relevant provisions of the 1993 Act are scattered throughout it. The logical place to start is section 39, which is the first section in B Chapter II, the chapter entitled "Individual right of tenant of flat to acquire new lease".

[7] Section 39 is headed "Right of qualifying tenant of flat to acquire new lease" and subsection (1) provides that:

> (1) This Chapter [that is Chapter II] has effect for the purpose of conferring on a tenant of a flat, in the circumstances mentioned in subsection (2), the right, exercisable subject to and in accordance with this Chapter, to acquire a new lease of the flat on payment of a premium determined in accordance with this chapter.

[8] Subsection (2) provides that:

> C (2) Those circumstances are that on the relevant date for the purposes of this Chapter —
> (a) the tenant has for the last two years been a qualifying tenant of the flat; …

[9] Subsection (3) provides that:

> references in this Chapter to a qualifying tenant of a flat shall accordingly be construed by reference to [the] provisions [of Chapter I].

[10] I will come back to those subsections shortly. Subsection (8) provides that:

> D (8) In this Chapter "the relevant date", in relation to a claim by a tenant under this Chapter, means the date on which notice of the claim is given to the landlord under section 42.

That is, in the present case and for present purposes, 5 July 2007.

[11] There are two sections in Chapter I that are pertinent. First, in section 5, which is headed "Qualifying tenants", there is a definition reading as follows in subsection (1):

> (1) Subject to the following provisions of this section, a person is a qualifying tenant of a flat for the purposes of this Chapter [I would add in parentheses and for Chapter II, as we have seen] if he is a tenant of the flat E under a long lease…

[12] Subsection (3) of section 5 provides that: "No flat shall have more than one qualifying tenant at any one time." That subsection has a crucial role in the argument, as I shall explain soon.

[13] Subsection (4) fills that out by beginning "Accordingly" and subsection (4)(a) reads:

> (4)(a) Where a flat is for the time being let under two or more leases to which subsection (1) applies, any tenant under any of those leases which is superior to that held by any other such tenant shall not be a qualifying tenant of the flat for the purposes of this Chapter;…

F [14] That, too, has a role in the argument, as I shall explain, in particular in the distinction to be drawn between: (a) a vertical hierarchy, under which there may be leases and subleases or agreements for leases and subleases between a freeholder, a head tenant and a subtenant; to be contrasted with (b), what I would call the horizontal hierarchy, in which a leasehold interest or subleasehold interest might be transferred by assignment between one tenant and a successor tenant or one subtenant and a successor subtenant.

[15] Also in Chapter I is section 7, headed "Meaning of 'long lease'", which says, in subsection (1), that:

> (1) In this Chapter "long lease" means (subject to the following provisions of this section) —
> (a) a lease granted for a term of years certain exceeding 21 years, whether or not it is (or may become) terminable before the end of that term by notice given by or to the tenant or by re-entry, forfeiture or otherwise; G

The lease in this case, as I have mentioned, was just under 32 years when granted in 1986 and therefore substantially more than the 21-year limit in section 7.

[16] I ought to mention a few other provisions of the 1993 Act. First of all, there is section 40(1), part of Chapter II, which defines a landlord:

> (1) In this Chapter "the landlord", in relation to the lease held by a qualifying tenant of a flat, means the person who is the owner of that interest in the flat H which for the time being fulfils the following conditions, namely —
> (a) it is an interest in reversion expectant (whether immediately or not) on the termination of the tenant's lease, and
> (b) it is either a freehold interest or a leasehold interest whose duration is such as to enable that person to grant a new lease of that flat in accordance with this Chapter,
> and is not itself expectant (whether immediately or not) on an interest which fulfils those conditions.

[17] That sort of definition is found in other statutes, in particular the Landlord and Tenant Act 1954 (the 1954 Act), which is the subject of a number of the authorities cited before me. Of course, in contrast with J the way in which the statute defines a landlord in respect of a lease held by a qualifying tenant of a flat, there is no definition of what constitutes a tenant who qualifies under section 31.

[18] Section 42 of the Act, in Chapter II, is headed "Notice by qualifying tenant of claim to exercise right" and begins:

> (1) A claim by a qualifying tenant of a flat to exercise the right to acquire a new lease of the flat is made by the giving of the notice of the claim under this section.

The section goes on to specify to whom the notice must be given, including the landlord, and what the notice must contain.

[19] Section 46 is headed "Proceedings relating to validity of K tenant's notice" and provides, in subsection (1), that:

> (1) Where —
> (a) the landlord has given the tenant a counter-notice under section 45 which… contains such a statement as is mentioned in subsection (2)(b) of that section, and
> (b) the court is satisfied, on an application made by the landlord, that on the relevant date the tenant had no right under this Chapter to acquire a new lease of his flat,
> the court shall by order make a declaration to that effect.

That is the provision on which the claimant, The Wellcome Trust, relies L to give this court the power to make the declarations sought by it.

[20] Finally, I would set out part of section 101, which deals with the general interpretation of Part I of the 1993 Act, which includes both Chapter I and Chapter II, and subsection (2) of which states that:

> (2) In this Part "lease" and "tenancy" have the same meaning, and both expressions include (where the context so permits) —
> (a) a sub-lease or sub-tenancy, and
> (b) an agreement for a lease or tenancy (or for a sub-lease or sub-tenancy), but do not include a tenancy at will or at sufferance…

Argument M

[21] I am very grateful indeed to the parties and in particular counsel, Mr Anthony Radevsky, for The Wellcome Trust, and Mr Andrew Skelly, for Ms Baulackey, for their very concise and helpful presentation of the matter before me. The argument concentrated on the issue of whether Ms McLaughlan became a tenant for the purposes of the Act, in particular section 39 of the Act, when she acquired the lease from her predecessor or whether she became a tenant only on registration of her title as obtained some four months or so later in 2005.

[22] Ms Baulackey did not raise any issue based on her having been treated as a matter of fact as a tenant by her exclusive occupation of the flat and the payment of rent or service charge or any other matters or conduct that would give rise to an estoppel on the part of the

landlord, preventing The Wellcome Trust from denying that she had become a tenant on her acquisition of the lease prior to registration. I therefore summarise the arguments on the limited legal issue that I bear in mind.

[23] On behalf of the claimant, The Wellcome Trust, the submission was made that Ms McLaughlan did not become a qualifying tenant on the assignment of the lease from her predecessor unless and until she became the registered proprietor. This was a case of registered land and section 27 of the Land Registration Act 2002 (the 2002 Act) applies.

[24] Section 27 of the 2002 Act provides that:

> (1) If a disposition of a registered estate or registered charge is required to be completed by registration, it does not operate at law until the relevant registration requirements are met.
>
> (2) In the case of a registered estate, the following are the dispositions which are required to be completed by registration —
>
> (a) a transfer;

[25] The claimant's submission was that: (i) until registration took place, Ms McLaughlan was the equitable owner; (ii) she acquired that equitable status on exchange of contracts; and (iii) as between her and her predecessor, there was a relationship of trustee and beneficiary. However, in *law*, the completion of her purchase and the execution of the transfer by her predecessor made no difference to her status. She remained no more than an equitable owner and it was only on registration that a change in legal ownership was effected.

[26] It was next submitted that the tenant would not be concerned, for the purposes of the 1993 Act and, indeed, other Acts, with any beneficial ownership of the landlord's interest but would deal only with the legal owner, the legal owner, namely the registered landlord, being the only party that can serve notices and otherwise effect statutory rights. The same should apply with regard to the tenant, the landlord not being concerned with any beneficial ownership of a tenant's interest but only with the legal ownership of that interest. So while between the lessee and her predecessor there would be an equitable assignment on the exchange of contracts or the execution of a transfer, only registration effects the legal interest and their relationships with third parties.

[27] It was suggested that that was emphatically highlighted by cases such as *Gentle* v *Faulkner* [1900] 2 QB 267, which led complainants and others subsequently to make sure that equitable assignments would be treated in the same way as parting with possession. Absent some express reference in equitable assignments, they would not be treated as having legal effect for the purposes of general covenants.

[28] A number of other cases were cited as examples of this principle, largely in the context of who is to constitute a landlord for the purposes of the 1954 Act: for example, *Biles* v *Caesar* [1957] 1 WLR 156 and *Pearson* v *Alyo* [1991] 1 EGLR 114*. Other cases appear to support a similar line of reasoning, at least as concerns landlords: see, in particular, *Smith* v *Express Dairy Co* [1954] JPL 45 and *Brown & Root Technology Ltd* v *Sun Alliance & London Assurance Co Ltd* [2001] Ch 733†, which concerned the lessee executing a transfer of the lease but that transfer not being registered at the Land Registry.

[29] Finally, in respect of the arguments as laid out by the claimant, The Wellcome Trust, they drew attention to section 101(2) of the 1993 Act, which I have rehearsed, suggesting that that constitutes an exception to what it would submit was the general rule requiring the registration of the tenant's interest in order to constitute its qualification as a tenant. That exception would allow a tenant that becomes a tenant in equity under a specifically enforceable provision of a lease to count as a qualifying tenant under the Act, reflecting the principle in *Walsh* v *Lonsdale* (1882) LR 21 Ch D 9.

[30] The point that was made on behalf of the claimant was that that is an exception to the generality as set out in the specific statutory provision at section 101(2), and was an exception that merely reflected established law to the effect that an agreement for a lease was as good as a lease and did not give rise to any conflict with section 5(3) of the 1993 Act because it did not give rise to the risk that there might be two people simultaneously being regarded as a qualifying tenant. In contrast, the claimant says, if the assignment of a lease without or prior to registration were to be treated as giving rise to a tenancy for this purpose as far as the assignee was concerned, that problem would arise. So, too, would it arise if, for example, a registered tenant were an express trustee, not by virtue of the sale and assignment of the lease but simply under a trust holding for the benefit of another.

[31] So it is said by the claimant that, by contrast with section 101(2), where the disposal of vertical interests would not give rise to a conflict with section 5(3), a horizontal transaction between a registered tenant and either a purchaser or some other beneficiary would give rise to such a problem were the statute to be construed such as to allow a purchaser to qualify as a tenant following mere purchase rather than registration.

[32] On behalf of the defendant, the starting point for the submissions was, as one might expect, entirely different. What was said was that the 1993 Act did not give any special definition to the word "tenant" or "tenancy" so as to limit those who would count as such a person to registered owners. Attention was drawn to the fact that the requirement for the tenant to have been for the last two years prior to the relevant date of giving notice a qualifying tenant of the flat suggested that there had to be a continuous period during which the tenant counted as such, and that it might well be that there would be a delay or registration gap during which the successor tenant had done everything by way of taking an agreement of sale and purchase of the lease and paying the price, but prior to registration. It was suggested that there was no reason why, with regard to that period, that same person should not be treated as having qualified.

[33] The defendant's submission was that a tenant for this purpose should be treated as any person with exclusive possession or occupation of the flat dealing with the normal incidents of tenancy, such as the payment of ground rent and so on.

[34] Some authority was relied on to suggest that this was right, in particular *Scribes West Ltd* v *Relsa Anstalt (No 3)* [2004] EWCA Civ 1744* and *Roberts* v *Church Commissioners for England* [1971] 3 WLR 566. It was suggested that *Brown & Root* was to be distinguished and *Scribes* was to be preferred, even though *Scribes* appeared to turn simply on the interpretation of a statutory phrase, namely section 141(2) of the Law of Property Act 1925, that provided that covenants might be enforced by "a person... entitled... to... income": note the emphasis there on persons entitled, as opposed to persons who have their status by ownership, whether legal or otherwise.

[35] With regard to *Roberts*, it was suggested that this case is helpful in showing what is required to constitute a long tenancy "in law or equity", notwithstanding, again, that that case was concerned with a statutory definition in which the phrase "in law or in equity" specifically appears.

Conclusions

[36] Although I have been attracted by some aspects of the arguments on both sides, in the end I consider that the right answer in this matter is clear. I accept the submissions of the claimant, The Wellcome Trust. Looking at the statute and the statutory provisions on their face, it appears to me that the system set out in Chapter II of the 1993 Act is one that: (a) depends very much on the requirement that there be only a single qualifying tenant at any one time; and (b) suggests very strongly that it is the legal owners on both sides, the landlord and the tenant (subject of course to specific explanations, exceptions or provisos), that are those who are to take part in the process and have the rights and concomitant obligations under the process.

[37] I am not impressed by the suggestion made at one stage on behalf of the defendant that the record of parliamentary debates indicated that the 1993 Act would replace residential requirements for tenants seeking an enfranchisement, or, rather, new leases, with the qualification of a two years' ownership condition in order to address the danger from short-term speculators and therefore, the defendant would

* Editor's note: Also reported at [1990] 25 EG 69

† Editor's note: Also reported at [1997] 1 EGLR 39; [1997] 18 EG 123

* Editor's note: Reported at [2005] 1 EGLR 22; [2005] 09 EG 190

infer, was not intended to distinguish between those who had had two years since they purchased the flat and those who had been legal owners of the flat for two years. I do not consider that that purpose, if it was a purpose, is needed to resolve any ambiguity in the 1993 Act or is or was the subject of a clear expression by the promoter of the 1993 Act. I am not persuaded that any debate with regard to the purpose of the 1993 Act assists in any way with regard to the relevant issue.

[38] What assists more, in my judgment, are two things: first, the analogies with landlord and other rights with regard to transfers and transferees in the various cases to which I have very briefly referred, and, second, the good sense of equating the system under the 1993 Act with other statutory regimes in which notification between legal owners gives rise to rights.

[39] With regard to the section 101(2) point, I am entirely satisfied that the fact that the 1993 Act provides for agreements for leases or subleases does not in any way derogate from the general principle that a tenant becomes a tenant only on acquiring the legal ownership of the lease, that is, on registration. I consider that to construe otherwise would actually damage the overall structure of the 1993 Act in giving rise to a risk that there would be more than one qualifying tenant under section 5(3). The distinction, as I have sought to adumbrate it, between vertical transactions and horizontal transactions can, I think, be well explained and maintained in that context.

[40] There was some debate as to whether what mattered by way of (a) continuity of ownership over the two years and (b) service of notices could be dealt with by some construction of the 1993 Act that required the tenant to be registered at the time of serving the notice, even though not necessarily registered for the whole of the previous two years — if, for example, for some small part of that period, the tenant had completed his or her purchase but there was a small registration gap before becoming the legal owner.

[41] For my part, however, again I do not consider that that sort of debate assists. At the end of the day, the tenant has to be a qualifying tenant for a period of two years prior to the relevant date. That seems to me a very simple test, or at least intended to be a very simple test, and it is one that is met with sufficient clarity, in my judgment, if registration is required throughout that period.

[42] For those reasons, without, I fear, doing justice to some of the more subtle points that were made in the submissions before me, I am minded to grant the declaration sought. I will hear counsel as to the precise form of order and with any other arguments consequential on that decision.

Claim allowed.

NEGLIGENCE

Joyce *v* Bowman Law Ltd

CHANCERY DIVISION

18 February 2010

VOS J

[2010] EWHC 251 (Ch)

Negligence — Licensed conveyancer — Damages — Property advertised for sale with option to purchase additional land — Purchaser's conveyancer failing to include option in conveyance — Claim for negligence and breach of contract — Quantum of damages — Whether award to be made for loss of chance to develop both areas of land together

The defendant firm of licensed conveyancers acted for the claimant on the purchase of a dilapidated cottage on a 0.12ha plot at a price of £360,000. The sales particulars for the property stated that the purchaser would have the option to purchase for £20,000, within 12 months of completing the cottage purchase, an additional 0.09ha of land at the lower end of the garden, in the event that the vendor's own application for planning permission for the erection of one dwelling on that land was unsuccessful.

Contracts were exchanged on the sale of the cottage and completion took place simultaneously. The sale contract did not contain an option to purchase in the form in which it had been advertised; instead of a buyer's option, under which the claimant could require a sale to him, it contained a seller's option, enabling the vendor to require the claimant to purchase. The vendor subsequently refused to sell the additional land to the claimant.

In a claim against the defendant, alleging breach of contract and negligence, the claimant contended that he was entitled to damages for loss of the profit that he would have made had he been able to redevelop the cottage plot and the additional land together, by demolishing the existing cottage and building a large detached house. The defendant admitted negligence in: (i) failing to appreciate the nature of the option contained in the contract; (ii) erroneously advising the claimant that he would be able to purchase the additional land should the vendor fail to obtain planning permission within 12 months of completion; and (iii) advising the claimant to enter into the contract when it contained a seller's rather than a buyer's option. However, it disputed the recoverability of loss of profits and submitted that the claimant had failed to mitigate his loss by carrying out a smaller development on the cottage plot.

***Held*: The claim was allowed in part. The claimant was entitled to recover consequential losses based on the profit that he would have made from his planned development. He was not confined to damages based on any reduction in value of the property as a result of the buyer's option not having been granted. The defendant had known that the claimant intended to develop the property for profit and that the additional land played a role in that development. Such knowledge amounted to knowledge of special circumstances outside the ordinary course of things, in addition to the matters of which the contracting parties were reasonably assumed to know at the time of contracting, in which a breach would be liable to cause more loss to the claimant, within the second rule in *Hadley* v *Baxendale* (1854) 9 Exch 341. It was not necessary for the defendant to be aware of the precise scale of the intended development. On the evidence, there was a real and substantive, as opposed to speculative, chance that had the matter of the option been brought to the parties' attention before completion, the vendor would have conceded a satisfactory buyer's option, the claimant would have exercised that option and he would have obtained planning permission and funding for his intended development. Evaluating those chances as a percentage, there was: (i) an 85% of obtaining the buyer's option; (ii) a 100% chance of the claimant exercising that option; (iii) a 40% chance of obtaining planning permission; and (iv) an 85% chance of obtaining funding. The percentage prospects of any future event that depends on the hypothetical actions of third parties are cumulative and must be multiplied together, at least where they relate to separate considerations that do not overlap. Since the above percentages already reflected any overlap, they were therefore to be multiplied together. That produced 28.9%, which was rounded up to 29%. Assuming that the new house would not have been completed and ready for sale before mid-2008, its value at that date would have been £1.135m, with a basic profit of £375,000 after the deduction of purchase and build costs. A deduction for failure to mitigate the loss should be made in the sum of £245,000, that being the profit that the claimant might have made by undertaking a smaller project without the additional land, leaving a balance of £130,000. Overall, the claimant was entitled to recover 29% of that balance for the loss of chance of making the additional profit had he acquired the additional land; accordingly, he was entitled to damages of £37,700.**

The following cases are referred to in this report.

Allied Maples Group Ltd v *Simmons & Simmons* [1995] 1 WLR 1602; [1995] 4 All ER 907, CA
Banco de Portugal v *Waterlow & Sons Ltd* [1932] AC 452
County Personnel (Employment Agency) Ltd v *Alan R Pulver & Co* [1987] 1 WLR 916; [1987] 1 All ER 289; [1986] 2 EGLR 246, CA
Dodd Properties (Kent) Ltd v *Canterbury City Council* [1980] 1 WLR 433; [1980] 1 All ER 928; [1980] 1 EGLR 15; (1979) 253 EG 1335, CA
Hadley v *Baxendale* (1854) 9 Exch 341
Hanif v *Middleweeks (a firm)* [2000] Lloyd's Rep PN 920, CA
Jenmain Builders Ltd v *Steed & Steed* [2000] PNLR 616, CA
Pilkington v *Wood* [1953] Ch 770; [1953] 3 WLR 522; [1953] 2 All ER 810
Smith New Court Securities Ltd v *Citibank NA; sub nom Smith New Court Securities Ltd* v *Scrimgeour Vickers (Asset Management) Ltd* [1997] AC 254; [1996] 3 WLR 1051; [1996] 4 All ER 769, HL
Victoria Laundry (Windsor) Ltd v *Newman Industries Ltd* [1949] 2 KB 528; [1949] 1 All ER 997; (1949) 65 TLR 274, CA

This was a claim by the claimant, Paul Joyce, against the defendant, Bowman Law Ltd, for damages for breach of contract and negligence in the conduct of a conveyancing transaction.

Simon Williams (instructed by Ross & Craig) appeared for the claimant; Eva Ferguson (instructed by Beachcroft LLP, of Bristol) represented the defendant.

Giving judgment, **VOS J** said:

Introduction

[1] The defendant, Bowman Law Ltd (Bowman), a firm of licensed conveyancers, acted for Mr Paul Peter Anthony Joyce (Mr Joyce), the claimant, when he purchased Warren Cottage, Severals Road, Bepton,

A West Sussex (the property), for £360,000 from Ms Dorothea Elizabeth Bannister Chesters (the vendor) on 2 September 2005. A somewhat dilapidated cottage (Warren Cottage), dating back between 200 and 300 years, stood on the property in a plot of 0.12ha, or 0.30 acres.

[2] The property had been advertised with an option for the buyer to purchase the lower end of the existing garden (the additional land) for £20,000, within 12 months, if the vendor's application for planning permission to erect one dwelling on that land was unsuccessful. The additional land extended to 0.09ha, or 0.22 acres. The property and the additional land, therefore, together amounted to 0.21ha, or 0.52 acres.

[3] Contracts for the sale of the property were exchanged (the
B contract) and completion (completion) took place simultaneously on 2 September 2005. At that time, Mr Joyce thought that the contract contained an option in the form in which it had been advertised. It fact, it contained a seller's option instead.

[4] On 3 February 2010, Bowman admitted negligence in failing to appreciate the nature of the option that the contract contained and in wrongly advising Mr Joyce that if the vendor failed to obtain planning permission within 12 months of completion, he (Mr Joyce) would be able to purchase the additional land. Bowman also admitted that it was negligent in permitting Mr Joyce to enter into the contract when, in fact, it contained a seller's option rather than a buyer's option.

C [5] The question that arises in these proceedings is what, if any, damages Mr Joyce is entitled to recover from Bowman for its negligence.

Background facts

[6] The sales particulars for the property, which were "subject to contract", included the following wording:

> A picturesque character cottage adjacent to woodland and offering scope for modernisation and extension (subject to the usual consents)…
> Important Note
> (Subject to Contract)
> The lower end of the garden is being retained by the vendor and a planning
> D application for the erection of one dwelling is to be submitted. Should this prove unsuccessful, both on application or any subsequent appeal that might be considered, then the purchaser will have the option within 12 months of completing the purchase of Warren Cottage to purchase the additional area of garden at the price of £20,000…

[7] As it stood, Warren Cottage had a total floor area of $182m^2$ and an external footprint of $142m^2$, both including garages and outbuildings, and a height of between 6.5m and 7m. The footprint without the garage was $118m^2$, and the upper floor had an area of $40m^2$.

[8] Mr Joyce made an offer of £385,000 for the property on 5 June 2005, expressly indicating that he wanted to take up the option to
E purchase the additional land if the vendor was unable to obtain planning permission. On 6 June 2005, King & Chasemore, the vendor's agent, prepared a memorandum of sale recording the sale "subject to contract" and including the following: "option for the purchasers to acquire remainder of the garden for £20,000 (see sales particulars)".

[9] Mr Joyce was first represented in the purchase transaction by George Ide Phillips LLP (GIP). GIP was replaced by Bowman in mid-August 2005. On 16 August 2005, MacDonald Oates LLP, the solicitor for the vendor, asked Bowman whether the option contained in draft special condition 5 was acceptable to Mr Joyce. In due course, Mr Joyce commented on the period of the option (asking that the
F period be reduced from two years to the one year advertised in the sales particulars), but the contract was exchanged and completed with the terms of the clause otherwise unchanged. Special condition 5 (the option clause) provided in part as follows:

> (i) 5.1 In this Clause:… 5.1.5 "the Option Notice" means the Notice served by the Buyer pursuant to clause 5.3
>
> (ii) 5.2 In consideration of [£1] paid by the Seller to the Buyer… the Buyer grants to the Seller the Option to require the Buyer to buy the freehold interest in the Additional Property [the "Additional Land"] at the Additional Purchase Price [£20,000].

[10] Mr Joyce raised the purchase price of the property by mortgaging it to Northern Rock Building Society to raise £306,720. The balance of
G the purchase price of the property came from Birmingham Midshires Building Society and was charged on Mr Joyce's existing family home, Yew Cottage, Charlton Road, East Dean, Chichester, West Sussex PO18 0JA (Yew Cottage). Mr Joyce contributed no cash towards the purchase of the property from his own resources.

[11] At some time during August 2005, Mr Joyce instructed Mr Barrie F Morse MRICS FCIOB FASI, of Wilkins, a firm of chartered surveyors, project managers and construction cost consultants, to advise him in connection with Warren Cottage. The date of Mr Joyce's first meeting with Mr Morse is rather unclear, but it was certainly before the contract was concluded on 2 September 2005.

H [12] Mr Joyce claims that, at around this time, he prepared a sketch plan of the large house, with dimensions noted of 12m by 18m, that he wanted to build on the property (the sketch plan) once he had been able to exercise the option to acquire the additional land.

[13] Shortly before the contract was executed, Mr Joyce negotiated a reduction in the contract price from £385,000 to £360,000 on the ground that he had discovered that a sand quarry landfill site was proposed close to the property.

[14] In the period leading up to completion, Mr Joyce dealt mostly with Mr Keith Curran (Mr Curran), an experienced conveyancer then employed by Bowman. Mr Joyce says that he also spoke to
J Mr Simon Bowman, the principal of Bowman. Mr Bowman now admits that he met Mr Joyce briefly on the morning of completion, and Mr Joyce undoubtedly met Mr Curran that morning to sign the contract and to discuss the transaction. Mr Joyce phoned Mr Curran from the airport (whence he was leaving for France) at lunchtime that day to confirm that the transaction had gone through.

[15] As I have said, the contract that was exchanged and completed on 2 September 2005 contained the option clause allowing the vendor to require Mr Joyce to buy the additional land for £20,000 during the period of one year ending on 2 September 2006. There was, as I have said, no option allowing Mr Joyce to require the vendor to sell the
K additional land to him.

[16] An exchange of correspondence immediately after the sale between Mr Eddie Lintott, of King & Chasemore, and Mr Martin Chesters, the vendor's nephew, who was handling the sale on her behalf, shows that both Messrs Lintott and Chesters were relieved that the sale had gone through.

[17] It appears that, on 3 September 2005, while Mr Joyce was in the south of France), he e-mailed Mr Morse as follows:

> very nice to meet you on Friday, e-mail as promised with full address of the [Property]… Can you please arrange TPOs orders on the trees asap as I dont
> L have the luxury of time… I will forward you a file on [Yew Cottage] asap and send it to you re the new planning application.

[18] On 5 September 2005:

(i) Mr Morse replied to Mr Joyce explaining how he was progressing the tree preservation orders (TPOs) on the additional land (which Mr Joyce had instructed him to obtain in order to obstruct the vendor's intended application for planning permission), and saying that a survey would be necessary and that "[t]he latter information will be required when we prepare a scheme to extend the present house [on the property]".

M (ii) Mr Morse also sent Mr Joyce a letter explaining his charges in detail. Almost the entire letter is premised on the proposed extension to the existing Warren Cottage to make it into a five/six bedroom house.

[19] On 9 September 2005, Mr Morse wrote again to Mr Joyce enclosing his proposed plans for such an extension, and saying that retaining the original portion of the building incurred problems, "low ceiling heights, poor thermal qualities and damp ingress", and that "[i]t may be prudent to consider demolishing the whole of the [property] (upon receipt of planning approval) and erect a purpose-designed house to comply with modern standards. You would also save VAT, as new properties are zero-rated."

[20] On 9 September 2005, Mr Joyce e-mailed Mr Morse as follows:

> think we may have a misunderstanding wanted to object to the planning on the extra piece of garden. Seems like a major project, as regards to the house if I wanted a modern one I would have looked elsewhere as it stands if I built new with the price I paid I would be looking at another project. If the way forward is to obtain outline planning thats fine, but I'm not ready to build… .

[21] I should say at once that I do not accept Mr Joyce's explanation for this e-mail, which was that he had his wires crossed with a proposed extension project at Yew Cottage. It seems clear to me that this e-mail expresses Mr Joyce's honestly held views on his proposals in respect of the property at the time it was written. In essence, Mr Joyce had instructed Mr Morse to submit an application for planning permission for an extension to Warren Cottage and also to obtain TPOs on the additional land. At that stage, Mr Joyce had certainly not decided that he wanted to demolish the existing house and build a new one.

[22] On 19 September 2005, Mr Morse submitted a planning application to Chichester District Council (CDC) to alter and extend Warren Cottage to a floor area of around 250m^2 (the planning application), including a new garage, but it did not involve the additional land.

[23] On 2 November 2005, Mr Andrew Runciman, the planning officer of CDC, faxed Mr Morse expressing his concerns about the scale and size of the proposed extension and suggesting that the application should be withdrawn and that "following discussions with the writer, a more appropriate scheme is submitted, smaller in scale and in keeping with the setting and character of the area". The planning application was refused on 14 November 2005 on the grounds that:

> the proposed extensions would result in development which by virtue of their size, bulk and mass are considered to relate unsympathetically to the existing building and unacceptably increase its visual prominence within the sensitive landscape of the locality resulting in development that would be seriously detrimental to the character and appearance of the Sussex Downs Area of Outstanding Natural Beauty… .

[24] In November 2005, Mr Joyce instructed Mr James Lacey, of Vail Williams LLP, property and planning consultant, to advise in respect of the property.

[25] On 10 November 2005, MacDonald Oates wrote to Bowman alleging that the contract provided that Mr Joyce would erect a fence along the boundary with the additional land (which it did not) and saying that the vendor wanted to do so anyway at her own expense. The letter required Mr Joyce to remove a chicken cage, which straddled the boundary, by 18 November 2005. A fence was duly erected.

[26] Mr Joyce was unhappy about these events, which made him feel that something was wrong with the option that he thought he had obtained, and he and/or Mr Graham Thompson, his uncle acting on his behalf, contacted Mr Bowman.

[27] On 17 November 2005, Mr Bowman wrote to Mr Joyce saying that he would register "the option against [the vendor's] unregistered title". He continued: "I am confident that this tactic will prevail and secure you the benefit of the registered option." Mr Bowman was suggesting registering a caution to protect a buyer's option that did not exist, but by doing so (as he eventually did), he was (on his own assessment) reassuring Mr Joyce.

[28] In December 2005, Mr Joyce consulted Mr David Bennett, a solicitor with GIP, concerning the option, and was told for the first time that the option clause in the contract did not operate in his favour.

[29] On 16 December 2005, the vendor applied for planning permission to build a new two-bedroom dwelling on the additional land. On 18 January 2006, Smiths Gore wrote on behalf of someone supposedly called "Mr Graham Joyce", who was said to be Mr Joyce's father, making detailed objections to the vendor's application, relying on the TPOs that Mr Joyce had procured. In fact, Mr Joyce's father died long ago, but Mr Thompson told me that he sometimes uses Mr Joyce's father's name, and had done so on this occasion.

[30] On 23 December 2005, Vail Williams wrote to Mr Runciman concerning a proposal to replace the existing dwelling with a new house having a footprint of 7m by 15m. On 6 January 2006, Mr Runciman, of CDC, wrote to Vail Williams saying that: "the principle of a replacement dwelling on the above site is in my opinion something which could be supported by officers and would not conflict with Local Plan policies… A footprint for the new house of approximately 7m by 14m is probably on the upper limit of what would be considered appropriate for this location, a smaller footprint would be preferred."

[31] In January 2006, Mr Joyce consulted another firm of solicitors, Thomas Eggar, in Chichester, and it again advised him that the option clause in the contract did not operate in his favour.

[32] The vendor's application for planning permission was eventually rejected on 13 February 2006.

[33] On 17 February 2006, Vail Williams submitted plans dated January 2006 to Mr Runciman, along the lines that he had suggested, for his consideration (the January 2006 plans). They related to a new four-bedroom house with a footprint of 7m by 15m and a floor area of 210m^2. This proposal also did not involve the additional land. It was not, however, pursued after 20 February 2006, when Mr Joyce seems to have instructed Vail Williams to submit a larger scheme.

[34] In February/March 2006, Mr Joyce sought advice on the option from another firm of solicitors in Brighton. Once again, he was told that the option clause was a vendor's option and that he had a claim against Bowman.

[35] On 31 May 2006, Bowman wrote again to Mr Joyce concerning the option. He sought to reassure Mr Joyce that all was well. He wrote, somewhat enigmatically: "As you are aware at the time of the purchase the seller was looking to obtain planning permission on [the additional land] and it was agreed that if their planning application failed you would have the option to buy that land for the benefit of the cottage."

[36] In June 2006, Mr Joyce consulted yet another solicitor on the option. This was the fourth time he had asked the same question. That firm, Green Wright Chalton Annis, gave the same advice as the three previous firms he had consulted.

[37] On 31 July 2006, the vendor applied again for planning permission to build on the additional land. This application was refused on 17 November 2006, and the vendor's appeal was dismissed on 29 November 2007. I am told that the vendor has since died aged around 98, but that her estate has continued to deal with the additional land through Mr Chesters.

[38] On 25 August 2006, Bowman purported to serve a notice on the vendor's solicitor exercising Mr Joyce's supposed option to purchase the additional land for £20,000 pursuant to the option clause. Prior to that, Mr Joyce had provided Bowman with £20,370 at Mr Bowman's request.

[39] On 3 September 2006, Mr Joyce completed the sale of his previous family home, Yew Cottage, at a price of £585,000. He received some £42,472 from the proceeds of sale. It was at this point that Mr Joyce moved with his family to Warren Cottage, despite its dilapidated state.

[40] On 4 September 2006, the vendor's solicitor wrote to Bowman pointing out that the option clause in the contract was not exercisable by Mr Joyce, but only by the vendor.

[41] On 6 September 2006, Bowman wrote to Mr Joyce, again somewhat enigmatically saying: "However, having studied the option more closely it is ambiguous in that clause 5.1.5 the option notice is defined as a Notice that has to be served by yourself as the buyer whereas in clause 5.3 the indication is that the option is exercisable by the seller" and "I am sure that the seller will indicate that the option is for his benefit and there is clearly a legal argument to be made in your favour." He concluded by saying that his firm did not undertake litigation and could not give a definitive answer as to how the contract would be construed.

[42] On 6 November 2006, Mr Joyce took still further legal advice on the option. This time he met Mr John Bridger, a partner at Preston Redman, who had been recommended by Mr Curran, who told me that he was keen to help Mr Joyce sort out the problem created by his mistake if he could. Mr Curran also attended the meeting. Mr Joyce told Mr Bridger that Mr Bowman had told him that the "intention of the parties was of paramount importance". Mr Bridger told Mr Joyce that that was "nonsense", and said unequivocally that the option did

A not operate in his favour and that he had a claim for damages against Bowman. Mr Curran admitted his error in the course of this meeting.

[43] On 25 January 2007, Preston Redman wrote to the vendor's solicitor saying that Mr Joyce was minded to instruct it to seek rectification and asking whether the vendor would anyway be prepared to sell the additional land to Mr Joyce for £20,000. Preston Redman had informed Mr Joyce that it had sent this letter two months earlier, but it eventually turned out that it had mistakenly failed to do so. The vendor's solicitor rejected the intimated rectification claim and said that Mr Joyce's complaint was against Bowman. It also said that the additional land was not then for sale because the vendor intended to B appeal the rejection of planning permission.

[44] On 1 February 2007, Mr Bowman wrote to Mr Joyce recording that Mr Joyce had telephoned him to say that he intended to sue Bowman for negligence, and returning the £20,350 that Mr Joyce had paid him to exercise the option.

[45] The claim form in these proceedings was issued against Bowman on 23 May 2008.

[46] In March 2009, Mr Morse prepared plans (the March 2009 plans) for a new larger dwelling with a footprint of just less than 12m by 18m (192m^2 plus a garage of 54m^2) to be built in place of the existing Warren Cottage. The March 2009 plans showed a total floor area of C 370m^2 for the house and 94m^2 for the garage block (totalling 464m^2).

[47] Despite Mr Joyce's attempts to persuade the vendor to sell the additional land to him, the vendor's estate has recently sold the additional land to a third party for an unknown sum.

Overview of the damages claim

[48] Mr Joyce claims damages for breach of the contract, by which he retained Bowman to act for him on the purchase of the property, and in tort for damages for the negligent advice he was given.

[49] Mr Joyce has not, however, adduced any evidence to show that the property was worth less than he paid as a result of the absence of the D buyer's option he had wanted, nor has he argued that the buyer's option he expected was inherently valuable in the sense that the £20,000 option fee was below the market value of the additional land.

[50] Instead, Mr Joyce has claimed the profit (put originally in the particulars of claim at £620,000) that he says he would have made had he been able to redevelop the property and the additional land together, by building a large detached house (eventually as shown in the March 2009 plans) in place of the dilapidated Warren Cottage. He originally estimated the value of that new house at £1.4m, from which he deducted the purchase price of £380,000 (including the option fee) E and £400,000 for estimated development costs.

[51] The values of the property and the additional land based on the March 2009 plans are now agreed, having been valued by the single joint expert instructed by the parties, Mr Kieran Godkin MRICS. The present-day value would be £1.025m, and the value in the fourth quarter of 2007, when Mr Joyce says he would have been able to sell the new property, was £1.155m. Mr Godkin also estimated the costs of constructing the new house, according to the March 2009 plans, at £375,000. No VAT arises on that sum (or at least on the bulk of that sum) because it would be a new building. Thus, Mr Joyce's basic claim is for £400,000 as at the end of 2007 (£1.155m less £380,000 for the F purchase and £375,000 for the development cost), or £270,000 if the values are taken as at today.

[52] Mr Simon Williams, counsel for Mr Joyce, accepted that these sums need to be reduced, first to take account of any failure to mitigate (although he said that there was in fact no such failure) and, second, to take account of the chances that a buyer's option in suitable form would not have been negotiated, that such an option would not have become exercisable, that Mr Joyce might not have been able to fund the redevelopment and of the possibility that planning permission would not have been granted. He invited me to find that there was no failure to mitigate and to take a single percentage for all these uncertainties. He put that percentage at 75%. In the result, therefore, Mr Joyce contends that since he would have been able to develop the property and the additional land, absent Bowman's negligence, by late 2007, he should G recover damages of £300,000 (75% x £400,000).

[53] Ms Eva Ferguson, counsel for Bowman, argued in response that none of these alleged consequential losses is recoverable as a matter of law because they were not within the scope of Bowman's duty of care, were not caused by the negligence, were too remote and, as a matter of fact, there was no real and substantial chance that Mr Joyce would have obtained or been able to exercise a buyer's option, or obtained planning permission for, or been able to fund the development of the property and the additional land based on the March 2009 plans as he maintains he would have. Even if all that were wrong, Ms Ferguson said that Mr Joyce has failed to mitigate his loss, in that he failed to develop H the property by building a smaller house along the lines of the January 2006 plans, which she said he could have done even without purchasing the additional land. Had he done so, he would have made a profit of £245,000 (£815,000 being the agreed value for the fourth quarter of 2007, less the purchase price of £360,000 and less £210,000, being the construction costs of the smaller property estimated by Mr Godkin). Bowman contends that Mr Joyce should give credit for this profit that he could have made. Finally, Ms Ferguson argued that the percentages for the uncertainties that I have recorded above should be assessed individually and multiplied together. She says, therefore, that if Mr Joyce is entitled to anything, he cannot have more than £400,000 J less £245,000 multiplied by a very small percentage (assuming, which she also contested, that the end of 2007 is the right date at which to assess Mr Joyce's loss).

Loss of a chance

[54] The principles in *Allied Maples Group Ltd* v *Simmons & Simmons* [1995] 1 WLR 1602, *per* Stuart Smith LJ, at p1614D, and Hobhouse LJ, at p1621A, are common ground between the parties. It is useful, however, to set out these *dicta* as follows:

(i) Stuart Smith LJ said:

> the plaintiff must prove as a matter of causation that he has a real or substantial K chance as opposed to a speculative one. If he succeeds in doing so, the evaluation of the chance is part of the assessment of the quantum of damage, the range lying somewhere between something that just qualifies as real or substantial on the one hand and near certainty on the other. I do not think that it is helpful to seek to lay down in percentage terms what the lower and upper ends of the bracket should be.

(ii) Hobhouse LJ said:

> The plaintiffs have satisfied the court that the loss they have suffered is not nominal. They are not obliged to prove more than that they have lost something of substance. This they have done by showing that they had a measurable chance of negotiating significantly better terms. They are entitled to an L assessment of their damages.
>
> I agree with Stuart-Smith LJ that the correct approach is that summarised by Lord Reid in *Davies* v *Taylor* [1974] AC 207, 213:
>
> "You can prove that a past event happened, but you cannot prove that a future event will happen and I do not think that the law is so foolish as to suppose that you can. All that you can do is evaluate the chance. Sometimes it is virtually 100 per cent: sometimes virtually nil. But often it is somewhere in between".

[55] Ms Ferguson also relies on a *dictum* of Mance LJ at paras 33 and 34 of *Hanif* v *Middleweeks (a firm)** as showing that the percentage prospects of any future event that depends on the hypothetical actions of M third parties are cumulative and should be multiplied together, at least in so far as they relate to separate considerations that do not overlap. That case, when both the judgments of Mance and Roch LJJ are considered, makes it clear that a mathematical approach is not always appropriate. There were, in that case, three putative defences to an action that the claimant had lost the chance of bringing. The Court of Appeal held that the judge ought not to have disregarded the first two defences (which had high probabilities of failing) and taken only the percentage that he assessed for the defence that was most likely to succeed. In essence, however, *Hanif* shows that it is permissible, indeed appropriate, to take

---

* Editor's note: Reported at [2000] Lloyd's Rep PN 920

A a less than entirely mathematical approach where the factors affecting the cumulative future events overlap or are affected by the same considerations. I will consider how these propositions affect the facts in this case under issue 4 below.

Issues

[56] Against this background, it seems to me that the following main issues arise for determination:

(1) Is Mr Joyce entitled to recover consequential losses based on the building profit that would have been made?

(2) What did Bowman know about Mr Joyce and his intention to
B develop the property?

(3) Had Mr Joyce been properly advised, was there a real and substantial chance that:

(a) Mr Joyce would have been granted the buyer's option?

(b) Mr Joyce would, in the event, have been able to exercise that option, had it been granted?

(c) Mr Joyce would have been able to obtain permission for the March 2009 plans, had the buyer's option been granted and exercised?

(d) Mr Joyce would have been able to obtain the funding for the development of the March 2009 plans?

(4) What were the percentage chances that Mr Joyce would have
C succeeded in surmounting these four hurdles?

(5) What would Mr Joyce's profit have been from building out the March 2009 plans and when should they be assessed?

(6) Has Mr Joyce failed to mitigate his loss?

(7) To what damages is Mr Joyce entitled?

Witnesses of fact

[57] Mr Joyce was not an entirely satisfactory witness. In some respects, however, he had a good recollection of what had occurred, and I bear in mind that this project was and remains extremely important to him. I am sure that he has turned over the events in his mind over many years and has, in some ways, "obsessed" about the option problem with
D which he was faced. This has led him, I think, to persuade himself that certain things happened when I am now fairly sure they did not. None the less, I did not regard him as deliberately untruthful, and in some important respects I have preferred his recollection to that of Mr Bowman and Mr Curran. The stark contrast between their evidence and that of Mr Joyce was that they were talking years after the event about one of many hundreds of transactions on which they had been engaged, while Mr Joyce was talking about something that formed a major part of his life. Mr Joyce's recollection of the events that occurred was therefore far better.

[58] There are two matters that Mr Joyce gave evidence on that I do
E not accept:

(i) First, he told me expressly that he had, right from the beginning, wanted a newly built house with a footprint that was 12m by 18m.

(ii) Second, as I have already mentioned, he explained his 9 September 2005 e-mail to Mr Morse saying he did not want to build a new house at all, by saying that he or they had got their wires crossed.

[59] Both these pieces of evidence are classic examples of the phenomenon that I have described above. The first suggestion of a newly built house came from Mr Morse, probably when they first viewed the site and certainly in his 9 September 2005 letter. It was at first rejected by Mr Joyce. Moreover, Mr Joyce did not think of
F building a new house as large as 12m by 18m until later. What is clear, however, is that he wanted, from before completion, to extend Warren Cottage as much as he possibly could and that he was hoping to produce something similar to the property next door to Warren Cottage, called Woodlands (which was extended to a footprint of 171m^2, and a floor area of 335m^2, both including a garage block). Thus, by the time Mr Joyce bought the property, he had both the comparison with Woodlands and a five/six-bedroom extended house in mind. This, I think, is what he told Mr Morse and what Mr Morse, in due course, passed on to Vail Williams. Mr Joyce may have staked out an extended Warren Cottage when he first visited the property, but not with any specific dimensions in mind. Nor do I think that Mr Joyce prepared the sketch plan before completion.

[60] As for the way in which Mr Joyce's plans developed, they G
are explained largely by the documentation that I have recited above and do not need to be repeated here. Mr Joyce left matters largely in the hands of Mr Morse and Vail Williams, and, as he told me, he quickly became absorbed by the option problem — probably from early to mid-November 2005 onwards. I accept that Mr Joyce was considering a staged approach to the planning, in that he hoped and expected that he would get planning (initially pursuant to the planning application) and that, when he later acquired the additional land, he would be able to extend the extent of his existing permission. I do not think that he had made any decision not to build a new house based on the January 2006 plans at the time, but I accept that he was happy to H
receive Mr Runciman's letter of 6 January 2006, which effectively gave the green light to a fairly substantial new development even without the additional land. He was still hoping at that stage to be able to extend the permission when he acquired the additional land. Of course, he knew by then that the option was not in his favour, but I am sure that he wanted to believe it would all be alright, and it was for that reason that he believed Mr Bowman's reassuring messages concerning the registration of the caution and the intention of the parties being the most important thing. Moreover, it might, at that stage, have turned out alright, since Mr Chesters might have wanted to sell the additional land for £20,000 even without an option requiring him to do so. It was for J
that reason that Mr Joyce, somewhat misguidedly in my view, thought that he should do whatever he could to stop the vendor from obtaining planning permission. His aggressive conduct in obtaining TPOs and opposing the vendor's planning application was probably not as clever as he thought it was. It seems to have engendered a hostility in Mr Chesters that resulted in his being unwilling to sell the additional land to Mr Joyce at any price. Had Mr Joyce played his cards more cautiously, it might have turned out that the vendor's planning application would anyway have been refused and Mr Chesters would have been happy to sell to him.

[61] As Mr Joyce told me, he put the planning applications for the K
property on the back burner in early 2006, as he became more and more concerned about the option. I accept that he did not want to spend more money on advisers than he needed to, until he had sorted out the option position.

[62] I shall deal later on with what Mr Joyce told Mr Bowman and Mr Curran before the contract was entered into.

[63] Before leaving Mr Joyce's evidence, however, I should consider whether he was, as he half-heartedly contends, a "property developer". In my judgment, he was not a commercial property developer in the sense that that term is often understood, but he was a private individual who was buying the property with the objective of turning L
a profit. Mr Joyce had, in September 2005, renovated and extended one property at Yew Cottage and invested in another. He was one of many such private individuals who wanted to try to take advantage of the (then) rising property market by buying, improving and selling on properties in which he and his family could sometimes, but not always, live along the way. As one would expect of such an individual, all the mortgages that he obtained were in the domestic retail market, and he never took any kind of commercial loan. However, that does not mean that he was not buying the property to make a profit. In my judgment, he undoubtedly was.

M

[64] Mr Thompson, Mr Joyce's uncle, sought to support Mr Joyce's evidence in a number of respects. The two most important aspects on which this support was given were in respect of Mr Joyce's suggestions that he had always wanted to build a new house with a footprint of 12m by 18m and in saying that Bowman was well aware of Mr Joyce's development plans. I did not find Mr Thompson a particularly impressive witness, and I think that he was mistaken to suggest that 12m by 18m had been mentioned before completion. That said, Mr Thompson was obviously a great help to Mr Joyce and was much involved with the project. He was therefore to a large extent involved in the same process as his nephew, namely of making himself believe that some things had happened that had not. Mr Thompson did,

however, give some important evidence concerning his contact with Bowman, to which I will return.

[65] Mr Morse, Mr Joyce's architect or designer, was a reasonably reliable witness. At the outset in cross-examination, he told me that he understood that his brief was primarily to maximise the development of the site. He said that his preference was the possibility of two smaller units, but when he first looked at the property before the contract, "the discussion then was bouncing ideas off each other as to what could be done and what could be achieved". He said that he had suggested demolition and replacement on 9 September 2005, but the size of 12m by 18m was not fixed, even later on. His January 2006 plans (based on a footprint of 7m by 15m) were as close as possible as he could achieve to meet his client's wishes at that stage. He supported Mr Joyce's evidence in several important respects, all of which I accept. First, he said that Mr Joyce had told him from the beginning that he wanted, in effect, to emulate what had been done at Woodlands. Second, he endorsed what Mr Joyce had said about the "stepped" approach to the planning process. Third, he supported Mr Joyce in thinking that "the Additional Land would be a planning point in his favour as the size of the plot can determine the size of the build". Finally, Mr Morse was clear that Mr Joyce wanted a big property from the outset — he described it in his initial letter as five to six bedrooms.

[66] I turn then to the witnesses for Bowman.

[67] Mr Curran was the conveyancer in charge of Mr Joyce's transaction. There is no doubt that his negligence that has been admitted in this case was astonishing. I am constrained to wonder whether Mr Curran can ever have read the option clause, since he admitted that he knew that Mr Joyce wanted a buyer's option and yet the option clause quite obviously gives him nothing of the sort. However, all this is now water under the bridge in the light of Bowman's admissions, save in one important respect. It seems to me that the crass nature of Mr Curran's breach of duty gives a hint as to his credibility and carefulness in giving his evidence. I am afraid I am unable to accept that he gave any proper attention to Mr Joyce's case, and I therefore view his evidence with the greatest caution. I simply do not accept that he has any real recollection of what happened prior to completion, beyond what the papers have revealed to him after the event. I accept Mr Joyce's recollection of his taciturn nature in the meeting on 2 September 2005.

[68] Both Mr Curran and Mr Bowman told me that it was the firm's practice not to make or keep any attendance notes. I am bound to say that I found this evidence extraordinary. I cannot understand how a professional firm can possibly offer a proper service without noting down the oral instructions that they are given, and important events that occur in meetings and telephone discussions. The absence of any note of any discussion with Mr Joyce in this case has been an important factor in the decision that I have reached as to what actually occurred between Mr Joyce and Bowman before the contract.

[69] Mr Bowman, the managing director of Bowman, was a cool customer. He was keen to answer questions as briefly as possible. On certain crucial matters, I found his evidence evasive. His witness statement was exiguous and inaccurate. It made no attempt to explain his undoubted relationship with Mr Thompson, Mr Joyce and Mr Richard Criddle, the mortgage broker who introduced them to him, or indeed to give any background to the transaction. He simply denied having any contact with Mr Joyce "regarding the file". I have no doubt that that evidence was false. He had extensive contact with Mr Joyce after the transaction and, as I shall explain later, some contact with both Mr Joyce and Mr Thompson before the transaction.

[70] He gave but one insight into how his firm worked, when he told me that all his clients have his mobile telephone number and that they use it, in effect, to bypass the system and to make sure that the conveyancer handling the case actually calls them back. The implication was that they would not otherwise be able to get hold of the conveyancer.

[71] His evidence, which I am unable to accept, was that Mr Joyce did not make clear to him what he was going to do with Warren Cottage. He accepted, however, that he knew that Mr Joyce was remortgaging Yew Cottage and did not think that Mr Joyce was moving to Warren Cottage at that time.

[72] Mr Bowman also accepted that it was fairly clear with hindsight that the option was a seller's option and not a buyer's option.

Evidence of the planning consultants

[73] Both experts did their best to assist the court, and demonstrated real expertise in their fields. Bearing that in mind, it is slightly surprising that they were so far apart in their assessments of whether the March 2009 plans would have been likely to receive planning permission: their evaluations varied between a 5% chance on appeal because of the possibility of a maverick inspector to a "strong possibility" of the plans being approved on appeal.

[74] Mr David Lander MRTPI, of Bowyer Planning, gave evidence for Mr Joyce. He gave his evidence carefully. Both he and Mr Tetlow reviewed the planning policies found in the structure plan and the local plan. I do not intend to set out that analysis here. Suffice it to say that Mr Lander thought that the most important issue in this case was whether the second criterion set out in the CDC's policy H12 could be satisfied. That criterion was that the property "does not detract from the rural character and appearance of the existing dwelling and/or the surrounding area particularly in Areas of Outstanding Natural Beauty where proposals should not detrimentally increase the bulk of the building visible from public vantage points".

[75] Mr Lander thought that this criterion could be satisfied, although he was closely questioned about the height and bulk of the building envisaged in the March 2009 plans and its proximity to Severals Road. Although he accepted that it would be closer to the road than any of his comparable properties nearby (Woodlands, Loxley Chase or Broadland Court) and that it would have covered 11.7% of the plot, whereas those other properties covered only 2%, 7.7% and 8.1% of their plots respectively, he thought that it could be effectively screened. His main point was that the additional land provided screening as one approached the property from the south up Severals Road, and that the existing Warren Cottage already presented a bulky profile from heavily wooded area to the north of the property, and the new building would be further back from the northern boundary than the existing property. He also drew a distinction between a relatively narrow flank wall without habitable windows close to the road and the southern elevation that was the main rear of the house to the rear garden providing the principal amenity space.

[76] Mr Lander explained also that policy H12 says that, in calculating the scale of any permissible increase, one should exclude garages, completely glazed conservatories and outbuildings. He said that CDC uses that interpretation because they are evidently concerned to restrict the spread of development.

[77] Mr Robin Tetlow MSc Dip Surv FRTPI FRICS FCIH FRSA, of Tetlow King Planning, gave evidence for Bowman. He too gave his evidence carefully.

[78] He told me that he and Mr Lander did not differ much on the planning policies that would be applicable, although he thought that one needed to look at the policies relating to the area of outstanding natural beauty and the wider policies on character and quality before one looked at the local policy H12. He agreed that the latter policy was the principal one.

[79] Mr Tetlow said that the house shown in the March 2009 plans would be more noticeable than the existing Warren Cottage because there was a significant difference in the height. He estimated the height of the new house at 10m, although he thought that the same conclusion would apply even if the height were 9.5m, as Mr Lander estimated. He accepted that there was scope for boundary planting, but said that the proposed house was big and bulky and difficult to screen.

[80] Mr Tetlow accepted that the main issue in planning terms was the harm that the proposed development would do, but he defended stoutly his conclusion that there was only a 2% chance of obtaining planning permission from CDC for the March 2009 plans, and only a 5% chance of success on appeal. He placed considerable reliance on the large footprint on a relatively small site and the fact that the

building was very close to both the west boundary and the eastern road boundary.

[81] He said that the advice that Vail Williams gave regarding achieving a height of 7.5m was good planning advice, and he thought that something along the lines of the January 2006 plans would have been approved, but he thought that it was stretching imagination to think that Mr Runciman could have recommended a much larger scheme. He did not think that an appeal would have succeeded, even though one in three appeals do succeed across the country.

[82] Finally, Mr Tetlow said that he thought that the size of the plot was irrelevant in this case. The March 2009 plans would have been rejected with or without the additional land. He said that the March 2009 plans represented over-intensive development.

Factual findings

[83] As the issues that I have described above demonstrate, I need to evaluate, to some extent, not only what the parties knew and understood but also what could have been expected to happen, in order properly to assess damages in this case.

[84] Mr Joyce had not decided what exactly he wanted to do with the property by the time he bought it. Undoubtedly, he wanted to develop it into as large a building as possible, but his initial objective was to extend and renovate the existing Warren Cottage, rather as had happened to Woodlands. He was uncertain, in my view, about whether he would eventually live in it, let it or sell it on. It was Mr Morse that suggested demolishing the old Warren Cottage and building afresh. Mr Joyce was initially unwilling to consider rebuilding, as his 9 September 2006 e-mail clearly shows, but he quickly came round, preferring one large new house to the two smaller ones that Mr Morse favoured. He always intended to make a "stepped approach" to the planning authorities, expanding the size of the development once the additional land was acquired.

[85] Having seen Mr Joyce and Mr Thompson, I am quite satisfied that they would have been talking volubly about their plans for the property to anyone who would listen. I do not accept that Mr Bowman was the kind of person who would not listen. It seemed to me that he was putting on a show in the witness box. In reality, he has built up his clientele by being approachable, as his revealing comment about his clients having his mobile telephone number demonstrated. I am quite satisfied, therefore, that Mr Joyce, and indeed Mr Thompson, spoke to Mr Bowman before completion about Mr Joyce's plans. I also find that Mr Joyce spoke to Mr Curran about his plans at the 2 September 2005 meeting, if not before.

[86] I am quite satisfied also that the option promised in King & Chasemore's sales particulars was important to Mr Joyce. Mr Joyce and Mr Morse understood from the beginning that the additional land would make it easier to obtain permission to extend Warren Cottage. Moreover, as soon as a new build was suggested, they thought that it would enable them to build a bigger house. If permission had been obtained for the January 2006 plans, they would have sought to increase the footprint and put rooms in the roof, once the option was exercised.

[87] As for the option, the conduct of the vendor (not herself, but through her agent) was surprising. They had promised a buyer's option, and then drafted a seller's option. It is extraordinary that they were not discovered in what seems to have been a try-on. Indeed, there is a further remarkable irony in that Mr Joyce noticed that they had extended the option to two years, and required that it come back to one year. The two years for the seller's option as drafted was actually rather beneficial to him, but his insistence on one year for what he thought was a buyer's option does show the importance he attached to obtaining the additional land in a short timescale.

[88] Had Bowman or Mr Joyce noticed what had happened, I think that the vendor and MacDonald Oates, her solicitor, would have been in a weak position. It is true that MacDonald Oates had been pressing for early exchange and completion and that it had expressed frustration and mistrust for Mr Joyce, but Mr Joyce would have had the upper hand in any negotiations about the option, had their attempt to move the goalposts been detected. Moreover, the post-completion correspondence between Mr Lintott and Mr Chesters shows how happy they were with the deal. I am therefore satisfied that there was a real and substantive prospect that, had the matter been brought to attention, the vendor would have conceded an option of the kind intimated in the sales particulars.

[89] The next problem is to try to predict what kind of buyer's option that would have been. This is a little more difficult. The wording of the sales particulars is not clear. The real issue is whether the buyer's option would have been watertight, in the sense that it could be exercised automatically unless planning permission had been granted to the vendor within a year from completion, or whether it would have given the vendor room for manoeuvre by, for example, pursuing an appeal and thereby defeating the option. I think that Mr Joyce would most likely have been able to negotiate a watertight option because the other purchasers had, by late August 2005, left the scene and, as Mr Chesters' 6 September 2005 letter reveals, he was reassured to have the put option he had negotiated: in other words, he would have wanted to be able to sell the additional land for £20,000 if the planning application failed. Moreover, it seems to me that the result might have been matching options with both the option clause and a suitable buyer's option included.

[90] Had matters progressed as Mr Joyce had wanted and expected, he would probably have been granted a satisfactory buyer's option and would have exercised it on 2 September 2006. This leaves a further difficult question, namely what planning permission would he have applied for and what would he have obtained.

[91] One thing is clear as to planning. Mr Joyce would have stood a good chance of obtaining permission to build on a 7m by 14m or 15m footprint along the lines of the January 2006 plans, even without the additional land. Both experts confirmed this. It is hard to speculate, but Mr Joyce probably failed to pursue this smaller project for a variety of reasons, including the fact that: (a) he became focused on the option (or lack of it) and on pursuing Bowman rather than the building project; and (b) he did not have huge resources easily to build the project out, and wanted something that would give him a large profit. The fact that Mr Joyce moved his family to Warren Cottage in September 2006, when he sold Yew Cottage, was not really a contributing factor, since he could have remained living in Warren Cottage although the new house was built in the garden, only demolishing the old house once the new one was completed. Moreover, although Mr Joyce was clearly never flushed with cash, I think he could have obtained a mortgage advance to build the new house, whether based on the January 2006 or the March 2009 plans, had he sought to do so in September 2006 or thereabouts. At that time, the property market remained strong, and I am sure that Mr Thompson would have assisted him to obtain finance if there was otherwise a problem.

[92] The final and most important question, then, is whether Mr Joyce would have obtained planning permission for a large house on a 12m by 18m footprint even if he had been able to purchase the additional land in September 2006. I have looked carefully at the correspondence with Mr Runciman, of CDC, and heard the careful expert evidence that I have described above. I have no doubt that Mr Runciman would have opposed such a proposal, and that other factors would have been against it, for example:

(i) The property is situated in an area of outstanding natural beauty.

(ii) Even with the additional land, the site was very small, at only just over half an acre.

(iii) The planners were sensitive about height as well as footprint.

(iv) The planners were also keen to see a sensitive development of the cottage type. Mr Joyce's proposed house would have been bulky and somewhat obtrusive. It would have been larger than the adjoining Woodlands that Mr Joyce was, on his own analysis, originally seeking to emulate.

[93] Despite all these factors, I have concluded that Mr Joyce would have had a real chance of obtaining planning permission for something like the March 2009 plans. He might well have had to have made some changes, and even reduced it marginally in size, but I think that he had a real possibility of building something rather large on the entire site,

including the additional land. I will not conceal my own aesthetic view, which is that such a building would have been brash and bulky and really rather out of character, but having seen the photographs of Loxley Chase and Broadland Court, I am not convinced that the chances of success would have been anything like as low as Mr Tetlow thought. I confess to agreeing with Mr Tetlow, which was, I think, the thrust of his evidence, that the chances *should* have been as low as he said. However, I do not think that they actually were. The true position falls somewhere between the views of the two experts.

First issue: Is Mr Joyce entitled to recover consequential losses based on the building profit that would have been made?

[94] In *County Personnel (Employment Agency) Ltd* v *Alan R Pulver & Co* [1987] 1 WLR 916*, Bingham LJ said, at pp925B-926D, as to the assessment of damages in a solicitor's negligence case:

> The principles to be applied in assessing damages in this case are, in my judgment, these.
>
> (1) The overriding rule was stated by Lord Blackburn in *Livingstone* v *Rawyards Coal Co* (1880) 5 App Cas 25, 39, and has been repeated on countless occasions since: the measure of damages is
>
> "that sum of money which will put the party who has been injured, or who has suffered, in the same position as he would have been in if he had not sustained the wrong for which he is now getting his compensation or reparation."
>
> As Megaw LJ added in *Dodd Properties (Kent) Ltd* v *Canterbury City Council* [1980] 1 WLR 433, 451:
>
> "In any case of doubt, it is desirable that the judge, having decided provisionally as to the amount of damages, should, before finally deciding, consider whether the amount conforms with the requirement of Lord Blackburn's fundamental principle. If it appears not to conform, the judge should examine the question again to see whether the particular case falls within one of the exceptions of which Lord Blackburn gave examples, or whether he is obliged by some binding authority to arrive at a result which is inconsistent with the fundamental principle."
>
> (2) On the authorities as they stand the diminution in value rule appears almost always, if not always, to be appropriate where property is acquired following negligent advice by surveyors. Such cases as *Philips* v *Ward* [1956] 1 WLR 471; *Pilkington* v *Wood* [1953] Ch 770; *Ford* v *White & Co* [1964] 1 WLR 885 and *Perry* v *Sidney Phillips & Son* [1982] 1 WLR 1297, lay down that rule and illustrate its application in cases involving both surveyors and solicitors.
>
> (3) That is not, however, an invariable approach, at least in claims against solicitors, and should not be mechanistically applied in circumstances where it may appear inappropriate. In *Simple Simon Catering Ltd* v *Binstock Miller & Co* (1973) 228 EG 527 the Court of Appeal favoured a more general assessment, taking account of the "general expectation of loss". In other cases the cost of repair or reinstatement may provide the appropriate measure: the *Dodd Properties* case [1980] 1 WLR 433, 456, *per* Donaldson LJ. In other cases the measure of damages may properly include the cost of making good the error of a negligent adviser: examples are found in *Braid* v *WL Highway & Sons* (1964) 191 EG 433, and *G&K Ladenbau (UK) Ltd* v *Crawley de Reya* [1978] 1 WLR 266.
>
> (4) While the general rule undoubtedly is that damages for tort or breach of contract are assessed as at the date of the breach (see, for example, *Miliangos* v *George Frank (Textiles) Ltd* [1976] AC 443, 468, *per* Lord Wilberforce), this rule also should not be mechanistically applied in circumstances where assessment at another date may more accurately reflect the overriding compensatory rule. The *Dodd Properties* case… both affirms this principle and illustrates its application.
>
> (5) On the facts of the present case the diminution in value rule would involve a somewhat speculative and unreal valuation exercise…
>
> (6) Even after an appropriate measure has been found to reflect damage recoverable under the first limb of the rule in *Hadley* v *Baxendale* (1854) 9 Exch 341, there will be cases in which a plaintiff will not be adequately compensated unless he receives damages to reflect his loss under the second limb also.

[95] It is common ground that Mr Joyce can recover the damages that he seeks only under the second rule in *Hadley* v *Baxendale*†, at p354, namely that damages for breach of contract should be "such as may reasonably be supposed to have been in the contemplation of both parties at the time they made the contract as the probable result of the breach of it". As Asquith LJ put the matter in *Victoria Laundry (Windsor) Ltd* v *Newman Industries Ltd* [1949] 2 KB 528, at p539:

> But to this knowledge, which a contract-breaker is assumed to possess whether he actually possesses it or not [under the first rule] there may have to be added in a particular case knowledge which he actually possesses, of special circumstances outside the "ordinary course of things," of such a kind that a breach in those special circumstances would be liable to cause more loss. Such a case attracts the operation of the "second rule" so as to make additional loss also recoverable.

[96] In addition to her argument that Bowman had no knowledge of special circumstances such as to engage the second rule (to which I will turn under the second issue), Ms Ferguson has argued that Mr Joyce should be entitled only to damages based on the reduced value of the property as a result of the option not having been renegotiated. She relied in this regard on Mr Joyce being a property developer, as to which point I have made findings above, and on *Jenmain Builders Ltd* v *Steed & Steed**. In that case, the Court of Appeal rejected a developer's claim to assess damages against its negligent solicitor based on the profit that it would have made had it been able to buy and develop the intended property as it should have done. Chadwick LJ said, at pp625E-627A:

> This was a property with development potential. It is common ground that this property was no longer to be used as a village hall. It would have to be used for some other purpose; and there would have to be some development so that it could be used for that purpose. The question was: for what development could planning permission be obtained and how valuable would the property be on completion of that development? But those are the factors which a properly informed market will take into account in fixing the market value of property. The profit potential of the property is an element to be taken into account in fixing its market value. It is not suggested that there was anything special about this property to the appellants as purchasers. It is not suggested that there were not other developers in the market for property of this nature who could have made a proper assessment of the value of this property. The problem for the appellants in the present case is that they never sought to persuade the judge — and never adduced evidence to establish — that the market value of this property, Dukes Hall, was anything greater than the £67,500 which the Parish Council was seeking. It is for those reasons that the claim for loss of profits is one which the court could not entertain in this case.
>
> …
>
> In the present case, there is no evidence that these appellants would not have been able to purchase other property in the market which they could develop profitably with the use of the money which they did not lay out in the purchase of Dukes Hall. There is no evidence that the respondents, insofar as their duty lay in contract, were aware of any special circumstances which made it impossible for the respondents to employ their funds in the ordinary course of their business, or of any circumstances which suggested that this property was being sold at an under-value. Indeed, in the circumstances that they were acting for the vendors, the Parish Council, it would be most unlikely that they would regard the property as being sold at an under-value rather than at market price.

[97] In this case, as in *Jenmain*, Mr Joyce has not suggested that the value of what he bought without the option was less than he paid. Nor has he suggested that the option was worth more than the £20,000 he would have (putatively) to have paid to exercise it. What he says is that he has sustained consequential losses as a result of the loss of the chance of having and being able to exercise the option he wanted.

[98] Mr Joyce was in some measure locked into the transaction in this case, unlike the situation in *Jenmain*. Indeed, he was, in my judgment, locked in by the conduct of Bowman itself, even if no claim was made against Bowman in respect of its post-completion conduct: *cf Smith New Court Securities Ltd* v *Scrimgeour Vickers (Asset Management Ltd)* [1997] AC 254 in respect of damages for fraudulent misrepresentation. Even though Mr Joyce received repeated advice from December 2005 onwards that the option was not what he had expected and was not a buyer's option at all, he continued to believe that he might none the less be able to exercise it. He did so, in material part at least,

* Editor's note: Also reported at [1986] 2 EGLR 246

† Editor's note: Reported at (1854) 9 Exch 341

* Editor's note: Reported at [2000] PNLR 616

because Mr Bowman told him so in the way I have described above. It does not, therefore, easily lie in Bowman's mouth to argue that Mr Joyce should be deprived of damages for consequential losses properly flowing from the fact that he retained the property through at least to early 2007, when he finally realised that he would for certain be unable to obtain the additional land.

[99] I do not think that *Jenmain* is strictly applicable here. First, Mr Joyce was not a "property developer" with cash to invest commercially in the way envisaged in *Jenmain*. Second, he was in some measure locked into the property until early 2007. It does not lie in this contract-breaker's mouth to say that Mr Joyce ought to have been more sensible and sold the property instantly he became aware of the problem. Moreover, when it came to 2007, his circumstances had changed and he had moved into the property with his family. By then, if he had suffered any consequential loss, it was too late to rectify it by a sale.

[100] In these circumstances, I am satisfied that there is no reason in principle why Mr Joyce cannot recover consequential losses in addition to the difference in value (which latter he does not claim). As I have said, however, he will be able to recover such damages only if they fall within the second rule in *Hadley*.

Issue 2: What did Bowman know about Mr Joyce and his intention to develop the property?

[101] Ms Ferguson has argued that to come within the second rule Mr Joyce has to show that Bowman knew that:

(i) it was Mr Joyce's intention to develop the property for profit;

(ii) a specific development was contemplated; and

(iii) the additional land was necessary for that development.

[102] Ms Ferguson bolsters this submission by contending that if Mr Joyce had no fixed idea of what he wanted to develop, he could not have had any fixed idea about the importance of the additional land.

[103] I have already found that Mr Joyce spoke to Mr Bowman and Mr Curran about his plans before completion, but it seems to me that these conversations have to have been either at the time of the retainer or at least far enough ahead of the contract to enable Bowman to take account of its knowledge in performing its contractual duties.

[104] Mr Joyce was clear in his evidence that he had told Mr Bowman during telephone calls with him between 16 August 2005 and completion about his development plans and about the importance of the additional land to him. I accept that he did so. I do not, of course, accept that he mentioned a new building or something as specific as a new house with a footprint of 12m by 18m. However, I am sure that he would have mentioned the adjoining cottage, Woodlands, and a major extension of that kind in addition to his quite transparent intention to turn a profit, and that he did so when he first instructed Bowman or soon afterwards.

[105] Ms Ferguson argued that the mortgage application forms on Bowman's files showed that this was a domestic conveyancing transaction and that they had no reason to suppose anything else. However, this ignores the reality of the kind of people that Mr Thompson and Mr Joyce are. They are, without any disrespect being intended, "wheelers and dealers", and Mr Bowman must have known that. I have no doubt he was told what Mr Joyce was plainly excited about, namely his plan to develop the property and make a profit, in part by acquiring the additional land.

[106] Although it is unnecessary for the purposes of Mr Joyce's claim, I also accept that Mr Curran was told the same things on 2 September 2009, just before the contract was signed. I am sure that Mr Thompson also mentioned the same matters to Mr Bowman in his conversations with him during August 2006.

[107] I do not think that it was necessary for Bowman to be aware of the precise scale of the intended development in order to have the special knowledge necessary to bring Mr Joyce within the second rule. The fact that Mr Bowman knew of the intention to make a profit from developing the property and about the role to be played by the additional land (about which Mr Joyce was particularly concerned) is enough to give Bowman "knowledge… of special circumstances outside the 'ordinary course of things' of such a kind that a breach in those special circumstances would be liable to cause more loss".

Issue 3: Had Mr Joyce been properly advised, was there a real and substantial chance that Mr Joyce would: (i) have been granted the buyer's option?; (ii) in the event, have been able to exercise that option had it been granted?; (iii) have been able to obtain permission for the March 2009 plans had the buyer's option been granted and exercised?; and (iv) have been able to obtain the funding for the development of the March 2009 plans?

[108] I have already dealt above with these four questions. I am entirely satisfied, for the reasons that I have given, that there was a real and substantial, rather than a speculative, chance of Mr Joyce:

(i) being granted a buyer's option that he could have exercised after one year if planning permission had not by then been obtained;

(ii) actually exercising that option in September 2006 since planning permission would not have been obtained by the vendor;

(iii) being able to obtain permission for the March 2009 plans; and

(iv) obtaining the funding for the development of the March 2009 plans.

Issue 4: What were the percentage chances that Mr Joyce would have succeeded in surmounting the four hurdles?

[109] This issue is perhaps the most difficult because it requires an evaluation of all the evidence in the case. It is therefore very much a balancing exercise, and it is hard to set out in full detail everything that one has taken into consideration.

[110] I will deal first with my percentage evaluations of the chances of each of these four events occurring and then with the question of how I have obtained a single figure from those percentages.

[111] The chance of Mr Joyce obtaining a suitably watertight buyer's option was, in my judgment, 85%. The correspondence, some of which I have described above, is informative. I think that Mr Chesters wanted to sell, thought he was getting a good price and that had his scheme to replace the promised buyer's option with a seller's option been rumbled, he would have done what had been promised in the sales particulars and the sales memorandum. The fact that he had tried it on in the way he had would have put him in a weak negotiating position and would have made Mr Joyce (who was very keen to have the promised option) all the more astute to make sure that the option was watertight and that Mr Chesters could not wriggle out of it by, for example, appealing the planning decision within the year. I think it is fairly certain that Mr Joyce would have obtained the buyer's option he wanted.

[112] The chance of Mr Joyce actually exercising that option in September 2006 was also very high. We know that the vendor could not get planning permission and that Mr Joyce had found the £20,000. I think that the chance of his exercising the option, had he got it, was 100%.

[113] The third chance concerning planning permission is the most difficult. I have sought to explain my approach to the expert evidence above. I am by no means sure that Mr Joyce would have obtained the precise permission he wanted, but on the assumption that something close to what is shown in the March 2009 plans would have satisfied him and enabled him to go ahead, I judge the prospects of getting such a permission, at least on appeal, as being 40%.

[114] As for the chance of Mr Joyce obtaining the funding for the development of the March 2009 plans, I accept that Mr Joyce was by no means cash rich, but he did seem able to borrow when he wanted to, and I think it likely that Mr Thompson would have helped him out. The chances of funding such a development, had he got planning permission, would have been 85%, in my view.

[115] I have fed in the overlaps between these chances as I have progressed through them. For example, the second chance is assessed at 100% because it assumes that the first chance has already been realised. Likewise, the fourth chance is assessed at 85% on the assumption that the March 2009 plans were passed. Thus, I think it is appropriate here to undertake the mathematical exercise that was not precisely appropriate in *Hanif*. This gives 28.9%, which I propose to round up to 29%.

Issue 5: What would Mr Joyce's profit have been from building out the March 2009 plans, and when should they be assessed?

[116] As I have already mentioned, Mr Joyce's putative profit is largely agreed. However, it is dependent on timing. To obtain planning permission for the March 2009 plans, I think that Mr Joyce would have had to go to appeal. Mr Runciman was very opposed to the idea of such a grand development. That would have taken time. It is hard to say how much, but it would very likely have delayed the project by six to nine months. Taken with the seven-month development period, and some time to get funded and geared up, I doubt that Mr Joyce would have completed the new house and made it ready for sale before mid-2008. Thus, I need to take the figures agreed for that period, rather than those I have mentioned above.

[117] The value of the house at mid-2008 figures would have been £1.135m. That figure less £380,000 purchase costs and £375,000 build costs amounts to a basic profit of £380,000.

Issue 6: Has Mr Joyce failed to mitigate his loss?

[118] Mr Williams argued that Mr Joyce should not have his damages reduced because he failed to mitigate his loss. He argued that it was not reasonable for Mr Joyce to be forced to undertake a smaller project. I disagree. This is not a matter of weighing in fine scales how an injured party should behave (see *Banco de Portugal* v *Waterlow & Sons Ltd* [1932] AC 452, at p506), nor is it equivalent to requiring a claimant to embark on speculative and costly litigation: see *Pilkington* v *Wood* [1953] Ch 770, at p777. It is simply requiring Mr Joyce to do what he always wanted to do, namely develop the property to the greatest level possible. It has not been suggested that he would not have been able to develop the property profitably on the basis of the January 2006 plans, and I think that he must give credit for the profit he would have made had he done so. He could, in my judgment, have obtained permission without the additional land for the January 2006 plans by early 2007, and could have completed the building and got it ready for sale by the end of 2007. Had he done so, he would have made a profit of £245,000, as I have explained above.

Issue 7: What damages is Mr Joyce entitled to?

[119] For the reasons that I have explained above, therefore, it seems to me that Mr Joyce has lost a profit of £375,000, less the profit he could have made of £245,000, making £130,000. Mr Joyce is entitled to recover 29% of that balance for the loss of the chance that he would have been able to make the additional profit had he been able to acquire the additional land, which he could not acquire because of Bowman's negligence.

[120] Checking back, as suggested by Megaw LJ in *Dodd**, I am satisfied that this sum of £37,700 (£130,000 x 29%) conforms with the requirement of Lord Blackburn's fundamental principle, in that it puts Mr Joyce in the position that he would have been in had he not sustained the wrong, or had the contract been performed by Bowman and the buyer's option had been granted to him. It reflects the chance that he would have built out a bigger development on the scale of the March 2009 plans. The figures are not so exact that it is appropriate to make a further reduction for the possibility that the March 2009 plans might have had to be slightly reduced in size. Mr Joyce cannot, however, blame Bowman for the fact that he did not develop the property to the lesser extent that was available to him, and make the £245,000 profit he could have achieved from that process.

Conclusion

[121] I therefore award Mr Joyce damages of £37,700 (£130,000 x 29%) together with interest from 30 June 2008, by which time he should have realised that additional profit. I will hear counsel on the rate of interest applicable and costs.

Claim allowed in part.

---

* Editor's note: *Dodd Properties (Kent) Ltd* v *Canterbury City Council* [1980] 1 EGLR 15; (1979) 253 EG 1335

PUBLIC UTILITIES

Barratt Homes Ltd *v* Dwr Cymru
A Cyfyngedig (Welsh Water)

SUPREME COURT

[2009] UKSC 13

9 December 2009

LORD PHILLIPS, PRESIDENT, LORD SAVILLE, LORD WALKER, LADY HALE AND LORD CLARKE

Public utilities — Sewers and drains — Connection — Residential
B ***development — Planning permission granted for development — Water undertaker refusing connection to public sewer at developer's proposed location on ground of risk of overload to public sewerage system — Section 106 of Water Industry Act 1991 — Whether undertaker entitled to refuse connection at proposed location on ground of overloading — Whether entitled to insist on connection at alternative location***

The respondent was granted planning permission for a residential development of 98 houses and a school. The appellant water undertaker had objected to the planning application because it was concerned that the existing public sewerage system would be unable to cope with the
C demands of the development. The planning permission was subject to a condition that required approval, before development could start, of a scheme of foul water and surface water drainage, to be completed before the buildings were occupied.

The respondent served a notice on the appellant, under section 106 of the Water Industry Act 1991, of its proposal to connect to the sewer at point X, which was situated in the road immediately adjoining the northern boundary of the development site. By its reply under section 106(3), the appellant insisted that the connection should be made at point Y, which was located between 300m and 400m to the east of point X; that would require the construction of a new connecting sewer at the
D respondent's expense. The appellant was concerned that a connection at point X would overload the system and create a risk of overspill. Despite this, the local planning authority, at the respondent's request, discharged the planning condition relating to drainage.

The respondent brought proceedings to establish its entitlement to connect at point X. The claim was dismissed at first instance but allowed on appeal: see [2008] EWHC 1936 (QB) and [2008] EWCA Civ 1552; [2009] 1 EGLR 55; [2009] 11 EG 120. The Court of Appeal held that that the grounds on which connection could be refused, under section 106(4), were narrow and related solely to the mode of construction or the condition of the connecting drain, such that a water
E undertaker could not refuse a connection to its existing system, or insist on a connection at an alternative location, on the ground of overloading. The appellant appealed.

Held **(Lady Hale dissenting): The appeal was dismissed. The right to connect to a public sewer afforded by section 106 of the 1991 Act is absolute. The undertaker is not entitled to refuse a connection with a public sewer on the ground that the intended point of connection is unsatisfactory or that the additional discharge into the system would overload it. The burden of dealing with the consequences of any additional discharge, and the cost of any consequential works, fall directly on the undertaker, although the expense will be shared by all who pay sewerage charges, including the occupiers of the houses in any new development.** F
The "drain or sewer" referred to in section 106(4) is the private drain or sewer that the developer proposes to connect to the public sewer, and the words "mode of construction" cannot be extended to include the point of connection. Consequently, section 106 entitles a developer to connect its private drain or sewer to the public sewer subject only to the undertaker's rights to give notice: (i) under section 106(4), refusing a connection on the ground of deficiencies in the condition of the private drain or sewer; and (ii) under section 107, that it will itself make the connection. The undertaker is not entitled to select the point of connection on the ground of its dissatisfaction with the proposed point. G

Per curiam**: Although the 1991 Act affords no such right, a case could be made for deferring the right to connect to a public sewer in order to give the sewerage undertaker a reasonable opportunity to ensure that the public sewer would be able to accommodate the increased load. The only way of achieving such a deferral is through the planning system, and it is desirable that both the undertaker and Office of Water Services be consulted as part of the planning process.**

The following cases are referred to in this report.

Ainley v *Kirkheaton Local Board* (1891) 60 LJ Ch 734

Barratt Homes Ltd v *Dwr Cymru Cyfyngedig (Welsh Water)* [2008] EWCA Civ 1552; [2009] 1 EGLR 55; [2009] 11 EG 120; [2009] JPL 1095; [2008] EWHC 1936 (QB) H

Beech Properties Ltd v *GE Wallis & Sons Ltd* [1977] 1 EGLR 142; (1976) 241 EG 685; [1976] JPL 429, Ch

Brown v *Dunstable Corporation* [1899] 2 Ch 378

Charles v *Finchley Local Board* (1883) LR 23 Ch D 767, Ch

Glossop v *Heston and Isleworth Local Board* (1879) 12 Ch D 102, CA

Grampian Regional Council v *Secretary of State for Scotland* [1983] 1 WLR 1340; [1983] 3 All ER 673; (1984) 47 P&CR 540; [1984] 2 EGLR 175; (1983) 271 EG 625; [1983] RVR 263; [1984] JPL 416, HL

McMonagle v *Westminster City Council* [1990] 2 AC 716; [1990] 2 WLR 823; [1990] 1 All ER 993, HL J

Marcic v *Thames Water Utilities Ltd*; *sub nom Thames Water Utilities Ltd* v *Marcic* [2003] UKHL 66; [2004] 2 AC 42; [2003] 3 WLR 1603; [2004] 1 All ER 135

Marleasing SA v *La Comercial Internacional de Alimentación SA* C-106/89 [1990] ECR I-4135, ECJ

Smeaton v *Ilford Corporation* [1954] Ch 450; [1954] 2 WLR 668; [1954] 1 All ER 923; (1954) 52 LGR 253, CA

Wilkinson v *Llandaff and Dinas Powis Rural District Council* [1903] 2 Ch 695, CA

This was an appeal by the appellant, Dwr Cymru Cyfyngedig (Welsh Water), from a decision of the Court of Appeal allowing an appeal by the respondent, Barratt Homes Ltd, against a decision of K Wyn Williams J dismissing a claim for declaratory and injunctive relief relating to a proposal, under section 106 of the Water Industry Act 1991, to connect a housing development to the public sewer.

Lord Pannick QC, David Holgate QC, Maurice Sheridan and Jessica Simor (instructed by Geldards LLP) appeared for the appellant; Anthony Porten QC, Steven Gasztowicz QC and Clare Perry (instructed by Darwin Gray, of Cardiff) represented the respondent.

Giving his opinion, **LORD PHILLIPS** (with whom **LORD SAVILLE**, **LORD WALKER** and **LORD CLARKE** agreed) said:

A *Introduction*

[1] This appeal concerns the right conferred by the Water Industry Act 1991 (the 1991 Act) on a property owner to connect its private drain or sewer to a public sewer for the purpose of discharging its sewage into that public sewer. The principal issue raised is whether it is the property owner or the sewerage undertaker that is entitled to determine the point at which the property owner's drain or sewer is to connect to the public sewer. This narrow issue of statutory construction conceals, however, wider and more fundamental issues that are less easily resolved. I propose, first, to resolve the narrow issue before commenting on these wider issues.

B [2] Llanfoist is a village near Abergavenny, in Monmouthshire. Its surface water and foul water drainage requirements are met by a public sewerage system that terminates in a waste-water treatment works (the treament works) around one-third of a mile to the east of the village and below it. This system is around 60 years old.

[3] Approximately midway between the village and the treatment works, at manhole SO29125900 (the CSO), the sewerage pipe that links the two reduces from a diameter of 225mm to one of 150mm and continues for a distance of 282m before it increases, at manhole SO29127901, to a diameter of 300mm for the final stretch to the treatment works. The narrow section, described as a "pipe bridge" C determines the capacity of the system, or at least all that part of it that lies upstream of manhole SO29127901.

[4] The respondent, Barratt, is in the process of building a substantial development of 98 houses and a primary school on a green-field site contiguous to the east of Llanfoist. It constructed a private sewer to receive the sewage from this development. It claimed a statutory right to connect its private sewer to the public sewer at a point of its own choosing, which was in the close vicinity of its development. This point of connection was not satisfactory to Welsh Water because it would overload the system upstream of manhole SO29127901. It claimed a statutory right to refuse connection at this point, offering instead D connection at manhole SO29127901, an option that would saddle Barratt with the cost of the link from its development to manhole SO29127901. Thus arose the narrow issue of the interpretation of the relevant provisions of the 1991 Act.

[5] At first instance, in a judgment delivered on 1 August 2008, Wyn Williams J found in favour of Welsh Water: [2008] EWHC 1936 (QB). His decision was reversed by the Court of Appeal on 28 November 2008: [2008] EWCA Civ 1552*. Barratt then proceeded to connect the development's sewer to the public sewer at the place of its choice. Welsh Water does not seek, by this appeal, to effect a physical reversal of what has taken place. It accepts that what has E taken place in this case is now water under the bridge. It is anxious to establish, however, that a sewerage undertaker has a right to refuse to permit connection to be made to one of its sewers when it considers that the proposed point of connection is not suitable. Should it establish this right of refusal, a further issue arises as to the effect of a statutory time limit for giving notice of refusal.

Water Industry Act 1991

[6] The law in respect of sewers has its origin in the reign of Henry VIII, but the modern law begins with the Public Health Act 1848. There followed a series of Acts that consolidated and amended the law, of which the 1991 Act is one. The provisions of that Act that F are directly relevant to this appeal can be traced back to the Victorian legislation. They provide as follows:

94 General duty to provide sewerage system

(1) It shall be the duty of every sewerage undertaker —

(a) to provide, improve and extend such a system of public sewers (whether inside its area or elsewhere) and so to cleanse and maintain those sewers and any lateral drains which belong to or vest in the undertaker as to ensure that that area is and continues to be effectually drained; and

(b) to make provision for the emptying of those sewers and such further provision (whether inside its area or elsewhere) as is necessary from time to time for effectually dealing, by means of sewage disposal works or otherwise, with the contents of those sewers. G

…

106 Right to communicate with public sewers

(1) Subject to the provisions of this section —

(a) the owner or occupier of any premises, or

(b) the owner of any private sewer which drains premises,

shall be entitled to have his drains or sewer communicate with the public sewer of any sewerage undertaker and thereby to discharge foul water and surface water from those premises or that private sewer.

…

(2) Subject to the provisions of Chapter III of this Part, nothing in subsection (1) above shall entitle any person — H

(a) to discharge directly or indirectly into any public sewer —

(i) any liquid from a factory, other than domestic sewage or surface or storm water, or any liquid from a manufacturing process; or

(ii) any liquid or other matter the discharge of which into public sewers is prohibited by or under any enactment; or

(b) where separate public sewers are provided for foul water and for surface water, to discharge directly or indirectly —

(i) foul water into a sewer provided for surface water; or

(ii) except with the approval of the undertaker, surface water into a sewer provided for foul water; or

(c) to have his drains or sewer made to communicate directly with a storm-water overflow sewer. J

(3) A person desirous of availing himself of his entitlement under this section shall give notice of his proposals to the sewerage undertaker in question.

(4) At any time within twenty-one days after a sewerage undertaker receives a notice under subsection (3) above, the undertaker may by notice to the person who gave the notice refuse to permit the communication to be made, if it appears to the undertaker that the mode of construction or condition of the drain or sewer —

(a) does not satisfy the standards reasonably required by the undertaker; or

(b) is such that the making of the communication would be prejudicial to the undertaker's sewerage system.

(5) For the purpose of examining the mode of construction and condition of a drain or sewer to which a notice under subsection (3) above relates a sewerage undertaker may, if necessary, require it to be laid open for inspection. K

In this judgment, I shall, where appropriate, refer to "the developer" as shorthand for the "owner or occupier" of premises who enjoys rights under section 106.

[7] Section 106(6) provides that any question as to the reasonableness of an undertaker's refusal to permit a communication to be made or of a requirement under subsection (5) may be referred for determination by the director of the Office of Water Services (Ofwat).

[8] Section 107 entitles the sewerage undertaker to give notice within 14 days of receipt of a notice under section 106(3) that the L undertaker intends to make the communication itself. In that event, the developer has to pay the reasonable cost of the work.

Point of connection

Submissions

[9] Mr Anthony Porten QC, for Barratt, submitted that the provisions of section 106 of the 1991 Act were clear. Subsection (1) gave a property owner the right to connect to a public sewer, subject only to such limitations as were imposed by other provisions of the section. That right was a right to connect at whichever point the property owner chose to do so. The only restrictions on that right were those set out M in subsection (4). Those restrictions were very limited. They gave the undertaker the right to refuse to permit the connection only on the ground of the inadequacy of the mode of construction or condition of the private drain or sewer that was to be joined to the public sewer. No objection could be made to the point of connection, however inconvenient that might be for the undertaker.

[10] Lord Pannick QC, for Welsh Water, submitted that the court should not accept this interpretation, for its consequences ran counter to the object of the legislation. That object was the protection of health and of the environment. Parliament cannot have intended that a property owner should be entitled to insist on a specific point of connection however great the harm that this would cause to the environment or to

* Editor's note: Reported at [2009] 1 EGLR 55; [2009] 11 EG 120

public health and however reasonable it might be to require the property owner to connect elsewhere.

[11] The potential harm identified by Lord Pannick was damage to the environment or to health as a result of the escape of foul water from the sewerage system. The overload on the system consequent on the point of connection chosen by Barratt had increased the risk of escape of foul water at the CSO. The CSO was intended to act as an escape point for sewage to a limited extent deemed acceptable in conditions of overload caused by exceptional rainfall in storm conditions. The additional loading on the system as a result of connecting Barratt's sewer upstream rather than downstream of the pipe bridge was calculated to lead to an escape of foul water beyond the limit that was acceptable. Such escape would result in Welsh Water committing criminal offences of strict liability under section 85 of the Water Resources Act 1991 and would infringe provisions of Directive 91/271/EEC concerning the collection, treatment and discharge of urban waste water (the directive) and the Urban Waste Water Treatment (England and Wales) Regulations 1994 (SI 1994/2841) (the 1994 Regulations) passed to give effect to the directive.

[12] Lord Pannick treated the facts of the present case as illustrative of the general effect of an interpretation of section 106 of the 1991 Act that permits a developer to select the point of connection between its sewer and a public sewer. That is the only relevance of the facts of this case to the issue of interpretation that is raised, and I shall defer a more detailed consideration of those facts to later in this judgment.

[13] Lord Pannick further submitted that the escape of waste water consequent on a property owner connecting to a public sewer at an inappropriate point could include pollution and risk to health, thereby infringing Articles 2, 3 or 8 of the European Convention on Human Rights. The court was bound, if possible, so to interpret section 106 of the 1991 Act as to avoid these consequences: see *Marsleasing SA* v *La Comercial Internacional de Alimentación SA* C-106/89 [1990] ECR 1-4135 and section 3 of the Human Rights Act 1998. Lord Pannick submitted that such interpretation could be achieved by reading the provisions of section 106 in a manner that implicitly incorporated express provisions in earlier legislation that the 1991 Act had replaced.

[14] Lord Pannick advanced two alternative ways of interpreting section 106 that produced the result for which he contended. The first involved reading "the mode of construction…; of the drain or sewer" in subsection (4) as embracing the point of connection. This interpretation was, he submitted, supported by the legislative history.

[15] Section 21 of the Public Health Act 1875 (the 1875 Act) provided:

> The owner or occupier of any premises within the district of a local authority shall be entitled to cause his drains to empty into the sewers of that authority on condition of his giving such notice as may be required by that authority of his intention so to do, and of complying with the regulations of that authority in respect of the mode in which the communications between such drains and sewers are to be made, and subject to the control of any person who may be appointed by that authority to superintend the making of such communications.

[16] The Public Health Act 1936 (the 1936 Act) replaced the provisions of the 1875 Act with provisions that more closely resemble those of the 1991 Act. Section 34 provided:

> (1) Subject to the provisions of this section, the owner or occupier of any premises, or the owner of any private sewer, within the district of a local authority shall be entitled to have his drains or sewer made to communicate with the public sewers of that authority, and thereby to discharge foul water and surface water from those premises or that private sewer:
>
> …
>
> (3) A person desirous of availing himself of the foregoing provisions of this section shall give to the local authority notice of his proposals, and at any time within twenty-one days after receipt thereof, the authority may by notice to him refuse to permit the communication to be made, if it appears to them that the mode of construction or condition of the drain or sewer is such that the making of the communication would be prejudicial to their sewerage system, and for the purpose of examining the mode of construction and condition of the drain or sewer they may, if necessary, require it to be laid open for inspection:
>
> Provided that any question arising under this subsection between a local authority and a person proposing to make a communication as to the reasonableness of any such requirement of the local authority, or of their refusal to permit a communication to be made, may on the application of that person be determined by a court of summary jurisdiction.

[17] Lord Pannick submitted that the legislature can have had no intention of restricting the rights of the local authority and that "mode of construction" in the 1936 Act should be given the same meaning as "mode in which the communications… are to be made" in the 1875 Act. The latter phrase was wide enough to embrace the point at which the communication should be made. The same interpretation should be given to "mode of construction" in section 106 of the 1991 Act.

[18] Alternatively, Lord Pannick submitted that section 106(1) did not confer any entitlement on a property owner to connect at any point of its choosing and that it was open to an undertaker to respond to a proposal under section 106(3) by identifying a location at which connection might be made, such a response being subject to dispute resolution under section 106(6).

[19] Lord Pannick relied in support of these submissions on observations by Walton J in *Beech Properties* v *GE Wallis & Sons Ltd* (1976) 241 EG 685*, to which I shall return.

Judgments below

[20] Wyn Williams J accepted the first of Lord Pannick's approaches to the construction of section 106, then advanced, on behalf of Welsh Water, by Mr Maurice Sheridan. In doing so, he relied on the judgment of Walton J in *Beech Properties.* He added that he considered that it would be objectionable to construe the statute in such a way as to preclude an undertaker from refusing a connection that would have potentially deleterious environmental consequences.

[21] In the leading judgment of the Court of Appeal, reversing the decision of the trial judge, Carnwath LJ held that section 34 of the 1936 Act, which was essentially reproduced in section 106 of the 1991 Act, provided only narrow grounds on which an undertaker could refuse connection. These related solely to the mode of construction or the condition of the connecting drain. This formulation was even narrower than under the 1875 Act, which permitted the authority to regulate the "mode of communication". Furthermore, the reason why Welsh Water objected to the point of connection was that that connection would overload the public sewer and there was clear authority that an undertaker could not resist connection on this ground.

[22] Lawrence Collins LJ agreed with the judgment of Carnwath LJ. Pill LJ also agreed. He held, in [54]:

> I am unable to conclude that the expression "mode of construction and condition of the drain or sewer" in section 106(4), repeated in section 106(5) of the 1991 Act, has any bearing upon the location of the communication with the public sewer contemplated in section 106(1)(b) and section 106(4). Mode of construction has nothing to do with location.

He added in the following paragraph that he would not accept the submission of Mr Porten that the owner or occupier could dictate the precise location of the connection.

> Circumstances may be such as to allow a modest discretion to the sewerage undertaker where good reason is shown, for example, that the precise location chosen by the applicant is not a feasible or sensible location at which to connect.

That was not this case. Welsh Water was seeking to dictate a communication situated around 300m from that requested and across land in third-party ownership and control.

Statutory scheme

[23] The right to connect to a public sewer afforded by section 106 of the 1991 Act and its predecessors has been described as an "absolute right". The sewerage undertaker cannot refuse to permit the connection

* Editor's note: Also reported at [1977] 1 EGLR 142

A on the ground that the additional discharge into the system will overload it. The burden of dealing with the consequences of this additional discharge falls directly on the undertaker, and the consequent expense is shared by all who pay sewerage charges to the undertaker. Thus, in *Ainley* v *Kirkheaton Local Board* (1891) 60 LJ Ch 734, Stirling J held that the exercise of the right of an owner of property to discharge into a public sewer conferred by section 21 of the 1875 Act could not be prevented by the local authority on the ground that the discharge was creating a nuisance. It was for the local authority to ensure that what was discharged into their sewer was freed from all foul matter before it flowed into any natural watercourse.

B [24] In *Brown* v *Dunstable Corporation* [1899] 2 Ch 378, at p390, Cozens-Hardy J described the right under section 21 as an "absolute right", adding that:

> This absolute right is no doubt subject to any regulations in respect of the mode of making connections and subject to the control of any person appointed to superintend the making of the connections; but no regulations can justify an absolute refusal to allow a connection to be made on any terms.

[25] In *Smeaton* v *Ilford Corporation* [1954] Ch 450, the corporation were the authority responsible for sewerage in Ilford. They were sued by the plaintiff in nuisance caused by the escape of sewage from a
C sewer. Upjohn J held that they were not liable. The nuisance was not caused by the corporation but arose because the corporation were bound by section 34 of the 1936 Act to permit occupiers or premises to make connections with the sewer and to discharge their sewage into it.

[26] *Smeaton* was cited with approval by the House of Lords in *Marcic* v *Thames Water Utilities Ltd* [2003] UKHL 66; [2004] 2 AC 42. Lord Nicholls of Birkenhead remarked, in [34], that Thames Water had no control over the volume of water entering its sewers. A sewerage undertaker was unable to prevent connections from being made to the existing system, and the ingress of water through those connections, even if this risked overloading the existing sewers.

D [27] It follows that the duty imposed on Welsh Water by section 94 of the 1991 Act requires it to deal with any discharge that is made into its sewers pursuant to section 106. It does not follow, however, that where a new development is constructed, Welsh Water is obliged, at its own expense, to construct a sewer to accept the sewage from the development if one does not already exist. Section 98 entitles a developer, among others, to requisition a public sewer or a lateral drain linking with a public sewer in order to service the buildings being constructed, but on terms that it meets the costs of so doing. Section 101 provides that the place or places where the public sewer and drain are to be located are to be agreed between the requisitioner and the undertaker
E or, in default of agreement, to be determined by Ofwat.

[28] Sections 102 and 103 of the 1991 Act make provision for a sewerage undertaker to adopt private sewers, lateral drains and disposal works. Section 104 makes provision for a party that is constructing or that proposes to construct a sewer, lateral drain or disposal works to enter into an agreement with a sewerage undertaker under which the undertaker will adopt the works at or after their completion.

[29] Section 112 of the 1991 Act provides:

> **Requirement that proposed drain or sewer be constructed so as to form part of general system**
>
> (1) Where —
>
> F (a) a person proposes to construct a drain or sewer; and
>
> (b) a sewerage undertaker considers that the proposed drain or sewer is, or is likely to be, needed to form part of a general sewerage system which that undertaker provides or proposes to provide, the undertaker may require that person to construct the drain or sewer in a manner differing, as regards material or size of pipes, depth, fall, direction or outfall or otherwise, from the manner in which that person proposes, or could otherwise be required by the undertaker, to construct it.
>
> (2) If any person on whom requirements are imposed under this section by a sewerage undertaker is aggrieved by the requirements, he may within twenty-eight days appeal to [Ofwat].

Any additional cost that this involves has to be paid by the undertaker.

[30] Section 113 of the 1991 Act provides:

> **Power to alter drainage system of premises in area** G
>
> (1) Where any premises have a drain or sewer communicating with a public sewer or a cesspool, but that system of drainage, though sufficient for the effectual drainage of the premises —
>
> (a) is not adapted to the general sewerage system of the area; or
>
> (b) is, in the opinion of the sewerage undertaker for the area, otherwise objectionable, the undertaker may, at its own expense, close the existing drain or sewer and fill up the cesspool, if any, and do any work necessary for that purpose.
>
> (2) The power conferred on a sewerage undertaker by subsection (1) above shall be exercisable on condition only that the undertaker first provides, in a position equally convenient to the owner of the premises in question, a drain
> or sewer which — H
>
> (a) is equally effectual for the drainage of the premises; and
>
> (b) communicates with the public sewer.

[31] The scheme of the legislation, as reflected in the above provisions and as affecting a developer, can be summarised as follows:

(i) Where the connection of a development to a public sewer requires consequential works to accommodate the increased load on the public sewer, the cost of those works falls exclusively on the undertaker.

(ii) Where works are carried out, whether by or on the requisition of the developer, that will be used exclusively by the development, the
costs of such works fall exclusively on the developer. J

(iii) In specified circumstances, the undertaker is entitled to require the developer to carry out the works in a manner other than that proposed by the developer, or to alter the works carried out by the developer. In either case, the undertaker has to bear the costs involved.

(iv) Costs that are borne by the undertaker are passed on to all who pay sewerage charges. These include those who occupy the houses in the development.

Natural meaning of section 106

[32] It is plain from section 106(5) that the "drain or sewer" referred to in section 106(4) is the private drain or sewer that the developer
proposes to connect to the public sewer, and Lord Pannick accepted that this was so. K
I agree with the Court of Appeal that it is impossible to extend the natural meaning of the "mode of construction" of the existing drain or sewer so as to include the point at which it is proposed to connect that drain or sewer to the public sewer. Lord Pannick argued that one reason why this extension of "mode of construction" should be made was that it was unlikely that the "mode of construction" of the private sewer or drain would be of concern to the undertaker were that phrase to be given its natural meaning. As to this, we received no evidence as to why the condition or mode of construction of the private drain or sewer should be of concern to the undertaker, but I note that
section 114 gives the undertaker a right to open a private drain or sewer for inspection if, *inter alia*, there are reasonable grounds for believing that "any such drain or private sewer is so defective as to admit subsoil water". L
I see no justification for approaching section 106(4) on the premise that the condition or mode of construction of the private drain or sewer is unlikely to be of concern to the undertaker.

[33] The provisions of section 106(4) of the 1991 Act contrast with the equivalent provisions in respect of sewerage in Scotland set out in section 12 of the Sewerage (Scotland) Act 1968:

> (3) The owner of any premises who proposes to connect his drains or
> sewers with the sewers or works of a local authority, or to alter a drain or sewer connected with such sewer or works in such a manner as may interfere with them, shall give to the authority notice of his proposals, and within 28 days of the receipt by them of the notice the authority may refuse permission for the connection or alteration, or grant permission for the connection or alteration, subject to such conditions as they think fit, and any such permission may in particular specify the mode and point of connection and, where there are separate public sewers for foul water and surface water, prohibit the discharge of foul water into the sewer reserved for surface water, and prohibit the discharge of surface water into the sewer reserved for foul water. M
>
> (4) A local authority shall forthwith intimate to the owner their decision on any proposals made by him under subsection (3) above, and, where permission is refused, or granted subject to conditions, shall inform him of the reasons for their decision and of his right of appeal under subsection (5) below.

(5) If a person to whom a decision has been given under subsection (4) above is aggrieved by the decision or any conditions attached thereto, he may appeal to the Secretary of State who may confirm the decision and any such conditions either with or without modification or refuse to confirm it.

This merely underlines the fact that "mode of construction" does not naturally embrace the "point of connection". No explanation was offered to us as to why those who drafted the Scottish Act chose different language from that of the 1991 Act.

[34] So far as Lord Pannick's alternative approach to construction is concerned, I can see no basis, if the wording of section 106 is given its natural meaning, for inferring that it confers a right on the part of the undertaker to refuse permission to communicate with a public sewer on the ground that the intended point of connection is not satisfactory.

Beech Properties v *Wallis*

[35] The issue in this case was whether a vendor of property had satisfied an obligation to provide the purchaser with the right to run foul and surface water from the land sold to a public sewer. The vendor contended that this obligation was satisfied by the right of the purchaser to connect a 12in diameter pipe to a 9in diameter public sewer at a particular location, pursuant to section 34 of the 1936 Act. Walton J held, essentially because of uncertainty as to this right, that the condition was not satisfied. His judgment contained the following observations, at p146F-H:

> However, it does appear to me that, wide as the words of subsection (1) may be, and for the moment ignoring the opening qualification, they do not confer upon an individual the right to connect his sewer to the water authority's sewer at any point which he may choose. In most cases, of course, the matter will be quite academic. There will be the water authority's sewer, going along the road; a new house is built in the road; and quite obviously and clearly the owner will expect to have a right to drain into that sewer, and it would be very difficult, assuming that there are no problems under the proviso to subsection (1), to imagine a set of circumstances where the water authority would be entitled to say that he must not connect to that sewer but to some other sewer. Even so, if the new house was built at a crossroads and there were available sewers in both roads, I can see no reason why the owner should be entitled to drain into the sewer of his choice if the water authority required him to drain into the other, which might, for example, well be a relief sewer expressly provided for the district because the other sewer was approaching capacity. Similarly, I see no reason why the owner is entitled to connect at point X rather than an adjacent point Y, if the water authority requires him to connect at Y.

[36] This passage sounds eminently sensible, but the judge gave no satisfactory explanation as to how the authority's option to select the point of connection could be derived from section 34. This decision cannot sustain the weight placed on it by the trial judge and by Lord Pannick.

Requirements of European law and Human Rights Convention

[37] Lord Pannick submitted that if the words of section 106 did not naturally bear the meaning for which he contended, they should be so interpreted as to carry that meaning none the less in order to avoid the infringement of the directive, the 1994 Regulations and the Human Rights Convention. Although he relied on the facts of the present case as illustrating his thesis, much of the argument focused on a rather different scenario. Mr Porten argued that Pill LJ had erred in suggesting that an undertaker enjoyed a "modest discretion" to refuse to connect at the precise location chosen by an applicant where this was not a feasible or sensible location at which to connect. He submitted that a developer's "proposals" under section 106(3) could specify the precise point of connection and that the undertaker had no right to insist on deviation from that point by so much as 1m.

[38] Lord Pannick seized on this *reductio ad absurdum* as demonstrating that the construction for which Barratt contended could not possibly be correct. The scenario postulated is indeed absurd. It is impossible to conceive of any reason why a developer should not be prepared, indeed eager, to co-operate with the sewerage undertaker in selecting the point of connection that is most suitable, provided that this is within reasonable proximity of the development. In the present case, the evidence placed before us shows that Barratt was prepared to contemplate any one of a number of manholes in the vicinity of its development as the connection point. It is, I believe, significant that, in almost a century and a half since the 1875 Act was passed, this is the first occasion on which the English courts have been required to resolve a dispute between property owner and sewerage undertaker as to the point of connection of a private sewer or drain to a public sewer. The 1875 Act permitted local authorities to make regulations in respect of the "mode in which the communications between such drains and sewers are to be made". There is no evidence that any regulations relevant to the issues raised on this appeal were ever made. Nor is there any evidence that suggests that the change to the single ground for refusing a connection made by the 1936 Act led to any practical difficulties.

[39] Pill LJ did not identify the source of the "modest discretion" that he suggested would exist on the part of an undertaker to object to the "precise" point of connection selected by the developer should this prove not feasible or sensible. I suggest that section 108(1) of the 1991 Act probably provides the answer. This requires the developer, before commencing the work of making the communication, to give reasonable notice to any person directed by the undertaker to superintend the carrying out of the work and to afford such person all reasonable facilities for superintending the carrying out of the work. The subsection is silent as to the powers of the superintendent, but his role can be traced back to section 21 of the 1875 Act, which provided that the making of the communication should be "*subject to the control* of any person who may be appointed by that authority to superintend the making of such communications" (my emphasis). It is at least arguable that section 108 of the 1991 Act implicitly confers on the undertaker's superintendent the power to control the making of the connection and thus to insist that the precise point of communication is one at which it is technically feasible and sensible to make the connection. There is a lacuna in the Act in that the powers of the superintendent are not spelt out and no machinery is provided for resolving any dispute between the superintendent and the developer. Once again, this may reflect the fact that the possibility of a dispute between the supervisor and the developer is one that exists in theory rather than in practice.

[40] I now turn from the unlikely scenario of a dispute as to the precise point of connection to the situation that has led to the dispute in the present case.

Real problem

[41] The real problem that is demonstrated by the facts of this case arises out of the "absolute right" conferred by section 106 of the 1991 Act on the owner or occupier of premises to connect those premises to a public sewer without any requirement to give more than 21 days' notice. Although this might create no problem in the case of an individual dwelling-house, it is manifestly unsatisfactory in respect of a development that may, as in the present case, add 25% or more to the load on the public sewer. The public sewer may well not have surplus capacity capable of accommodating the increased load without the risk of flooding unless the undertaker has received sufficient advance notice of the increase and has been able to take the necessary measures to increase its capacity.

[42] This problem is accentuated by the fact that the budgets of sewerage undertakers and the charges that they are permitted to make have to be agreed by Ofwat and that this process takes place at five-year intervals, so that forward planning may have to be carried out five years in advance. This is not a problem that arises because, if it be the case, the developer has the right to select the point of connection. It is fortuitous that in this case there was spare capacity in the final short section of Welsh Water's sewer that led to the treatment works. In many cases, there will be no alternative point of connection that will avoid overload on the public sewer. Welsh Water has presented this appeal as though the problem to be addressed relates to the point of connection, whereas in truth the problem relates to the right of a developer, on no more than 21 days' notice, to connect to a public sewer that lacks the relevant capacity.

[43] The Court of Appeal suggested that the practical answer to this problem lies in the fact that the building of a development

A requires planning permission under the Town and Country Planning Act 1990. The planning authority can make planning permission conditional on there being in place adequate sewerage facilities to cater for the requirements of the development without ecological damage. If the developer indicates that it intends to deal with the problem of sewerage by connecting to a public sewer, the planning authority can make planning permission conditional on the sewerage authority first taking any steps necessary to ensure that the public sewer will be able to cope with the increased load. Such conditions are sometimes referred to as *Grampian* conditions after the decision of the House of Lords in *Grampian Regional Council* v *Secretary of State for Scotland* [1983] B 1 WLR 1340*. Thus, the planning authority have the power, which the sewerage undertaker lacks, of preventing a developer from overloading a sewerage system before the undertaker has taken steps to upgrade the system to cope with the additional load.

[44] Mr David Holgate QC, whose expertise in the field of planning led Lord Pannick to delegate to him this area of the case, sought to persuade us that planning law did not provide a satisfactory answer to the problem. He demonstrated that there are some projects that have a major effect on sewerage that are not subject to any planning control. Further, the planning authority may not always take the right decision so far as demands on the sewerage system are concerned. Article 10 C of the Town and Country Planning (General Development Procedure) Order 1995 (SI 1995/419) sets out a wide range of bodies that must be consulted by the local planning authority on an application in respect of a development such as Barratt's. They include the Health and Safety Executive, highway authorities, the Environment Agency, English Heritage, Natural England, the Countryside Council for Wales and the National Assembly for Wales, but not sewerage undertakers.

[45] If conditions of planning permissions are to provide the answer to the problem of the connection of private sewers to public sewers that are not adequate to bear the additional load, it would seem essential that there should be input to planning decisions from both the relevant sewerage undertaker and Ofwat. In the present case, there was input D from each, but in the submission of Welsh Water the county council's planning department made an erroneous decision. Before looking briefly at what occurred, it is instructive to note that Welsh Water and Ofwat were approaching the situation from different viewpoints.

[46] In 1997, an appeal was made to Ofwat, purportedly under section 106 of the 1991 Act, by the Post Office against a refusal by Yorkshire Water to allow a connection to a sewer in Sheffield on the sole ground of lack of capacity in the sewer. Ofwat ruled that this was not a valid ground for refusing the connection. Subsequently, on 28 November, Ofwat sent a letter to all sewerage undertakers about this E decision. It included the following passages:

> The key issue, which the Director was required to consider when making his recent determination, is whether the Act allows companies to refuse, or impose conditions upon, a connection of a surface water drain to its public combined sewer on the grounds of limited capacity in the latter.
>
> The Director concluded in his determination that the company was not able to refuse a connection solely on the grounds of lack of capacity. The Act refers only to the condition or construction of the private drain or sewer which is to be connected. This cannot, in the Director's view, extend to a consideration of the additional flows to be discharged into the public sewer, except in very specific circumstances. For example, if the additional flows were to be discharged at such high pressure as to potentially cause damage to the receiving sewer. F The Director also considers that companies are not able to make connection conditional upon works, by the person requesting the connection, designed to reduce flows and therefore address capacity problems in the companies' own systems.
>
> ...
>
> The Director also acknowledges that it is not in anybody's interest for new connections to lead to flooding from the public sewers. Although there is no specific provision in the Act to allow conditions to be imposed as to the timing of the connection, there may be circumstances in which it would be desirable to seek a deferment of the connection date to allow the company time to carry out necessary works to prevent flooding. However, if the company has had warning of a development and ought reasonably to have foreseen a likely connection G (for example, if it is included in the local structure plan), but fails to act, then a deferment condition is unlikely to be defensible. In this context, the companies' duty under Section 94 of the Act to provide, improve and extend the system of public sewers so as to ensure that the area is effectually drained is relevant.
>
> Finally, all of the comments above regarding rights of connection assume a situation in which there are no specific planning conditions upon a development specifying the nature of the connection or works to be completed prior to making the connection. There may be cases in which a planning condition would prohibit making a connection to a particular sewer, or place conditions upon that connection. There are mechanisms by which developers may appeal against such planning conditions, in which the Director has no role.

[47] Despite this advice, Mr Ian Wyatt, the new business manager of H Welsh Water, made it clear in a statement in these proceedings that Welsh Water believed that fairness required that a developer such as Barratt should bear any costs caused by the connection of the development's private sewer to a public sewer. Welsh Water had not budgeted for the cost of upgrading its system to cope with the demands that Barratt proposed to make on it by connecting at its chosen point. Upgrading involved replacing the pipe bridge with a pipe of larger diameter at a cost of around £200,000. Welsh Water's attitude throughout has been that Barratt should pay for this to be done or, alternatively, requisition Welsh Water under section 98 of the 1991 Act to build a parallel sewer to link the development to the public sewer at SO29127901, again at J Barratt's expense.

Facts in this case

[48] In 1999, a pre-deposit draft of Monmouthshire County Council's (MCC) unitary development plan was sent to Welsh Water for purposes of consultation. This made provision for, *inter alia*, the Llanfoist development. Welsh Water's response was that it objected to this proposed development because its sewerage system was already overloaded and improvements to it were not included in its relevant development programme.

[49] On 18 August 2005, Barratt applied to MCC for planning K permission for a development of 120 dwellings. Welsh Water was consulted and, on 14 September 2005, objected to this development for the same reason given in 1999. It added, however, that it might be possible for the developer to fund the accelerated provision of replacement infrastructure or to requisition a new sewer under sections 98 to 101 of the 1991 Act. Barratt revised its planning application, reducing the number of dwellings to 98 but adding a primary school. On 14 May 2007, MCC granted planning permission, subject to a number of conditions, which included:

> 10. No development shall take place until a scheme of foul drainage, and surface water drainage has been submitted to, and approved, by the Local L Planning Authority and the approved scheme shall be completed before the building(s) is/are occupied.

[50] Meanwhile, negotiations proceeded between Barratt and Welsh Water under the common assumption that if the development were to proceed, Barratt would have to fund either the upgrading of the public sewer to accommodate the increased load or the construction of a new sewer to link with the public sewer at manhole SO29127901.

[51] On 29 May 2007, Barratt served a notice under section 106 of the 1991 Act, on a standard form provided by Welsh Water, of its intention to make a foul water connection to the public sewer "on or M after June 07" at SO29131302, this being a manhole in close proximity to the development. A parallel application was made in respect of surface water. Welsh Water replied on 26 June 2007, as follows:

> Thank you for your application to connect the foul and surface water flows from the above-proposed development into the public sewerage systems.
>
> We are in a position to approve the connections, however, the foul water connection must be made into or downstream of manhole SO29127901, as shown on the attached plan (ref ConF1).
>
> Please note that if you encounter problems with third party landowners you may requisition, under Sections 98 to 101 of the Water Industry Act 1991, one of the following: —
>
> • A new sewer from the boundary of your site to this point of adequacy, or,

* Editor's note: Also reported at [1984] 2 EGLR 175; [1983] 271 EG 625

• The necessary improvement works as identified in the hydraulic assessment dated November 2006.

It is now accepted that this somewhat confusing letter is to be treated as a refusal of Barratt's proposal.

[52] Discussions continued between Barratt and Welsh Water on the premise that, in one way or another, Barratt would be funding the cost of dealing with Welsh Water's capacity problem. However, on 11 September 2007, Barratt wrote to Welsh Water, referring to its letter of 26 June, asserting that Welsh Water had no right, under section 106, to set the point of connection and asking Welsh Water to approve the connection. Welsh Water's response on 26 September was to contend that Barratt had served a requisition notice under section 98 and that this precluded any right to connect under section 106.

[53] On 25 January 2008, Ofwat, which had been kept informed of these developments, wrote to Welsh Water with a copy to Barratt, stating that there was no impediment on a developer pursuing simultaneously rights under sections 98 and 106. This letter concluded with the following statement:

> In any case, it is apparent that the application under section 106 of the Act by Barratt Homes was made on 29 May 2007, received by Welsh Water on 30 May 2007 and the company did not respond to the application until 26 June 2007. The response on 26 June 2007 was outside the statutory 21 days provided under section 106(4) and the company was not, therefore, entitled to refuse the application as made.
>
> That being the case, please confirm by 1 February, that Barratt Homes' proposal for connection as notified on 29 May 2007 can proceed. It is for Barratt Homes to confirm with the Planning Authority that it can satisfy the planning condition No 10.

[54] This letter was, I suspect, something of a bombshell. If so, it was as nothing compared to the next development. Barratt, with the aid of Ofwat's letter and an opinion from Mr Porten, the content of which has not been disclosed, persuaded MCC to treat condition 10 as discharged. The present proceedings followed.

Conclusions on the point of connection

[55] On its natural construction, section 106 of the 1991 Act gives the developer the right to connect its private drain or sewer to a public sewer subject only to the right of the sewerage undertaker to give notice: (i) refusing permission to make the communication on the ground of deficiencies in the condition of the private drain or sewer (section 106(4)); and (ii) that it will make the connection itself: section 107. The section confers no express right on the sewerage undertaker to select the point of connection or to refuse permission to make the communication on the ground that the point of connection proposed by the developer is open to objection. Lord Pannick has argued that, despite its natural meaning, the section must be interpreted as conferring such a right if the operation of the relevant provisions of the 1991 Act are not to be rendered "insensible, absurd or ineffective to achieve its evident purpose" — the phrase used by Lord Bridge of Harwich as justifying the disregard of particular words or phrases in a statute in *McMonagle* v *Westminster City Council* [1990] 2 AC 716, at p726E.

[56] I have not been persuaded by this argument. The lengthy history of the right to communicate with a public sewer does not suggest that the point of connection has ever given difficulty in practice. The facts of this case do not illustrate that section 106 gives rise to a problem with the point of connection. It illustrates the more fundamental problem that can arise as a result of the fact, accepted by Lord Pannick, that no objection can be taken by a sewerage undertaker to a connection with a public sewer on the ground of the lack of capacity of the sewer.

[57] As Ofwat has pointed out, although the 1991 Act affords no such right, there is a case for deferring the right to connect to a public sewer in order to give a sewerage undertaker a reasonable opportunity to make sure that the public sewer will be able to accommodate the increased loading that the connection will bring. The only way of achieving such a deferral would appear to be through the planning process. Some difficult issues of principle arise however:

• Is it reasonable to expect the sewerage undertaker to upgrade a public sewerage system to accommodate linkage with a proposed development regardless of the expenditure that this will involve?
• How long is it reasonable to allow a sewerage undertaker to upgrade the public sewerage system?
• Is it reasonable to allow the sewerage undertaker to delay a planned upgrading of a public sewer in the hope or expectation that this will put pressure on the developer to fund the upgrading?

[58] The facts of this case suggest that a sewerage undertaker may well take a different view from Ofwat as to how these questions should be answered. Be that as it may, it would seem desirable that the sewerage undertaker and Ofwat should at least be consulted as part of the planning process. I would endorse the comment made by Carnwath LJ, in [48], that more thought may need to be given to the interaction of planning and water regulation systems under the modern law to ensure that the different interests are adequately protected.

[59] These comments are an aside from the narrow issue of statutory interpretation raised in respect of the point of connection. For the reasons that I have given, I would endorse the judgments of the Court of Appeal in holding that a sewerage undertaker has no right to select the point of connection or to refuse a developer the right to connect with a public sewer because of dissatisfaction with the proposed point of connection.

Twenty-one-day limit

[60] Section 106(4) of the 1991 Act provides that the sewerage undertaker has 21 days from the receipt of a notice under section 106(3) in which to give notice of refusal to permit the communication to be made. The issue arises as to whether this time limit results in an absolute bar on giving such a notice once it has expired. In the light of my conclusion that the right of a sewerage undertaker to refuse permission to connect under section 106 of the 1991 Act arises only where there is reason to question the condition of the private drain or sewer that is to be connected, this issue is of limited importance, and of no significance at all on the facts of this case.

[61] A similar issue arises in respect of section 107(1), which gives the sewerage undertaker 14 days in which to give notice that it intends itself to make the communication.

[62] In the Court of Appeal, both Carnwath LJ and Pill LJ inclined to the view that the 21-day time limit was not mandatory, but refrained from deciding the point. I take the opposite view. Notices given under sections 106(4) and 107(1) remove a right to connect that is otherwise vested in the developer. Under the provisions of sections 107 and 109 respectively, it is a criminal offence to cause a drain or sewer to communicate with a public sewer after a notice has been given under section 106(4) or section 107(1). In these circumstances, it seems to me that the time limits in those two subsections must be strictly applied.

[63] For the reasons that I have given, I would dismiss this appeal.

Giving the second opinion, **Lady Hale** (dissenting) said:

[64] It is curious that it should have taken so long for a dispute of this sort to reach the courts. One might have thought that developers and sewerage undertakers were quite frequently at odds with one another about how best to accommodate a new housing development within the sewerage system and how the costs should be borne. However, there is no English or Welsh case directly in point. Wyn Williams J reached one conclusion on the meaning of the legislation and the Court of Appeal reached another. Most members of this court agree with the Court of Appeal, but the legislative history of the matter leads me to disagree.

[65] Section 106 of the 1991 Act can be traced back to section 21 of the 1875 Act and before that to section 8 of the Sanitary Act 1866 (the 1866 Act). The 1875 Act consolidated, with amendments, the patchwork of public health legislation that began with the Public Health Act 1848 (the 1848 Act). The 1848 Act, together with the Local Government Act 1858, provided for the setting up of local boards of health with a variety of powers dealing with sewers and drains, road cleaning, water supply and the like. Under those Acts, the local boards had the duty of effectually draining their districts. There was no right to connect to their sewers without their consent. However, the drive was

to get new and existing houses to connect. The board could direct how any new house built within 100ft of a sewer was to connect to it and could require old houses within the same distance to connect. However, local boards did not cover the entire country. The Sewage Utilization Act 1865 set up sewerage authorities in other areas and gave them all the powers of the local boards. Section 8 of the 1866 Act gave owners or occupiers of premises within the district of a sewerage authority the conditional right to cause its drains to empty into the authority's sewers in almost identical terms to section 21 of the 1875 Act. The Public Health Act 1872 rationalised the administration by dividing the whole of England and Wales (apart from the metropolis) into urban and rural sanitary districts. The metropolis was included in 1874 and the entire legislative scheme consolidated in the 1875 Act.

[66] Section 21 of the 1875 Act provided that the owner or occupier of any premises within the district of a local authority "shall be entitled to cause his drains to empty into the sewers of that authority", subject to giving the authority such notice as they required of his intention to do so and by "complying with the regulations of that authority in respect of the mode in which the communications between such drains and sewers are to be made" and subject to superintendence of its making.

[67] In *Ainley* v *Kirkheaton Local Board* (1891) 60 LJ Ch 734, the plaintiff was already connected to the authority's sewer, but they wanted to cut him off because the sewer emptied into an open stream and proceedings had been taken against the authority for fouling the stream. Stirling J held that the owner's right to drain into the existing sewers was not affected by the authority's obligation under section 17 of the 1875 Act not to allow its sewers to convey untreated sewage into a natural stream or watercourse. It was for the authority to provide sufficient sewers and to treat the sewage before discharging it into the stream.

[68] This case was followed in *Brown* v *Dunstable Corporation* [1899] 2 Ch 378, in which Cozens-Hardy J held that he could not grant an injunction to prevent the authority from allowing new connections to a sewer. Following *Ainley*, in preference to *Charles* v *Finchley Local Board* (1883) LR 23 Ch D 767, at 390, he held that the:

> absolute right is no doubt subject to any regulations in respect of the mode of making connections and subject to the control of any person appointed to superintend the making of the connections; but no regulations can justify an absolute refusal to allow a connection to be made on any terms… It is obvious that under this by-law the surveyor can only prescribe the manner of connection. He cannot refuse to allow any connection.

[69] In *Wilkinson* v *Llandaff and Dinas Powis Rural District Council* [1903] 2 Ch 695, the main issue was whether a roadside surface water drain was a "sewer" within the meaning of the 1875 Act. If it was, the authority had to keep it clean. One of the arguments against it being a sewer was that section 21 would then give everyone the right to connect his or her own drains into it. Romer LJ bluntly observed, at p702, that "it does not follow that, because this channel is a "sewer" within the definition of the Act, it can be used by any inhabitants of the district for sewage or faecal matter". Stirling LJ (as he had become) thought, at p703, that the argument was an exaggeration of the effect of section 21:

> Sect 21 does not provide that every owner or occupier of premises within the district of a local authority shall be entitled as of right to connect every drain which he has with every sewer belonging to the local authority. That is not the meaning of the section. All that is given by that section to the owner and occupier is a right to have the drain connected or made to communicate with the sewers of the local authority, subject to compliance with certain conditions — amongst others, that he is to comply "with the regulations of the local authority in respect of the mode in which the communication" with the sewers is to be made. So that, in my opinion, the local authority may define by regulation the particular sewer with which the communication is to be made.

[70] Each party in this case can get something from these three authorities. For the developer, the fact that continuing an existing connection or allowing a new one would cause a nuisance to the public or to a private individual was not by itself a reason to stop up or prohibit the connection. For the undertaker, on the other hand, the "mode in which the communication… is to be made" could be regulated and this could cover the time and the place at which the connection was to be made.

[71] The 1875 Act was consolidated with other enactments and some amendments in the 1936 Act. Section 21 of the 1875 Act became section 34 of the 1936 Act. Once again, the owner or occupier of any premises or the owner of any private sewer within the district of a local authority was entitled to have its drains or sewer made to communicate with the public sewers of that authority. This was subject to various restrictions in the section itself, and to the requirement, in section 34(3), that a person wanting to avail himself of this right should give notice to the local authority and "at any time within 21 days after receipt thereof, the authority may by notice to him refuse to permit the communication to be made, if it appears to them that the mode of construction or condition of the drain or sewer is such that the making of the communication would be prejudicial to their sewerage system…". Disputes concerning the reasonableness of any refusal could be determined by a magistrates' court.

[72] Lord Pannick has referred us to the report that led up to the 1936 Act: Cmd 5059 of 1936. There is nothing in that report to suggest that the change in language, from the "mode in which the communications between such drains and sewers are to be made" to the "mode of construction or condition of the drain or sewer", was intended to cut down the existing scope of the local authority's power to control the place and manner of the connection. Yet one would expect such a significant change to be flagged up in any report proposing consolidation with amendments. It would be very strange if parliament had intended to make such a change. The public interest in ensuring that connections were made in ways that were not prejudicial to the sewerage system remained the same. There were no other means available of doing so. It could not have been contemplated, for example, that the developer could knock a big hole into an existing sewer and simply stick its own perfectly sound drain through it without making good.

[73] It would also be strange if parliament had legislated for such a change in England and Wales while leaving the position in Scotland, under section 110 of the Public Health (Scotland) Act 1897, the same as it had been in England and Wales under section 21 of the 1875 Act. Further, that parliament should later re-enact and clarify that provision in section 12(1) of the 1968 Act, which provided that the Scottish local authorities could "specify the mode and point of connection". It is inexplicable why provisions that began in the same legislation covering the entire UK should diverge in this respect. It is much more likely that parliament intended them to mean the same thing.

[74] Then came the well-known case of *Smeaton* v *Ilford Corporation* [1954] Ch 450. The local authority's Victorian sewers were overloaded and, from time to time, sewage erupted from a manhole near the plaintiff's house and overflowed into his premises. Despite section 31 of the 1936 Act providing that a local authority shall so discharge their functions "as not to create a nuisance", the plaintiff's claim in nuisance failed. The local authority were not causing or adopting the nuisance. Upjohn J explained, at pp464-465:

> It is not the sewers that constitute the nuisance; it is the fact that they are overloaded. That overloading, however, arises not from any act of the defendant corporation but because, under section 34 of the Public Health Act 1936, subject to compliance with certain regulations, they are bound to permit occupiers of premises to make connexions to the sewer and to discharge their sewage therein… Nor, in my judgment, can the defendant corporation be said to continue the nuisance, for they have no power to prevent the ingress of sewage into the sewer.

[75] The real problem in such a case, as both Lord Nicholls of Birkenhead and Lord Hoffmann pointed out in *Marcic* v *Thames Water Utilities Ltd* [2003] UKHL 66; [2004] 2 AC 42, is that "Every new house built has an absolute right to connect" ([34]) and the undertaker has a duty "to accept whatever water and sewage the owners of property in their area choose to discharge": [53]. The overflow is not caused by any failure to clean or maintain the existing sewers but by a failure to

build new or bigger ones. Moreover, there is a long line of authority, dating back to *Glossop* v *Heston and Isleworth Local Board* (1879) 12 Ch D 102, that the authority's duty to provide sufficient sewers effectually to drain the area is to be enforced through the statutory scheme and not by private action.

[76] The decisions in *Smeaton* and *Marcic* were predicated on the authority's or undertaker's duties to allow connections and to accept sewage, but they did not decide what that duty entailed. The only other relevant observations to which we have been referred are in *Beech Properties Ltd* v *GE Wallis & Sons Ltd* [1977] 1 EGLR 142, in which the question was whether a condition in a contract for the sale of land had been performed. Part of this depended on whether the purchaser would have the right to connect to the public sewer at a particular point. Walton J, at p146D, thought it "obvious" that the right given by section 34 of the 1936 Act "is not an absolute, but a qualified, right". He continued, at p146F-H:

> wide as the words of subsection (1) may be… they do not confer upon an individual the right to connect his sewer to the water authority's sewer at any point which he may choose. In most cases, of course, the matter will be quite academic. There will be the water authority's sewer, going along the road; a new house is built in the road; and quite obviously and clearly the owner will expect to have a right to drain into that sewer… Even so, if the new house was built at a crossroads and there were available sewers in both roads, I can see no reason why the owner should be entitled to drain into the sewer of his choice if the water authority required him to drain into the other, which might, for example, well be a relief sewer expressly provided for the district because the other sewer was approaching capacity. Similarly, I see no reason why the owner is entitled to connect at point X rather than an adjacent point Y, if the water authority requires him to connect at Y.

[77] So, we have three propositions for which there is respectable authority going back over many years and that are not inconsistent with one another. The first is that the sewerage authority or undertaker cannot refuse to allow an owner or occupier to connect at all. It must allow some sort of connection even if the system is already overloaded or will thereby become so overloaded that a nuisance will result. The second is that the authority or undertaker is not liable for nuisances that result from such overloading. The remedy lies in the statutory procedures to oblige them to build more sewers. However, the third is that all courts that have addressed themselves specifically to the point at issue here, namely the place at which and manner in which a particular connection is to be made, have expressed the view that the authority or undertaker can refuse to agree to the developer's proposals.

[78] There is no material difference between the 1936 and the 1991 Acts for this purpose. The 1936 Act provided that disputes between developers and authorities should go to a magistrates' court. The 1991 Act provides that a developer that argues that an authority's refusal is unreasonable can take the dispute to Ofwat, which is a much more appropriate body to resolve such matters. The 1936 Act provided that a local authority could refuse on the ground that "the making of the communication would be prejudicial to their sewerage system" and section 106(4)(b) provides the same. This is obviously capable of including the deleterious effects of connecting at point A rather than point B. This, too, may help cast some light on the meaning of the words "mode of construction or condition": it is easier to think of ways in which the place and manner of making the connection would be deleterious to the system than of ways in which the physical condition of the developer's drain would be so.

[79] In the light of the historical development of this difficult legislation, therefore, I would hold that the words "mode of construction or condition" do cover the way in which it is proposed to connect that private drain or sewer to the public sewer, including the place. Whether the undertaker's reasons for refusing to allow the proposed connection are reasonable is another matter, which, in my view, it is for Ofwat to resolve. If that were the only issue in the case, therefore, I would have allowed this appeal.

Appeal dismissed.

RENTCHARGES

A

Shand and another *v* Morgoed Estates Ltd

BIRMINGHAM COUNTY COURT

21 December 2009

HH JUDGE DAVID COOKE

Rentcharges — Enforcement — Terre-tenant — First rentcharge reserved by vendor of land on sale to developer — Second rentcharges created on sale of developed plots — Identity of terre-tenant as party liable to pay chief rent under first rentcharge — Whether terre-tenant holder of second rentcharge or current freehold owner of affected land B

By a number of conveyances, various parcels of land were conveyed to a developer. Rentcharges were reserved over the land in favour of the vendor; the respondents were the successor in title to those rentcharges. The appellant held a second layer of rentcharges as the successor in title to the developer, which had, after developing the land for residential housing, sold the individual plots and reserved the second rentcharges out of those conveyances. On taking a transfer of the second rentcharges, the appellant covenanted with the transferor C to pay the chief rent due under the respondents' rentcharges, but the appellant gave no such covenant directly to the respondents.

The respondents brought proceedings against the appellant to recover arrears of chief rent. The appellant argued that it had no direct liability to the respondents for the chief rent because there was neither privity of contract nor privity of estate between them. It contended that, in the absence of a direct covenant with the respondents, only the terre-tenant, being a freeholder in actual possession and enjoyment of the land or in receipt of its profits, was liable to pay the chief rent; the appellant did not hold a freehold interest.

Allowing the claim, the district judge held that since a rentcharge was a legal interest in land and the appellant was in possession of the D rent thereunder, its position was equivalent to being in possession of the land and, as such, it was the terre-tenant. The appellant appealed.

Held: **The appeal was allowed. The action in debt to enforce a rentcharge lies against the terre-tenant, which must be both the party with the freehold interest in the property and the party entitled to possession or to "pernancy of the profits" of that property, namely the right to receive the profits from the land. The two tests, of holding a freehold interest in the land and of being the pernor of the profits, are both necessary and cumulative. If a freeholder is entitled to immediate possession, it is no objection E that it is not in actual occupation of the profits, or has not actually received such rents and profits as exist or that they are insufficient to pay the chief rent. However, in order to be liable, the party charged must be the freeholder of land. The owner of a second rentcharge is not the freeholder but only the party to whom the freeholder pays the second rent, and, as such, it cannot be the terre-tenant. The appellant was not the terre-tenant and was not liable for the chief rent under the respondents' rentcharge; the respondents would have to pursue their entitlement to rent against the freehold occupiers of the land for the time being.**

The following cases are referred to in this report.

Cundiff v *Fitzsimmons* 91910] 1 KB 513, KB
Swift v *Kelly* (1889) 24 LR Ir 478, Ir CA
Thomas v *Sylvester* (1873) LR 8 QB 368 F
Varley v *Leigh* (1848) 2 Exch 446

This was an appeal by the appellant, Morgoed Estates Ltd, from a decision of District Judge Brown allowing a claim by the respondents, Diana Shand and Christopher Alderley, for chief rent under a rentcharge.

Mr Jonathan Roberts, of Morgoed Estates Ltd, appeared for the appellant; William Hanbury (instructed by Blackhurst Budd, of Blackpool) represented the respondents.

Giving judgment, **HH JUDGE DAVID COOKE** said:

[1] This is an appeal by the defendant in the original action, Morgoed G Estates Ltd, against the decision of District Judge Brown given in a reserved judgment on 25 August of this year. The district judge himself gave permission for this appeal.

[2] I summarise the facts from the district judge's judgment (which is at p25 of the bundle). The claimants, the respondents to this appeal, are the successors in title of the grantee of various rentcharges in Manchester and its surrounding area. Those rentcharges were created by a number of conveyances in the early part of the 20th century. One Mr George Frederick Derbyshire, the grantor of those rentcharges, that is, the purchaser of the various parcels of land, developed the land for residential housing and sold off the individual properties, reserving H further rentcharges for himself out of each house sold. The defendant, the appellant before me, is the successor in title to Derbyshire. Therefore, there are two layers of rentcharges: those reserved out of the conveyance to Derbyshire in favour of the claimant's predecessors in title and those reserved out of the conveyances of the individual housing plots in favour of Derbyshire. Hereafter, I will call the rentcharge granted by Derbyshire in favour of the original vendor the chief rent, and the rentcharge reserved by Derbyshire out of the individual houses sold the second rent.

[3] The respondents seek to recover from the appellant arrears of chief rent. The appellant does not dispute that the arrears exist or that J the respondents have the benefit of and are entitled to enforce payment of the chief rent. However, the appellant argues that it has no direct liability for the chief rent because there is neither privity of contract nor privity of estate between these parties.

[4] Now, as has been pointed out before me, in the conveyance to the appellant, by which it acquired the second rents that it holds, the appellant covenanted to the vendors of the second rents to pay the chief rent. The vendors of the second rents (or their predecessors in title) will in turn have given covenants to the present householders to pay the chief rent, the expectation being that the householders would be responsible for the payment of the second rents and the second rent K holder would pay the chief rent.

[5] The question here, however, is whether the second rent holder is directly liable to the holder of the chief rent. It arises in circumstances in which, as Mr William Hanbury, for the respondents, accepted, there is no direct covenant by the appellant enforceable by the respondents for the payment of the chief rents, and although the appellant holds a legal interest in the property, namely a rentcharge, it is not a freehold interest in the property. Indeed, the present appellant has never held a freehold interest in the property. The appellant submits that only a freeholder can be liable to pay the chief rent.

[6] The district judge was referred, as I have been, to a number of textbooks and to one particular case, *Thomas* v *Sylvester**. I have had the benefit of additional case authority that was not before the district judge, but the textbook authorities are the same in both cases. First, there is an extract from *Halsbury's Laws of England* (vol 39(2)), in para 877, which is headed "Nature of Remedy", which says:

> Since the abolition of real actions, a rentcharge in fee may be recovered in an action of debt from the terre-tenant (that is, the person in actual possession and enjoyment of the land) entitled to a freehold interest, even where he is terre-tenant of part only of the land charged with the rent.

A number of authorities are cited for the proposition that the rentcharge is recoverable from a terre-tenant entitled to a freehold interest. The first is *Varley* v *Leigh* (1848) 2 Exch 446, although it is noted that, in that case, the terre-tenant had personally covenanted to pay the rentcharge, which is not applicable in this case. The second is the case that was referred to the district judge, *Thomas*, and the note goes on to say "where the decision was independent of covenant". I will refer to that case in due course.

[7] The next paragraph, 878, is headed "Who may be liable". It says:

> In order that the terre-tenant be made liable in an action of debt he must be "pernor of the profits" of the land (that is, the person who receives the profits), but he may be made liable whether he has or has not received profits equal to the rentcharge claimed. If the freeholder has an immediate right to take possession, he has sufficient "pernancy of the profits" to render him liable. Accordingly a mortgagee who had never entered might formerly be liable where the mortgage was by conveyance of the fee simple.

The authority for the first proposition, that the terre-tenant must be the pernor of the profits of the land, is an Irish case: *Swift* v *Kelly* (1889) 24 LR Ir 478. The district judge was referred to that case, but, unfortunately, he was not able to be provided with a report of it. I do now have a copy of that case, so I have the advantage over the district judge in being able to consider it.

[8] Second, there is an extract from *Megarry & Wade: The Law of Real Property* (in para 31-028) that is to similar effect. Third, there is an extract from an old textbook of 1884, Harrison's *The Law Relating to Chief Rents and Other Rentcharges*. The respondents refer particularly to a paragraph at p181 from the original text, which is as follows:

> In the case of a first chief rent payable by the owner of a "second" rent, the action will apparently not lie against the latter directly, there being no privity. Here, as in other cases, the person to be sued is the pernor of the profits of the land upon which the rent is charged, who must recoup himself by means of the indemnity (if any) which he possesses under the deed creating the second rent.

That paragraph seems to commence by referring to a first chief rent being "payable by" the owner of a second rent, but it goes on to say that no action may be maintained against the second rent holder directly. Of course, the appellant relies on that.

[9] On the previous page in the same work, there is also this section, which seems to me to be relevant:

> All real actions being now abolished… the remedy by action of debt has become applicable to all cases of rent in arrear.

The "rent" referred to is a rentcharge:

> The action is properly brought against the pernor of the profits of the land, the liability arising by reason of the receipt of the profits, and it therefore lies against an assignee, and not merely against the original grantor.

[10] All those passages, it seems to me, rely, first, on the fact that the party to be made liable is what is referred to as a terre-tenant, that is to say, the holder for the time being of the freehold interest in the land. They say, second, that the terre-tenant, in order to be liable, must be the pernor of the profits of the land, that is to say, the party that is entitled to receive the rents and profits of the land. These seem to me to be references to two separate conditions. The first condition being that the party sought to be made liable is the freeholder, and the second is that, as freeholder, it is also the pernor of the profits. That is less clearly brought out in the extract from Harrison, but from the passage at p180 it does seem to me that the author there is referring to an "assignee [that is to say, an assignee of the freehold] and not merely the original grantor" of the rentcharge, which of course would have been the freeholder at the time of the grant. So where he passes on to refer to the pernor of the profits, it seems to me that he is referring to an assignee of the freehold that is also the pernor of the profits.

[11] The district judge's conclusion is set out at p29 of the bundle, which, as I say, he based primarily on the textbooks that he was referred to, having only one of the cases. He says, in [12] of his judgment:

> The starting point must be that a rentcharge is a legal interest in the land. (Law of Property Act 1925 Section 1(2)(*b*)). As a result it must be the case that the defendant as owner of the second rent has a legal interest in the properties to which it relates. Is that enough to make him the terre tenant and does he thereby have pernancy of the profits of the land? To assist me further I have made reference to the general definitions section of the Law of Property Act 1925. The following provisions are relevant: —
>
> Section 205 (1) (ix): "Land" includes land of any tenure, and mines and minerals and other corporeal hereditaments, also a manor, an advowson and a rent and other incorporeal hereditaments and an easement right.
>
> …
>
> (xxiii) possession includes receipt of rent and profits or the right to receive the same, if any…
>
> (xxiv) rent includes a rent service or a rentcharge.

He goes on to say:

> Thus the owner of the rentcharge;
>
> (1) Has a legal interest;
>
> (2) That interest is an interest in land;
>
> (3) He is in possession of the rent, which is the equivalent to being in possession of the land.
>
> As a result it seems to me that it must follow that the owner of the second rent is the terre-tenant as he must be treated as in possession of the land.

He then notes that that is a conclusion opposite to that reached by Harrison.

[12] I now refer to the case authorities that I have been provided with, which are the foundation of the statements that are made by the authors of *Halsbury's* and *Megarry & Wade*. *Thomas* was decided in 1873. The case note opens by recording that the plaintiff, as freeholder, had conveyed the relevant property to the defendant's predecessors in title subject to the payment of a rentcharge. There were three judges in the case, Blackburn, Quain and Archibald JJ. The principal citations were from the judgment of Quain J, but all three judges were agreed in the result. Quain J says, at p372 of the report:

> Putting aside the remedy by real action, would debt have lain [that is to say, would an action of debt have been maintainable] against an assignee who is in possession of the same estate in land as the grantor of the rent? That appears to be decided in *Sir W Loringes case*…

A little further on, he says:

> The action was against the person who is called the pernor of the land, provided he had the same estate as the grantor. I apprehend that the reason is that the land is the debtor, as is stated by Wilson J in *Mills* v *Auriol*. If a man comes into possession of land as tenant in fee, he is the pernor of the profits of the land, and was liable to a real action for the non-payment of a rentcharge created by a former tenant in fee; if this be so, since real actions are abolished, an action of debt will lie.

[13] In both those passages, Quain J is clearly referring to an action being maintained against the person who is, for the time being, the freeholder of the property. That is the person who is referred to as the tenant in fee. It is clear that he is referring to a case in which the person sought to be charged has the same estate as the grantor, that is to say, the freehold estate out of which the rentcharge was granted. Having that estate, he was liable for the payment of the rentcharge by a real action, but since the abolishing of real actions he is made liable in an action for debt.

---

* Editor's note: Reported at (1873) LR 8 QB 368

A [14] *Swift* is an Irish case that, as I say, was referred to the district judge but was not available for him to read. That is an appeal decision. It is a decision made prior to the partition of the UK, and I think Mr Roberts is right to say that it is an authority, therefore, that is binding on this court, but even if not binding it would be highly persuasive, being made by an appeal court at a time when these matters were rather more frequently before the courts than I suspect they are at the moment.

[15] The Chancellor Lord Ashbourne said, starting at p479:

> I am satisfied to treat as settled that where an owner of fee-simple land is in possession, and the land is charged with a rentcharge, the arrear can be
> B recovered from the owner by a personal action.

A little further on, he says:

> But in the present case there is a term of 100 years…

And a little further down:

> What is the effect of this term, having regard to the fact that the liability of the owner to be sued at law is based upon the ground of his pernancy of the profits?

He goes on to hold that, in the circumstances of that case, the defendant,
C who was the freeholder, was not to be treated as the pernor of the profits by reason of the existence of the term of years. However, it seems to be clear in that passage that the discussions start from the proposition that the person to be charged must be the owner of the land in fee simple, and the question in that case was whether he was also the pernor of the profits. The second judge was Chief Baron Palles, and, at p481, he says:

> Each of the Judges in the Court below based his judgment that the defendant was *terre-tenant,* not upon the ground that he had the estate of the grantor, but that he had that estate, *and* was pernor of the profits. Gibson J expressly states that "privity of estate and possession or pernancy of profits must co-exist.

D (Emphasis in the original.)
At p482, he says:

> As, however, the defendant was admittedly in possession, and but for the existence of the term of ninety-nine years, would have had the entire of the fee which was vested in the grantor at the time of the grant, he would unquestionably have been pernor of the profits but for the existence of the term. Thus, the case depends upon the effect of the interposition of the term between the rentcharge and the freehold estate of the defendant…

He goes on to conclude that the defendant is not the pernor of the profits.

E [16] Finally, I was referred to extracts from the decision of Naish LJ, beginning at p489. Concurring in the result arrived at by other members of the court, he says at p490:

> *Thomas* v *Sylvester* may be taken as establishing that if there were no term, and if the freehold reversioner were in possession in the pernancy of the rents and profits, an action of debt would lie against him.
>
> Does the interposition of the term create a distinction?

He goes on to conclude that it does, and, at p491, he says:

> It is said, however, that here he is in possession — in the pernancy of rents and profits. I ask, however, in what capacity this possession and pernancy is
> F enjoyed at law? Clearly as tenant at will, or, it may be, bailiff of the termor; and if it be a possession *under*, and not *over*, the term, I do not see how *Thomas* v *Sylvester* rules the case.

In the next paragraph, he goes on to say:

> I am therefore of opinion that the possession of the defendant here is wholly different from that of the defendant in *Thomas* v *Sylvester*. In the latter case he had at law large and extensive rights, and, having these, was held liable to a personal action. Here there are no corresponding rights — a fact arising from a particular remedy and security — namely, a trust term having been created with the consent, or on the stipulation, it must be taken, of the jointress, for the purpose of protecting the jointure.

The "jointure" refers to the rentcharge.

[17] Finally, by way of authority Mr Hanbury was able to obtain a G
copy of *Cundiff* v *Fitzsimmons* [1911] 1 KB 513, in which a rentcharge was sought to be recovered against a mortgagee. It is relevant to note that at the time of that action, being before the 1925 property law reforms, the mortgage in that matter was held under a conveyance of the fee simple and not as it would be now by grant of a term of years or charge by way of legal mortgage. At pp517-518, Darling J says:

> It is conceded by both parties that the person who is liable to pay the chief rent is the terre tenant, that is to say the person who, having a freehold interest in the land, has the pernancy or possession of such rents and profits as there are. Mr Hughes says that the defendants were not the terre tenants because
> although they were owners in fee simple of the land, they had never entered into H
> possession of the land and therefore were not pernors of the profits.

At the foot of p518, he says:

> That seems to shew conclusively that the terre tenant or person who is liable to pay the rent-charge need not be in actual possession of the land and that the owner will be terre tenant even where the land is derelict.

Over the page:

> I come then to the conclusion that the mortgagees here are the terre tenants, that they were in possession in the sense that they had the legal possession, and the mere fact that they had not the actual occupation did not excuse them from J
> liability to pay the chief rent.

So that again was a situation in which the person sought to be charged held the freehold interest of the property, and the question was whether that person was also the pernor of the profits.

[18] The case authorities therefore seem to me to make it clear that the action in debt lies against the party that is referred to as the terre-tenant, which must be both the party with the freehold interest in the property and also the party entitled to possession of the property or to the pernancy of the profits. No doubt, a freeholder in actual occupation is in the pernancy of profits, and the cases that I have been referred to,
Swift and *Cundiff,* discuss whether, in the circumstances, a freeholder is K
to be regarded as the pernor of the profits in those particular examples. What they hold is that the freeholder is not necessarily the pernor of the profits if, as in *Swift*, he does not have the immediate rights of freeholder by virtue of the interposition of a term of years, a 100-year term. However, if he is entitled to immediate possession, it is not an objection that he is not actually in occupation of the profits or has not actually received such rents and profits as there are or that they are not sufficient to pay the chief rent. That is the authority of *Cundiff.* They all start from the proposition that to be liable the party in charge must in the first instance be the freeholder.

[19] In my judgment, the district judge regrettably allowed himself L
to be led astray by the argument over possession and pernancy or entitlement to receive the rents and profits. His conclusion was that the owner of the second rent is the terre-tenant, since it must be treated as being in possession of the land for the reasons that he gave. Where he went wrong, it seems to me with respect, is that he did not treat the two tests as being necessary and cumulative, that is to say, that the terre-tenant must have the freehold interest in the land, that is, the same estate as the original grantor of the chief rent, and it must also be the pernor of the profits. He conflated the two and treated the question of who is the terre-tenant as depending only on the entitlement to rents
and profits. That said, it may be open to question whether the receipt M
of a second rent would be regarded as sufficient to make a party the "pernor" of the profits, given the amplification of that expression that was given in *Swift* and the emphasis on the existence of the wide range of rights normally available to the freeholder. However, that, I think, is probably an academic question because it seems to me *ex hypothesi* if the party sought to be charged is the owner of a second rent, it is not the freeholder, it is the party to whom the freeholder pays the second rent. It cannot, in my judgment, ever be the terre-tenant.

[20] For those reasons, in my judgment, the appellant is not and has never been the terre-tenant and is not liable to pay the chief rent.

[21] That puts the claimants, the respondents to this appeal, in the position that they must pursue their entitlement to rent against those

A who are the terre-tenants, that is, the present freehold occupiers. The district judge felt that that would cause difficulty and confusion, but it seems to me that the chief rent holder is, in fact, in a very privileged position, it being the case on authority that it is entitled to recover the entirety of the chief rent from the present freeholder of any part of the land. The confusion and difficulty therefore falls on those freeholders, if there are many of them as there appear to be here, who must, if called on to pay, exercise their rights of contribution or indemnity against the B
other present freeholders or against the appellant, which has given an indemnity to those freeholders or its predecessors.

[22] However, for those reasons, in my view, the appeal must be allowed and the district judge's order set aside.

Appeal allowed.

TOWN AND VILLAGE GREENS

A
R (on the application of Lewis) *v* Redcar and Cleveland Borough Council and another

SUPREME COURT

[2010] UKSC 11

3 March 2010

LORD HOPE, DEPUTY PRESIDENT, LORD RODGER, LORD WALKER,
B LORD BROWN AND LORD KERR

Town and village greens — Registration — Commons Act 2006 — User as of right — Application to register part of golf course as green under section 15 of 2006 Act — Inspector finding user by local inhabitants not as of right because inhabitants deferring to golf course use — Test of use as of right — Whether deference precluding use from being as of right

In 2007, the appellant applied to the first respondent council to register part of a local golf course (the disputed land) as a town green He applied under section 15 of the Commons Act 2006 on the ground of 20 years' use by local habitants as of right for sports and pastimes.
C The first respondents owned the land but let it to a golf club. The second respondent was the first respondents' development partner for a proposed residential and leisure development on land that included the disputed land.

The first respondents refused the application on the recommendation of an inspector. The inspector found that the local inhabitants had not used the land as of right, within the meaning of section 15, because they had overwhelmingly deferred to the extensive use of the land by the members of the golf club, such that they did not give the appearance of asserting a right; for example, recreational users refrained from walking on the playing areas while play was in progress.

D A judicial review of that decision was dismissed on the ground that this deference prevented a finding that the local inhabitants' user was as of right: see [2008] EWHC 1813 (Admin). That approach was upheld by the Court of Appeal: see [2009] EWCA Civ 3; [2009] 1 EGLR 79; [2009] 15 EG 100. The appellant appealed.

***Held*: The appeal was allowed. The use of land by local inhabitants for sports and pastimes will be "as of right" if the land has been enjoyed openly and in the way in which a person rightfully entitled to do so would use it and that use was not by force, stealth or the licence of the owner. The focus is on the quality of the user relied on and the issue of deference relates to**
E **that matter; there is no additional requirement that it would have appeared to a reasonable landowner that users were asserting a right to use the land for the lawful sports and pastimes in which they were indulging. Deference to the landowners' use of the land is not inconsistent with the inhabitants' use being as of right. Where the recreational use by local inhabitants has coexisted with the owner's use during the 20-year period relied on to establish the right, the rights of each will continue to coexist after registration subject to give and take on both sides. Registration does not enlarge the rights of local inhabitants so as effectively to prevent the landowner from using the land for its own purposes, owing to the principle of equivalence between the user relied on to establish a right and the way in which that right, once established, can** F
be exercised. In the instant case, the inspector had misdirected himself. The recreational use of the disputed land met the test under section 15. The local inhabitants' use was peaceful, open and not based on any licence from either the first respondents or the golf club. Courtesy and common sense dictated that they interfered with the golfers' progress over the course as little as possible. That was not inconsistent with the use of the land as of right for lawful sports and pastimes. The first respondents would be ordered to register the disputed land as a town green.

The following cases are referred to in this report.

Bridle v *Ruby* [1989] 1 QB 169; [1988] 3 WLR 191; (1987) 56 P&CR 155, CA G

Bright v *Walker* (1834) 1 CM&R 211

Costagliola v *English* (1969) 210 EG 1425

Cumbernauld and Kilsyth District Council v *Dollar Land (Cumbernauld) Ltd* (1993) SC (HL) 44, HL, (1992) SLT 1035, Ct Sess

Dalton v *Henry Angus & Co; Commissioners of HM Works and Public Buildings* v *Henry Angus & Co; sub nom Angus & Co* v *Dalton* (1881) LR 6 App Cas 740

Fitch v *Fitch* (1797) 2 Esp 543

Gardner v *Hodgson's Kingston Brewery Co Ltd* [1903] AC 229, HL

Henderson v *Volk* (1982) 35 OR (2d) 379

Hollins v *Verney* (1884) LR 13 QBD 304, CA; (1883) LR 11 QBD 715, QB

Humphreys v *Rochdale Metropolitan Borough Council* unreported 18 June 2004, Ch H

Mercer v *Woodgate* (1870) LR 5 QB 26

New Windsor Corporation v *Mellor; sub nom Bachelors, Acre, New Windsor, Berkshire, Re* [1975] Ch 380; [1975] 3 WLR 25; [1975] 3 All ER 44; (1975) 73 LGR 337; 30 P&CR 429, CA

Oxfordshire County Council v *Oxford City Council* [2006] UKHL 25; [2006] 2 AC 674; [2006] 2 WLR 1235; [2006] 4 All ER 817; [2006] 2 EGLR 95; [2004] EWHC 12 (Ch); [2004] Ch 253; [2004] 2 WLR 1291; [2004] 2 P&CR 19; [2004] 1 EGLR 105; [2004] 2 PLR 65

Pye (JA) (Oxford) Ltd v *United Kingdom* 44302/02 [2005] 3 EGLR 1; [2005] 49 EG 90; [2006] RVR 188; (2006) 43 EHRR 3, ECtHR

R (on the application of Beresford) v *Sunderland City Council; sub nom R* v *Sunderland City Council ex parte Beresford* [2003] UKHL 60; [2004] 1 AC 889; [2003] 3 WLR 1306; [2004] 1 All ER 160; [2004] 2 P&CR 23; [2004] 1 EGLR 94; [2004] 1 PLR 85; [2004] JPL 1106 J

R (on the application of Godmanchester Town Council) v *Secretary of State for the Environment, Food and Rural Affairs; R (on the application of Drain)* v *Secretary of State for the Environment, Food and Rural Affairs* [2007] UKHL 28; [2008] 1 AC 221; [2007] 3 WLR 85; [2007] 4 All ER 273; [2008] 1 P&CR 12; [2007] 3 EGLR 119; [2007] JPL 1691

R (on the application of Laing Homes Ltd) v *Buckinghamshire County Council* [2003] EWHC 1578 (Admin); [2004] 1 P&CR 36; [2003] 3 EGLR 69; [2003] 3 PLR 60; [2004] JPL 319

R (on the application of Lewis) v *Redcar and Cleveland Borough Council* [2009] EWCA Civ 3; [2009] 1 WLR 1461; [2009] 4 All ER 1232; [2009] 1 EGLR 79; [2009] 15 EG 100, [2008] EWHC 1813 (Admin); [2008] PLSCS 212; [2008] NPC 97 K

R v *Oxfordshire County Council, ex parte Sunningwell Parish Council* [2000] 1 AC 335; [1999] 3 WLR 160; [1999] 3 All ER 385; (1999) 79 P&CR 199; [1999] 2 EGLR 94; [1999] 31 EG 85; [2000] JPL 384, HL

R v *Suffolk County Council, ex parte Steed; sub nom Steed* v *Suffolk County Council* (1996) 75 P&CR 102; [1997] 1 EGLR 131; [1997] 10 EG 146, CA, (1995) 70 P&CR 487; (1996) 71 P&CR 463, QB

Rhins District Committee of Wigtownshire County Council v *Cuninghame; sub nom Rhins District Committee of the County Council of Wigtownshire* v *Cuninghame* (1917) 2 SLT169, OH

Sturges v *Bridgman* (1879) 11 Ch D 852, CA

A *White* v *Taylor (No 2)* [1969] 1 Ch 160; [1968] 2 WLR 1459; [1968] 1 All ER 349

This was an appeal by the appellant, Kevin Lewis, from a decision of the Court of Appeal upholding a decision of Sullivan J, sitting in the Administrative Court, dismissing a claim against the respondents, Redcar and Cleveland Borough Council and Persimmon Homes (Teesside) Ltd, for judicial review of a decision of the first respondents on an application for registration of a town or village green under the Commons Act 2006.

Charles George QC, Jeremy Pike and Cain Ormondroyd (instructed by Irwin Mitchell, of Sheffield) appeared for the appellant;
B George Laurence QC and Rodney Smith (instructed by the legal department of Redcar and Cleveland Borough Council) appeared for the first respondents; Ross Crail (instructed by Ward Hadaway, of Newcastle upon Tyne) represented the second respondent.

Giving the first opinion, **Lord Walker** said:

Issue

[1] Section 15 of the Commons Act 2006 (the 2006 Act), so far as relevant to this appeal, provides as follows:

> 15. Registration of Greens
> C (1) Any person may apply to the commons registration authority to register land to which this Part applies as a town or village green in a case where subsection (2), (3) or (4) applies.
> (2) This subsection applies where —
> (a) a significant number of the inhabitants of any locality, or of any neighbourhood within a locality, have indulged as of right in lawful sports and pastimes on the land for a period of at least 20 years; and
> (b) they continue to do so at the time of the application.
> …
> (4) This subsection applies (subject to subsection (5)) where —
> (a) a significant number of the inhabitants of any locality, or of any neighbourhood within a locality, indulged as of right in lawful sports and
> D pastimes on the land for a period of at least 20 years;
> (b) they ceased to do so before the commencement of this section; and
> (c) the application is made within the period of five years beginning with the cessation referred to in paragraph (b).
> …
> (7) For the purposes of subsection (2)(b) in a case where the condition in subsection (2)(a) is satisfied —
> (a) where persons indulge as of right in lawful sports and pastimes immediately before access to the land is prohibited as specified in subsection (6), those persons are to be regarded as continuing so to indulge; and
> (b) where permission is granted in respect of use of the land for the purposes of lawful sports and pastimes, the permission is to be disregarded in
> E determining whether persons continue to indulge in lawful sports and pastimes on the land "as of right".

The application relevant to this appeal was expressed to be made under section 15(4). It was suggested in argument that (because of the "deeming" provision in subsection (7)) it was also or alternatively made under subsection (2). In any case, it was a valid application, and neither subsection (5) nor subsection (6) is in point.

Issue

[2] The general issue for the court is whether a piece of open land next to the sea in Redcar ought to have been registered as a town
F green under section 15. For at least 80 years before 2002, the land in question (the disputed land) formed part of a golf course in regular use by members of Cleveland Golf Club, whose trustees were tenants of the course. The inspector who held a public inquiry found as a fact that when local residents using the disputed land for recreation encountered members of the golf club playing golf, the former "deferred" to the latter. In these circumstances, the legal issue for the court can be more particularly stated as whether the legal consequence of this deference was that the local residents were not indulging in recreation "as of right" within the meaning of the 2006 Act.

[3] During the past decade, there have been three important decisions of the House of Lords dealing with different aspects of the law (as it stood before the 2006 Act) as to town and village greens: see *R* v *Oxfordshire County Council, ex parte Sunningwell Parish Council* G
[2000] 1 AC 335*, *R (on the application of Beresford)* v *Sunderland City Council* [2003] UKHL 60; [2004] 1 AC 889† and *Oxfordshire County Council* v *Oxford City Council* [2006] UKHL 25; [2006] 2 AC 674‡. In none of these appeals did the House of Lords have to decide the point now at issue, although both sides have placed reliance on some passages in their lordships' opinions. The 2006 Act (which is still not fully in force) makes important changes in the law, but does not directly affect the issue of deference.

Facts

[4] The appellant, Mr Kevin Lewis, is one of five local residents H
who made the application for registration of the disputed land under section 15 of the 2006 Act. The first respondents, Redcar and Cleveland Borough Council, have a dual capacity, being both the registration authority and the freehold owners of the disputed land. The second respondent, Persimmon Homes (Teesside) Ltd, is an interested party. It has, since 2003, been the borough council's development partner in the Coatham Links coastal regeneration project. The project is for a mixed development for residential and leisure purposes on a site extending to 14ha. The disputed land forms an important, and possibly indispensable, part of the development site. The appeal is therefore of great importance to the parties, as well as raising a point of law of general public interest. J

[5] Redcar is on the south side of the Tees estuary. The disputed land is part of an area known as Coatham Common or Coatham Links (Coatham was originally a separate village but is now part of Redcar). On the south (landward) side of the disputed land, there is a mainly residential area. To the east, is the site of the former clubhouse and a leisure centre (the clubhouse site is not included in the disputed land but was included in the earlier application mentioned below). To the west is more open land still used as a golf course. To the north is the beach and the North Sea. The disputed land formerly included the tees, fairways and greens of the first and 18th holes and a small practice area.

[6] The inspector's report, dated 14 March 2006, described the K
boundaries in more detail and contained (paras 6 and 7) this further description of the disputed land (referred to as "the Report Land"):

> The character of the Report Land is typical of coastal sand dunes, with irregular sand hills covered in rough grass. The dunes are noticeably higher on the northern side. There is a flatter area along the southern side, particularly west of the Church Street access. The former tees, greens and fairways of the golf course are no longer obvious. The Report Land is crossed by numerous informal paths of which the most well used run alongside and close to the southern and northern boundaries. A number of photographs show the general nature of the land.
> There are some fairly new signs erected by [the borough council] on the L
> Report Land. The gist of the signs is that they give the public temporary permission to use the Report Land for recreation pending its redevelopment. I call these signs "the permissive signs".

The footpath near the southern boundary is a public footpath.

[7] Mr Lewis and his fellow applicants applied for registration of the disputed land on 8 June 2007, soon after section 15 of the 2006 Act came into force on 5 April 2007. It was not the first application that had been made in respect of the disputed land. An earlier application had been made by another group of local residents on 1 March 2005. It was therefore considered under the earlier law, that is, the Commons
Registration Act 1965 (the 1965 Act) as amended by the Countryside M
and Rights of Way Act 2000 (the 2000 Act). This earlier application was the subject of a public inquiry held by Mr Vivian Chapman QC, as an inspector appointed by the borough council as registration authority. The inquiry was held over several days in December 2005 and January 2006. Mr Chapman produced a lengthy report, dated 14 March 2006, recommending that the application should be refused, and the borough council accepted his recommendation. An application for

---

* Editor's note: Also reported at [1999] 2 EGLR 94; [1999] 31 EG 85
† Editor's note: Also reported at [2004] 1 EGLR 94
‡ Editor's note: Also reported at [2006] 2 EGLR 95

A leave for judicial review of that decision was refused on the papers by Collins J on 22 August 2006 and was not renewed.

[8] When the second application was made in 2007, it was rightly thought that it was unnecessary, and would be a waste of time and money, to hold a second public inquiry since it would be directed to the same factual issues. Mr Chapman did, however (in connection with the first application), make a second report, dated 9 June 2006, addressing the decision of the House of Lords in *Oxfordshire* (he advised that it made no difference to his conclusions and that in any case it was not open to the borough council to reopen their decision).

[9] The relevant findings of fact are therefore in Mr Chapman's B report, dated 14 March 2006, on the first application. The crucial findings are in paras 171, 172 and 175. These paragraphs are set out in full, in [8] and [9], in the judgment of Dyson LJ in the Court of Appeal [2009] EWCA Civ 3; [2009] 1 WLR 1461*, but they are of such central importance that they need to be set out again. Paragraph 171 dealt with use of the disputed land by golfers:

> 171. I find that, from as far back as living memory goes (at least as far back as the 1920s), the Report Land was continuously used as part of the Cleveland Golf Club links. The only exception is that the golfing was suspended during World War II. Golfing use ceased in 2002. I find that the club was a popular C one and that the golf links were well used nearly every day of the year. In the years before 2002, the report land was used for the club house, the first and eighteenth holes and for a practice ground. There is some evidence that the precise configuration of the course changed somewhat over the years. The club house, tees, fairways, greens and practice ground did not, however, take up the whole of the Report Land and there were substantial areas of rough ground beside and between these features.

[10] Paragraph 172 dealt with use by non-golfers (that is, local residents):

> 172. I find that from as far back as living memory goes, the open parts of the report land have also been extensively used by non golfers for informal D recreation such as dog walking and children's play. Some of the walking has been linear walking in transit. Thus the informal paths running east-west have been used by caravan residents to get access to the centre of Redcar with its shops and public houses. Also, there is evidence of people taking a short cut south-north from Church Street to the gap in the fence in Majuba Road. However I am satisfied that the open parts of the report land have been extensively used by non golfers for general recreational activities apart from linear walking. I prefer the evidence on this point of the applicants' witnesses and of Mr Fletcher to the evidence of the objector's other witnesses that such use was occasional and infrequent…

[11] Paragraphs 173 and 174 concluded that the local people who E used the land for informal recreation came primarily from the Coatham area of Redcar. Then, para 175 dealt with the relationship between the two types of use:

> I find that the relationship between the golfers and the local recreational users was generally cordial. There was evidence of only a few disputes. Only Squadron Leader Kime seems to have caused problems by actively asserting a right to use the Report Land and the golf club appears to have tried to avoid any formal dispute with him. In my judgment, the reason why the golfers and the local people generally got on so well was because the local people (with the exception of Squadron Leader Kime) did not materially interfere with the use of the land for playing golf. Many of the applicants' witnesses emphasised F that they would not walk on the playing areas when play was in progress. They would wait until the play had passed or until they were waved across by the golfers. Where local people did inadvertently impede play, a shout of "fore" would be enough to warn them to clear the course. I find that recreational use of the Report Land by local people overwhelmingly deferred to golfing use.

[12] Paragraph 221 (in the part of the report applying the law to the facts as found) referred to the decisions of Sullivan J in *R (on the application of Laing Homes Ltd)* v *Buckinghamshire County Council* [2003] EWHC 1578 (Admin); [2004] 1 P&CR 36† and HH Judge Howarth in *Humphreys* v *Rochdale Metropolitan Borough Council* unreported 18 June 2004: G

> Leaving aside the public footpath, I consider that the reasoning in [*Laing Homes*] and [*Humphreys*] squarely applies to the report land in the present case. Use of the report land as a golf course by the Cleveland Golf Club would have been in breach of Inclosure Act 1857 section 12 and Commons Act 1876 section 29 if the report land had been a town or village green. It was a use which conflicted with the use of the report land as a place for informal recreation by local people. It was not a use which was with a better view to the enjoyment of the report land as a town or village green. The overwhelming evidence was that informal recreational use of the report land deferred to its extensive use as a golf course by the Cleveland Golf Club. Accordingly, use of the report land by local people was not as of right until use as a golf course ceased in 2002. H

Mr Chapman concluded (para 223) that (apart from use of the public footpath) recreational user of the disputed land was not as of right before 2002 because it deferred to extensive use of the land by the golf club, and that user as of right was not continuing because of the permissive signs erected in 2003.

[13] It is convenient, at this point, to dispose of the matter of the signs. They were contentious earlier but are no longer a live issue. There were two sets of signs: warning signs erected by the golf club in 1998 and the permissive signs erected by the borough council in 2003. The warning signs read "Cleveland Golf Club. Warning. It is dangerous to J trespass on the golf course." The inspector found, in para 176:

> Although these were vandalised several times after which the golf club gave up trying to maintain them, I am satisfied that they were in place long enough for regular users of the report land to know of them. Indeed it seems that they caused a stir locally because of the implication that local people using Coatham Common were trespassers.

The inspector treated them as material to the outcome of both applications, but on judicial review of the second application Sullivan J ([2008] EWHC 1813 (Admin)*, in [11] to [23]) held that the wording was too ambiguous to alter the character of the residents' use of the K land, and that conclusion has not been challenged by the respondents. The permissive signs erected in 2003 were fatal to the first application but not to the second application because of the change in the law made by section 15 of the 2006 Act.

Course of the second application

[14] Mr Chapman advised the borough council, in an opinion dated 12 June 2007, that the application made on 6 June 2007 was bound to fail on two of the same grounds on which the first application failed, that is, the deference issue and the 1998 warning notices. He recommended that the application should be summarily dismissed, subject to any new points raised by the applicants. Various points were raised, but in L three further opinions, dated 29 July, 13 October and 18 October 2007, Mr Chapman maintained his advice that the application should be rejected. On 19 October 2007, the borough council, by their general purposes and village greens committee, accepted Mr Chapman's advice and resolved to reject the application for registration.

[15] On 18 July 2008, Sullivan J, at a "rolled up" hearing, granted the applicants permission to apply for judicial review of the borough council's decision, but dismissed the substantive application. He did so on the ground that the local residents' deference to the golfers had prevented their user being "as of right" before 2002. He relied on [82] of his own judgment in *Laing Homes* and on [57] of Lord Hoffmann's M opinion in *Oxfordshire*. He granted leave to appeal, commenting that "deference is judge-made law, judge-made by me".

[16] The Court of Appeal (Laws, Rix and Dyson LJJ) unanimously dismissed the appeal in reserved judgments handed down on 15 January 2009: see [2009] EWCA Civ 3; [2009] 1 WLR 1461†. Dyson LJ gave the principal judgment and Rix LJ added a concurring judgment. Both judgments put the decision squarely on the ground of deference excluding user as of right (although Dyson LJ denied

* Editor's note: Also reported at [2009] 1 EGLR 79; [2009] 15 EG 100

† Editor's note: Also reported at [2003] 3 EGLR 69

* Editor's note: Reported at [2008] NPC 97

† Editor's note: Also reported at [2009] 1 EGLR 79; [2009] 15 EG 100

A that there was any "principle of deference"). The provisions of two Victorian statutes relating to greens (section 12 of the Inclosure Act 1857 (the 1857 Act) and section 29 of the Commons Act 1876), which had formed part of the grounds of decision in *Laing Homes,* were not relied on in the Court of Appeal. In short, all the subsidiary issues have disappeared and this court is faced with the single issue of deference. It is not, however, a simple issue.

As of right

[17] The concept of user "as of right" is found (either in precisely those words or in similar terms) in various statutory provisions dealing
B with the acquisition by prescription of public or private rights. Section 5 of the Prescription Act 1832 (the 1832 Act) makes it sufficient to plead enjoyment "as of right", while section 2 refers to a way "actually enjoyed by any person claiming right thereto without interruption for the full period of 20 years". Section 31 of the Highways Act 1980 refers to use of a way being "actually enjoyed by the public as of right and without interruption for the full period of 20 years". Section 22(1A) of the 1965 Act, as substituted by the 2000 Act, refers simply to inhabitants indulging in lawful sports and pastimes "as of right" for at least 20 years

[18] Both *Sunningwell* and *Beresford* were concerned with the
C meaning of "as of right" in the 1965 Act. In *Sunningwell*, Lord Hoffmann discussed the rather unprincipled development of the English law of prescription. He explained that, by the middle of the 19th century, the emphasis shifted from fictions: see pp350H-351A:

> to the quality of the 20-year user which would justify recognition of a prescriptive right or customary right. It became established that such user had to be, in the Latin phrase, *nec vi, nec clam, nec precario*: not by force, nor stealth, nor the licence of the owner. (For this requirement in the case of custom, see *Mills* v *Colchester Corporation* (1867) LR 2 CP 476, 486.) The unifying element in these three vitiating circumstances was that each constituted a reason why it would not have been reasonable to expect the owner to resist the exercise
> D of the right in the first case, because rights should not be acquired by the use of force, in the second, because the owner would not have known of the user and in the third, because he had consented to the user, but for a limited period.

Lord Hoffmann pointed out that for the creation of a highway there was an additional requirement that an intention to dedicate it must be evinced or inferred: as to that aspect see *R (on the application of Godmanchester Town Council)* v *Secretary of State for the Environment, Food and Rural Affairs* [2007] UKHL 28; [2008] 1 AC 221*.

[19] In *Sunningwell*, the villagers had used around 10 acres of glebe land for dog-walking, children's games and similar activities. This use
E seems to have coincided with the land being let for grazing by horses, but the report gives little detail about this. The inspector (as it happens, Mr Chapman) advised against acceptance of the registration because although the witnesses had said that they thought they had the right to use the glebe, they did not say that they thought the right was confined to villagers (as opposed to the general public). Lord Hoffmann held (and the rest of the Appellate Committee agreed) that this was an error. The decision of the Court of Appeal in *R* v *Suffolk County Council, ex parte Steed* (1998) 75 P&CR 102† was overruled. That was the context in which Lord Hoffmann stated in a passage, at pp352H-353A, relied on by the respondents:

> F My Lords, I pause to observe that Lord Blackburn [in *Mann* v *Brodie* (1885) 10 App Cas 378, 386, as to dedication of a highway] does not say that there must have been evidence that individual members of the public using the way believed there had been a dedication. He is concerning himself, as the English theory required, with how the matter would have appeared to the owner of the land. The user by the public must have been, as Parke B said in relation to private rights of way in *Bright* v *Walker* (1834) 1 CM&R 211, 219, "openly and in the manner that a person rightfully entitled would have used it." The presumption arises, as Fry J said of prescription generally in *Dalton* v *Angus & Co* 6 App Cas 740, 773, from acquiescence.

---

* Editor's note: Also reported at [2007] 3 EGLR 119

† Editor's note: Also reported at [1997] 1 EGLR 131; [1997] 10 EG 146

[20] The proposition that "as of right" is sufficiently described by G
the tripartite test *nec vi, nec clam, nec precario* (not by force, nor stealth, nor the licence of the owner) is established by high authority. The decision of the House of Lords in *Gardner* v *Hodgson's Kingston Brewery Co Ltd* [1903] AC 229 is one of the clearest: see Lord Davey, at p238, and Lord Lindley, at p239. Other citations are collected in *Gale on the Law of Easements* (18th ed) 2008, in paras 4-80 and 4-81. The proposition was described as "clear law" by Lord Bingham of Cornhill in [3] of *Beresford.* The opinion of Lord Rodger of Earlsferry ([55]) is to the same effect. So is that of Lord Scott of Foscote ([34]), although with a cautionary note as to the difference between the acquisition of public and private rights. H

Laing Homes

[21] The respondents' case is that although Sullivan J, in his judgment in *Laing Homes*, was indeed the first judge to speak in terms of "deference" shown by local residents, he was not striding into entirely unknown and uncharted territory. Earlier authorities (including those mentioned in the passage of Lord Hoffmann's opinion in *Sunningwell* quoted in [19] above) suggest that although the local residents' private beliefs as to their rights are irrelevant, the same is not true of their outward behaviour on the land in question as it would appear to a reasonable owner of the land. It is relevant, on this J
argument, to look at what might today be called the residents' attitude or body language: this thought is elaborated in an imaginary example given by JG Riddall, *Miss Tomkins and the Law of Village Greens* [2009] Conveyancer and Property Lawyer 326. I propose to look next at *Laing Homes*, and then to consider how far the respondents can claim much more long-established roots for the doctrine of deference that *Laing Homes* articulates.

[22] *Laing Homes* was concerned with three adjoining fields (the application area), extending in all to 38 acres, on the edge of Widmer End in Buckinghamshire. This land, together with three smaller fields not affected by the application for registration, had been acquired by K
Laing Homes, a housebuilder, and held in its "land bank" since 1963. The land was subject to a grazing licence from 1973 to 1979, when the farmer stopped using it for grazing because of repeated troubles with trespassers. In the course of time, footpaths were established round the three fields in the application area (cutting some corners) and these were officially recognised as public footpaths in June 2000. An application for registration of the application area was made in August 2000. The registration authority's decision to register the land as a village green was challenged by way of judicial review on various grounds (including human rights grounds on which Sullivan J did not find it necessary to rule). L

[23] In his judgment, Sullivan J listed, in [50], the four main grounds on which Laing Homes was attacking the inspector's report (and the registration based on it). The first ground was that there was insufficient evidence of the use of the whole of the application area for lawful sports and games over the 20-year period. The second was the inspector's conclusion that the use of the fields for an annual hay crop (from around 1980 until the early 1990s) was not incompatible with the establishment of village green rights. Sullivan J considered the second ground first. He discussed it at some length and differed from the inspector. He did so primarily on the view that he took of the perception of a reasonable landowner, although he was also influenced by the point (no longer M
relied on) as to the Victorian statutes: [86]:

> Like the Inspector, I have not found this an easy question. Section 12 [of the 1857 Act] acknowledges that animals may be grazed on a village green. Rough grazing is not necessarily incompatible with the use of the land for recreational purposes: see *Sunningwell.* If the statutory framework within which s22(1) [of the 1965 Act] was enacted had made provision for low-level agricultural activities to coexist with village green type uses, rather than effectively preventing them once such a use has become established, it would have been easier to adopt the Inspector's approach, but it did not. I do not consider that using the three fields for recreation in such a manner as not to interfere with [the farmer's] taking of an annual hay crop for over half of the 20-year period, should have suggested to Laings that those using the fields believed that they

were exercising a public right, which it would have been reasonable to expect Laings to resist.

[24] I have to say that I am rather puzzled by Sullivan J's summary of the evidence concerning hay-making and the discussion of it, both by the inspector in paras 56 and 57 and by the judge himself in [59] to [63]. There is a detailed description of the local residents keeping off the fields for a few days in spring when they were harrowed, rolled and fertilised, and again for a few days during hay-making. However, there are only the most passing references by the judge (in [59] and [111]) to the further need for people to keep off the fields for many weeks while the crop was growing if it was to be worth the farmer's while to get it in. The length of this period would vary with the quality of the land and the seasonal weather, but would usually, I imagine, be of the order of three months. The evidence was that the farmer generally got well more than 2,000 bales of hay from the application area. So it seems that the local residents must, in general, have respected the hay crop.

[25] The puzzle is partly explained by Sullivan J's consideration of the first ground (evidence of use of the entire application area) that follows in [88] to [111]. In [111], the judge commented that there was an overlap between the two grounds because the existence of public footpaths round the three fields (cutting some corners) provided an alternative explanation of the local residents' use of the fields. It seems likely that they used the perimeter paths and kept off the hay while it was growing, although their dogs may not have done, as the judge discussed at some length: [103] to [110].

[26] There are some *dicta* concerning *Laing Homes* in Lord Hoffmann's opinion in *Oxfordshire*. Lord Rodger and I expressed general agreement with Lord Hoffmann, but did not comment on this point. Lord Hoffmann observed, in [57]:

> No doubt the use of the land by the owner may be relevant to the question of whether he would have regarded persons using it for sports and pastimes as doing so "as of right". But, with respect to the judge, I do not agree that the low-level agricultural activities must be regarded as having been inconsistent with use for sports and pastimes for the purposes of section 22 if in practice they were not.

[27] There was some discussion in the course of argument of what Lord Hoffmann meant by the first sentence of this passage. In the Court of Appeal, Dyson LJ, in [45], took it to mean inconsistency between competing uses manifested "where the recreational users adjust their behaviour to accommodate the competing activities of the owner (or his lessees or licensees)". I am rather doubtful about that. I think it just as likely that Lord Hoffmann had in mind, not concurrent competing uses of a piece of land, but successive periods during which recreational users are first excluded and then tolerated as the owner decides. An example would be a fenced field used for intensive grazing for nine months of the year but left open for three months when the animals are indoors for the worst of the winter.

]28] Whether or not that is correct, I see great force in the second sentence of the passage quoted. Taking a single hay crop from a meadow is a low-level agricultural activity compatible with recreational use for the late summer and from then until next spring. *Fitch* v *Fitch* (1797) 2 Esp 543 is venerable authority for that. That is not to say that *Laing Homes* was wrongly decided, although I see it as finely balanced. The residents of Widmer End had gone to battle on two fronts, with the village green inquiry in 2001 following a footpaths inquiry two or three years earlier, and some of the evidence concerning their intensive use of the footpaths seems to have weakened their case as to sufficient use of the remainder of the application area.

Earlier authorities

[29] I have already referred to *Fitch*, the case concerning cricket and hay-making at Steeple Bumpstead, in Essex. The report is brief, but what Heath J is reported as having said, at p544, is a forthright declaration of the need for coexistence between concurrent rights:

> The inhabitants have a right to take their amusement in a lawful way. It is supposed that because they have such a right, the plaintiff should not allow the grass to grow; there is no foundation in law for such a position. The rights of both parties are distinct, and may exist together. If the inhabitants come in an unlawful way, or not fairly, to exercise the right they claim of amusing themselves, or to use it in an improper way, they are not justified under the custom pleaded, which is a right to come into the close to use it in the exercise of any lawful games or pastimes, and are thereby trespassers.

[30] Against that, Mr George Laurence QC relied on the general proposition that if the public (or a section of the public) is to acquire a right by prescription, it must by its conduct bring home to the landowner that a right is being asserted against it, so that the landowner has to choose between warning off the trespassers or eventually finding that they have established the asserted right against it. That was in line with what Lord Hoffmann (in *Sunningwell*, at pp350-351, quoted in [18] above) called "the unifying element" in the tripartite test: why it would not have been reasonable to expect the owner to resist the exercise of the right.

[31] The first of the old authorities relied on by Mr Laurence was *Bright* v *Walker* (1834) 1 CM&R 211, at p219, a case on a private right of way in which Parke B spoke of use of a way "openly and in the manner that a person rightfully entitled would have used it". I read that reference to the manner of use as emphasising the importance of open use, rather than as prescribing an additional requirement. On its facts, the case raised as much of an issue as to *vi* as to *clam*, since gates had been erected and broken down during the relevant period. The point of law in the case turned on the peculiarity that the freehold owner of the servient tenement was a corporation sole.

[32] The next case relied on (another case concerning a claim to a private way) was *Hollins* v *Verney* (1884) 13 QBD 304 (there is a fuller statement of the facts in the first instance report: see (1883) LR 11 QBD 715. Lindley LJ, giving the judgment of the Court of Appeal, observed, at p315:

> No user can be sufficient which does not raise a reasonable inference of such a continuous enjoyment. Moreover, as the enjoyment which is pointed out by the statute is an enjoyment which is open as well as of right, it seems to follow that no actual user can be sufficient to satisfy the statute, unless during the whole of the statutory term (whether acts of user be proved in each year or not) the user is enough at any rate to carry to the mind of a reasonable person who is in possession of the servient tenement, the fact that a continuous right to enjoyment is being asserted, and ought to be resisted if such right is not recognised, and if resistance to it is intended. Can an user which is confined to the rare occasions on which the alleged right is supposed in this instance to have been exercised, satisfy even this test? It seems to us that it cannot: that it is not, and could not reasonably be treated as the assertion of a continuous right to enjoy; and when there is no assertion by conduct of a continuous right to enjoy, it appears to us that there cannot be an actual enjoyment within the meaning of the statute.

[33] The second sentence of this passage begins with "Moreover", suggesting that Lindley LJ was adding to the requirement that the use should be continuous. However, the passage as a whole seems to be emphasising that the use must be openly (or obviously) continuous (the latter word being used three more times in the passage). The emphasis on continuity is understandable because the weight of the evidence was that the way was not used between 1853 and 1866 or between 1868 and 1881. It was used exclusively, or almost exclusively, for carting timber and underwood that was cut on a 15-year rotational system. The use relied on was too sparse for any jury to find section 2 of the 1832 Act satisfied.

[34] In *Bridle* v *Ruby* [1989] 1 QB 169, the plaintiff established a right of way by prescription despite his personal belief that he had such a right by grant. Ralph Gibson LJ said, at p178E-F:

> The requirement that user be "as of right" means that the owner of the land, over which the right is exercised, is given sufficient opportunity of knowing that the claimant by his conduct is asserting the right to do what he is doing without the owner's permission. If the owner is not going to submit to the claim, he has the opportunity to take advice and to decide whether to question the asserted right. The fact that the claimant mistakenly thinks that he derived the right, which he is openly asserting, from a particular source, such as the conveyance to him of his property, does not by itself show that the nature of the user was

A materially different or would be seen by the owner of the land as other than user as of right.

That the claimant's private beliefs are generally irrelevant, in the prescription of either private or public rights, was finally confirmed by the House of Lords in *Sunningwell*: see [18] and [19] above.

[35] The last authority calling for mention on this point is *Cumbernauld and Kilsyth District Council* v *Dollar Land (Cumbernauld) Ltd* (1992) SLT 1035 (Court of Session); (1993) SC (HL) 44 (House of Lords). In the Court of Session, the Lord President (Lord Hope), after considering several authorities, observed, at p1041G:

> B The significance of these passages for present purposes is that, where the user is of such amount and in such manner as would reasonably be regarded as being the assertion of a public right, the owner cannot stand by and ask that his inaction be ascribed to his good nature or to tolerance.

Lord Hope's reference to the manner of use must, I think, be related to the unusual facts of the case, set out in detail at pp1037-1038. The issue was whether there was a public right of way over an extensive walkway in a new town, designed to separate pedestrian from vehicular traffic. It gave access to the town centre, where there were numerous shops (whose tenants no doubt had private rights of way for themselves and their customers). However, the walk was also used for access to public
C places such as the railway station, the church, a health centre and a swimming pool. It was held that the use of the way "had the character of general public use of a town centre pedestrian thoroughfare": see p1042K. The House of Lords upheld this decision. It is worth noting that Lord Jauncey of Tullichettle stated, at p47H:

> There is no principle of law which requires that there be conflict between the interest of users and those of a proprietor.

Deference or civility?

[36] In the light of these and other authorities relied on by Mr Laurence, I have no difficulty in accepting that Lord Hoffmann
D was absolutely right, in *Sunningwell*, to say that the English theory of prescription is concerned with "how the matter would have appeared to the owner of the land" or, if there were an absentee owner, to a reasonable owner that was on the spot. However, I have great difficulty in seeing how a reasonable owner would have concluded that the residents were not asserting a right to take recreation on the disputed land simply because they normally showed civility (or, in the inspector's word, deference) towards members of the golf club who were playing golf. It is not as though the residents took to their heels and vacated the land whenever they saw a golfer. They simply acted (as all the members of the court agree, in much the same terms) with
E courtesy and common sense. However, courteous and sensible though they were (with occasional exceptions), the fact remains that they were regularly, in large numbers, crossing the fairways as well as walking on the rough, and often (it seems) failing to clear up after their dogs when they defecated. A reasonably alert owner of the land could not have failed to recognise that this user was the assertion of a right and would mature into an established right unless the owner took action to stop it; as the golf club tried to do, ineffectually, with the notices erected in 1998.

[37] There is, in my opinion, a significant difference on this point between the acquisition of private and public rights. As between
F neighbours living in close proximity, what I have referred to as "body language" may be relevant. In a Canadian case of that sort, *Henderson* v *Volk* (1982) 35 OR (2d) 379, at p384d, Cory JA, delivering the judgment of the Court of Appeal of Ontario, observed:

> It is different when a party seeks to establish a right-of-way for pedestrians over a sidewalk. In those circumstances the user sought to be established may not even be known to the owner of the servient tenement. In addition, the neighbourly acquiescence to its use during inclement weather or in times of emergency such as a last minute attempt to catch a bus, should not too readily be accepted as evidence of submission to the use.
>
> It is right and proper for the courts to proceed with caution before finding that title by prescription or by the doctrine of lost modern grant was established in a case such as this. It tends to subject a property owner to a burden without compensation. Its ready invocation may discourage acts of kindness and G good neighbourliness; it may punish the kind and thoughtful and reward the aggressor.

[38] That is, if I may say so, obviously good sense. However, I do not think that it has any application to a situation, such as the court now faces, in which open land owned by a local authority is regularly used, for various different forms of recreation, by a large number of local residents. The inspector's assessment did, in my opinion, amount to an error of law. He misdirected himself as to the significance of perfectly natural behaviour by the local residents.

Rights after registration H

[39] Mr Laurence made some forceful submissions as to what the position would have been on a double hypothesis: namely that the disputed land had been registered as a town green and that it had continued to be let to the golf club after its registration. In those circumstances, he said, the fortunes of the golfers and the local residents would be dramatically reversed: instead of being all "give" by the residents it would be all "take", to the point at which the golf club would no longer be able to function. There was, he said, a massive mismatch between what the residents would have done in order to gain the rights and what they would be in a position to do after the green had been registered. This lack of symmetry was a reason, he argued, J for doubting the soundness of the reasoning on which the appellant's case rested.

[40] These submissions raise two distinct questions. The first is a question of law concerning the effect of registration of a green. The second is a speculative question of predicting the behaviour of a group of people in an eventuality that cannot now arise.

[41] I would spend little time on the second question. Like other members of the court, I am sceptical about the notion that the local residents' attitude towards the golfers, if the green were to be registered in circumstances where it was still being used by the golf club, would suddenly turn from friendly civility to vindictive triumphalism. Many K of them must have friends or neighbours who are members of the golf club; some are even members themselves. However, I would accept that the question of law needs to be considered on the footing that it is at least possible that relations between the two groups might become rather more strained.

[42] Here it is necessary to come back to *Oxfordshire*. The proceedings in that case were not judicial review proceedings. They were initiated by the registration authority, by a claim form under CPR 8, for guidance on a pending application for registration; the first instance judgment is reported at [2004] EWHC 12 (Ch); [2004] L Ch 253*. In the House of Lords, both Lord Scott of Foscote and Baroness Hale of Richmond, regarded some of the questions raised as unnecessary, academic and inappropriate: see Lord Scott, in [91] to [103], and Baroness Hale, in [131] to [137]. The questions to which they most strongly objected were whether: (i) when a green was registered, the relevant inhabitants had legal rights to take recreation on it; and (ii) land registered as a green fell within the scope of what had been referred to as the Victorian statutes.

[43] Lord Hoffmann, while recognising these concerns, thought that it would be appropriate to answer the questions because Oxford City Council had a real interest in the question: see [45]: M

> But the interest of the city council in these questions is concrete in the most literal sense. They wish to build houses on the land. If registration creates no rights and the land does not fall within the Victorian statutes, they will be able to do so.

So, Lord Hoffmann proceeded to answer them, and Lord Rodger and I expressed general agreement with his opinion.

[44] Lord Hoffmann noted, in [46], that registration is conclusive evidence of the matters registered, but:

---

* Editor's note: Also reported at [2004] 1 EGLR 105

In the case of a town or village green, the registration states simply that the land is a green. No other information is prescribed.

The position under the 2006 Act will be similar once it has come fully into force. The only rights specifically registrable in respect of a town or village green will be rights of common: see section 2(2) and section 3(4). However, section 3(5) enables regulations to be made requiring or permitting other information to be included in the register. Regulations have been made — the Commons Registration (England) Regulations 2008 (SI 2008/1961) — but they do not require or permit specific rights of recreation to be registered. The extensive management provisions in Part 2 of the Act apply to town or village greens only if they are subject to rights of common, and deal with the regulation of rights of common. This seems to be in line with what Lord Hoffmann said in *Oxfordshire*, in [48], that although the 1965 Act was intended to be followed by further legislation with regard to the management of commons, it was by no means clear that parliament contemplated further legislation as to rights over greens.

[45] I must set out at some length what Lord Hoffmann said about rights after registration in [49] to [51]:

> 49. So one has to look at the provisions about greens in the 1965 Act like those of any other legislation, assuming that Parliament legislated for some practical purpose and was not sending commons commissioners round the country on a useless exercise. If the Act conferred no rights, then the registration would have been useless, except perhaps to geographers, because anyone asserting rights of recreation would still have to prove them in court. There would have been no point in the conclusive presumption in section 10. Another possibility is that registration conferred such rights as had been proved to support the registration but no more. So, for example, if land had been registered on the strength of a custom to have a bonfire on Guy Fawkes Day, registration would confer the right to have a bonfire but no other rights. But this too would make the registration virtually useless. Although the Act provides for the registration of rights of common, it makes no provision for the registration of rights of recreation. One cannot tell from the register whether the village green was registered on the basis of an annual bonfire, a weekly cricket match or daily football and rounders. So the establishment of an actual right to use a village green would require the inhabitants to go behind the registration and prove whatever had once satisfied the Commons Commissioner that the land should be registered.
>
> 50. In my view, the rational construction of section 10 is that land registered as a town or village green can be used generally for sports and pastimes. It seems to me that Parliament must have thought that if the land had to be kept available for one form of recreation, it would not matter a great deal to the owner whether it was used for others as well. This would be in accordance with the common law, under which proof of a custom to play one kind of game gave rise to a right to use the land for other games: see *Sunningwell* [2000] 1 AC 335, 357A-C.
>
> 51. This does not mean that the owner is altogether excluded from the land. He still has the right to use it in any way which does not interfere with the recreational rights of the inhabitants. There has to be give and take on both sides.

Lord Hoffmann then, in [54] to [57], dealt with the Victorian statutes, as I have already mentioned.

[46] Lord Scott, thinking it right to express a limited view on this issue, disagreed: see [105]:

> But I do not agree that registration can authorise local inhabitants to enjoy recreative user of the land that is different in kind from the 20 years' user that has satisfied the statutory criteria for registration or that would diminish the ability of the landowner to continue to use the land in the manner in which he has been able to use the land during that 20-year period. I do not accept that a tolerant landowner who has allowed the local inhabitants to use his grass field for an annual 5 November bonfire for upwards of 20 years must, after registration, suffer his field to be used throughout the year for all or any lawful sports and pastimes with the consequential loss of any meaningful residual use that he could continue to make of the field.

[47] Having reconsidered the general agreement that I expressed in *Oxfordshire*, I find that I agree with almost all that Lord Hoffmann said in the paragraphs that I have quoted. He had already, in *Sunningwell* at p357D, explained that "sport or pastime" denotes a single composite class and recognised that "dog walking and playing with children [are], in modern life, the sort of informal recreation which may be the main function of a village green". The only point on which I differ from Lord Hoffmann is the point that Lord Scott picked up in [105]: the notion that a custom to have an annual bonfire on Guy Fawkes' day could be a sufficient basis for registration of a green. Such a right might have been established as a stand-alone custom, but would, to my mind, be far too sporadic to amount to continuous use for lawful sports and pastimes (quite apart from the fact that most bonfires are now illegal on environmental grounds). Once that special case is eliminated, I see little danger, in normal circumstances, of registration of a green leading to a sudden diversification or intensification of use by local residents. The alleged asymmetry between use before and after registration will in most cases prove to be exaggerated. Golfers and local residents can coexist without much friction even when the latter have established legal rights.

Conclusion

[48] Disparaging references are sometimes made to the "village green industry" and to applications for registration being used as a weapon of guerrilla warfare against development of open land. The House of Lords has, both in *Beresford* and *Oxfordshire*, expressed some doubt over the extension of town or village green protection to land very different (both in size and appearance) from a traditional village green. However, in the 2006 Act, parliament has made it easier, rather than more difficult, to register a green. There is also the prospect (as Lord Hope mentions in [56] of his judgment) of further legislation, which might possibly make provision for the management of greens on lines comparable to those proposed for commons in Part 2 of the 2006 Act. As it is, district councils have the power, under section 1 of the Commons Act 1899, to make by-laws for the preservation of order on commons, which are defined (in section 15) as including town and village greens. Even without such regulation, conflicts over competing uses (whether as between the owner and the local residents or between different interest groups among the local residents) are capable of resolution by the "constant refrain in the law of easements that 'between neighbours there must be give as well as take'": see Gray and Gray, *Elements of Land Law* (5th ed) 2009, in para 5.2.72, citing Megarry J in *Costagliola* v *English* (1969) 210 EG 1425, at p1431.

[49] For these reasons, I would allow the appeal and order that the borough council should register the disputed land as a town green under section 1 of the 2006 Act (if then in force in Redcar and Cleveland) or under the applicable transitional provisions.

Giving the second opinion, **Lord Hope** said:

[50] This appeal relates to an application by Mr Lewis for judicial review of a decision of the general purposes and village greens committee of Redcar and Cleveland Borough Council, on 19 October 2007, to reject an application to register part of the land in Redcar known as Coatham Common as a town or village green under the 2006 Act. On 18 July 2008, Sullivan J dismissed the application but granted permission to appeal: see [2008] EWHC 1813 (Admin). On 15 January 2009, the Court of Appeal (Laws, Rix and Dyson LJJ) dismissed the appeal: see [2009] EWCA Civ 3; [2009] 1 WLR 1461. The applicant now appeals to this court. The interested party, Persimmon Homes (Teesside) Ltd, seeks to develop the land for housing and leisure activities. It supports the case for the local authority, as it did in the courts below.

[51] As Lord Walker has explained, the land is owned by the local authority. Until 2002, it was part of the land that formed the links of Cleveland Golf Club. It comprised the first and 18th holes of the golf course and a practice ground. There were also substantial areas of rough ground beside and between these features. It was also used by the local inhabitants for informal recreation such as walking their dogs, children's games and picnics. They did not interfere with or interrupt play by the golfers. They would wait until the play had passed or until they were waved through by the golfers. The relationship between the golfers and the local inhabitants was cordial. The two activities appear to have coexisted quite happily during this period. The details are set out in the report by Mr Chapman (the inspector). He was appointed by the local authority to hold an inquiry following an application by

Mr Lewis and a number of other local inhabitants to register an area of land, which included the clubhouse, as a town or village green under the 1965 Act. He was asked to provide a further report following a second application to register the area with which this case is concerned that was made after the 2006 Act came into force. His comments in a series of further opinions on the relationship between the golfers and local inhabitants confirmed his earlier conclusions that the local inhabitants deferred to the golfers and that the deferral to golfing use precluded use of the land by the local inhabitants as of right for recreational purposes. The relevant findings have been quoted in full by Lord Walker: see [9] to [11] above.

[52] On 18 January 2008, these judicial review proceedings were commenced. Sullivan J agreed with Mr Chapman's conclusion that the recreational use of the land was not "as of right" because it deferred to the use of the land by the golf club. Asking himself how the matter would have appeared to the golf club, he said that it would not be reasonable to expect the club to resist the recreational use of the land by local users if their use of the land did not in practice interfere with its use by the golf club: see [41]. The Court of Appeal agreed with that approach: see Dyson LJ, in [54], and Rix LJ, in [64] and [65]. Rix LJ said that if it were otherwise, there would be no way of resolving questions that would subsequently arise, given that registration does not confer qualified or limited rights but the unqualified right to use the land generally for sports and pastimes. He envisaged questions as to whether, if a right of registration were to be assumed, the local inhabitants had a right of walking on the golf greens themselves during play or of playing golf as though they were members of the club.

Issues

[53] As Lord Walker has explained, the question is whether the land ought to have been registered. In an attempt to focus their arguments more precisely, the parties were agreed that it raised the following issues:

(1) Where land has been extensively used for lawful sports and pastimes *nec vi, nec clam, nec precario* for 20 years by the local inhabitants, is it necessary, under section 15(4) of the 2006 Act, to ask the further question whether it would have appeared to a reasonable landowner that users were asserting a right to use the land for the lawful sports and pastimes in which they were indulging?

(2) If the answer to (1) is "yes", does the mere fact that local inhabitants did not prevent the playing of golf by walking in front of the ball (or seeking to prevent the playing of strokes by golfers) preclude the use from being "as of right" under section 15(4)?

(3) If the answer to (2) is "no", did the local authority (and the inspector) err in law in concluding that the inhabitants' use was not "as of right", given what the inspector described as "overwhelming evidence" that recreational use of the land by local people deferred to the golfing use?

[54] This presentation was not, as it turned out, particularly helpful. As counsel recognised, issues (2) and (3) fall to be taken together because they are both directed to the question of deference. I agree with Lord Brown that the critical question, which none of these issues addresses, is what are the respective rights of the local inhabitants and the owner of the land once it has been registered? It is a remarkable fact that the statute gives no guidance on this issue. In *R (on the application of Laing Homes Ltd)* v *Buckinghamshire County Council* [2003] EWHC 1578 (Admin); [2004] 1 P&CR 36, in [27] to [29], referring to what Carnwath J said in *R* v *Suffolk County Council, ex parte Steed* (1995) 70 P&CR 487, Sullivan J said that this was not the original intention. The 1965 Act was intended to be a two-stage legislative process. As a first step, the registers would establish the facts and provide a definitive record of which land was and was not common land or a town or village green. In the second stage, parliament would deal with the consequences of registration by defining what rights the public had over the land that had been registered.

[55] In *New Windsor Corporation* v *Mellor* [1975] Ch 380, at p392, Lord Denning MR said that he hoped that the second-stage legislation would not be long delayed. However, here we are, 45 years after the passing of the 1965 Act. Parliament has still not said what these rights are. In *Oxfordshire County Council* v *Oxford City Council* [2006] UKHL 25; [2006] 2 AC 674, in [48], Lord Hoffmann said that although there were indications that further legislation on rights over common land was in prospect, it was by no means clear that parliament contemplated further legislation concerning rights over village greens. It has been left to the courts to try to work this out for themselves. As Lord Hoffmann put in [49], one has to look at the provisions on greens like those of any other legislation and assume that parliament legislated for some practical purpose. I think that one must assume too that it was parliament's intention that practical common sense would be the best guide to the way in which the public right was to be exercised once the land had entered the register.

[56] In answer to a series of written questions by Lord Greaves, the Parliamentary Under-Secretary of State for the Department for Environment, Food and Rural Affairs (Defra), Lord Davies of Oldham said that the government proposes to consult in the spring of 2010 as to whether changes are needed to the existing framework: see *Hansard* (HL) Written Questions, 15 January 2010, Qs 961-964. This initiative appears to have been prompted by a research report that was received by Defra into the registration of new town and village greens, which has identified particular concerns as to its use in respect of land that is subject to proposals for residential development. I hope that the opportunity will be taken to look at the consequences of registration as revealed by the developing case law as well as how the registration system is working.

Previous authority

[57] I agree with Lord Walker that in none of the three decisions of the House of Lords to which he refers (see [3], above) was it necessary for the House to address the question of deference which lies at the heart of this case. *R* v *Oxfordshire County Council, ex parte Sunningwell Parish Council* [2000] 1 AC 335 was concerned with the registration of a glebe that was used predominantly by the villagers for informal recreation. The diocesan board had obtained planning permission to build two houses on part of the glebe, and it objected to registration. However, the inspector found that it had been tolerant of harmless public use of the land for informal recreation. In *R (on the application of Beresford)* v *Sunderland City Council* [2003] UKHL 60; [2004] 1 AC 889, the land was an open, flat area of grass that was used by the local inhabitants for ball games and other lawful pastimes. The council cut the grass from time to time, but they did not use it in any other way that might have interfered with its use by the locals. In *Oxfordshire*, in [125], the land was described by Lord Walker as an overgrown, rubble-strewn, semi-submerged area, sandwiched between the canal and the railway in north-west Oxford, hardly the ideal site to focus close attention on the critical issue that is before us in this case.

[58] The only passages in these three cases that might be taken as suggesting that the rights acquired by the local inhabitants would be enlarged over those of the owner once the land had been registered, as Rix LJ assumed would happen in this case, are to be found in Lord Hoffmann's speech in *Oxfordshire*. In [51], he said of the effect of registration:

> This does not mean that the owner is altogether excluded from the land. He still has the right to use it in any way which does not interfere with the recreational rights of the inhabitants.

In [59], where he distinguished *Oxfordshire* from the decision of the European Court of Human Rights in *JA Pye (Oxford) Ltd* v *United Kingdom* 44302/2 [2005] 3 EGLR 1*, there is a subtle change of language. He said:

> In the present case, first, the owner retains his title to the land and his right to use it in any way which does not prevent its use by the inhabitants for recreation and, secondly, the system of registration under the 1965 Act was introduced to preserve open spaces in the public interest.

* Editor's note: Also reported at [2005] 49 EG 90

I think that the first passage, in which Lord Hoffmann uses the words "interfere with", goes some way to supporting the idea that after registration the rights of local inhabitants predominate. The second passage, on the other hand, does not. "Preventing" the use of the land for recreation would of course defeat the point of registration completely.

[59] Lord Scott of Foscote was obviously very troubled in *Oxfordshire* by the idea that the public would acquire much broader, more intrusive rights over the land after registration and the management problems that this might give rise to: see [85]. However, his objections were, as I read them, based on an assumption as to the effect of the registration as a town or village green on places such as a dense wood in which people wandered to pick bluebells or to look for mushrooms: see [76]. His dissent casts some light on what he thought was at issue in that case. However, I do not think that it can be used to elevate what Lord Hoffmann said, in [51], to a ruling on the point that, on the facts of that case, did not arise.

[60] The only case that directly addresses the question of deference is *Laing Homes*, in which Sullivan J quashed the resolution that the land should be registered. As Dyson LJ observed in the Court of Appeal, in [30], the concept of deference as a bar to the creation of a new town or village green is Sullivan J's creation. The land in that case consisted of three adjacent fields that Laing Homes held as part of its land bank. It granted a grazing licence to a farmer, Mr Pennington, who, for a few years at the start of the 20-year period, kept cattle on the fields until he had to give this up because of problems with members of the public, whose use of the perimeters of the fields resulted in the paths that they had established there being registered as public footpaths. For more than half of that period, Mr Pennington used the land for taking an annual crop of hay. The question was whether this use of the land, or the growing of any other crop, was inconsistent with the right to use the land for recreation that was contended for by the local inhabitants.

[61] After referring to passages in Lord Hoffmann's speech in *Sunningwell* concerning the extent of the user by the public that was needed to establish that the land was being used by it as of right, Sullivan J said, in [82]:

> If the starting point is, "how would the matter have appeared to Laings?" it would not be reasonable to expect Laings to resist the recreational use of their fields so long as such use did not interfere with their licensee, Mr Pennington's use of them for taking an annual hay crop.

In [84], he said that so long as the local inhabitants' recreational activities did not interfere with the way in which the owner had chosen to use its land, there would be no suggestion to it that they were exercising or asserting a public right to use it for lawful sports and pastimes. In [85], he said:

> I do not believe that Parliament could have intended that such a user for sports or pastimes would be "as of right" for the purposes of s22 [of the 1965 Act]. It would not be "as of right", not because of interruption or discontinuity, which might be very slight in terms of numbers of days per year, but because the local inhabitants would have appeared to the landowner to be deferring to his right to use his land (even if he chose to do so for only a few days in the year) for his own purposes.

[62] In [86], he added these words:

> Like the Inspector, I have not found this an easy question. Section 12 acknowledges that animals may be grazed on a village green. Rough grazing is not necessarily incompatible with the use of the land for recreational purposes: see *Sunningwell*. If the statutory framework within which s22(1) was enacted had made provision for low-level agricultural activities to co-exist with village green type uses, rather than effectively preventing them once such a use has become established, it would have been easier to adopt the Inspector's approach, but it did not. I do not consider that using the three fields for recreation in such a manner as not to interfere with Mr Pennington's taking of an annual hay crop for over half the 20-year period, should have suggested to Laings that those using the fields believed that they were exercising a public right, which it would have been reasonable to expect Laings to resist.

[63] This passage suggests that Sullivan J was approaching the case on the assumption that registration was inconsistent with the continued use of the land by Mr Pennington for taking the annual hay crop. In other words, registration would bring non-interference to an end. The public right to use the fields for recreational purposes would make it impossible for them to be used for growing hay. His approach has also been taken as indicating that in cases where the land has been used by a significant number of inhabitants for 20 years for recreational purposes *nec vi, nec clam, nec precario*, there is an additional question that must be addressed: would it have appeared to a reasonable landowner that the inhabitants were asserting a right to use the land for the recreational activities in which they were indulging? I am not sure that Sullivan J was really saying that there was an additional question that had to be addressed. However, if he was, I would respectfully disagree with him on both points.

Section 15 questions

[64] The application in this case was made under section 15(4) of the 2006 Act, which provides that a person may apply for registration of land as a town or village green where "a significant number of the inhabitants of any locality, or of any neighbourhood within a locality, indulged as of right in lawful sports and pastimes on the land for a period of at least 20 years" if they ceased to do so before the commencement of that subsection, so long as the application is made within a period of five years beginning with the date of the cessation. The words that I have set out in quotation marks appear in each of subsections (2), (3) and (4) of section 15. The definition of the phrase "town or village green" in section 22(1) of the 1965 Act, as amended by section 98 of the 2000 Act, has been repeated throughout this section, with the addition of the words "a significant number".

[65] The theory on which these provisions are based is known to the common law as prescription: see Lord Hoffmann's explanation in *Sunningwell*, at pp349-351, of the background to the definition of "town or village green" in section 22(1) of the 1965 Act. As the law developed in respect of private rights, the emphasis was on the quality of the user for the 20-year period that would justify recognition of a prescriptive right:

> It became established that such user had to be, in the Latin phrase, *nec vi, nec clam, nec precario*: not by force, nor stealth, nor the licence of the owner… The unifying element in these three vitiating circumstances was that each constituted a reason why it would not have been reasonable to expect the owner to resist the exercise of the right in the first case, because rights should not be acquired by the use of force, in the second, because the owner would not have known of the user and in the third, because he had consented to the use, but for a limited period. So in *Dalton* v *Henry Angus & Co* (1881) 6 App Cas 740, 773 Fry J (advising the House of Lords) was able to rationalise the law of prescription as follows:
>
> "the whole law of prescription and the whole law which governs the presumption or inference of a grant or covenant rest upon acquiescence."

Section 2 of the 1832 Act made it clear that what mattered was the quality of the user during the 20-year period. It had to be by a person "claiming right thereto". It must have been enjoyed openly and in the manner that a person rightfully entitled would have used it, not by stealth or by licence: see *Bright* v *Walker* (1834) 1 CM&R 211, at p219, *per* Parke B. In *Gardner* v *Hodgson's Kingston Brewery Co Ltd* [1903] AC 229, at p239, Lord Lindley said that the words "as of right" were intended to have the same meaning as the older expression *nec vi, nec clam, nec precario*.

[66] Referring then to section 1(1) of the Rights of Way Act 1932 (the 1932 Act), Lord Hoffmann said in *Sunningwell*, at p353D-F:

> The words "actually enjoyed by the public as of right and without interruption for a full period of 20 years" are clearly an echo of the words "actually enjoyed by any person claiming right thereto without interruption for the full period of 20 years" in section 2 of the Act of 1832. Introducing the Bill into the House of Lords (HL Debates), 7 June 1932, col 637, Lord Buckmaster said that the purpose was to assimilate the law on public rights of way to that of private rights of way. It therefore seems safe to assume that "as of right" in the Act of 1932 was intended to have the same meaning as those words in section 5 of the Act of 1832 and the words "claiming right thereto" in section 2 of that Act.

A He concluded, at p354, that there was no reason to believe that "as of right" in section 22(1) of the 1965 Act was intended to mean anything different from what those words meant in the Acts of 1832 and 1932. The same can be said of the meaning of those words in section 15 of the 2006 Act.

[67] In the light of that description, it is, I think, possible to analyse the structure of section 15(4) in this way. The first question to be addressed is the quality of the user during the 20-year period. It must have been by a significant number of inhabitants. They must have been indulging in lawful sports and pastimes on the land. The word "lawful" indicates that they must not be such as will be likely to cause injury or
B damage to the owner's property: see *Fitch* v *Fitch* (1797) 2 Esp 543. Moreover, they must have been doing so "as of right": that is to say, openly and in the manner that a person rightfully entitled would have used it. If the user for at least 20 years was of such amount and in such manner as would reasonably be regarded as being the assertion of a public right (see *Beresford*, in [6] and [77]), the owner will be taken to have acquiesced in it unless it can claim that one of the three vitiating circumstances applied in its case. If it does, the second question is whether that claim can be made out. Once the second question is out of the way, either because it has not been asked or because it has been answered against the owner, that is an end of the matter. There is no
C third question. The answer to the first issue (see [4] above) is: no.

[68] Mr Charles George QC, for the appellant, said that there was only one simple test: was the use caught by any of the three vitiating circumstances? Mr Laurence confirmed that it was common ground that the use of the land for recreation in this case was *nec vi, nec clam, nec precario*, but he said that this did not exhaust the issue. The unifying principle was one of reasonableness. He said that if it was not reasonable to expect the owner to resist what the users were doing, no harm could come to the owner from its omission to resist or complain. In this case, as the inspector held, the local inhabitants overwhelmingly deferred to the golfers. As Dyson LJ said in the Court of Appeal, the
D user of the local inhabitants was extensive and frequent, but so too was the use by the golfers: the greater the degree of deference, the less likely it was that it would appear to the reasonable owner that the locals were asserting any right to use the land: see [48] and [49].

[69] I agree with Mr George that all the authorities show that there are only three vitiating circumstances: *Gardner*, at p238, *per* Lord Davey, and p239, *per* Lord Lindley; *Sunningwell*, at p350, *per* Lord Hoffmann; *Beresford*, in [3], *per* Lord Bingham of Cornhill, in [16], *per* Lord Scott of Foscote, in [55], *per* Lord Rodger of Earlsferry; Riddall and Trevelyan *Rights of Way: A Guide to Law and Practice* (4th ed) 2007, at pp41-47. There is no support there for the proposition
E that there is an additional requirement. However, that does not answer Mr Laurence's point, which was really and quite properly directed to the first question as to the quality of the user that is relied on. That, as has been said, is the critical question in this case.

Deference

[70] In [175] of his report, the inspector said that he found that the relationship between the golfers and the local recreational users was generally cordial. This was because local people (with the exception of Squadron Leader Kime) did not materially interfere with the use of the land for playing golf. They would wait until the play had passed or
F until they had been waved on by the golfers. When local people did inadvertently impede play, the golfers' shout of "fore" was enough to warn them to clear the course. The inspector asked himself whether this indicated deference to the golfers. Following what Sullivan J said in *Laing*, in [85], he understood that the use would not be "as of right" if the local inhabitants would have appeared to the owners to be deferring to the owners' right to use their land for their own purposes. That approach is based on the judge's assumption, which the Court of Appeal endorsed, that the effect of registration would be to enlarge the right of the local inhabitants in a way that would effectively prevent the golfers from using the land for their own purposes.

[71] I do not find anything in the words used in section 15(4) of the 2006 Act that supports that approach. On the contrary, the theme that runs right through all the law on private and public rights of way G
and other similar rights is that of an equivalence between the user that is relied on to establish the right, on the one hand, and the way in which the right may be exercised once it has been established, on the other. In *Dalton* v *Henry Angus & Co* (1881) LR 6 App Cas 740, at p774, Fry J, having stated, at p773, that the entire law of prescription rests on acquiescence, said that it involved, among other things, the abstinence by the owner from any interference with the act relied on "for such a length of time as renders it reasonable for the Courts to say that he shall not afterwards interfere to stop *the act* being done" (my emphasis). In other words, one looks to the acts that have been acquiesced in. It is those acts, and not their enlargement in a way that makes them more H
intrusive and objectionable, that he afterwards cannot interfere to stop. This is the basis for the familiar rule that a person who has established by prescriptive use a right to use a way as a footpath cannot without more use it as a bridleway or for the passage of vehicles.

[72] In *White* v *Taylor (No 2)* [1969] 1 Ch 160, at p192F, Buckley LJ said that the user must be shown to have been "of such a character, degree and frequency as to indicate an assertion by the claimant of a continuous right, *and of a right of the measure of the right claimed*". (again, my emphasis). That was a case in which it was claimed, among other things, that sheep rights had been established by prescription at common law. However, I think that this observation is consistent with J
the approach that is taken to prescriptive rights generally. It has to be recognised, of course, that once the right to use the land for lawful sports and pastimes is established and the land has been registered, its use by the local inhabitants for those purposes is not restricted to the sports or pastimes that were indulged in during the 20-year period. Lord Hoffmann said, in *Oxfordshire*, in [50], that the rational construction of section 10 of the 1965 Act, which did not require the rights of recreation as such to be registered, was that land registered as a town or village green can be used generally for sports and pastimes:

> It seems to me that Parliament must have thought that if the land had to be K
> kept available for one form of recreation, it would not matter a great deal to the owner whether it was used for others as well. This would be in accordance with the common law, under which proof of a custom to play one kind of game gave rise to a right to use the land for other games: see the *Sunningwell* case [2000] 1 AC 335, 357A-C.

As he put it in the passage referred to in *Sunningwell*, as long as the activity can properly be called a sport or pastime, it falls within the composite class. This approach indicates that while the principle of equivalence tells one in general terms what the land may be used for, there may be some asymmetry as to the manner of its use for that L
purpose before and after it has been registered. However, it does not follow that, where the use for recreation has coexisted with the owner's use of the land during the 20-year period, the relationship of coexistence is ended when registration takes place.

[73] In *Fitch*, where the inhabitants had the right to play lawful games and pastimes on the plaintiff's close that he used for growing grass for hay, the jury was told that the rights of both parties were distinct and might coexist together. However, the inhabitants could not use the close in the exercise of their right in a way that was not fair or was improper. Referring to that case in *Oxfordshire*, in [51], Lord Hoffmann said that there had to be give and take on both sides. M
Mr Rodney Stewart Smith, following Mr Laurence QC, did not agree. He said that it was fundamental to his argument that the concept of give and take had no place in rights of the kind that were established by registration under the 2006 Act. He submitted that these rights were unqualified and unlimited. He said that *Fitch* did not support the idea of give and take, and he sought to contrast rights of the kind that follow registration with those of the kind discussed in *Mercer* v *Woodgate* (1870) LR 5 QB 26, where there was dedication of the right of way to the public subject to the owner's right to plough the soil in the due course of husbandry. Cockburn CJ said, at p30, that there would be great injustice and hardship to hold that there had been an absolute dedication where the owner had clearly intended only a limited

A dedication. Blackburn J said, at p31, that he could see no objection in law to such a partial dedication.

[74] I agree that care needs to be taken in drawing conclusions from cases concerning the creation of a right of way by dedication. However, the concepts of partial dedication and the coexistence of rights on both sides appear to me to be capable of being applied generally. Lord Hoffmann would not have mentioned give and take in *Oxfordshire* had he thought it had no application to town and village greens. If it were otherwise, it would in practice be very difficult, if not impossible, to obtain registration in cases where the owner is putting its land to some use other than, perhaps, the growing and cutting of grass
B for hay or silage. There being no indication in the statute to the contrary, I would also apply these concepts to the rights created by registration as a town or village green.

[75] Where, then, does this leave deference? Its origin lies in the idea that, once registration takes place, the landowner cannot prevent use of the land in the exercise of the public right that interferes with its use of it: see *Laing*, in [86]. So it would be reasonable to expect it to resist use of its land by the local inhabitants if there were reason to believe that its continued use of the land would be interfered with when the right was established. Deference to its use of the land during the 20-year period would indicate to the reasonable landowner that there was no reason to
C resist or to object to what was taking place. However, once one accepts, as I would do, that the rights on either side can coexist after registration subject to give and take on both sides, the part that deference has to play in determining whether the local inhabitants indulged in lawful sports or pastimes as of right takes on an entirely different aspect. The question is whether the user by the public was of such amount and in such manner as would reasonably be regarded as being the assertion of a public right. Deference by the public to what the owner does on its land may be taken as an indication that the two uses can in practice coexist.

[76] Of course, the position may be that the two uses cannot sensibly coexist at all. However, it would be wrong to assume, as the inspector
D did in this case, that deference to the owner's activities, even if it is as he put it overwhelming, is inconsistent with the assertion by the public to use of the land as of right for lawful sports and pastimes. It is simply attributable to an acceptance that where two or more rights coexist over the same land there may be occasions on which they cannot practically be enjoyed simultaneously: see Rowena Meager *Deference & user as of right: an unholy alliance*, Rights of Way Law Review, October 2009, 147, at p152. If any of the local inhabitants were to exercise his right by way of all take and no give in a way to which legitimate objection could be taken by the landowner, he could no doubt be restrained by an injunction: Philip Petchey, *R (on the application of Lewis)* v *Redcar*
E *and Cleveland Borough Council, Rights of Way Law Review*, March 2009, 139, at p143. In my opinion, the inspector misdirected himself on this point. The question, then, is whether the council's decision that was based on his recommendation can be allowed to stand if the facts are approached in the right way.

[77] The facts of this case, as described by the inspector, show that the local inhabitants (except for Squadron Leader Kime) were behaving when they were using the land for sports and pastimes in the way in which people normally behave when they are exercising public rights over land that is also used as a golf course. They recognise that golfers have as much right to use the land for playing golf as they do
F for their sports and pastimes. Courtesy and common sense dictate that they interfere with the golfer's progress over the course as little as possible. There will be periods of the day, such as early in the morning or late in the evening, when the golfers are not yet out or have all gone home. During such periods the locals can go where they like without causing inconvenience to golfers. When golf is being played, gaps between one group of players and another provide ample opportunity for crossing the fairway while jogging or dog-walking. Periods of waiting for the opportunity are usually short and rarely inconvenience the casual walker, rambler or bird-watcher. I cannot find anything in the inspector's description of what happened in this case that was out of the ordinary. Nor do I find anything that was inconsistent with the use of the land as of right for lawful sports and pastimes.

Conclusion G

[78] For these reasons, and those given by everyone else with which I agree, I would allow the appeal and make the order that has been proposed by Lord Walker.

Giving the third opinion, **Lord Rodger** said:

[79] I agree with the judgment of Lord Walker. In view of the importance of the issue, I add some observations of my own.

[80] As Lord Walker has explained, until 2002 an area of land (the disputed land) in the Coatham district of Redcar formed part of a golf course on which members of Cleveland Golf Club played. The club was a tenant of the council, which owned the land. Then, in 2002, the course H
was reconfigured and the club gave up its tenancy of the disputed land. The following year, the council entered into an agreement with Persimmon Homes (Teesside) Ltd for a mixed residential and leisure development on an area of land of which the disputed land formed an important part.

[81] In March 2005, a group of residents applied to have the disputed land registered as a village green. In March 2006, the inspector recommended against registration. In June 2007, Mr Lewis and his fellow applicants put in a fresh application under section 15 of the 2006 Act. Again, the inspector recommended against registration and the matter has now led to the present appeal. J

[82] This sequence — a proposal to develop an area of open land, followed by an application to register the land as a village green in order to stop the development — is very familiar. The House of Lords dealt with three such cases in the space of a few years and newspaper articles refer to many other examples. However, the fact that the disputed land was used by the golf club during the period of 20 years that the applicants rely on to justify its registration as a village green has prompted much heart-searching as to what the position would have been had the land been registered as a village green while the club was still in occupation and its members were still wanting to play on the land. Would registration have enabled the dog walkers of Redcar to K
take over and in effect extinguish the rights of the golfers to play on that part of their course?

[83] However interesting the point of law may be, in a case such as this the issue is more than just a little unreal. The fact of the matter is that if the golf club had remained as a tenant after 2002, the golfers would have continued to hack their way over the disputed area and the dog walkers would have continued to make their way across the course. It is a fair bet that in that happy state of affairs no one would have dreamed of applying to have the land registered as a village green. It was only the prospect of the development on this open space, when the golf club was no longer using it, that prompted the application for L
registration with a view to stopping the development in its tracks. So, in the real world, the dog walkers and golfers will never actually have to coexist on the disputed land if it is registered as a village green.

[84] If, however, in some imaginary parallel universe, the two groups had been required to coexist after registration, then, like Lord Walker, I find it hard to imagine that there would, in practice, have been many problems. The pre-existing situation suited the local inhabitants well enough: doubtless, some of them were themselves members of the club and played on the land. In any event, the golf club must have kept the grass cut and the area looking presentable. If the inhabitants had previously shown no inclination to break out the croquet hoops or to set M
up butts or cricket stumps or to dance around a maypole on the disputed land, it seems unlikely that registration would have suddenly brought on the urge. Indeed, too many developments of these kinds would probably have upset the dog walkers almost as much as the golfers. In all likelihood, therefore, things would have gone on much as before, with a bit of give and take on both sides. I would therefore particularly associate myself with what Lord Walker says in [47] of his judgment.

[85] Under section 15 of the 2006 Act, registration of land as a village green requires that a significant number of the inhabitants of any locality, or of any neighbourhood in a locality, have indulged as of right in lawful sports and pastimes on the land for a period of at least 20 years. Since *R* v *Oxfordshire County Council, ex parte*

A *Sunningwell Parish Council* [2000] 1 AC 335, it has been settled law that dog-walking and playing with children count as lawful sports and pastimes. Since both activities can and do take place on almost any and every open space close to centres of population, the scope for applying to register land as a village green is correspondingly wide. Owners of land are taken to be aware of this chapter of the law and of the need to take appropriate preventive steps if they see a risk of circumstances arising in which an application could be made and their land becomes registered as a village green. If they fail to do so, they are treated as having acquiesced in the inhabitants indulging in sports and pastimes on their land "as of right".

B [86] Here, the evidence shows that, as far back as living memory goes, many local inhabitants used the disputed land for informal recreation such as dog-walking and children's play. However, the courts below have held that they were not doing so "as of right".

[87] The basic meaning of that phrase is not in doubt. In *Sunningwell*, Lord Hoffmann showed that the expression "as of right" in the 1965 Act was to be construed as meaning *nec vi, nec clam, nec precario*. The parties agree that the position must be the same under the 2006 Act. The Latin words need to be interpreted, however. Their sense is perhaps best captured by putting the point more positively: the user must be peaceable, open and not based on any licence from the owner C of the land.

[88] The opposite of "peaceable" user is user that is, to use the Latin expression "*vi*". However, it would be wrong to suppose that user is "*vi*" only where it is gained by employing some kind of physical force against the owner. In Roman law, where the expression originated, in the relevant contexts *vis* was certainly not confined to physical force. It was enough if the person concerned had done something that he was not entitled to do after the owner had told him not to do it. In those circumstances, what he did was done *vi*: see, for instance, D43.24.1.5-9, Ulpian 70 *ad edictum*, commenting on the word as used in the interdict *quod vi aut clam*.

D [89] English law has interpreted the expression in much the same way. For instance, in *Sturges* v *Bridgman* (1879) 11 Ch D 852, at p863, where the defendant claimed to have established an easement to make noise and vibration, Thesiger LJ said:

> Consent or acquiescence of the owner of the servient tenement lies at the root of prescription, and of the fiction of a lost grant, and hence the acts or user, which go to the proof of either the one or the other, must be, in the language of the civil law, *nec vi nec clam nec precario*: for a man cannot, as a general rule, be said to consent to or acquiesce in the acquisition by his neighbour of an easement through an enjoyment of which he has no knowledge, actual or constructive, *or which he contests and endeavours to interrupt*, or which he E temporarily licenses.

(Emphasis added.)

If the use continues despite the neighbour's protests and attempts to interrupt it, it is treated as being *vi* and so does not give rise to any right against him. Similarly, in *Dalton* v *Henry Angus & Co* (1881) LR 6 App Cas 740, at p786, Bowen J equated user *nec vi* with peaceable user and commented that a neighbour "without actual interruption of the user, ought perhaps, on principle, to be enabled by continuous and unmistakeable protests to destroy its peaceable character, and so to annul one of the conditions upon which the presumption of right is raised: *Eaton* v *Swansea Waterworks Co* (1851) 17 QB 267". The contrary F view, that the only manner in which enjoyment of window lights could be defeated before the 1832 Act was by a physical obstruction of the light, "was not the doctrine of the civil law, nor the interpretation which it placed upon the term '*non vi*'...".

[90] In short, as *Gale on the Law of Easements* (18th ed) 2008, in para 4-84, suggests, user is peaceable (*nec vi*) only if it is neither violent nor contentious.

[91] In *Sunningwell*, at pp350-351, Lord Hoffmann found that the unifying element in the three vitiating circumstances was that each constituted a reason why it would not have been reasonable to expect the owner to resist the exercise of the right. In the case of *nec vi*, he said that this was "because rights should not be acquired by the use of force". If by "force" Lord Hoffmann meant only physical force, G I would respectfully disagree. Moreover, some resistance by the owner is an aspect of many cases where use is *vi*. Assuming, therefore, that there can be *vis* where the use is contentious, a perfectly adequate unifying element in the three vitiating circumstances is that they are all situations in which it would be unacceptable for someone to acquire rights against the owner.

[92] If, then, the inhabitants' use of land is to give rise to the possibility of an application being made for registration of a village green, it must have been peaceable and non-contentious. This is at least part of the reason why, as Lord Jauncey observed in the context of a claim to a public right of way in *Cumbernauld and Kilsyth District* H *Council* v *Dollar Land (Cumbernauld) Ltd* (1993) SC (HL) 44, at p47: "There is no principle of law which requires that there be conflict between the interest of users and those of a proprietor."

[93] In this case, the local inhabitants' use of the disputed land for recreation was peaceable, open and not based on any licence from the council or the golf club. So prima facie, the inhabitants did everything that was necessary to bring home to the council, if they were reasonably alert, that the inhabitants were using the land for recreation "as of right".

[94] However, the council argue that, since there were competing interests, the inhabitants' use of the land was peaceable only because J they "overwhelmingly" deferred to the golfers' simultaneous use of the same land. Had they not done so, it would have become contentious. However, because they routinely deferred to the golfers, the inhabitants did not do "sufficient to bring home to the reasonable owner of the application site that they were asserting a right to use it": *cf* [2009] EWCA Civ 3; Dyson LJ, [2009] 1 WLR 1461, in [49]. In other words, the reasonable owner of the disputed land would have inferred from the behaviour of the inhabitants that they were not asserting a right over the land and so would have seen no need to take any steps to prevent such a right from accruing.

K [95] On closer examination, the starting point for this argument must be that the owner of the land is entitled to infer from the inhabitants' behaviour in deferring to the golfers that they are aware of the legal position. However, that starting point is inherently implausible. To adapt what Lord Sands said in connection with a public right of way in *Rhins District Committee of Wigtownshire County Council* v *Cuninghame* (1917) 2 SLT 169, at p172, people walk their dogs or play with their children on the disputed land because they have been accustomed to see others doing so without objection. The great majority know nothing about the legal character of their right to do so and never address their minds to the matter. Moreover, to draw an inference based L on the premise that the inhabitants are aware of the legal position is hard to reconcile with the decision in *Sunningwell*, at pp355-356, that the subjective views of the inhabitants as to their right to indulge in sports and pastimes on the land are irrelevant. It would therefore have been far from reasonable for the council to infer that the inhabitants' behaviour towards the golfers was based on some understanding of the legal position. It would have been equally unreasonable for the council to go further and conclude that the inhabitants were deferring to the golfers because of a conscious decision on their part to respect what they perceived to be the superior rights of the owners of the land.

[96] Such a conclusion might just conceivably have been plausible M and legitimate had there been no other explanation for the inhabitants' behaviour. However, that is far from so. The local inhabitants may well have deferred to the golfers because they enjoyed watching the occasional skilful shot or were amused by the more frequent duff shots, or simply because they were polite and did not want to disturb the golfers, who experience shows almost invariably take their game very seriously indeed. A reasonable landowner would realise that any of these motives was a more plausible explanation for the inhabitants' deference to the golfers than some supposed unwillingness to go against a legal right that they acknowledged to be superior. In my view, the inspector misdirected himself on this aspect of the case.

[97] I would accordingly allow the appeal and make the order proposed by Lord Walker. I confess that I view the outcome with little

enthusiasm. The idea that this land should be classified and registered as a village green, when it was really just an open space that formed part of a golf course, is unattractive to say the least. It is hard to imagine that those who devised the registration system ever contemplated that it would produce such a result. However, given the established case law and given also that parliament has not amended the law despite the known problems, the result is unavoidable.

Giving the fourth opinion, **Lord Brown** said:

[98] I would formulate the critical question for the court's determination on this appeal very differently from any of those identified in the statement of facts and issues. The critical question to my mind is what are the respective rights of the landowner (the owner) and the local inhabitants (the locals) over land once it is registered as a town or village green?

[99] Take the facts of this case, as already sufficiently recounted by other members of the court, but assume that the land here in question, instead of becoming vacant in 2002 and subject now to development proposals, remained in use by the owner (as, for convenience, I shall call the Redcar & Cleveland Golf Club, the actual owner's licensee) as the first and 18th holes (and practice green) of its golf course. Suppose then that the local inhabitants, having themselves made such use of the land as the inspector records, "deferring" to the golfers in the way he describes, successfully applied for its registration as a town green, what then would be the consequences with regard to the owner's own continuing rights? Would the owner remain entitled to use the land for golf with the locals continuing to "defer" to the golfers? Or would the balance shift entirely, the locals' rights being substantially enlarged by registration, the owner's effectively extinguished?

[100] So, far from this question begging that as to the right to registration (the ultimate question at issue here), it seems to me one that necessarily should be resolved before it can sensibly be decided what must be established in order to have the land registered. Indeed, I may as well say at once that were it the law that, on registration, the owner's continuing right to use its land as it has been doing becomes subordinated to the locals' rights to use the entirety of the land for whatever lawful sports and pastimes they wish, however incompatibly with the owner continuing in its, I would hold that more is required to be established by the locals merely than use of the land for the stipulated period *nec vi, nec clam, nec precario*. If, however, as I would prefer to conclude, the effect of registration is rather to entrench the previously assumed rights of the locals, precluding the owner from thereafter diminishing or eliminating such rights but not at the expense of the owner's own continuing entitlement to use the land as it has been doing, I would hold that no more is needed to justify registration than what, by common consent, is agreed to have been established by the locals in the present case.

[101] This is not merely because, in my opinion, no other approach would meet the merits of the case. Also it is because, to my mind, on the proper construction of section 15 of the 2006 Act, the only consequence of registration of land as a green is that the locals gain the legal right to continue to "indulge" in lawful sports and pastimes on it (which previously they have done merely as *if* of right), no more and no less. To the extent that the owner's own previous use of the land prevented its indulgence in such activities in the past, it remains restricted in its future use of the land. The owner's previous use *ex-hypothesi* would not have been such as to have prevented the locals from satisfying the requirements for registration of the land as a green. No more should the continuance of the owner's use be regarded as incompatible with the land's future use as a green. Of course, in so far as future use by the locals would *not* be incompatible with the owner continuing in its previous use of the land, the locals can change, or indeed increase, their use of the land; they are not confined to the same "lawful sports and pastimes", the same recreational use as they had previously enjoyed. However, they cannot disturb the owner so long as it wishes only to continue in its own use of the land.

[102] Is there, then, anything in the case law that precludes our deciding, as I have already indicated I would prefer to decide, that registration does not carry with it a right in future to use the land inconsistently with such use as the owner itself has been making and wants to continue making of it? The respondents here urge that the decision of the House of Lords in *Oxfordshire County Council* v *Oxford City Council* [2006] UKHL 25; [2006] 2 AC 674 is just such a case. They so submit notwithstanding that the land there was disused scrubland of which the owner made no use whatever so that no question arose as to possibly conflicting uses or the respective rights of owners and locals following registration. For my part, I simply cannot regard *Oxfordshire* as having decided the particular question I am addressing here. The respondents rely on passages in Lord Hoffmann's speech such as that, following registration, "[The owner] still has the right to use it in any way which does not interfere with the recreational rights of the inhabitants" ([51]) and "[T]he owner retains his title to the land and his right to use it in any way which does not prevent its use by the inhabitants for recreation": [59]. To my mind, however, these are not inconsistent with the position that I have suggested arises on registration and, indeed (also in [51]), Lord Hoffmann states: "There has to be give and take on both sides."

[103] True it is that, in a partially dissenting opinion, Lord Scott, in [105], himself appears to have understood the other members of the committee to have decided that registration of land as a green "bring[s] about a diminution of the landowner's property rights, not simply by establishing the local inhabitants' right to go on doing what they had been doing for the last 20 years but by depriving the landowner of the right to go on doing what he has been doing for the last 20 years". Lord Scott did "not agree [inferentially, with the majority view] that registration can authorise local inhabitants to enjoy recreative user of the land that is different in kind from the 20 years' user that has satisfied the statutory criteria for registration or that would diminish the ability of the landowner to continue to use the land in the manner in which he had been able to use the land during that 20-year period… [or] that a tolerant landowner who has allowed the local inhabitants to use his grass field for an annual 5 November bonfire for upwards of 20 years must, after registration, suffer his field to be used throughout the year for all or any lawful sports and pastimes with the consequential loss of any meaningful residual use that he could continue to make of the field". That, however, was in the context of Lord Scott's view ([106]) that registration of the land there in question would (or at least should) entitle the locals only to "recreative rights of user… commensurate with the nature of the user that had led to that result and would not necessarily extend to the right to use the land for all or any lawful sports or pastimes [for instance, clay pigeon shooting or archery contests]". It is important to note, moreover, that all of this was concerned with the first of the 10 issues before the House as to which it was held (*per* the headnote) that "registration gave rise to rights for the relevant inhabitants to indulge in lawful sports and pastimes, such rights extending (Lord Scott of Foscote dissenting) to sports and pastimes generally and not merely that use which had been the basis for registration, the landowner retaining the right to use the land in any way which did not interfere with those rights".

[104] I repeat, the position arising on registration when both the owner and the locals are using land in theoretically conflicting ways but in fact harmoniously simply did not arise in *Oxfordshire*, and I for my part would decline to treat that case as though it has decided how such an issue should be resolved.

[105] I would therefore hold that in this different situation the owner remains entitled to continue its use of the land as before. If, of course, as in *Oxfordshire* itself, it has done nothing with its land, it cannot complain that, on registration, the locals gain full and unqualified recreational rights over the land. However, that is not the position that I am considering here.

[106] In short, on the facts of this case, had the use of the land as part of a golf course continued, the locals would, in my opinion, have had to continue "deferring" to the golfers. By this, I understand the inspector to have meant no more than that the locals (with the single exception of Squadron Leader Kime) recognised the golfers' rights to play (in this sense only, the locals "overwhelmingly deferred to golfing use"), both locals and golfers sensibly respecting the use being made of the

A land by the other, neither being seriously inconvenienced by the other, sometimes the locals waiting for the golfers to play before themselves crossing, sometimes the golfers waiting for the walkers to cross before playing. It is not unique for golf courses to embrace at least some common land, and there are innumerable courses crossed by public footpaths. Both walkers and golfers are generally sensible and civilised people and common courtesy dictates how to behave. Harmonious coexistence is in practice easily achievable. For my part, and in the light of my own experience both as a golfer and a walker for more than six decades, I do not read the inspector's findings as indicating (to quote Sullivan J) [2008] EWHC 1813 (Admin), in [40], that "there was B overwhelmingly 'give' on the part of the local users and 'take' on the part of the golfers".

[107] This being so, I see no good reason whatever to superimpose on the conventional tripartite test for the registration of land that has been extensively used by local inhabitants for recreational purposes a further requirement that it would appear to a reasonable landowner that the users were asserting a right to use the land for the lawful sports and pastimes in which they were indulging. As Lord Walker has explained, there is nothing in the extensive jurisprudence on this subject to compel the imposition of any such additional test. Rather, as Lord Hope, Lord Walker and Lord Kerr make plain, the focus must always be on C the way in which the land has been used by the locals and, above all, the quality of that user.

[108] I too, therefore, would allow this appeal.

Giving the fifth opinion, **Lord Kerr** said:

[109]] For the reasons given by Lord Hope, Lord Rodger, Lord Walker and Lord Brown, with all of whom I agree, I too consider that this appeal should be allowed. I venture to offer a few words of my own because my conclusion that the appeal should be allowed represents a change from the view that I initially held and because I can well understand why the Court of Appeal and Sullivan J dismissed the D application for judicial review.

[110] The critical question in this case centres on the meaning to be given to the words "as of right" in section 15 of the 2006 Act. It is not possible to give a literal interpretation to the words since, clearly, the right cannot vest in the local inhabitants until the period of 20 years has elapsed. They cannot be considered to have indulged in sports and pastimes by dint of a right until the right has come to fruition: see Lord Bingham in *R (on the application of Beresford)* v *Sunderland City Council* [2003] UKHL 60; [2004] 1 AC 889, in [3]. It is also clear that they do not need to believe that they have a right: see below. As Lord Walker said in *Beresford*, in [72], it has sometimes E been suggested that the meaning of the statutory formula is closer to "as if of right": see, for instance, Lord Cowie in *Cumbernauld and Kilsyth District Council* v *Dollar Land (Cumbernauld) Ltd* (1992) SLT 1035, at p1043.

[111] Using this formulation, the question is what does "as if of right" mean. Does it simply mean openly indulging in the pastimes etc without force or under licence or does it connote something more? Clearly, it cannot be construed to mean "as if they believed they had the right". The House of Lords so held in *R* v *Oxfordshire County Council, ex parte Sunningwell Parish Council* [2000] 1 AC 335. Does it mean that they *acted* as though they had the right? If so, how is F that to be judged? Does it mean that they gave every indication that they had the right to indulge in the pastimes and sports? According to Mr George, the only exception to the tripartite test arises where the users expressly represent that they are not asserting any right at all. In those circumstances, according to him, they are either benefiting from the implied permission of the owner or they are covertly allowing the necessary period to elapse, in which case they fall foul of the requirement that the use of the land should not be secret.

[112] The question that has troubled me is "what if the inhabitants' engagement in the pastimes and sports is not on foot of an express representation that they are not asserting a right but on the basis of an unspoken understanding by all concerned that they are not doing so"? Is there a reason why, as a matter of principle, there should be any different legal outcome? It appears to me that there is none. If the owner G of the land and those who recreate on it share the appreciation that no right is being asserted, no right is acquired. Therefore, as Lord Hope has said (in [69] of his judgment), one must focus on the manner in which the local inhabitants have used the land or, as he has put it, "the quality of the user that is relied on".

[113] The use of the word "deferring" in the context of the inhabitants' use of land is potentially misleading. In common parlance, "deferring to an owner's use of his lands" can easily be understood to mean no more than the ordinary courteous and civilised acknowledgement of the entitlement of the owner to make use of the land. Such civility does not necessarily import an acceptance of any lack of entitlement on the H part of the users to continue to indulge their recreations with a view to the acquisition of a right under section 15. However, if deference takes the form of acceptance that the users are not embarked on a process of accumulating the necessary number of years of use of the lands or if it evinces an intention not to embark on such a process, this must surely have significance in respect of the question of whether the inhabitants have indulged in the activities "as of right".

[114] It is for this reason in particular that I am in emphatic agreement with Lord Hope in his view that one must focus on the way in which the land has been used by the inhabitants. Have they used it as though they had the right to use it? This question does not require any examination J of whether they believed that they had the right. That is irrelevant. The question is whether they acted in a way that was comparable to the exercise of an existing right. Posed in that way, one can understand why the Court of Appeal considered that the examination of the relevant question partook of an inquiry as to the outward appearance created by the use of the lands by the inhabitants. On that basis also one can recognise the force of Mr Laurence's argument that it was necessary to show not only that the lands had been used *nec vi, nec clam, nec precario* but also that it was reasonable to expect the landowner to resist the use of the land by the local inhabitants. The essential underpinning of both these assertions, however, was the view that the registration K of the land as a village or town green had the inexorable effect of the enlargement of the inhabitants' rights and the commensurate diminution of the right of the owner to maintain its pre-registration level of use, if that interfered with the inhabitants' extended use of the land.

[115] For the reasons that Lord Hope and Lord Walker have given, the view that this was the effect of the relevant authorities in this area may now be discounted. For my part, I find it unsurprising that this view formerly held sway. Mr Laurence (without direct demur from Mr George) informed us that it was the universal opinion of all who practised in this field that the inevitable consequence of the decision in *Oxfordshire County Council* v *Oxford City Council* [2006] UKHL L 25; [2006] 2 AC 674 was that local inhabitants acquired unrestricted rights of recreation after registration. Passages from the speech of Lord Hoffmann in that case — particularly in [51] — appeared to lend support for the notion that general, unrestricted rights of recreation over the entire extent of the lands followed on registration. Moreover, the speech of Lord Scott certainly seemed to imply that he apprehended that this was the outcome of the decision by the majority. Whatever may have been the position previously, however, it is now clear that, where it is feasible, co-operative, mutually respecting uses will endure after the registration of the green. Where the land has been used by both the inhabitants and the owner over the pre-registration period, the M breadth of the historical user will be, if not exactly equivalent to, at least approximate to that which will accrue after registration.

[116] On that basis, I am content to accept and agree with the judgments of Lord Hope, Lord Walker and Lord Brown that no overarching requirement concerning the outward appearance of the manner in which the local inhabitants used the land is to be imported into the tripartite test. The inhabitants must have used it as if of right, but that requirement is satisfied if the use has been open in the sense that they have used it as one would expect those who had the right to do so would have used it; that the use of the lands did not take place in secret; and that it was not on foot of permission from the owner. If the use of the lands has taken place in such circumstances, it is unnecessary

A to enquire further as to whether it would be reasonable for the owner to resist the local inhabitants' use of the lands. Put simply, if confronted by such use over a period of 20 years, it is *ipso facto* reasonable to expect an owner to resist or restrict the use if it wishes to avoid the possibility B of registrations.

Appeal allowed.

Secretary of State for Environment, A Food and Rural Affairs *v* Meier and others

SUPREME COURT

[2009] UKSC 11

1 December 2009

LORD RODGER, LORD WALKER, LADY HALE, LORD NEUBERGER AND B LORD COLLINS

Trespass — Possession order — Extent of order — Travellers setting up unauthorised camp in wood — Order covering other woods in area — Whether permissible to grant possession order against trespassers in respect of land not occupied by them where move to such land anticipated — Whether injunctive relief appropriate

The appellant travellers lived and travelled in motor vehicles. In 2007, they established an unauthorised camp in Hethfelton Wood, near Wool, Dorset, one of numerous woods managed by the Forestry Commission on behalf of the respondent owner. The respondent brought proceedings for possession against the appellants and "persons C unknown". It sought possession not only of Hethfelton Wood but also of other woodland managed by the Commission in Dorset covering an area approximately 25 miles by 10 miles; this was with a view to prevent the travellers from moving to another area of woodland if possession of Hethfelton were ordered. The respondent also applied for injunctions against the same parties restraining them from re-entering Hethfelton Wood or entering the other woods.

The recorder granted an order for possession in respect of Hethfelton Wood. He refused to make a wider order for possession in respect of the other woods or to grant the injunction sought, primarily on the ground that the Commission had not, prior to commencing the D proceedings, considered the matters suggested by 2004 government guidance on managing unauthorised camping. The relief sought by the respondent was subsequently granted on appeal. The Court of Appeal held, following *Drury* v *Secretary of State for the Environment, Food and Rural Affairs* [2004] EWCA Civ 200; [2004] 3 EGLR 85; [2004] 37 EG 142, that the court had the power in appropriate cases to grant a possession order in respect of land of which the trespassers were not yet in occupation. The appellants challenged the correctness of that view in an appeal to the Supreme Court. The appropriateness of the injunctive relief was also in issue on the appeal.

***Held*: The appeal was allowed in part. (1) A possession claim E against trespassers involves the party entitled to possession seeking recovery of the land; the nature of the claim is reflected in the terms of the relevant court order. There is no legitimate basis for making a wider, precautionary order for possession in respect of land that the trespassers have not occupied, but are threatening to occupy. A wider order for possession would require the trespassers to do the impossible, namely to deliver up possession of land that they did not occupy, and would purport to return to the landowner something that it had not lost, namely possession of land of which it was already in possession. Although the concept of occupying part of a property in the name of the whole is well established and an order for possession can be made in respect of the whole of land of which the trespasser occupies part, that reasoning cannot legitimately be F extended to apply to land that is wholly distinct or miles distant from the occupied land. It does not justify the conclusion that a wider order for possession can be made in respect of separate pieces of land, only one of which is occupied by trespassers, simply because they happen to be in the same ownership. The trespassers did not occupy or possess such land in any conceivable way and the landowner enjoyed uninterrupted possession of it: *Drury* wrongly decided. (2) Where a trespass is threatened, and particularly where it is being committed and has been committed in the past, an injunction to restrain the threatened trespass will usually be appropriate unless there are good reasons to the contrary. Although G the court should not make orders that it did not intend or would be unable to enforce, it might be appropriate to grant an injunction even in cases where there was little prospect of enforcing it by imprisonment or sequestration if it might have a real deterrent effect on the particular trespassers. The 2004 guidance does not preclude the granting of an injunction to restrain travellers from trespassing on other land. There were no grounds for setting aside the injunction granted in the instant case.**

The following cases are referred to in this report.

Bloomsbury Publishing Group plc v *News Group Newspapers Ltd* [2003] EWHC 1205 (Ch); [2003] 1 WLR 1633; [2003] 3 All ER 736 H

Connors v *United Kingdom* 66746/01 [2004] 4 PLR 16; [2004] HLR 52, (2005) 40 EHRR 9, ECtHR

Doherty v *Birmingham City Council*; *sub nom Birmingham City Council* v *Doherty* [2008] UKHL 57; [2008] 3 WLR 636; [2009] 1 All ER 653; [2009] 1 P&CR 11; [2008] HLR 45

Drury v *Secretary of State for the Environment, Food and Rural Affairs*; *sub nom Secretary of State for the Environment, Food and Rural Affairs* v *Drury* [2004] EWCA Civ 200; [2004] 1 WLR 1906; [2004] 2 All ER 1056; [2004] 3 EGLR 85; [2004] 37 EG 142

Fourie v *Le Roux* [2007] UKHL 1; [2007] 1 WLR 320; [2007] 1 All ER 1087

Gledhill v *Hunter* (1880) LR 14 Ch D 492

Hampshire Waste Services Ltd v *Persons unknown* [2003] EWHC 1738 (Ch); J [2003] 42 EG 126 (CS); [2004] Env LR 9

Hemmings v *Stoke Poges Golf Club Ltd* [1920] 1 KB 720, CA

Henderson v *Squire* (1869) LR 4 QB 170; 10 B&S 183; 38 LT QB 73; 19 LT 601

McPhail v *Persons, Names Unknown*; *Bristol Corporation* v *Ross* [1973] Ch 447; [1973] 3 WLR 71; [1973] 3 All ER 393; 72 LGR 93, CA

Manchester Airport plc v *Dutton* [2000] 1 QB 133; [1999] 3 WLR 524; [1999] 2 All ER 675; (2000) 79 P&CR 541; [1999] 1 EGLR 147

Manchester Corporation v *Connolly* [1970] 1 Ch 420; [1970] 2 WLR 746; [1970] 1 All ER 961; (1970) 21 P&CR 154, CA

Masri v *Consolidated Contractors International Co SAL* [2008] EWCA Civ 303; [2009] QB 450; [2009] 2 WLR 621; [2008] 2 All ER (Comm) 1099; K [2008] 2 Lloyd's Rep 128

Ministry of Agriculture, Fisheries and Food v *Heyman* (1990) 59 P&CR 48, QB

R v *Wandsworth County Court, ex parte Wandsworth London Borough Council* [1975] 1 WLR 1314; [1975] 3 All ER 390; [1975] 74 LGR 62

Secretary of State for the Environment, Food and Rural Affairs v *Meier* [2008] EWCA Civ 903; [2009] 1 WLR 828; [2009] 1 All ER 614; [2009] HLR 16

South Bucks District Council v *Porter*; *Chichester District Council* v *Searle*; *Wrexham County Borough Council* v *Berry*; *Herstmere Borough Council* v *Harty* [2003] UKHL 26; [2003] 2 AC 558; [2003] 2 WLR 1547; [2003] 3 All ER 1; [2003] 2 PLR 101; [2003] JPL 1412

South Cambridgeshire District Council v *Persons Unknown* [2004] EWCA Civ 1280, [2004] 4 PLR 88; [2005] JPL 680

A *Thompson* v *Elmbridge Borough Council* [1987] 1 WLR 1425; (1987) 86 LGR 245; (1987) 19 HLR 526, CA

University of Essex v *Djemal* [1980] 1 WLR 1301; [1980] 2 All ER 742; (1981) 41 P&CR 340, CA

Wykeham Terrace, Brighton, Re; sub nom Territorial Auxiliary & Volunteer Reserve Association for the South East, ex parte Wykeham Terrace, Brighton; Territorial Association v *Hales* [1971] Ch 204; [1970] 3 WLR 649

This was an appeal by the appellants, Sharon Horie and Lesley Rand, against a decision of the Court of Appeal granting a possession order and injunctive relief to the respondent, the Secretary of State for
B Environment, Food and Rural Affairs, requiring the appellants to give up occupation of land and restraining them from occupying other land also owned by the respondent in the area.

Richard Drabble QC and Marc Willers (instructed by Community Law Partnership, of Birmingham) appeared for the appellants; John Hobson QC and John Clargo (instructed by Whitehead Vizard, of Salisbury) represented the respondent.

Giving the first opinion, **Lord Rodger** said:

[1] If a group of people comes onto my land without my permission, I shall want the law to provide a speedy way of dealing with the
C situation. If the people leave but come back repeatedly, depending on the evidence I shall be able to obtain an interlocutory and final injunction against them returning. However, they may come onto my land and set up camp there. Again, depending on the evidence, I shall be able to obtain an injunction (interlocutory and final) against them remaining and also against them coming back once they leave as required by the injunction. Similarly, if the evidence shows that, once they leave, they are likely to move and set up camp on other land that I own, the court can grant an injunction (interlocutory and final) against them doing that. If authority is needed for all this, it can be found in the judgment of Lord Diplock in the Court of Appeal in *Manchester Corporation* v *Connolly* [1970] Ch 420.

D [2] Of course, it is quite likely that I will not know the identities of at least some of the trespassers. If so, Wilson J regarded an injunction as "useless" since "it would be wholly impracticable for the claimant to seek the committal to prison of a probably changing group of not easily identifiable travellers, including establishing service of the injunction and of the application": *Drury* v *Secretary of State for the Environment, Food and Rural Affairs* [2004] EWCA Civ 200; [2004] 1 WLR 1906*, at p1912, in [19]. That may well have been an unduly pessimistic assessment. Certainly, claimants have used injunctions against unnamed defendants. Moreover, Sir Andrew Morritt V-C was satisfied that the procedural problems could be overcome. Admittedly,
E the circumstances in the first of his cases, *Bloomsbury Publishing Group plc* v *News Group Newspapers Ltd* [2003] EWHC 1205 (Ch)†, were very different from a situation involving trespassers. However, trespassing protesters were the target of the interlocutory injunction that he granted in *Hampshire Waste Services Ltd* v *Persons unknown* [2003] EWHC 1738 (Ch)‡. Similarly, in *South Cambridgeshire District Council* v *Persons unknown* [2004] EWCA Civ 1280§, the Court of Appeal (Brooke and Clarke LJJ) granted an injunction against persons unknown "causing or permitting hardcore to be deposited, caravans, mobile homes or other forms of residential accommodation to be stationed, or existing caravans or other mobile homes to be occupied
F on land" adjacent to a gypsy encampment in rural Cambridgeshire. Brooke LJ commented, in [8]: "There was some difficulty in times gone by against obtaining relief against persons unknown, but over the years that problem has been remedied either by statute or by rule." See the discussion of such injunctions by Dr Jillaine Seymour, *Injunctions Enjoining Non-Parties: Distinction without Difference* (2007) 66 CLJ 604-623.

* Editor's note: Also reported at [2004] 3 EGLR 85; [2004] 37 EG 142

† Editor's note: Reported at [2003] 1 WLR 1633

‡ Editor's note: Reported at [2004] Env LR 9

§ Editor's note: Reported at [2004] 4 PLR 88

[3] The present case concerns travellers who set up camp on G
the Forestry Commission's land at Hethfelton. Lord Neuberger has explained the circumstances. The identities of some but not all of those involved were known to the Commission. So, the defendants included "persons unknown". Despite this, the Commission sought an injunction against all the defendants, including those described as "All persons currently living on or occupying the claimant's land at Hethfelton". The recorder declined to grant an injunction on the view that it would be disproportionate. However, the Court of Appeal ([2008] EWCA Civ 903*), by a majority, reversed the recorder on this point and granted an order that:

> The respondents, and each of them, be restrained from entering upon, trespassing H
> upon, living on, or occupying the parcels of land set out in the Schedule hereto, and, for the avoidance of doubt, the 4th respondent shall mean "those people trespassing on, living on, or occupying the land known as Hethfelton Wood on any date between 13th February 2007 and 3rd August 2007 save for those specifically identified as 1st, 2nd, 3rd, 5th and 6th respondents".

In my view, for the reasons given by Lord Neuberger, the majority was right to grant the injunction. In any event, Mr Richard Drabble QC, who appeared for the travellers, did not suggest that this injunction had been incompetent or defective for lack of service or in some other respect. Even Wilson LJ, who dissented on the injunction point in the Court of
Appeal, did not go so far as to suggest that it was inherently useless: J
he simply took the view that it added nothing of value to the order for possession and, therefore, the recorder would have been entitled to exercise his discretion to refuse it on that basis: [76].

[4] This brings me to the order for possession that lies at the heart of the appeal. If people not only come onto my land but oust me from it, I can bring an action for recovery of the land. That is what the Commission did in the present case: it raised an action in Poole County Court for the recovery of "land at Hethfelton nr Wool and all that land described on the attached schedule all in the County of Dorset". In effect, the Commission was asking for two things: to be put back
into possession of the land on which the appellants were camped at K
Hethfelton and to be put into possession of the other specified areas of land that it owned, but on which it anticipated that the appellants might well set up camp once they left Hethfelton.

[5] The Court of Appeal granted an order for possession in respect both of the land at Hethfelton and of the other parcels of land situated some distance away. With regard to the competency of granting an extended order of this kind, the court was bound by the decision in *Drury*. The central issue in the present appeal is whether that case was rightly decided. In my view, it was not.

[6] Most basically, an action for recovery of land presupposes that
the claimant is not in possession of the relevant land: the defendant L
is in possession without the claimant's permission. This remains the position even if, as the Court of Appeal held in *Manchester Airport plc* v *Dutton* [2000] 1 QB 133†, the claimant no longer needs to have an estate in the land. See *Megarry & Wade: The Law of Real Property* (7th ed) 2008, in para 4-026. To use the old terminology, the defendant has ejected the claimant from the land; the claimant says that he has a better right to possess it and he wants to recover possession. That is reflected in the form of the order that the court grants: "that the claimant do forthwith recover" the land, or, more fully, "that the said AB do recover against the said CD possession" of the land: see
Cole on Ejectment (1857), at p786, form 262. The fuller version has M
the advantage of showing that the court's order is not *in rem*; it is *in personam*, directed against and binding only the defendant. Of course, if the defendant refuses to leave and the court grants a writ of possession requiring the bailiff to put the claimant into possession, in principle the bailiff will remove all those who are on the relevant land irrespective of whether or not they were parties to the action: *R* v *Wandsworth County Court, ex parte Wandsworth London Borough Council* [1975] 1 WLR 1314. So, in that way, non-parties are affected.

* Editor's note: Reported at [2009] 1 WLR 828

† Editor's note: Also reported at [1999] 1 EGLR 147

However, if anyone on the land has a better right than the claimant to possession, he can apply to the court for leave to defend. If he proves his case, he will be put into possession in preference to the claimant. However, the original order for possession will continue to bind the original defendant: see Stamp J's lucid account of the law in *Re Wykeham Terrace, Brighton* [1971] Ch 204, at pp209D-210B.

[7] *Wykeham Terrace* and *Manchester Corporation* showed the need for some reform of the procedures used in actions for the recovery of land. The twin problems of unidentifiable defendants and the lack of any facility for granting an interim order for possession were tackled by a new Order 113, the provisions of which, with some alteration of the details, have been incorporated into the current CPR 55. In the present case, no issue arises concerning the wording of CPR 55. However, I would certainly not interpret "occupied" in CPR 55.1(b) as preventing the use of the special procedure in a case such as *University of Essex* v *Djemal* [1980] 1 WLR 1301, where some protesters were excluding the university from one part of its campus but many students and members of staff were legitimately occupying other parts.

[8] The intention behind the relevant provisions of CPR 55 remains the same as with Order 113: to provide a special, fast procedure in cases that involve only trespassers and to allow the use of that procedure even when some or all of the trespassers cannot be identified. These important, but limited, changes in the rules cannot have been intended, however, to go further and to alter the essential nature of the action itself: it remains an action for recovery of possession of land from people who are in wrongful possession of it. I should add that, in the present case, the defendants do not dispute that they are — or, at least, were at the relevant time — in possession, rather than mere occupation, of the Commission's land at Hethfelton. Wonnacott, *Possession of Land* (2006), at p27, points out that defendants rarely dispute this. However, here, in any event, the appellants' possession is borne out by their offer to co-operate to allow the Commission's ordinary activities on the land not to be disrupted. This is inconsistent with the Commission being in possession. So, the preconditions for an action for recovery of land are satisfied.

[9] By contrast, the Commission was at all relevant times in undisturbed possession of the parcels of land listed in the schedule to the Court of Appeal's order. That being so, an action for the recovery of possession of those parcels of land is quite inappropriate. The only authority cited by the Court of Appeal in *Drury* for granting such an order was the decision of Saville J in *Ministry of Agriculture, Fisheries and Food* v *Heyman* (1990) 59 P&CR 48. However, in that case, the defendant trespassers were not represented and so the point was not fully argued.

[10] Saville J referred to the decision of the Court of Appeal in *University of Essex*, which I have just mentioned. That decision is clearly distinguishable, however. The defendant students, who had previously taken over and been removed from certain administrative offices of the University of Essex, had been occupying another part of the university buildings known as level 6. The Court of Appeal made an order for possession extending to the whole property of the university — in effect, the entire campus. This was justified because the university's right to possession of its campus was indivisible: "If it is violated by adverse occupation of any part of the premises, that violation affects the right of possession of the whole of the premises": p1305C-D, *per* Shaw LJ. In *Heyman*, by contrast, the ministry's right to possession of its land at Grovely Woods was not violated in any way by the trespassers' adverse possession of its other land two or three miles away at Hare Wood. In my view, *Heyman* was wrongly decided and did not form a legitimate basis for the Court of Appeal's decision in *Drury*.

[11] Mummery LJ described Wilson J's approach in *Drury* as "pragmatic": see p1916, in [35]. Further, of course, the common law does evolve by making pragmatic incremental developments. However, if they are to work, they must be consistent with basic principle and they must make sense.

[12] I would not put undue emphasis on the supposed practical difficulties in providing for adequate service by attaching notices to stakes, etc, on these remoter areas of land. Doubtless, adequate arrangements could be worked out if extended orders were otherwise desirable. The real objection is that the Court of Appeal's extended order that "the [Commission] do recover the parcels of land set out in the Schedule hereto" is inconsistent with the fundamental nature of an action for recovering land because there is nothing to recover: the Commission was in undisturbed possession of those parcels of land. Moreover, the law is harmed rather than improved if a court grants orders that lay defendants, knowing the facts, would rightly find incomprehensible. How, the appellants could well ask, can the Commission "recover" parcels of land that it already possesses? How, too, are the appellants supposed to comply with the order? Only a lawyer could understand and explain that the order "really" means that they are not to enter and take over possession of the other parcels of Commission land. This is, of course, what the injunction already says, in somewhat old-fashioned, but tolerably clear language.

[13] Doubtless, the wording could in theory be altered, but this would really be to change the nature of the action and turn the order into an injunction, so creating parallel injunctions, one leading to the possible intervention of the bailiff and the other not.

[14] The claimed justification for granting an extended order for possession of this kind is indeed that it is the only effective remedy against travellers such as the present appellants, since it can ultimately lead to them being removed by a bailiff under a warrant for possession. Moreover, unless the Commission can obtain an extended order, it will be forced to come back to court for a new order each time the appellants move to another of its properties. An injunction is said to be a much weaker remedy in a case such as the present since, if the appellants fail to comply with it, all that can be done is to seek an order for their sequestration or committal to prison. Sequestration is an empty threat, the argument continues, against people who have few assets, while committal to prison might well be inappropriate in the case of defendants who are women with young children.

[15] Plainly, the idea of the Commission having to return to court time and again to obtain a fresh order for possession in respect of a series of new sites is unattractive. However, the scenario presupposes that the appellants would, with impunity, disobey the injunction restraining them from entering the other parcels of land. So, this point is linked to the contention that the injunction would not work.

[16] I note in passing that there is actually no evidence that these appellants would fail to comply with the injunction in respect of the other parcels of land. So there is no particular reason to suppose that the Court of Appeal's injunction will prove an ineffective remedy in this case. On the more general point concerning the alleged ineffectiveness of injunctions in cases of this kind, *South Bucks District Council* v *Porter* [2003] UKHL 26; [2003] 2 AC 558 is of some interest. There, the council wanted to obtain an injunction against gypsies living in caravans in breach of planning controls because an injunction was thought to be a potentially more effective weapon than the various enforcement procedures under the planning legislation. This is in line with the thinking behind the application for an injunction in *South Cambridgeshire*, which I mentioned in [2].

[17] Admittedly, if the present appellants did fail to comply with the injunction, sequestration would not be a real option because they are unlikely to have any substantial assets. Moreover, of course, there are potential difficulties in a court trying to ensure compliance with an injunction by committing to prison defendants who are women with young children. Nevertheless, as Lord Bingham of Cornhill observed in *South Bucks* in [32], in connection with a possible injunction against gypsies living in caravans in breach of planning controls:

> When granting an injunction the court does not contemplate that it will be disobeyed… Apprehension that a party may disobey an order should not deter the court from making an order otherwise appropriate: there is not one law for the law-abiding and another for the lawless and truculent.

Taking that approach, we should, in my view, be slow to assume that an injunction is a worthless remedy in a case such as the present and that only the intervention of a bailiff is likely to be effective. If that is indeed the considered consensus of those with experience in the field,

A consideration may have to be given to changing the procedures for enforcing injunctions of this kind.

[18] However, any such reform would raise far-reaching issues that are not for this court. In particular, travellers are by no means the only people without means whose unlawful activities the courts seek to restrain by injunction and where the assistance of a bailiff might be attractive to claimants. Especially when parliament has intervened from time to time to regulate the way in which the courts should treat travellers, the need for caution in creating new remedies is obvious. At the very least, the matter is one for the Master of the Rolls and the Rules Council, who have the leisure and facilities to consider B the issues.

[19] For these reasons, I would allow the appellants' appeal to the extent proposed by Lord Neuberger.

Giving the second opinion, **Lord Walker** said:

[20] I agree with all the other members of the court that this appeal should be allowed to the extent of setting aside the wider possession order. In *Drury* v *Secretary of State for the Environment, Food and Rural Affairs* [2004] EWCA Civ 200; [2004] 1 WLR 1906, the Court of Appeal went too far in trying to achieve a practical solution. The decision cannot be seen as simply an extension of *University of Essex* C v *Djemal* [1980] 1 WLR 1301, in which the facts were very different. I respectfully agree with the observations on injunctive relief made by Lord Rodger at the end of his judgment.

Giving the third opinion, **Lady Hale** said:

[21] Two questions are before us. First, can the court grant a possession order in respect of land, no part of which is yet occupied by the defendant, because of the fear that she will do so if ejected from land that she does occupy? Second, should the court grant an injunction against that feared trespass? The Court of Appeal unanimously answered the first question in the affirmative, following the reasoning D of that court in *Drury* v *Secretary of State for the Environment, Food and Rural Affairs* [2004] EWCA Civ 200; [2004] 1 WLR 1906 and the decision of Saville J in *Ministry of Agriculture, Fisheries and Food* v *Heyman* (1990) 59 P&CR 48. The majority also answered the second question in the affirmative; Wilson LJ dissented, but only because he thought the wider possession order a sufficient remedy in the circumstances.

[22] The approach in *Drury* and *Heyman* was rightly described by Mummery LJ in *Drury* as "pragmatic" ([35]), depending as it did on the comparative efficacy of possession orders and injunctions. A possession order gives the claimant the right to call on the bailiffs E or the sheriff physically to remove the trespassers from its land, which is what it wants. An injunction can be enforced by imposing penalties only on those who disobey. Mummery LJ considered it a "legitimate, incremental development" of the ruling of the Court of Appeal in *University of Essex* v *Djemal* [1980] 1 WLR 1301 that a possession order can cover a greater area of the claimant's land than that actually occupied by the trespassers.

[23] The situation in *University of Essex* was very like the situation in this and no doubt many other cases. The University of Essex consists (mainly) of some less than beautiful buildings erected in the 1960s on a beautiful campus at Wivenhoe Park, near Colchester. The students F had occupied a small part of the university buildings. The university wanted an order covering the entirety of the university premises. The judge had given them an order covering only the part actually occupied by the students. The Court of Appeal made the wider order sought by the university, holding that there was jurisdiction to cover "the whole of the owner's property in respect of which *his right of occupation has been interfered with*": *per* Buckley LJ, at p1304E, emphasis supplied. Shaw LJ reasoned that the right of the university to possession of the site and buildings was "indivisible. If it is violated by adverse occupation of any part of the premises, that violation affects the *right of possession* of the whole of the premises": p1305D, emphasis supplied. These were extempore judgments in a case in which the students had already decided to call off their direct action, but it will be noted that Buckley LJ spoke of interference with a right of occupation, G while Shaw LJ spoke of violation of a right of possession.

[24] The appellants in this case are occupying part only of Hethfelton Wood. We can, I think, assume that the Commission is occupying the remainder. It is carrying on its forestry work as best it can; indeed, one of its problems is that it is impeded from doing so because of the risk of harm to the vehicles and their occupants. Yet Mr Drabble, for the appellants, has never resisted an order covering the whole of Hethfelton Wood, nor does he invite us to disagree with *University of Essex*. Being a sensible man, he recognised that we would be disinclined to hold that if trespassers set up camp in a large garden the householder could obtain an order enabling them to be physically removed only from that part of H the garden that they have occupied, even if it is clear that they will then simply move their tents to another part of the garden.

[25] The questions raised by this case and *University of Essex* should be seen as questions of principle rather than of pragmatism or procedure. Still less should they be answered by reference to the forms of action that were supposedly abolished in 1876. The underlying principle is *ubi ius, ibi remedium*: where there is a right, there should be a remedy to fit the right. The fact that "this has never been done before" is no deterrent to the principled development of the remedy to fit the right, provided that there is proper procedural protection for those against whom the remedy may be granted. So the questions are: what is the right to be J protected and what is the appropriate remedy to fit it?

[26] If we were approaching this case afresh, without the benefit and burden of history, we might think that the right to be protected is the right to the physical occupation of tangible land. A remedy should be available against anyone who does not have that right and is interfering with it by occupying the land. That remedy should provide for the physical removal of the interlopers if need be. The scope of the remedy actually granted in any individual case should depend on the scope of the right, the extent of the actual and threatened interference with it and the adequacy of the procedural safeguards available to those at risk of physical removal. K

[27] In considering the nature and scope of any judicial remedy, the parallel existence of a right of self-help against trespassers must not be forgotten because the rights protected by self-help should mirror the rights that can be protected by judicial order, even if the scope of self-help has been curtailed by statute. No civil wrong is done by turning out a trespasser using no more force than is reasonably necessary: see *Hemmings* v *Stoke Poges Golf Club* [1920] 1 KB 720. In *Cole on Ejectment* (London, Sweet, 1857), a comprehensive textbook written after the Common Law Procedure Act 1852, there is considerable discussion (in Chapter VII) of the comparative merits L of self-help and ejectment. Any person with a right to enter and take possession of land might choose simply to do that rather than to sue in ejectment. However, this was not advised where the right of entry was not clear and beyond doubt or where resistance was to be expected. The effect of the criminal statutes against forceable entry was "by no means clear": whether no force at all or only reasonable force might be used against the trespasser. Cole was not as sanguine as was Lord Denning MR in *McPhail* v *Persons, Names Unknown* [1973] Ch 447, at p456. Lord Denning took the view that the statutes against forceable entry did not apply to the use of reasonable force against trespassers. Those statutes have now been replaced by section 6 of the Criminal Law M Act 1977. This prohibits the use or threat of violence against person or property for the purpose of securing entry to any premises without lawful excuse. However, it also provides that a right to possession or occupation of the premises is no excuse, although there is now an exception for a "displaced residential occupier" or "protected intending occupier". This does not include the Commission, although it is not impossible that it would be able to evict the travellers without offending against the criminal law. However, in any event, the use of self-help, even if it can be lawfully achieved, is not encouraged because of the risk of disorder that it may entail.

[28] Lord Denning considered that the statutes of forceable entry did not apply because the trespassing squatters in *McPhail* were not

A in possession of the land. He quoted *Pollock on Torts* (15th ed) 1951, at p292:

> A trespasser may in any case be turned off land before he has gained possession, and he does not gain possession until there has been something like acquiescence in the physical fact of his occupation on the part of the rightful owner.

A trespasser who merely interferes with the right to possession or occupation of the property may also be ejected with the use of reasonable force: one does not need to go to court, or even call the police, to eject a burglar or a poacher from one's property.

B [29] Although *Cole* contemplated that self-help might be used against a tenant that had wrongfully continued in occupation after the end of its tenancy, tenants are clearly now in a different position from squatters. Lord Denning thought that the statutes of forceable entry did apply to protect them (although *Cole* says that the authorities on which he relied had later been overruled). Most, but not all, residential tenants are now protected by statute against eviction otherwise than by court order. This is a complicated area that need not concern us now since we are dealing with people who have never been granted any right to be where they are.

[30] However, Lord Denning's basic point is important here. C "In a civilised society, the courts should themselves provide a remedy which is speedy and effective: and thus make self-help unnecessary": *McPhail*, at p457C. It seems clear that the right of self-help has never been limited to those who have actually been dispossessed of their land: in fact, on one view, it is limited to those who have not been so dispossessed. There is no reason in principle why the remedy of physical removal from the land should be available only to those who have been completely dispossessed. It should not depend on the niceties of whether the person wrongfully present on the land was or was not in "possession" in whatever legal sense the word is being used. Were the students in *University of Essex* in possession of the university's D premises at all? Lord Denning, supported by Sir Frederick Pollock, would not think so: see *McPhail*, at p456F. Were these new travellers in possession of Hethfelton Wood at all? Again, Lord Denning would not think so. They had parked their vehicles there, but the work of the Commission was going on around them as best it could.

[31] If we accept that the remedy should be available to a person whose possession or occupation has been interfered with by the trespassers, as well as to a person who has been totally dispossessed, a case such as *University of Essex* becomes completely understandable, as does the order for possession of the whole of Hethfelton Wood in this case. Nor need we be troubled by the form of the order, that the claimant E "recover" the land. His occupation of the whole has been interfered with and he may recover his full control of the whole from those who are interfering with it.

[32] As is obvious from the above, a great deal of confusion is caused by the different meanings of the word "possession" and its overlap with occupation. As Mark Wonnacott pointed out in his interesting monograph, *Possession of Land*, Cambridge University Press, 2006, at p1, the term "possession" is used in three quite distinct senses in English land law: "first, in its proper, technical sense, as a description of the relationship between a person and an estate in land; secondly, in its vulgar sense of physical occupation of tangible land" (the third sense F need not concern us here). Possession, in its first sense, he divides into a relationship of right — the right to the legal estate in question — and a relationship of fact — the actual enjoyment of the legal estate in question; a person might have the one without the other. Possession of a legal estate in fact may often overlap with actual occupation of tangible land, but they are conceptually distinct: a person may be in possession of the headlease if he collects rents from the subtenants, but he will not be in physical occupation of tangible land.

[33] The modern action for the possession of land is the successor to the common law action of ejectment (and some statutory remedies developed for use in the county and magistrates' courts in the 19th century). The ejectment in question was not the ejectment sought by the action but the wrongful ejectment of the right holder. Its origins lay in the writ of trespass, an action for compensatory damages rather G than the recovery of the estate. However, the common law action to recover the estate was available only to freeholders and not to term-holders (tenants). So, the judges decided that this form of trespass could be used by tenants to recover their terms. Trespass was a more efficient form of action than the medieval real actions, such as novel disseisin, so this put tenants in a better position than freeholders. As is well known, the device of involving real people as notional lessees and ejectors was used to enable freeholders to sue the real ejectors. These were then replaced by the fictional characters John Doe and Richard Roe. Eventually, the medieval remedies were (mostly) abolished by the Real Property Limitation Act 1833; the H fictional characters of John Doe and Richard Roe by the Common Law Procedure Act 1852; and the forms of action themselves by the Judicature Acts 1873-75: see AWB Simpson, *A History of the Land Law*, Oxford, Clarendon Press (2nd ed) 1986, Chapter VII.

[34] The question for us is whether the remedy of a possession action should be limited to deciding disputes about "possession" in the technical sense described by Wonnacott. The discussion in *Cole on Ejectment* concentrates on disputes between two persons, both claiming the right to possession of the land, one in occupation and the other not. Often, these are between landlords and tenants who have remained in possession when the landlord thinks their time is up. However, it is clear J that, in reality, what was being protected by the action was the right to physical occupation of the land, not the right to possession of a legal estate in land. The head lessee, who was merely collecting the rents, would not be able to bring an action that would result in his gaining physical occupation of the land unless he was entitled to it.

[35] It seems clear that the modern possession action is there to protect the right to physical occupation of land against those who are wrongfully interfering with it. The right protected, to the physical occupation of the land, and the remedy available, the removal of those who are wrongfully there, should match one another. The action for possession of land has evolved out of ejectment, which itself evolved K out of the action for trespass. There is nothing in CPR 55 that is inconsistent with this view, far from it. The distinction is drawn between a "possession claim", which is a claim for the recovery of *possession of land* (CPR 55.1(a)), and a "possession claim against trespassers", which is a claim for the *recovery of land* that the claimant alleges is "occupied only by a person or persons who entered or remained on the land without the consent of a person entitled to possession of that land…". The object is to distinguish between the procedures to be used where a tenant remains in occupation after the end of its tenancy and the procedures to be used where there are squatters or others who have never been given permission to enter or remain on the land. That, to my L mind, is the reason for inserting "only": not to exclude the possibility that the person taking action to enforce his right to occupy is also in occupation of it. There is then provision for taking action against "persons unknown". However, the remedy in each case is the same: an order for physical removal from the land.

[36] It was held in *R* v *Wandsworth County Court, ex parte Wandsworth London Borough Council* [1975] 1 WLR 1314, that a bailiff executing a possession warrant is entitled to evict anyone found on the premises whether or not they were party to the judgment. However, there is nothing to prevent the order from distinguishing between those who are and those who are not lawfully there, provided M that some means is specified of identifying them. No one would suggest that an order for possession of Hethfelton Wood would allow the removal of Commission workers or picnickers who happened to be there when the bailiffs went in. In principle, court orders should be tailored to fit the facts and the rights they are enforcing, rather than the other way round.

[37] This does not, however, solve the principal question before us. What is the extent of the premises to which the order may relate? As Mummery LJ suggested in *Drury*, in [31], the origin was in an action to recover a term of years. The land covered by the term would be defined in the grant. It would not extend to all the land anywhere in the lawful possession of the claimant. Equally, however, as discussed

A earlier, the remedy can be granted in respect of land to which the claimant is entitled even though the trespasser is not technically in possession of it. This suggests that the scope may be wider than the actual physical space occupied by the trespasser, who may well move about from time to time. In any event, the usual rule is that possession of part is possession of the whole, thus begging the question of how far the "whole" may extend. It was suggested during argument that it might extend to all the land in the same title at the Land Registry. This could be seen as the modern equivalent of the "estate" from which the claimant had been unlawfully ousted. However, this is artificial when a single parcel of land may well be a combination of several different
B registered titles.

[38] The main objection to extending the order to land some distance away from the parcel that has actually been intruded on is one of natural justice. Before any coercive order is made, the person against whom it is made must have an opportunity of contesting it, unless there is an emergency. In the case of named defendants, such as the appellants here, this need not be an obstacle. They have the opportunity of coming to court to contest the order, both in principle and in scope. The difficulty lies with "persons unknown". They are brought into the action by the process of serving notice not on individuals but on the land. If it were to be possible to enforce the physical removal of "persons unknown" from
C land on which they had not yet trespassed when the order was made, notice would also have to be given on that land. That might be thought an evolution too far. Whatever else a possession order may be or have been, it has always been a remedy for a present wrongful interference with the right to occupy. There is an intrusion and the person intruded on has the right to throw out the intruder.

[39] Thus, although I would translate the modern remedy into modern terms designed to match the remedy to the rights protected, and would certainly not put too much weight on the word "recover", I would hesitate to apply it to quite separate land that has not yet been intruded on. The more natural remedy would be an injunction against
D that intrusion, and I would not be unduly hesitant in granting that. We should assume that people will obey the law and, in particular, the targeted orders of the court, rather than that they will not. We should not be too ready to speculate about the enforcement measures that might or might not be appropriate if it is broken. However, the main purpose of an injunction would be to support a very speedy possession order, with severely abridged time limits if it is broken.

[40] However, I would not see these procedural obstacles as necessarily precluding the "incremental development" that was sanctioned in *Drury*. Provided that an order can be specifically tailored against known individuals who have already intruded on the claimant's
E land, are threatening to do so again and have been given a proper opportunity to contest the order, I see no reason in principle why it should not be so developed. It would be helpful if the rules provided for it, so that the procedures could be properly thought through and the forms of order properly tailored to the facts of the case. The main problem at the moment is the "scattergun" form of the usual order (although it is not one prescribed by the rules).

[41] It is for that reason, and that reason alone, that I would allow this appeal to the extent of setting aside the wider possession order made in the Court of Appeal.

Giving the fourth opinion, **Lord Neuberger** said:

F [42] There is an acute shortage of sites in this country to satisfy the needs of travellers, people who prefer a nomadic way of life. Thus, in the county in which the travellers in this case pitched their camp, Dorset, it has been estimated that more than 400 additional pitches are required. The inevitable consequence is that travellers establish their camps on land that they are not entitled to occupy, normally as trespassers and almost always in breach of planning control. Proceedings seeking to prevent their occupation have led to human rights issues being raised before domestic courts: for instance, in the House of Lords, *Doherty* v *Birmingham City Council* [2008] UKHL 57* and before the European Court of Human Rights in *Connors* v *United Kingdom* 66746/01 (2005) G
40 EHRR 9. The present appeal, however, raises issues of purely domestic law, namely the permissible physical ambit of any possession order made against trespassing travellers and the appropriateness of granting an injunction against them.

Facts and procedural history

[43] Travellers often set up their camps in wooded areas. Many woods and forests in this country are managed by the Forestry Commission (the Commission) and owned by the Secretary of State for the Environment, Food and Rural Affairs. The functions of the Commission are "promoting the interests of forestry, the development H
of afforestation and the production and supply of timber and other forest products…": section 1 of the Forestry Act 1967. The Commission runs its woods and forests commercially, although it affords members of the public relatively free and unrestricted access to such areas.

[44] All undeveloped land in the UK is susceptible to unauthorised occupation by travellers and much of such land is vested in public bodies. However, land managed by the Commission is particularly vulnerable to incursion by travellers. As the recorder who heard this case at first instance said, "[g]iven the public access that it affords to its land and its needs for access for forestry vehicles, it is not protected and barricaded in the same way as much of the other land in private and J
local authority ownership in Dorset is now protected".

[45] In 2004, the Office of the Deputy Prime Minister issued its *Guidance on Managing Unauthorised Camping* (the 2004 guidance). This suggests that local authorities and other public bodies distinguish between unauthorised encampment locations that are "unacceptable" (for instance, because they involve traffic hazard or public health risks) and those that are "acceptable". It further recommends that the "management of unauthorised camping must be integrated" and states that "each encampment location must be considered on its merits". The 2004 guidance also indicates that specified welfare enquiries should be undertaken in respect of the travellers and their families in any K
unauthorised encampment before any decision is made as to whether to bring proceedings to evict them. The Secretary of State has accepted throughout these proceedings that the Commission should comply with the terms of the 2004 guidelines before possession proceedings are brought against any travellers on land that it manages, and that failure to do so may invalidate such proceedings.

[46] One of the woods managed by the Commission is Hethfelton Wood (Hethfelton), near Wool, where, at the end of January 2007, a number of new travellers established an unauthorised camp. After the Commission had carried out the enquiries recommended by the 2004 guidance, the Secretary of State issued the current proceedings, L
a possession claim against trespassers within CPR 55.1(b) and an application for an injunction, in Poole County Court, on 13 February 2007. The original defendants were Ms Natalie Meier, Mr Robert and Mrs Georgie Laidlaw, Ms Sharon Horie and "Persons Names Unknown". Ms Meier travels and lives in a vehicle with her two children, having done so since 2002. Mr Laidlaw sadly died before the hearing and, unsurprisingly in the circumstances, Mrs Laidlaw appears to have played no part in the proceedings. Ms Horie has pursued a nomadic way of life since around 1982 and lives in vehicles together with her three children. Ms Lesley Rand (who has been a traveller since around 1996 and lives together with her severely disabled nine-year-old M
daughter in a specially adapted vehicle) and Ms Kirsty Salter (who was pregnant at the time and has been a traveller for 10 years) were subsequently added as defendants.

[47] Two of the appellants had previously been encamped on another area of woodland some five miles from Hethfelton, called Moreton Plantation (Moreton), which was also managed by the Commission. Following the issue of possession proceedings in respect of Moreton, a compromise was agreed on 9 January 2007, which provided that the Secretary of State should recover possession on 29 January 2007. It was on that day that a number of the appellants moved from Moreton to Hethfelton. Some of the other appellants had previously occupied another wood managed by the Commission, Morden Heath (Morden),

* Editor's note: Reported at [2008] 3 WLR 636

which had also been subject to proceedings brought by the Secretary of State, which had resulted in a possession order that was due to be executed on 5 February 2007. In anticipation of the execution of that order, those other appellants moved from Morden to Hethfelton.

[48] In the claim form in the instant proceedings, the Secretary of State sought possession not only of Hethfelton but also of "all that land described on the attached schedule all in the county of Dorset". That schedule set out more than 50 separate woods that were owned by the Secretary of State and managed by the Commission and were marked on an attached plan. The number of woods of which possession was sought in addition to Hethfelton was subsequently reduced to 13, and the plan showed that those 13 woods (the other woods) were spread over an area of Dorset around 25 miles east to west and 10 miles north to south. In the injunction application, the Secretary of State sought an order against the same appellants (including "Persons Names Unknown") restraining them "from re-entering [Hethfelton] or from entering [the other woods]". Copies of the claim form seeking possession were served on the named appellants and at Hethfelton in accordance with the provisions of CPR 55.6, together with copies of the injunction application.

[49] The evidence established that all the occupiers of the camp at Hethfelton were new travellers living and travelling in motor vehicles, mostly with children and often with animals. The evidence also indicated that the camp was relatively tidy and did not involve any anti-social conduct on the part of any of the occupants. However, the presence of children and animals caused the Commission to avoid the use of heavy plant or the carrying out of substantial work, which might otherwise have occurred, in the surrounding area. The Commission's evidence showed that other areas in Dorset managed by the Commission in addition to Hethfelton, including Moreton and Morden, had been occupied by travellers as unauthorised camps, sometimes by one or more of the named appellants.

[50] The claim came before Mr Recorder Norman, who gave a full and careful judgment on 3 August 2007. He had to resolve three issues. The first was whether to grant an order for possession against the defendants in respect of Hethfelton. The second was whether to grant an order for possession in respect of any or all of the other woods. The third was whether to grant an injunction restraining the defendants from entering onto all or any of the other woods.

[51] The recorder decided to grant an order for possession against the appellants in respect of Hethfelton. However, he refused to make any wider order for possession or to grant the injunction sought by the Secretary of State. Although he accepted that he had jurisdiction to make such orders, he considered it inappropriate to do so primarily because the Commission had failed to consider the matters suggested by the 2004 guidance before the current proceedings were begun and because it was not prepared to assure the recorder that consideration would be given to that guidance before any wider order for possession or any injunction was enforced. Paragraph 1 of the order that was drawn up to reflect this decision provided that "[t]he claimant do forthwith recover the land known as Hethfelton Wood".

[52] The appellants did not appeal against this order for possession. However, the Secretary of State appealed against the recorder's refusal to grant an order for possession in respect of the other woods (which I will refer to as a "wider order for possession") and the injunction, and the Court of Appeal allowed the appeal: [2008] EWCA Civ 903; [2009] 1 WLR 828. The order made by the Court of Appeal ordered that the Secretary of State "do recover" the other woods and that each of the appellants "be restrained from entering upon, trespassing upon, living on, or occupying" any of the other woods.

[53] In her judgment, Arden LJ followed and applied the reasoning of the Court of Appeal in the earlier decision of *Drury* v *Secretary of State* [2004] EWCA Civ 200; [2004] 1 WLR 1906, under which it had been held that an order for possession, at least when made pursuant to a possession claim against trespassers, could, in appropriate cases, extend to land not forming part of or contiguous with or even near the land actually occupied by the trespassers. She concluded that the evidence demonstrated that at least some of the appellants had set up unauthorised encampments on woods managed by the Commission in Dorset and that there was a substantial risk that at least some of the appellants would move onto other such woods once an order for possession was made in respect of Hethfelton.

[54] Arden LJ also said, in disagreement with the recorder, that any failure on the part of the Commission to consider the matters recommended by the 2004 guidance before issuing the proceedings for possession of the other woods did not justify refusing to make such a wider order. This was essentially on the basis that if there were any such failure, it could be considered at the time the wider order for possession was sought to be enforced. Pill and Wilson LJJ agreed. Arden LJ also considered that, for the same reasons, the recorder had been wrong to refuse the injunction sought by the Secretary of State, and again Pill LJ agreed. However, Wilson LJ dissented on this point on the ground that the recorder had been entitled to refuse an injunction on the additional ground that he had mentioned, namely that, had he made a wider order for possession, it would have been disproportionate to grant an injunction as well.

[55] The instant appeal is brought by Ms Horie and Ms Rand, and it raises two principal issues. The first is the extent to which an order for possession can be made in favour of a claimant in respect of land not actually occupied by a defendant. The second concerns the circumstances in which an injunction restraining future trespass can and should be granted; this raises two points: (a) whether an injunction against travellers is generally appropriate; and (b) the point on which the Court of Appeal differed from the recorder, namely the effect of the 2004 guidance. I will consider these two issues in turn and then briefly review the implications of my conclusions.

An order for possession of land not occupied by the defendants

[56] In *Drury*, the facts were similar to those here, except that the Court of Appeal held that there was no evidence establishing that the travellers in that case had occupied, or threatened to occupy, other property managed by the Commission. Accordingly, the order for possession was in the normal form, limited, as with the order made by the recorder in this case, to the wood occupied by the travellers. However, the Court of Appeal decided that an order for possession could be granted, not merely in respect of land that the defendant occupied but also in respect of other land that was owned by the claimant and which the defendant threatened to occupy.

[57] The essence of the Court of Appeal's reasoning was that: (a) the law recognises that an anticipated trespass can give rise to a right of action; (b) an injunction would be of limited, if any, real use; (c) in those circumstances, the law should provide another remedy; (d) a wider order for possession would be of much more practical value than an injunction; (e) such an order for possession was justified by previous authority and in the light of the court's jurisdiction to grant *quia timet* injunctions; and (f) accordingly, such an order could be made; but (g) it should be made only in relatively exceptional circumstances: see in [20] to [24], [34] to [36] and [42] to [46], *per* Wilson J, Mummery LJ and Ward LJ respectively.

[58] Particularly with the advent of the Civil Procedure Rules, it is clear that judges should strive to ensure that court procedures are efficacious, and that where there is a threatened or actual wrong there should be an effective remedy to prevent or remedy it. Further, as Lady Hale points out, so long as landowners are entitled to evict trespassers physically, judges should ensure that the more attractive and civilised option of court proceedings is as quick and efficacious as legally possible. Accordingly, the Court of Appeal was plainly right to seek to identify an effective remedy for the problem faced by the Commission as a result of unauthorised encampments, namely that when a possession order is made in respect of one wood the travellers simply move on to another wood, requiring the Commission to incur the cost, effort and delay of bringing a series or potentially endless series of possession proceedings against the same people.

[59] None the less, however desirable it is to fashion or develop a remedy to meet a particular problem, courts have to act within the law, and their ability to control procedure and achieve justice is not

A unlimited. Judges are not legislators, and there comes a point where, in order to deal with a particular problem, court rules and practice cannot be developed by the courts, but have to be changed by primary or secondary legislation — or, in so far as they can be invoked for that purpose, by practice directions. In my view, it is simply not possible to make the sort of enlarged or wider order for possession that the Court of Appeal made in this case, following (as it was, I think, bound to do) the reasoning in *Drury*.

[60] The power of the county court for present purposes derives from section 21(1) of the County Courts Act 1984 (the 1984 Act), which gives it "jurisdiction to hear and determine any action for the
B recovery of land". The concept of "recovery" of land was the essence of a possession order both before and after the procedure was recast by section 168ff of the Common Law Procedure Act 1852, although, until the Supreme Court of Judicature Act 1875, the action lay in ejectment rather than in recovery of land: see *per* Lord Denning MR in *McPhail* v *Persons, Names Unknown* [1973] Ch 447, at pp457-458. None the less, the change of name did not involve a change of substance, and the essence of an order for possession, whether framed in ejectment or recovery, is that the claimant is getting back the property from the defendant whether by recovering the property from the defendant or because the claimant had been wrongly ejected by the defendant.
C As stated by Wonnacott, in *Possession of Land* (2006), at p22, "an action for recovery of land (ejectment) is an action to be put into possession of an estate in land. The complaint is that the claimant is not currently 'in' possession of it, and... wants... to be put 'in' possession of it.": see also Simpson, *A History of the Land Law* (2nd ed), at pp144-145, and *Gledhill* v *Hunter* (1880) LR 14 Ch D 492, at p496, *per* Sir George Jessel MR.

[61] As Sir George Jessel explained, an action for ejectment and its successor, recovery of land, were normally issued "to recover possession from a tenant" or former tenant. An action against a trespasser who did not actually dispossess the person entitled to possession was based
D on *trespass quare clausum fregit*, physical intrusion onto the land. None the less, where a trespasser exclusively occupies land so as to oust the person entitled to possession, the cause of action must be for recovery of possession. (Hence, if such an action is not brought within 12 years, the ousting trespasser will often have acquired title by "adverse possession".) Accordingly, in cases where a trespasser is actually in possession of land, an action for recovery of land, that is, for possession, is appropriate, as Lord Denning implicitly accepted in *McPhail*, at pp457-458.

[62] This analysis is substantially reflected in the provisions of the CPR and in the currently prescribed form of order for possession.
E CPR 55 is concerned with possession claims, and CPR 55.1 provides:

> (a) "a possession claim" means a claim for the recovery of possession of land (including buildings or parts of buildings);
>
> (b) "a possession claim against trespassers" means a claim for the recovery of land which the claimant alleges is occupied only by a person or persons who entered or remained on the land without the consent of a person entitled to possession of that land but does not include a claim against a tenant or sub-tenant whether his tenancy has been terminated or not;...

The special features of a possession claim against trespassers are that the defendants to the claim may include "persons unknown", such
F proceedings should be served on the land as well as on the named defendants and the minimum period between service and hearing is two days (or five days for residential property), rather than the 28 days for other possession claims: see CPR 55.3(4), 55.6, and 55.5(2) and (3).

[63] The drafting of CPR 55.1 is rather peculiar in that, unlike that in CPR 55.1(a), the definition in CPR 55.1(b) does not include the word "possession". Given that, since 1875, the cause of action has been for the recovery of land, the oddity, as Lord Rodger has pointed out, is the inclusion of the word "possession" in the former paragraph, rather than its exclusion in the latter. However, in so far as the point has any significance, the definition of "a possession claim", as with the definition of "land" in CPR 55.1(a), may well be carried into CPR 55.1(b). In any event, the important point, to my mind, is that a possession claim against trespassers involves the person "entitled to G
possession" seeking "recovery" of the land. Form N26 is the prescribed form of order in both a simple possession claim and a possession claim against trespassers: see CPR 4 PD, table 1. That form orders the defendant to "give the claimant possession" of the land in question. Although the orders at first instance (as drafted by counsel) and in the Court of Appeal direct that the claimant do "recover" the land in question from the defendants, that is the mirror image of ordering that the defendants "give" the claimant possession.

[64] The notion that an order for possession may be sought by a claimant and made against defendants in respect of land that is wholly detached and separated, possibly by many miles, from that occupied H
by the defendants, accordingly seems to me to be difficult, indeed impossible, to justify. The defendants do not occupy or possess such land in any conceivable way and the claimant enjoys uninterrupted possession of it. Equally, the defendants have not ejected the claimant from such land. For the same reasons, it does not make sense to talk about the claimant recovering possession of such land or to order the defendant to deliver up possession of such land.

[65] This does not mean that where trespassers are encamped in part of a wood an order for possession cannot be made against them in respect of the entirety of the wood (at least if there are no other occupants of the wood), just as much as an order for possession may J
extend to an entire house where the defendant is trespassing only in one room (at least if the rest of the house is empty).

[66] However, the fact that an order for possession may be made in respect of the entirety of a piece of property when the defendant is in occupation only of part and the remainder is empty does not appear to me to assist the argument in favour of a wider possession order as made by the Court of Appeal in this case. Self-help is a remedy still available, in principle, to a landowner against trespassers (other than former residential tenants). Where only part of its property is occupied by trespassers, a landowner, exercising that remedy through privately instructed bailiffs, would no doubt be entitled to evict the trespassers K
from the entirety of its property. Similarly, it seems to me, bailiffs (or sheriffs) who are required by a warrant (or writ) of possession to evict defendants from part of a property owned by the claimant would be entitled to remove the defendants from the entirety of that property. However, that does not mean that the bailiffs, whether privately instructed or acting pursuant to a warrant, could restrain the trespassers from moving onto another property, perhaps miles away, owned by the claimant.

[67] Further, the concept of occupying part of property (the remainder of which is vacant) effectively in the name of the whole is well established: see, for example, albeit in a landlord and tenant L
context, *Henderson* v *Squire* (1869) LR 4 QB 170, at p172. However, that concept cannot be extended to apply to land wholly distinct, even miles away, from the occupied land. So, too, the fact that one can treat land as a single entity if it is divided by a road or river (in different ownership from the land) seems to me to be an irrelevance: as a matter of law and fact, the two divisions can sensibly be regarded as a single piece of land. Accordingly, I have no difficulty with the fact that the possession order made at first instance in this case extended to the whole of Hethfelton, even though the appellants occupied only a part of it. M

[68] The position is more problematic where a defendant trespasses on part of land, the rest of which is physically occupied by a third party or even by the landowner. Particular difficulties in this connection are, to my mind, raised in respect of a wide order for possession in a claim within CPR 55.1(b). Such "a claim" may be brought "for the recovery of land which the claimant alleges is occupied only by a person or persons who entered or remained on the land without... consent...". Given that such a claim is limited to "land... occupied only by" trespassers, it is not immediately easy to see how it could be brought, even in part, in respect of land occupied by persons who are not trespassers. Moreover, it is fundamental that the court cannot accord a claimant more relief than he seeks (although it is of course possible,

in appropriate circumstances, for a claimant to amend to increase the extent of his claim, but that is not relevant here).

[69] The Court of Appeal in *University of Essex* v *Djemal* [1980] 1 WLR 1301 none the less decided that a university could be granted a possession order under RSC Order 113 r 1, which was (in respect of the issue in this case) in similar terms to CPR 55.1(b), in respect of its whole campus, against trespassers who were squatting in a relatively small part, even though the remainder of the campus was lawfully occupied by academics, other employees and indeed students. This was a thoroughly practical decision arrived at to deal with a fairly widespread problem at the time, namely student sit-ins. There was an obvious fear that if an order for possession was limited to the rooms occupied by the student trepassers they would simply move to another part of the campus.

[70] As already mentioned, given that there is the alternative remedy of self-help, the court should ensure that its procedures are as effective as lawfully possible. None the less, there is obviously great force in the argument that the fact that areas of the campus in that case were lawfully and exclusively occupied by academic staff, employees and students should have precluded a claim and an order for possession in respect of those areas, both in principle and in the light of the wording of RSC Order 113 r 1.

[71] However, this is not the occasion formally to consider the correctness of the decision in *University of Essex*, which was not put in issue by either of the parties, since the Secretary of State (like the Court of Appeal in *Drury*) relied on it, and the appellants were content to distinguish it. Accordingly, the implications of overruling or explaining the decision, which may be far-reaching in terms of principle and practice, have not been debated or canvassed.

[72] The Court of Appeal's conclusion in *Drury*, that the court could make a wider order for possession such as that in the instant case, rested very much on the reasoning in *University of Essex* and in the subsequent first instance decision of *Ministry of Agriculture, Fisheries and Food* v *Heyman* (1990) 59 P&CR 48, which represented an "incremental development of the ruling in [*University of Essex*]", as Mummery LJ put it in [35]. However, it seems to me that the decision in *Drury* was an illegitimate extension of the reasoning and decision in *University of Essex*. The fact that an order for possession can be made in respect of a single piece of land, only part of which is occupied by trespassers, does not justify the conclusion that an order for possession can be made in respect of two entirely separate pieces of land, only one of which is occupied by trespassers, just because both pieces of land happen to be in common ownership. As already mentioned, bailiffs, whether acting on instructions from a landowner exercising the right of self-help to evict a trespasser or acting pursuant to a warrant of possession, can remove the trespasser on part of a piece of property from the whole of that property, but they cannot prevent him from entering a different property, possibly many miles away. Similarly, while it is acceptable, at least in some circumstances, to treat occupation of part of property as amounting to occupation of the entirety of that property, one cannot treat occupation of one property as amounting to occupation of another entirely separate property, possibly miles away, simply because the two properties are in the same ownership.

[73] Having said all that, I accept that the notion of a wider, effectively precautionary order for possession, as made in *Drury*, has obvious attraction in practice. As the Court of Appeal explained in that case, the alternative to a wider possession order, namely an injunction restraining the defendant from camping in other woods in the area, would be of limited efficacy. An order for possession is normally enforced in the county court by applying for a warrant of possession under CCR Order 26, which involves the occupiers being removed from the land by bailiffs. (The equivalent in the High Court is a writ of possession executed by the sheriff under RSC Order 45 r 3.) This is a procedurally direct and simple method of enforcement. An injunction, however, "may be enforced", and that was treated by the court in *Drury* as meaning "may only be enforced", by sequestration or committal: see RSC Order 45 r 5(1) and, in respect of the county court, CCR 29 and section 38 of the 1984 Act. Given that the claimant's aim is to evict the travellers, those are unsatisfactory remedies compared with applying for a warrant of possession. They are not only indirect but are normally procedurally unwieldy and time-consuming and, in any event, they are of questionable value in cases against travellers, as explained in the next section of this opinion.

[74] There is also some apparent force as a matter of principle in the notion that the courts should be able to grant a precautionary wider order for possession. If judges have developed the concept of an injunction that restrains a defendant from doing something that he has not yet done, but is threatening to do, why, it might be asked, should they now not develop an order for possession that requires a defendant to deliver up possession of land that he has not yet occupied, but is threatening to occupy? The short answer is that a wider or precautionary order for possession, whether in the form granted in this case or in the prescribed form N26, requires a defendant to do something that he cannot do, namely to deliver up possession of land he does not occupy, and purports to return to the claimant something he has not lost, namely possession of land of which he already has possession.

[75] What the respondent is really seeking in the present case is an order that if the appellants go onto the other woods the respondent should be entitled to possession. That is really in the nature of declaratory or injunctive relief: it is not an order for possession. A declaration identifies the parties' rights and obligations. A *quia timet* injunction involves the court forbidding the defendant from doing something that he may do and that he would not be entitled to do. Both those types of relief are different from what the Court of Appeal intended to grant here, namely a contingent order requiring the defendant to do something (to deliver up possession) if he does something else (trespassing) that he may do and that he would not be entitled to do. I describe the Court of Appeal as intending to grant such an order because, as just explained, the actual order is in the form of an immediate order for possession of the other woods, which, as I have mentioned, is also hard to justify, given that the appellants were not in occupation of any part of them.

[76] Further, although it would be beneficial to be able to make a wider possession order because of the relative ease with which it could be enforced in the event of the appellants trespassing on other woods, such an order would not be without its disadvantages and limitations. An order for possession binds only those persons who are parties to the proceedings (and their privies), although the bailiffs (and sheriffs) are obliged to execute a warrant (or writ) of possession against all those in occupation: see *Re Wykeham Terrace, Brighton* [1971] Ch 204, at pp209-210, *R* v *Wandsworth County Court, ex parte Wandsworth London Borough Council* [1975] 1 WLR 1314, at pp1317-1319, *Thompson* v *Elmbridge Borough Council* [1987] 1 WLR 1425, at pp1431-1432, and the full discussion in *Wonnacott, op cit*, at pp146-152. It would therefore be wrong in principle for the court to make a wider order for possession against trespassers (whether or not named) in one wood with a view to its being executed against other trespassers in other woods. None the less, because the warrant must be executed against anyone on the land, there is either a risk of one or more of the occupiers of another wood being evicted without having the benefit of due process or room for delay while such an occupier applies to the court and is heard before a warrant is executed against him.

[77] Quite apart from this, a warrant of possession to execute an order for possession made in the county court in a claim for possession against trespassers can be issued without leave only within three months of the order: CCR Order 24, r 6(2). So, after the expiry of three months, a wider possession order does not obviate the need for the claimant to apply to the court before it can obtain possession of any land the subject of the order. Further, as pointed out by Wilson J in *Drury*, in [22], it seems rather arbitrary that only a person who owns land that is being unlawfully occupied can obtain a wider order for possession protecting all its land in a particular area.

[78] In conclusion on this issue, although there is considerable practical attraction in the notion that the court should be able to make the wide type of possession order that the Court of Appeal made in this case, following *Drury*, I do not consider that the court has such power. It is inconsistent with the nature of a possession order and with the

relevant provisions governing the powers of the court. The reasoning in the case on which it is primarily based, *University of Essex*, cannot sensibly be extended to justify the making of a wider possession order, and there are aspects of such an order that would be unsatisfactory. I should add that I have read what Lord Rodger has to say on this, the main issue, and I agree with him.

Should an injunction be refused as it will probably not be enforced?

[79] That brings me to the question of whether an injunction restraining travellers from trespassing on other land should be granted in circumstances such as the present. Obviously, the decision as to whether to grant an order restraining a person from trespassing will turn very much on the precise facts of the case. None the less, where a trespass to the claimant's property is threatened, and particularly where a trespass is being committed and has been committed in the past by the defendant, an injunction to restrain the threatened trespass would, in the absence of good reasons to the contrary, appear to be appropriate.

[80] However, as Lord Walker said during argument, the court should not normally make orders that it does not intend or will be unable to enforce. In a case such as the present, if the appellants had disobeyed an injunction not to trespass on any of the other woods, it seems highly unlikely that the two methods of enforcement prescribed by CCR 29 and section 38 of the 1984 Act (RSC Order 45 r 5(1) in the High Court) would be invoked. The appellants presumably have no significant assets apart from their means of transport, which are also their homes, so sequestration would be pointless or oppressive. Further, many of the appellants are vulnerable and most of them have young children, so imprisonment may very well be disproportionate. In *South Bucks District Council* v *Porter* [2003] UKHL 26; [2003] 2 AC 558, local planning authorities sought injunctions to restrain gypsies from remaining on land in breach of planning law and, in [32], Lord Bingham of Cornhill said that "[t]he court should ordinarily be slow to make an order which it would not… be willing, if need be, to enforce by imprisonment".

[81] On the other hand, in the same paragraph of his opinion, Lord Bingham also said that "[a]pprehension that a party may disobey an order should not deter the court from making an order otherwise appropriate". A court may consider it unlikely that it would make an order for sequestration or imprisonment if an injunction it was being invited to grant were to be breached, but it may none the less properly decide to grant the injunction. Thus, the court may take the view that the defendants are more likely not to trespass on the claimant's land if an injunction is granted because of their respect for a court order or because of their fear of the repercussions of breaching such an order. Or the court may think that an order of imprisonment for breach, while unlikely, would none the less be a real possibility, or it may think that a suspended order of imprisonment, in the event of breach, may well be a deterrent (although a suspended order should not be made if the court does not anticipate activating the order if the terms of suspension are breached).

[82] It was suggested in argument that if a defendant established an unauthorised camp in a wood that, in earlier proceedings, he had been enjoined from occupying, the court would be likely to be sympathetic to an application by the Commission to abridge even the short time limits in CPR 55.5(2). However, as Lord Rodger observed, if the court were satisfied that a defendant was moving from unauthorised site to unauthorised site on woods managed by the Commission, an abridgement of time limits might be thought to be appropriate anyway. Quite apart from this, if the only reason for granting an injunction restraining a defendant from trespassing in other woods was to assist the Commission in obtaining possession of any of those other woods should the defendant camp in them, it seems to me that this could be catered for by declaratory relief. For instance, the court could grant a declaration that the Commission is in possession of those other woods and the defendant has no right to dispossess it.

[83] In some cases, it may be inappropriate to grant an injunction to restrain a trespassing on land unless the court considers not only that there is a real risk of the defendants so trespassing but also that there is at least a real prospect of enforcing the injunction if it is breached. However, even where there appears to be little prospect of enforcing the injunction by imprisonment or sequestration, it may be appropriate to grant it because the judge considers that the grant of an injunction could have a real deterrent effect on the particular defendants. If the judge considers that some relief would be appropriate only because it could assist the claimant in obtaining possession of such land if the defendants commit the threatened trespass, a declaration would appear to me to be more appropriate than an injunction.

[84] In the present case, neither the recorder nor the Court of Appeal appears to have concluded that an injunction should be refused on the ground that it would not be enforced by imprisonment or because it would have no real value. Although it may well be that a case could have been (and may well have been) developed along those lines, it was not adopted by the recorder, and clearly did not impress the Court of Appeal. In those circumstances, it seems to me that it is not appropriate for this court to set aside the injunction unless satisfied that it was plainly wrong to grant it or that there was an error of principle in the reasoning that led to its grant. It does not appear to me that either of those points has been established in this case.

Effect of the 2004 guidance on the grant of an injunction

[85] The recorder considered that it was inappropriate to grant an injunction in favour of the Secretary of State because the Commission had not complied with the 2004 guidance in respect of the other woods before issuing the proceedings, and would not give an assurance that it would comply with the 2004 guidance before it enforced the injunction. The Court of Appeal considered that the injunction could none the less be granted, since the issue of the commission's compliance with the 2004 guidance could be considered before the injunction was enforced.

[86] As I have already mentioned, it has been conceded by the Secretary of State throughout these proceedings that the Commission is obliged to comply with the 2004 guidance, and that failure to do so may vitiate its right to possession against travellers trespassing on land that it manages. On that basis, there is some initial attraction in the appellants' argument that if the 2004 guidance ought to be complied with before the injunction is enforced it would be inappropriate to grant the injunction before the guidance was complied with. After all, now the injunction has been granted, the appellants would be in contempt of court and prone to imprisonment (once the appropriate procedures had been complied with) if they encamped on any of the other woods.

[87] However, I am of the opinion that the Court of Appeal was right to conclude that, even in the light of the Secretary of State's concession, the 2004 guidance did not present an obstacle to the granting of an injunction in this case. The guidance is concerned with steps to be taken with regard to existing unauthorised encampments: it is not concerned with preventing such encampments from being established in the first place. The recommended procedures in the 2004 guidance were relevant to the question of whether an order for possession should be made against the appellants in respect of their existing encampment in Hethfelton. However, quite apart from the fact that they are merely aspects of a non-statutory code of guidance, those recommendations are not directly relevant to the issue of whether the appellants should be barred from setting up a camp on other land managed by the Commission. Accordingly, I do not see how it could have justified an attack on the lawfulness of the Secretary of State seeking an injunction to restrain the appellants from setting up such unauthorised camps. At least on the basis of the concession to which I have referred, I incline to the view that the existence and provisions of the 2004 guidance could be taken into account by the court when considering whether to grant an injunction and when fashioning the terms of any injunction. However, I prefer to leave the point open because it was, understandably, not much discussed in argument before us.

[88] Even if the 2004 guidance was of relevance to the issue of whether the injunction should be granted, it seems to me that it could not be decisive. Otherwise, it would mean that such an injunction could never be granted because it would not be possible to carry out

up-to-date welfare enquiries in respect of defendants who might not move onto a wood that they were enjoined from occupying for several months or, conceivably, even several years after the order was made. As Arden LJ held, particularly bearing in mind that it purports to be no more than guidance, the effect and purpose of the 2004 guidance is simply not strong enough to displace the Secretary of State's right to seek the assistance of the court to prevent a legal right from being infringed. Further, the fact that welfare enquiries were made in respect of the appellants' occupation of Hethfelton by social services means that the more significant investigations required by the 2004 guidance had been carried out anyway.

[89] Following questions from Lady Hale, it transpired for the first time in these proceedings that, at the time of the issue of the claim, the Commission had (and has) a detailed procedural code that is intended to apply when there are travellers unlawfully on its land, and that this code substantially followed the 2004 guidance. It therefore appears that the Commission has considered the 2004 guidance and promulgated a code that takes its contents into account. On that basis, unless it could be shown in a particular case that the code had been ignored, it appears to me that the Commission's decision to evict travellers could not be unlawful on the ground relied on by the appellants in this case. However, it appears to me that failure to comply with non-statutory guidance would be unlikely to render a decision unlawful, although failure to have regard to the guidance could do so.

[90] Were the appellants to trespass onto land covered by the injunction, the Commission would presumably comply with its code before seeking to enforce the injunction. If it did not do so, then, if justified on the facts of a particular case, there may (at least if the Commission's concession is correct) be room for argument that, in seeking to enforce the injunction against travellers who have set up a camp in breach of an injunction, the Secretary of State was acting unlawfully. It is true that this means that, in a case such as this, a defendant who trespasses in breach of an injunction may be at risk of imprisonment before the Commission has complied with the 2004 guidance. However, where imprisonment is sought and where it would otherwise be a realistic prospect, the defendant could argue at the committal hearing that the injunction should not be enforced, even that it should be discharged, on the ground that the recommendations in the 2004 guidance have not been followed.

[91] Accordingly, on this point, I conclude that even assuming (in accordance with the Secretary of State's concession) that the Commission's failure to comply with the 2004 guidance may deter the court from making an order for possession against travellers, it should not preclude the granting of an injunction to restrain travellers from trespassing on other land. However, at least in a case where it could be shown that the claimant should have considered the 2004 guidance but did not do so, the guidance could conceivably be relevant to the question of whether an injunction should be granted (and, if so, on what terms) and, if the injunction is breached, to the question of whether it should be enforced (and, if so, how). In the event, therefore, the grant of an injunction was appropriate, as Arden and Pill LJ concluded (and the only reason Wilson LJ thought otherwise, namely the existence of the wider possession order, no longer applies).

Implications of this analysis

[92] As I have explained, the thinking of the Court of Appeal in *Drury* proceeded on the basis that an injunction restraining trespass to land could be enforced only by sequestration or imprisonment. In the light of the terms of RSC Order 45, r 5(1), this may very well be right. Certainly, in the light of the contrast between the terms of that rule and the terms of RSC Order 45, r 3(1) and CCR 26, r 16(1) (which respectively provide for writs and warrants of possession only to enforce orders for possession), it is hard to see how a warrant of possession in the county court or a writ of possession in the High Court could be sought by a claimant where such an injunction was breached.

[93] However, where, after the grant of such an injunction (or, indeed, a declaration), a defendant entered onto the land in question, it is, I think, conceivable that, at least in the High Court, the claimant could apply for a writ of restitution, ordering the sheriff or bailiffs to recover possession of the land for the benefit of the claimant. Such a writ is often described as one of the "writs in aid of" other writs, such as a writ of possession or a writ of delivery: see, for instance, RSC Order 46 r 1. Restitution is normally the means of obtaining possession against a defendant (or his privy) who has gone back into possession after having been evicted pursuant to a court order. It appears that it can also be invoked against a claimant who has obtained possession pursuant to a court order that is subsequently set aside (normally on appeal): see sc46.3.3 in *Civil Procedure*, vol 1, 2009. Historically at any rate, a writ of restitution could also be sought against a person who had gone into possession by force: see *Cole on Ejectment* (1857), pp692-694. So, there may be an argument that such a writ may be sought by a claimant against a defendant who has entered onto the land after an injunction has been granted restraining him from doing so, or even after a declaration has been made that the claimant is and the defendant is not entitled to possession. It may also be the case that it is open to the county court to issue a warrant of restitution in such circumstances.

[94] Whether a writ or warrant of restitution would be available to support such an injunction or declaration, and whether the present procedural rules governing the enforcement of injunctions against trespass on facts such as those in the present case are satisfactory, seem to me to be questions that are ripe for consideration by the CPR Committee. The precise ambit of the circumstances in which a writ or warrant of restitution may be sought is somewhat obscure, and could usefully be clarified. Further, if, as I have concluded, it is not open to the court to grant a wider order for possession, as was granted by the Court of Appeal in *Drury* and in this case, it appears likely that there may very well be defects in the procedural powers of the courts of England and Wales. Where a person threatens to trespass on land, an injunction may well be of rather little, if any, real practical value if the person is someone against whom an order for sequestration or imprisonment is unlikely to be made and an order for possession is not one that is open to the court. In addition, it seems to me that it may be worth considering whether the current court rules satisfactorily deal with circumstances such as those that were considered in *University of Essex*.

Disposal of this appeal

[95] Accordingly, it follows that, for my part, I would allow the appellants' appeal to the extent of setting aside the wider possession order made by the Court of Appeal, but dismiss their appeal to the extent of upholding the injunction granted by the Court of Appeal.

Giving the fifth opinion, **Lord Collins** said:

[96] At the end of the argument, my inclination was to the conclusion that, in *Drury* v *Secretary of State for the Environment, Food and Rural Affairs* [2004] EWCA Civ 200; [2004] 1 WLR 1906, the Court of Appeal had legitimately extended *University of Essex* v *Djemal* [1980] 1 WLR 1301 to fashion an exceptional remedy to deal with cases of the present kind. I was particularly impressed by the point that an injunction might be a remedy that was not capable of being employed effectively in cases such as this. However, I am now convinced that there is no legitimate basis for making an order for possession in an action for the recovery of wholly distinct land of which the defendant is not in possession.

[97] However, in my opinion, *University of Essex* represented a sensible and practical solution to the problem faced by the university, and was correctly decided. I agree, in particular, that it can be justified on the basis that the university's right to possession of its campus was indivisible, as Lord Rodger says, or that the remedy is available to a person whose possession or occupation has been interfered with, as Lady Hale puts it. Where the defendant is occupying part of the claimant's premises, the order for possession may extend to the whole of the premises. First, it has been pointed out, rightly, that the courts have used the concept of possession in differing contexts as a functional and relative concept in order to do justice and to effectuate the social purpose of the legal rules in which possession (or, I would

A add, deprivation of possession) is a necessary element: see Harris, "The Concept of Possession in English Law", in *Oxford Essays in Jurisprudence* (ed AG Guest, 1961) 69, at p72. Second, the procedural powers of the court are subject to incremental change in order to adapt to the new circumstances: see, for example, in respect of the power to grant injunctions, *Fourie* v *Le Roux* [2007] UKHL 1; [2007] 1 WLR 320, in [30], and *Masri* v *Consolidated Contractors International Co* B
SAL [2008] EWCA Civ 303; [2009] 2 WLR 621, in [182].

[98] I would therefore allow the appeal to the extent of setting aside the wider possession order.

Appeal allowed in part.

LANDS CHAMBER
LEASEHOLD ENFRANCHISEMENT

Sherwood Hall (East End Road) Management Co Ltd *v* Magnolia Tree Ltd

UPPER TRIBUNAL: LANDS CHAMBER

[2009] UKUT 158 (LC)

10 September 2009

Mr Norman Rose FRICS

Leasehold enfranchisement — Leasehold Reform, Housing and Urban Development Act 1993 — Acquisition of freehold — Premium — Deferment rate — Whether departure from generic rate in* Earl Cadogan *v* Sportelli *justified for 88-year reversion outside prime central London area — Whether additional value to be attributed in respect of potential for development of undeveloped land

The appellant applied to acquire the freehold interest in four blocks of flats in Barnet, London N2, under the Leasehold Reform, Housing and Urban Development Act 1993. The property comprised 36 two-bedroom flats in the four blocks, together with 12 lock-up garages, ancillary gardens and parking areas and a small area of undeveloped land to the rear of the site. The property fell to be valued as at August 2005 for the purposes of assessing the premium payable for the freehold.

The leasehold valuation tribunal (LVT) determined the premium at £248,825. In doing so, it applied a deferment rate of 5%, in accordance with *Earl Cadogan* v *Sportelli* [2007] 1 EGLR 153, in respect of the 35 flats that had unexpired terms of between 47.8 and 87.6 years. It attributed no reversionary value to the remaining flat, the lease of which had 177.6 years' unexpired, but attributed an additional £50,000 to the value of the undeveloped land.

On appeal, the appellant contended for a premium of £119,012. It argued that the LVT should have applied a higher deferment rate of 7% to the flats with an unexpired term of 88 years because *Sportelli* had been concerned with properties in the prime central London area (PCL) with reversions only up to 71 years. It submitted that a higher deferment rate should apply to properties outside the PCL with leases of 75 years' unexpired, a higher rate still to those with 80 years' unexpired and that reversions that were 100 years distant had no value because landlords in non-prime areas took a shorter-term view of investments than those in the PCL. The respondent argued that because the value of a reversion, discounted at the appropriate deferment rate, automatically declined as the length of the unexpired term increased, any further reduction by an increase in the deferment rate would involve double-counting.

The appellant further argued that the undeveloped land should be valued at nil, owing to the various problems and risks inherent in any development on that site.

***Held*: The appeal was allowed in part. (1) A party may call evidence seeking to demonstrate that a different deferment rate is appropriate in the case of longer-dated reversions. The risk premium, one of the several components of the deferment rate, involves a consideration of the risks of investment in long reversions in terms of volatility, illiquidity, deterioration and obsolescence. Whether one of those risks increases over time is a matter for evidence, and if the evidence demonstrates that the risk increases, that can, without any double-counting, be reflected in the yield as well as the period of deferment. However, the evidence adduced by the appellant did not support its contention that the circumstances of the appeal property justified an addition of 2% to the deferment rate for leases with 88 years' unexpired. Decisions of other LVTs were of little evidential value on the deferment rate, just as they were in respect of relativity. (2) The evidence showed that development of the undeveloped land would not be financially viable and its speculative value was therefore confined to the existing use value of the garage on the site at £10,000. Overall, the premium payable by the appellant was £205,000.**

The following cases are referred to in this report.

Arbib v *Earl Cadogan*; *Earl Cadogan* v *55/57 Cadogan Square Freehold Ltd*; *Earl Cadogan* v *Moussaieff*; *Day* v *32 Rosary Gardens (Freehold) Ltd*; *In re Earl Cadogan's Appeal* [2005] 3 EGLR 139; [2005] RVR 401, LT

Arrowdell Ltd v *Coniston Court (North) Hove Ltd* [2007] RVR 39, LT

Earl Cadogan v *Sportelli*; *Earl Cadogan* v *27/29 Sloane Gardens Ltd*; *Earl Cadogan* v *Grandeden Property Management Ltd*; *Howard de Walden Estates Ltd* v *Maybury Court Freehold Co Ltd*; *Bircham & Co Nominees (No 2) Ltd* v *Clarke* [2007] EWCA Civ 1042; [2008] 1 WLR 2142; [2008] 2 All ER 220; [2008] 1 EGLR 137; [2007] RVR 314, LRA/50/2005; [2007] 1 EGLR 153; [2006] RVR 382, LT

King & Queen Wharf Freehold Co Ltd v *Freehold Properties Ltd* LON/00BE/2008/0265, 19 June 2009, LVT

Maryland Estates Ltd v *63 Perham Road Ltd* [1997] 2 EGLR 198; [1997] 35 EG 94, LT

Nell Gwynn House Freehold Ltd v *NGH Properties Ltd* LON/00AW/OCE/2008/0086, 28 April 2009, LVT

Stokes v *Cambridge Corporation*; *sub nom Stokes* v *City Council of Cambridge* (1961) 13 P&CR 77; (1961) 180 EG 839, LT

This was an appeal from a decision of the leasehold valuation tribunal on a reference to determine the premium payable by the appellant, Sherwood Hall (East End Road) Management Co Ltd, to the respondent, Magnolia Tree Ltd, for the freehold of properties to be acquired on a leasehold enfranchisement application under the Leasehold Reform, Housing and Urban Development Act 1993.

Gary Cowen (instructed by Alan Edwards & Co) appeared for the appellant; Ellodie Gibbons, instructed by direct access, represented the respondent.

Giving judgment, **Mr Norman Rose FRICS** said:

Introduction

[1] This is an appeal by Sherwood Hall (East End Road) Management Co Ltd against a decision of the Leasehold Valuation Tribunal for the London Rent Assessment Committee (LVT), determining the price payable by the appellant for the freehold interest in 36 flats, arranged in four blocks and known as Sherwood Hall, East End Road, London N2 0TA, under the provisions of section 24 of the Leasehold Reform, Housing and Urban Development Act 1993 (the 1993 Act), at £248,825. Permission to appeal by way of rehearing was granted by the president on 3 March 2008. The appellant's case was that the price should be £119,012 and the respondent freeholder, Magnolia Tree Ltd, contended for a figure of £246,907.

[2] Mr Gary Cowen, of counsel, appeared for the appellant. He called one expert witness, Mr Bruce Maunder Taylor FRICS MAE. Counsel for the respondent, Ms Ellodie Gibbons, called Mr Eric Shapiro BSc

A (Est Man) FRICS IRRV FCIArb to give expert evidence. A factual witness statement by Mr Bulent Mehmet, company secretary of the respondent, was submitted and not challenged. By arrangement with the parties, I visited the site of the appeal property, unaccompanied, on 17 August 2009, and made an external inspection of the four blocks of flats and their grounds.

Facts

[3] In the light of the evidence and my site inspection, I find the following facts. The appeal property is located on the north side of East End Road in the London borough of Barnet, approximately
B midway between East Finchley underground station to the south-east and the North Circular Road to the north-west. East End Road forms the northern boundary of Hampstead Garden Suburb in a mixed inner London residential suburb, with local shops immediately adjacent to and opposite the appeal property and within easy distance of shopping facilities in the centre of the suburb in Lyttelton Road to the south and Temple Fortune and Brent Cross to the south-west.

[4] The appeal property comprises four blocks of flats, two containing 12 units each and two with six flats each, together with 12 lock-up garages, ancillary gardens and parking areas and a small undeveloped area at the rear of the site.

C [5] Access into the estate is via a front service road, entered at the western end of the East End Road frontage. This service road is also used for car parking. It leads to the central service road, which in turn gives access to the two rear blocks, the lock-up garages and the undeveloped site; it too is used for car parking. The estate was developed during the late 1920s or early 1930s. The flats are arranged on ground and two upper floors in facing brickwork beneath pitched tiled roofs, the second floors projecting forward slightly and faced with vertical tile hanging. All the flats contain similar accommodation, with a bay window to the principal room and with two flats on each landing. The entrance hall and stairs are basic and there are no lifts. Each ground-floor entrance
D provides front access and a secondary rear access to the gardens.

[6] Internally, the flats each comprise a reception room, kitchen, two bedrooms and bathroom/wc. The approximate dimensions are as follows:

Reception room	4.14m x 3.71m
Kitchen	2.9m x 2.31m
Bedroom	13.38m x 3.30m
Bedroom	23.28m x 2.57m
Bathroom/wc	

E

[7] The following matters are agreed. The valuation date is 1 August 2005. The ground rent income should be capitalised at 6%. The value of the long leasehold interest in each flat was £230,000 unimproved. These values should be uplifted by £1,000 to reflect the freehold value. The freehold value of each of the two garages included in a flat lease was £5,000 and that of each of the untied garages was £10,000. The existing lease values of flats 9 and 36 were £202,400.

Issues

[8] At the rehearing, there were two principal issues. First, the
F deferment rate for those flats with an unexpired lease term of 87.6 years. The appellant argued for a rate of 7% and the respondent supported the LVT's decision to adopt 5%. The LVT also applied a deferment rate of 5% to those flats with 47.8 and 66.6 years' outstanding, and that conclusion was not challenged. The lease of one flat had 177.6 years' unexpired. The LVT attached no reversionary value to that flat and that decision is also not disputed. The second issue relates to the development value, if any, to be attributed to part of the site for which planning permission was granted on appeal for the erection of two flats. The appeal decision was dated 27 March 2006, some eight months after the valuation date. The respondent agreed with the LVT's decision to attribute £50,000 to this land and the appellant contended for a nil value.

Deferment rate G

[9] Mr Shapiro considered that the LVT had been right to conclude that there was no reason to deviate from the decision of the Lands Tribunal (LT) in *Earl Cadogan* v *Sportelli* [2007] 1 EGLR 153 that a deferment rate of 5% should be adopted when valuing flats with unexpired lease terms of more than 20 years. Subsequent decisions of the LT, relating to properties in Greater London but outside the prime central London area (PCL) with which *Sportelli* was concerned, had applied the same deferment rate and there was no justification for applying a different rate in the case of the appeal property.

[10] Mr Maunder Taylor noted that, in *Sportelli*, the LT had been concerned with properties in the PCL area with reversions of up to a H maximum of 71 years. He considered that the deferment rate should start to increase with leases of 75 years' unexpired and increase further at 80 years. Reversions that were 100 or more years distant had no value. In support of this approach, he cited differences between the strategies of landlords in the PCL area and elsewhere and to the LT's observation in *Arbib* v *Earl Cadogan* [2005] 3 EGLR 139, in [169], that:

> there might be reason to increase the deferment rate also where there is more than 80 years unexpired.

[11] Mr Maunder Taylor attached significance to the fact that when the appeal property was offered for sale by auction in 2004, J the sale catalogue referred to the short leases as having a valuable reversion, but no such description was applied to the long leases. He also referred to LVT decisions in which a higher rate than the *Sportelli* 5% had been applied and to two settlements, which he said showed a similar pattern.

[12] I start by considering the weight to be given, in the context of the current appeal, to the guidance provided by the LT in *Sportelli*. As I have said, that case was concerned with a number of properties in the PCL area. The LT concluded that the generic deferment rate should be 4.75% ([79]) and that this should be increased by 0.25% for flats: [95]. In [85], the LT said: K

> Our conclusion is that the deferment rate is constant beyond 20 years. Below 20 years, we accept the view of Mr Dumas, Professor Lizieri and Mr Orr-Ewing that the rate would need to have regard to the property cycle at the time of valuation. Beyond 75 years, we see no reason on the evidence before us to conclude that the rate would be either higher or lower.

[13] In [123], the LT (Mr George Bartlett QC, president, HH Judge Michael Rich QC and Mr Paul Francis FRICS) said:

> The application of the deferment rate of 5% for flats and 4.75% for houses that we have found to be generally applicable will need to be considered in relation to the facts of each individual case. Before applying a rate that is different from this, however, a valuer or an LVT should be satisfied that there L are particular features that fall outside the matters that are reflected in the vacant possession value of the house or flat or in the deferment rate itself and can be shown to make a departure from the rate appropriate.

[14] The decision in *Sportelli* was appealed. In the Court of Appeal ([2007] EWCA Civ 1042; [2008] 1 WLR 2142*), Carnwath LJ agreed that this general guidance was appropriate. In [99] he said:

> I agree with the tribunal that an important part of its role is to promote consistent practice in land valuation matters. It was entirely appropriate for the tribunal to offer guidance as they have done in this case, and, unless and until the legislature intervenes, to expect leasehold valuation tribunals to follow M generally that lead.

[15] This approval by the Court of Appeal, however, was qualified. In [102], Carnwath LJ said:

> The tribunal's later comments on the significance of their guidance do not distinguish in terms between the PCL area and other parts of London or the country. However, there must in my view be an implicit distinction. The issues within the PCL were fully examined in a fully contested dispute between directly interested parties. The same cannot be said in respect of other areas. The judgment that the same deferment rate should apply outside the PCL area

---

* Editor's note: Also reported at [2008] 1 EGLR 137

was made, and could only be made, on the evidence then available. That must leave the way open to the possibility of further evidence being called by other parties in other cases directly concerned with different areas. The deferment rate adopted by the tribunal will no doubt be the starting point; and their conclusions on the methodology, including the limitations of market evidence, are likely to remain valid. However, it is possible to envisage other evidence being called, for example, on issues relevant to the risk premium for residential property in different areas. That will be a matter for those advising future parties, and for the tribunals, to consider as such issues arise.

[16] The property that forms the subject of the present appeal is outside the PCL area and the deferment rate in issue relates to leases with unexpired terms of approximately 88 years. The appellant accepts that, so far as leases with 66.6 and 47.8 years' unexpired are concerned, the *Sportelli* rate of 5% is applicable, despite the fact that the appeal property lies outside the PCL area. It seeks to depart from the *Sportelli* deferment rate when valuing the flats with 87.6 years' outstanding, on the ground that this period is outside the range of unexpired terms considered by the LT in *Sportelli*, namely 21.25 to 71.05 years. In the light of the comments of Carnwath LJ in [102], I am satisfied that it is indeed open to a party to call evidence seeking to demonstrate that a different deferment rate is appropriate in the case of longer-dated reversions, such as those with which I am here concerned.

[17] Mr Shapiro's view was that there could be no such evidence. He described the suggestion that the deferment rate could change depending on the length of the unexpired term as a fundamental error of valuation. He said that the present value of a reversion, discounted at the appropriate deferment rate, automatically declined as the length of term increased. To reduce that present value further by increasing the deferment rate would therefore constitute double-counting.

[18] I do not think that that approach is correct. It is clear from *Sportelli* that there are a number of components to the deferment rate — the risk-free rate, real growth and the risk premium. An assessment of the latter involves a consideration of the risks of investment in long reversions, namely volatility, illiquidity, deterioration and obsolescence. Whether one or more of these risks increases over a period of time is a matter for evidence. If the evidence demonstrated that there was an increased risk, it would not constitute double-counting to reflect it in the yield as well as in the period of deferment. Indeed, to ignore any increase in risk would itself amount to a valuation error.

[19] I therefore turn to Mr Maunder Taylor's evidence in support of an increase in the deferment rate. First, he said that many residential ground rent investments in the PCL area were owned by landlords, commonly known as the great estates. In the no-Act world, those landlords would sell relatively short lease extensions or hold their investments to term date. This was not the approach of landlords outside the PCL. In Mr Maunder Taylor's opinion, landlords of non-prime residential ground rent investments had a shorter-term view than those interested in prime property in prime locations.

[20] The identity of the hypothetical purchaser of a long-dated reversion was considered by the LT in *Sportelli* in the following terms:

> 76. It is, in our judgment, the combined effect of the other components, volatility and illiquidity, that must have the major impact on the risk premium. If the market was composed or contained a substantial number of people who intended to hold the reversion to term, the fluctuations in the residential property prices and the illiquidity of the investment would have very little influence. The investor would simply lock away the investment and the passage of time would iron out the fluctuations. The illiquidity would have no influence because the investment would not be sold. Mr Cullum's assessment was based, it appears to us, on the assumption of a market very much of this sort, and he identified pension funds and the great estates as the likely purchasers. We do not, however, accept that, in the market that we have to envisage, there would be any significant number of investors that would be looking to hold these very long-term assets throughout their lives. The attraction of the investment would be its relative security, the prospect of growth and the opportunity for both long-term retention and earlier sale. Tradeability would, we think, be important as one of its components, and it is this that would make the volatility of the housing market and the relative illiquidity of the investment significant factors in the mind of a purchaser.

Contrary to Mr Maunder Taylor's assumption, therefore, the *Sportelli* deferment rate did not reflect the requirements of the great estates.

[21] Mr Maunder Taylor also referred to the following evidence as to the purchasing strategy of investors in residential ground rents, given to the LT in *Maryland Estates Ltd* v *63 Perham Road Ltd* [1997] 35 EG 94* by Mr Bebbington, a director of the landlord company, at p199L:

> Mr Bebbington contrasted the auction market and the results (prices) achieved with those by private treaty sales involving serious investors and specialist brokers or agents. He said that investors did not capitalise and defer future increases in ground rent and that length of lease was seen as relevant only if it was less than 80 years unexpired. He put in evidence a schedule of private treaty sales of reversions during 1996 in various categories, for example ground rent only receivable, insurance commission also receivable and where in addition the freeholder manages the premises. This schedule of transactions negotiated by the Ground Rent Brokers, said to be the largest agent active in this specialised market, included both single reversions and "parcels" or portfolios. Mr Bebbington stressed that these were sales between investors and that at the point of sale full relevant information was available as to the position between the vendor and the lessees of flats; this situation contrasted markedly with the circumstances prevailing at auction.

Mr Maunder Taylor relied on Mr Bebbington's evidence to support his view that investors required a higher deferment rate where leases had more than 80 years' unexpired.

[22] In his decision in *Maryland*, the member (Dr T Hoyes FRICS) accepted Mr Bebbington's evidence. In stating his conclusions on the value of the freehold interest, he said, at p203E:

> The evidence therefore supports and favours the approach of Mr Angel [Mr Bebbington's expert] in annexure 2 save that, as Mr Bebbington said, the market pays no material regard to prospective, but somewhat distant, increases in ground rent. In the instant appeal their capitalisation and deferment produced only *de minimis* amounts.

[23] Mr Angel's annexure 2 contained the following valuation of the freeholder's interest:

Ground rent income		
4 x £75 pa =	£300	
Years' purchase @ 14.28%	7	
		£2,100
Reversion		
Flat 3 1 x £300 pa =	£300	
Years' purchase in perpetuity @ 14.28%	7	
Present value £1 in 56yrs @ 14.28%	0.0006	
		£1
Flats Reversion to	£230,000	
1, 2 and 4 PV £1 in 89 yrs @ 14.28%	0	
		£0
Insurance commission		
Premium £765.44		
Commission at 20%	£153.08	
6.0 Years' purchase (Note 2)	6.0	
		£918
Total =		£3,019
Freehold value =		
Say		£3,020

[24] In so far as any weight is to be attributed in this appeal to the evidence given in *Maryland*, it does not, in my view, support Mr Maunder Taylor's contention that a higher deferment rate is appropriate for leases with long unexpired terms. In that case, the present capital values of flats 1, 2 and 4, deferred 89 years, and the ground rent from flat 3, deferred 56 years, were *de minimis*, not because a higher deferment rate was applied to either but because, with a deferment rate of 14.28% in each case, the values produced by the calculation were vanishingly small.

[25] Mr Maunder Taylor also referred a number of post-*Sportelli* LVT decisions, relating to properties subject to long leases, that he considered to be of assistance. The first concerned 50 Wilton Crescent, London SW1, where he said that a deferment rate of 5.5%

* Editor's note: Also reported at [1997] 2 EGLR 198

was determined for a house, and the second concerned 220 Ladbroke Grove, London W10, where the LVT deferred the reversion to five flats at 5.25%. In oral evidence-in-chief, Mr Maunder Taylor confirmed that the rate determined for 50 Wilton Crescent was in fact 5%.

[26] Fifty Wilton Crescent was subject to a lease with 180 years' outstanding. The LVT said:

> *Sportelli* however, was not concerned with a freehold reversion 180 years in the future and it does seem to the Tribunal that a reversion so far distant would persuade an investor that a slightly higher return was appropriate.

[27] In the case of 220 Ladbroke Grove, the unexpired term was 123 years. It is plain from [42] of the LVT's decision that it decided to add 0.25% to the *Sportelli* starting point because of the property's location, not the length of the lease.

[28] In my judgment, neither of these decisions provides any support for Mr Maunder Taylor's contention that the circumstances of the appeal property justify an addition of 2% to the *Sportelli* rate for leases with 88 years' unexpired.

[29] In his supplementary report, Mr Maunder Taylor produced two further LVT decisions. The first related to a block of 429 flats plus commercial space, known as Nell Gwynn House, Sloane Avenue, London SW3, and the second to seven blocks of 150 flats and a small leisure complex, known as King and Queen Wharf, Rotherhithe Street, London SE16. The unexpired lease terms were a mixture of 92 and 120 years at Nell Gwynn House and 106 years in the case of King and Queen Wharf. In both cases, the freeholder contended for a deferment rate of 5%. Mr Maunder Taylor gave evidence for the nominee purchaser at both hearings. He deferred the reversions at King and Queen Wharf at 7%. For Nell Gwynn House, he contended for 7% for those leases with 91.68 years' unexpired and nil reversionary value where the unexpired term was 119.68 years.

[30] Although the King and Queen Wharf hearing took place first, the LVT decision on Nell Gwynn House* was published before that of King and Queen Wharf†. The Nell Gwynn House LVT said of Mr Maunder Taylor's evidence that "it may not be overwhelming but it was much better than none". It determined the deferment rate at 6.5% for the shorter leases, with no reversionary value attributed to the longer leases.

[31] Mr Maunder Taylor sent a copy of the *Nell Gwynn House* decision to the King and Queen Wharf LVT before the latter had finalised its decision. The LVT said this of *Nell Gwynn House*:

> In *Nell Gwynn House* Mr Maunder Taylor's arguments based on his analysis of auction sales met with a measure of success. That decision, by a particularly strongly constituted tribunal, is clearly one of very considerable importance. However, the actual finding by the Tribunal in that case is of little evidential value. Notwithstanding that Mr Maunder Taylor's evidence found favour with the Tribunal in that case, having considered his report and listened with great care to his oral evidence both in examination in chief and cross examination, we do not reach the same conclusion on the facts. Whereas the Tribunal in *Nell Gwynn House* did not find Mr Maunder Taylor's evidence *overwhelming*, following the very thorough and skilful cross examination by Mr Letman we have found it less than convincing. We have not had the opportunity to investigate what evidence the Tribunal had before it and how it had treated it. Moreover, as we have stated above, it is incumbent upon the expert to provide sufficient particulars to substantiate the reliability of the material upon which the opinion is based. In this instance, we were not persuaded as to the reliability of the material produced by Mr Maunder Taylor.

[32] Although Mr Maunder Taylor did not draw them to my attention in either of his written reports for the present appeal, it emerged that he had presented a similar argument at several further recent LVT hearings, but without success apart from in *Nell Gwynn House*.

[33] As the LVT indicated in its decision on King and Queen Wharf, decisions of other tribunals are of little or no evidential value. The position was considered by the LT (Mr George Bartlett QC, president, and Mr Norman Rose FRICS) in *Arrowdell Ltd* v *Coniston Court (North) Hove Ltd* [2007] RVR 39. In [37], the LT said:

> In our judgment leasehold valuation tribunal decisions on relativity are not inadmissible, but the mere percentage figure adopted in a particular case is of no evidential value. The reason for this is that each tribunal decision is dependent on the evidence before it, and thus, in order to determine how much weight should be attached to the figure adopted in a decision, it would be necessary to investigate what evidence the leasehold valuation tribunal had before it and how it had treated it. Such a process of investigation is potentially lengthy, and it is inherently undesirable that leasehold valuation tribunal hearings should resolve themselves into rehearings of earlier determinations.
>
> 38. It is certainly understandable that valuers negotiating the settlement of an enfranchisement claim should have regard to LVT decisions on relativity, since these might seem to them to be the best guide of the likely outcome if they were unable to reach agreement, even though, as Mr Pridell said, the decisions are disparate and fail to show any established pattern. But the decisions themselves can constitute no useful evidence in subsequent proceedings.

[34] I consider that those observations apply to LVT decisions on the deferment rate as much as to those on relativity. None of the recent LVT decisions that have considered Mr Maunder Taylor's theory of an increased deferment rate for long-dated reversions adds to the weight of the other evidence before me. Nor do I obtain assistance from the price that Mr Maunder Taylor negotiated for the freehold in Alexandra Court, Chase Road, London N14, where the freeholder's decision was influenced by personal circumstances.

[35] At the end of Mr Maunder Taylor's evidence, I questioned him on changes that had been made to the appellant's pleadings. In its statement of case dated 15 May 2008, the appellant said:

> 7. The subject property is in East Finchley and certainly not in PCL. The Applicant's expert witness, Bruce Maunder Taylor FRICS gave evidence before the LVT and will give evidence before the Lands Tribunal that the subject property has greater risks associated with volatility and illiquidity than PCL property. In addition, the property has greater risk of obsolescence and deterioration than property in PCL.
>
> 8. The applicant submits that the LVT ought to have adopted a deferment rate of 7% for the subject property.

[36] The statement of case was amended on 5 August 2008. The words in para 7 after "Lands Tribunal" were deleted and the following two paragraphs added:

> 5. It is accepted for the purposes of this appeal that for the leases in the block with less than 80 years unexpired, the relevant deferment rate ought to be 5%. The Appellant does not seek to overturn the LVT's decision on that one relatively narrow aspect.
>
> 6. However, the decision in *Sportelli* was expressly limited to leases with less than 75 years unexpired. The 1993 Act as amended by the Commonhold and Leasehold Reform Act 2002 expressly recognises that leases with in excess of 80 years unexpired have no marriage value. Mr Maunder Taylor will provide evidence that for leases in excess of 80 years unexpired, the appropriate deferment rate should not be 5% and that such leases ought to be treated differently to those with less than 80 years left unexpired.

[37] Paragraph 8 of the original statement of case was also amended to read:

> 7. The Applicant submits that the LVT ought to have adopted the following deferment rates: —
>
> (a) 5% for flats 9, 29 and 36 (as determined by the LVT and in respect of which, therefore, there is no appeal);
>
> (b) Nil for flats 25 and 26
>
> (c) Nil for flat 27 (as determined by the LVT and in respect of which, therefore there is no appeal); and
>
> (d) 7% for the remaining flats in the subject property.

[38] I asked Mr Maunder Taylor whether he accepted that the deferment rate for those flats with 66.6 and 47.8 years' unexpired should be 5%. He replied that he did not. In the light of legal advice received, his client had decided to accept the 5% rate, but he still considered that it should be 7%. The following exchange then took place:

---

* Editor's note: Reported at *Nell Gwynn House Freehold Ltd* v *NGH Properties Ltd* LON/00AW/OCE/2008/0086

† Editor's note: *Reported at King & Queen Wharf Freehold Co Ltd* v *Freehold Properties Ltd* LON/00BE/2008/0265

Q. If you are right that the deferment rate should be 7% instead of 5% for the shorter leases, does it follow that you would be asking for 9% for the longer leases?

A. No sir.

Q. I thought you said that there was a difference between the markets for shorter and longer reversions.

A. There is a different type of market. There is a point at which the thoughts and bids of those two different types of purchasers would come together.

[39] That final answer is irreconcilable with Mr Maunder Taylor's previous evidence. In para 6.1 of his first expert report, Mr Maunder Taylor said that:

> As far as concerns non-prime property, it is my opinion that the deferment rate should start to increase at 75 years unexpired, increase further at 80 years unexpired, and from 90 years unexpired upwards, reversionary value should be treated as nil.

In cross-examination, he said that he had carried out further research, and now considered that the nil value applied to leases of 100 years and more. Nowhere in his written or oral evidence was there any suggestion that the deferment rates for long and short leases would converge beyond 75 years.

[40] Mr Maunder Taylor's evidence on the significance of the price paid for the appeal property at auction was also troubling. In para 11.6 of his report, he said:

> It is my opinion that the auction price reflects the existence of the Leasehold Reform, Housing and Urban Development Act 1993 however, it is my opinion that the existence of the Act cannot possibly justify the difference between £95,000 in the with Act world and £248,500 in the no Act world.

[41] As Mr Maunder Taylor accepted in the course of cross-examination, that view was directly contrary to one that he had previously expressed in reply to a written question from the respondent's solicitor. It was given in the context of proceedings in Central London County Court arising from the respondent's decision to challenge the validity of the lessees' initial notices served under the 1993 Act on the ground that the prices proposed were wholly unrealistic. Those prices were based on a valuation report by Mr Maunder Taylor dated 31 December 2004, two weeks after the appeal property had been sold at auction. Mr Maunder's individual valuations of the four blocks totalled £47,100, half the auction price.

[42] The respondent's solicitor asked Mr Maunder Taylor:

> 10. Was the auction relevant and contemporaneous evidence of a comparable transaction affecting the above property?

On 28 July 2006, Mr Maunder Taylor replied:

> All evidence has some value in my opinion; the quality of the evidence must first be looked at and the circumstances behind it, and then the weight that should be given to such evidence can be properly assessed. In my opinion little weight, if any, should be placed on the auction result for the purposes of the statutory valuation.

[43] The statement in para 11.6 of Mr Maunder Taylor's report was also not supported by the enfranchisement price paid to the owner of 33 Addison Gardens, London W14 ODP. The freehold interest in that property was purchased at auction for £30,000 in September 2007. The nominee purchaser served an initial notice to purchase the freehold on 11 January 2008. The freeholder served a counternotice at £100,000, based on a valuation prepared by Mr Maunder Taylor. Following negotiations between Mr Maunder Taylor and the nominee purchaser's surveyor, the price was agreed at £79,600.

[44] The price determined by the LVT in respect of the appeal property and the price paid by the nominee purchaser of 33 Addison Gardens bore the same relationship to the auction price a multiple of approximately 2.5. When I pointed this out to Mr Maunder Taylor, he expressed the view that the value he had agreed for 33 Addison Gardens was exorbitant. That was a surprising suggestion, bearing in mind his acceptance that the nominee purchaser had been represented by a very experienced surveyor.

[45] I am satisfied that Mr Maunder Taylor's original opinion on the relevance of the auction price, quoted in [42] above, was right. The respondent completed the purchase of the freehold interest in the appeal property in January 2005. The decision of the LT in *Sportelli* was not published until 15 September 2006. It is clear that, prior to *Sportelli*, enfranchisement claims in this part of London were being settled based on a deferment rate, for reversions both long and short, of at least 7%. The then current valuation practice would have been of considerable significance to anyone bidding for the appeal property at auction, since the auction particulars disclosed that the lessees were interested in acquiring the freehold interest. The effect of *Sportelli* was that, for such properties, the starting point for assessing the deferment rate was reduced from at least 7% to 5%. This change has resulted in a substantial increase in the prices payable on enfranchisement. The increases are particularly marked where the unexpired terms of the existing leases are very long, because of the effects of compound interest on the calculation.

[46] Mr Maunder Taylor also pointed to the auction catalogue that would have been seen by the respondent before it agreed to purchase the appeal property on 14 December 2004. This referred to "valuable reversion in 2072" in the case of flats 9 and 36, having approximately 67 years' unexpired, and "valuable reversion in 2053" for flat 29 with approximately 48 years' unexpired. No such note appeared in respect of the remaining flats with 88 years or more outstanding on the lease. I do not attach any significance to this distinction. The auctioneers would have known that, in the real world, purchasers were interested in the possibility of realising marriage value that, it is agreed, did not exist where leases were significantly longer than 80 years. In the no-Act world, on the other hand, it is clear from *Sportelli* that marriage value is to be ignored when assessing the deferment rate, irrespective of the length of the unexpired term.

[47] Mr Maunder Taylor sought to rely on the LT's observation in *Arbib* that there may be reason to increase the deferment rate with leases of more than 80 years' unexpired. He did not refer to [123] of *Arbib*, which drew attention to the limitations of the evidence before the LT (HH Judge Michael Rich QC and Mr Peter Clarke FRICS) as follows:

> We must emphasise, however, that, as appears from such analysis, we are disappointed as to the quality and analysis of the evidence that has been put before us in the course of these appeals…

Since *Arbib* left room for further evidence and argument as to what deferment rate was right, the LT decided to collect up further cases raising the same issue for hearing together. The result was the hearing of *Sportelli* and other appeals. In the course of his judgment in the Court of Appeal in *Sportelli*, Carnwath LJ said, in [98]:

> The Lands Tribunal could hardly have done more to ensure that the issues were fully ventilated and exhaustively examined. They had already been discussed in detail in *Arbib*… I have already referred to the steps taken by the tribunal to bring together the present group of cases. Furthermore it is difficult to envisage a better qualified panel of experts for the purpose than those called in this case, or of specialist counsel on both sides of the argument.

Having heard such evidence, the LT decided, in [85], that:

> Beyond 75 years, we see no reason on the evidence before us to conclude that the rate would be either higher or lower.

Mr Maunder Taylor's selective quotation from *Arbib*, and his failure to state in terms that the LT had subsequently arrived at a different conclusion, having heard more detailed evidence, was misleading. I regret that, in my judgment, the inconsistencies and selective choice of material in Mr Maunder Taylor's evidence mean that I can place no reliance on his assertion that a higher deferment rate should be applied to 88-year reversions than to those with shorter unexpired terms. The appeal on this issue must therefore fail.

A *Development value*

[48] Prior to the auction, Mr Mehmet investigated the planning history of the appeal property. On 27 November 2004, he discovered that the only previous applications had related to trees. Immediately after the auction, Mr Mehmet instructed his architect, Foster Lomas, to prepare a scheme for the erection of new flats on the land at the rear of the site. Foster Lomas wrote to the local planning authority, the London borough of Barnet, on 12 January 2005, requesting a pre-planning application meeting. On 12 September 2005, the architect submitted a planning application for the erection of two two-bedroom flats totalling 157m^2. That application was refused in November 2005, contrary to B the recommendation of the planning officer. A revised application for two smaller flats was then submitted and approved. At the same time as it submitted the second application, the respondent appealed against the refusal of the first application, and the appeal was allowed on 27 March 2006.

[49] Mr Maunder Taylor considered that, at the valuation date, the market value of the undeveloped area was nil, for the following reasons. Enquiries of the local planning authority were likely to have resulted in an indication being given that planning permission would be refused. The leases of two of the existing flats would each have required a deed of variation permitting development of the land, C and there was no guarantee that such deeds would be entered into. Access to the completed development would be down an access way leading from East End Road to the garages and the development area. The management and maintenance of the access way, together with the right to collect service charges was part of the function of the management company, which was controlled by the lessees. If the lessees wished to frustrate the development, they could refuse to enter into an agreement regarding improvements to drainage, service charge contributions and membership of the management company. Finally, Mr Maunder Taylor's professional judgment was that the sale value of the two flats proposed was £400,000, which was insufficient to produce D a profit after all costs had been taken into account.

[50] Mr Shapiro produced a residual valuation of the site, assuming a gross development value for the two flats of £550,000, which, after deducting the agreed figure for costs and profit, produced a land value ripe for immediate development of £95,000. He said that had the site been purchased for £50,000, it would have produced a profit of £116,361 before allowing for the cost of buying out legal impediments. With the benefit of hindsight, it was known that planning permission was capable of being obtained on appeal. Although two leaseholders had an absolute indefeasible right to use the area proposed for the development, in view of the comparatively low value of the flats Mr Shapiro believed that the freeholder would have been able to secure the release of their rights E over the land for a realistic payment. In accordance with the principle in *Stokes* v *Cambridge Corporation* (1961) 13 P&CR 77*, the most the lessees could have extracted from the freeholder would be 30-40% of the development value. Mr Shapiro did not think that there would have been any difficulty with the management company because the two flats could, if necessary, be sold as separate freeholds with only rights of access over the estate roads. On planning, he considered that a purchaser would have had regard to the fact that the land was an overgrown brownfield site, not in a conservation area and unencumbered by tree preservation orders. The only reason planning permission might have been refused was because of objections from the adjoining leaseholder. Such objections F were indeed made to the inspector on appeal and rejected. The LVT had reflected all the risks by deducting 50% from its estimated unencumbered development value of £100,000. Mr Shapiro considered that this was a fair allowance and he, too, valued the land at £50,000.

[51] As is widely known, the LT is reluctant to accept residual valuations if there is any comparable evidence of land sales. No such comparable evidence has been produced in this case, and so the residual method must form the basis of my decision. The LVT arrived at its valuation of £100,000 as follows. It valued the two flats at £550,000 and deducted £450,000, being the agreed costs, and thus arrived at a residual development value of £100,000 with the benefit of planning permission G and assuming no hindrance to development. From that figure, the LVT deducted 50% to reflect the various risk factors and thus arrived at a value of £50,000. Mr Shapiro agreed with the LVT's gross development value of £550,000. The value of each of the existing unimproved flats in Sherwood Court had been agreed at £230,000. The proposed flats would be 30% larger and brand new. Mr Maunder Taylor pointed to certain deficiencies in the proposed flats, but he had no comparable evidence to justify his suggested value of £200,000 per flat. I find that the proposed flats would have been worth £275,000 each, giving a gross development value of £550,000 and a potential development value of £100,000 after deducting the agreed figure for costs and profit. H

[52] Mr Shapiro pointed out that, in *Arrowdell*, the LT had made a deduction of 50% to reflect the absence of planning permission. In that case, however, planning permission had recently been granted on appeal for a virtually identical development on the immediately adjoining site, and the planning officer had advised that refusal of permission for the development proposed "could not be upheld on appeal". In the case of the site at Sherwood Hall by contrast, there was no comparable planning history. Mr Mehmet did not say whether he had discussed the possibility of development with the local authority before the auction and, if he had, what view they had expressed. He produced a letter from his architect to the local planning authority dated 12 January 2005, enclosing information J for a pre-application meeting, but he provided no information as to the objections to the proposed development that, it appears from a letter dated 12 September 2005, were made by the council at that meeting.

[53] I do not agree with Mr Shapiro's suggestion that *Stokes* provides an appropriate yardstick for estimating the price likely to be required by the two lessees enjoying a right to prevent the development. In that case, the LT had assessed the ransom value of a strip of land in a single ownership at one-third of the increase in value of the claimant's land that would result from its release. One of the considerations that led to this conclusion was the LT's finding that there was "an inducement to the owner of the brown [ransom] strip to sell it as access" because K development of the claimant's land would expedite the re-zoning of other land, held together with the access strip, from allotment to industrial use. In the present case, there was no evidence to suggest that the two leaseholders would benefit in any way from the proposed development, nor that both could have been bought out at a price that would have made the development financially viable. It is significant, in my judgment, that the auctioneers did not even refer to the development potential of the site in their sale particulars prepared in December 2004. The LVT's decision recorded in [41] that garage 13 was located on the development land. Garage 13 is not tied to the lease of a particular flat and it is agreed that such garages are worth £10,000 each. Taking L all the evidence into account, I find that the speculative value of the development site did not exceed the existing use value of garage 13. I therefore assess the value of the development site at £10,000.

Conclusion

[54] The LVT valued the existing buildings at £198,827 and the development site at £50,000. This produced a total of £248,827, say £248,825. Before me, Mr Shapiro arrived at a slightly lower value, £196,907, for the existing buildings, based on a deferment rate of 5% throughout and utilising an exact unexpired term in each case and no rounding. I accept that figure. I determine that the price payable by M the appellant to the respondent for the freehold interest in the appeal property is £205,000, calculated as follows:

Existing buildings	£196,907
Development site	£10,000
Total	£206,907
Say	£205,000

[55] I am not aware of any circumstances that would justify the unusual step of awarding costs in an appeal of this nature. I make no order as to costs.

Appeal allowed in part.

* Editor's note: Also reported at *Stokes* v *City Council of Cambridge* (1961) 180 EG 839

Zuckerman and others *v* Trustees of the Calthorpe Estate

UPPER TRIBUNAL: LANDS CHAMBER

[2009] UKUT 235 (LC)

18 November 2009

MR NORMAN J ROSE FRICS

***Leasehold enfranchisement — Leasehold Reform, Housing and Urban Development Act 1993 — Lease extension — Premium — Deferment rate — Obsolescence — Growth rates in capital value of residential properties — Additional management burden associated with flats over houses — Whether evidence supporting departure from 5% rate in* Earl Cadogan *v* Sportelli**

The appellants applied, under Chapter II of the Leasehold Reform, Housing and Urban Development Act 1993, for lease extensions in respect of 11 two-bedroom flats in a 1970s development located within the Edgbaston conservation area of Birmingham. The existing lease of each flat was due to expire in September 2071. Since the surrender of a headlease in 1981, the appellants had held their properties directly from the respondent freeholders.

It was agreed, for the purposes of valuing the flats, that the unexpired term of the leases was 64 years, save in respect of one flat for which it was a few months less. On a reference to the leasehold valuation tribunal (LVT), to determine the premiums payable for the new leases, the only disputed matter was the deferment rate to be applied to the value of the extended lease. The LVT applied the 5% generic rate for flats laid down by the Lands Tribunal in *Earl Cadogan* v *Sportelli* [2007] 1 EGLR 153 and determined premiums of £9,843 for 10 of the flats and £9,887 for the flat with the shorter unexpired term.

On appeal, the appellants contended for a deferment rate of 6.25%. They submitted that a higher deferment rate was justified than the one applied in *Sportelli* to properties in the prime central London area (PCL), owing to: (i) a higher rate of obsolescence or deterioration of properties in the local area, arising from the lower market value of such properties when compared with the cost of their maintenance; (ii) lower growth rates in capital values in Birmingham than in the PCL area; and (iii) an increased gap between houses and flats resulting from recent legislative changes, such as those introduced by the Commonhold and Leasehold Reform Act 2002, which had placed greater significant management burdens on investors in flats.

***Held*: The appeal was allowed; a deferment rate of 6% was applied. (1) The 5% *Sportelli* rate provides the starting point when considering the appropriate deferment rate to be adopted. However, there was a striking difference between the value of the properties considered in *Sportelli* (between £740 and £1,100 psf) and that of the development in which the appellants' flats were located (£198.50 psf). It was likely to remain economically viable to repair high-value properties in the PCL area for considerably longer than similarly sized flats in the appellants' development. Consequently, although the individual flats might be let on full repairing terms, there was a greater risk of deterioration in the appellants' development than in PCL properties. That difference was not reflected in the respective vacant possession values. A purchaser of the freehold reversion to the development would require an increase of 0.25% in the risk premium to 4.75% to compensate for that difference. (2) Although there were deficiencies in the available statistical information, such as lenders' indices of price movements in residential properties, that information demonstrated a considerable difference between past rates of long-term price increases in the PCL area and those in the West Midlands. That information would persuade an investor that it could reasonably anticipate significantly slower long-term growth from residential properties in the West Midlands generally than in the PCL area. An investor would reduce its bid for the appellants' development accordingly. That reduction should be reflected by a further 0.5% increase in the risk premium to 5.25%. (3) The provisions of the Service Charge (Consultation Requirements) (England) Regulations 2003, introduced pursuant to the 2002 Act, are potentially serious for landlords in that they impose strict procedural requirements, and a failure to comply with these could leave a landlord unable to recover a large proportion of the costs of repairs. Since the market had become more aware of the dangers posed by the regulations than had been the case at the time of *Sportelli*, investors would require an additional management allowance of 0.5% to reflect the greater management problems associated with flats compared with houses. However, had the headlease remained in existence, it would not have been appropriate to depart from the *Sportelli* uplift of 0.25% in respect of that factor. (4) Applying the above adjustments to the *Sportelli* calculation, the appropriate deferment rate was 6%. The premium payable by the appellants was £8,100 for the 10 flats with the 64-year unexpired terms and £8,150 for the other.**

The following cases are referred to in this report.

Arbib v *Earl Cadogan; Earl Cadogan* v *55/57 Cadogan Square Freehold Ltd; Earl Cadogan* v *Moussaieff; Day* v *32 Rosary Gardens (Freehold) Ltd*; *In re Earl Cadogan's Appeal* [2005] 3 EGLR 139; [2005] RVR 401, LT
Camden London Borough Council v *Leaseholders of 37 Flats at 30-40 Grafton Way)* LRX/185/2006; *sub nom Re 30-40 Grafton Way* [2008] PLSCS 198, LT
Culley v *Daejan Properties Ltd* LRA/163/2007; *sub nom 5-8 Royston Court, North View, Eastcote, Middlesex, HA5 1PG* [2009] UKUT 168 (LC); [2009] 3 EGLR 165; [2010] L&TR 2
Daejan Investments Ltd v *The Holt (Freehold) Ltd* [2008] PLSCS 146, LT
Earl Cadogan v *Sportelli*; *Earl Cadogan* v *27/29 Sloane Gardens Ltd*; *Earl Cadogan* v *Grandeden Property Management Ltd*; *Howard de Walden Estates Ltd* v *Maybury Court Freehold Co Ltd*; *Bircham & Co Nominees (No 2) Ltd* v *Clarke* [2007] EWCA Civ 1042; [2008] 1 WLR 2142; [2008] 2 All ER 220; [2008] 1 EGLR 137; [2007] RVR 314, [2007] 1 EGLR 153; [2006] RVR 382, LT
Hildron Finance Ltd v *Greenhill Hampstead Ltd* [2008] 1 EGLR 179, LT
Mansal Securities Ltd, Re [2009] 2 EGLR 87; [2009] 20 EG 104, LT

This was an appeal from a decision of the leasehold valuation tribunal on a reference to determine the premium payable by the appellants, Mr Daniel Zuckerman, Ms Shelpa Patel, Mr Anthony and Mrs Lisa Collins, Mr Bindu Nair, Mr John Hutton, Mr Terence and Mrs Annette Clives, Mr Rajinder and Mrs Gita Soni, Mr Alan Adam, Dr Olga O'Klein, Mrs Joanna Good and Mr John Kail, to the respondents, the Trustees of the Calthorpe Estate, for new leases to be granted on a leasehold enfranchisement application under Chapter II of the Leasehold Reform, Housing and Urban Development Act 1993.

Douglas Readings (instructed by Hadgkiss Hughes & Beale, of Birmingham) appeared for the appellants; Anthony Radevsky (instructed by Boodle Hatfield) represented the respondents.

Giving judgment, **MR NORMAN ROSE FRICS** said:

Introduction

[1] This is an appeal by the leaseholders against a decision of the Leasehold Valuation Tribunal of the Midland Rent Assessment Panel (LVT), determining the premiums payable in respect of the grant of 11 lease extensions pursuant to Chapter II of Part I of the Leasehold Reform, Housing and Urban Development Act 1993 (the 1993 Act). Permission to appeal was granted by the LVT. The properties in question are flats 1, 2, 5, 6, 19, 21, 31, 33, 34, 35 and 36 Kelton Court, Carpenter Road, Edgbaston B15 2JX. The LVT decided that the premium payable by the leaseholder of flat 33 should be £9,887 and that the premium for each of the other flats should be £9,843.

[2] It was agreed at the LVT that the capitalisation rate to be applied to the ground rent was the same as the deferment rate to be applied to the value of the extended lease and the latter was the only point in dispute. The LVT decided that the deferment rate should be 5%, the figure suggested by the freeholders, the Trustees of the Calthorpe Estate. Before me, the LVT's decision was supported by the respondent freeholders and the appellants contended for a deferment

A (and capitalisation) rate of 6.5%, resulting in premiums of £7,569 (£7,594 in the case of flat 33).

[3] The appellants are:

Flat no	*Leaseholder(s)*
1	Mr D Zuckerman
2	Ms S Patel
5	Mr and Mrs A C Collins
6	Mr B Nair
19	Mr J W Hutton
21	Mr and Mrs T J Clives
31	Mr and Mrs R Soni
B 33	Mr A J Adam
34	Dr O'Klein
35	Mrs J M Good
36	Mr J Kail

[4] Mr Douglas Readings of counsel appeared for the appellants and called expert evidence from Mr Edward J Rutledge FRICS, a partner at Lawrence & Wightman, chartered valuation and building surveyors, of Birmingham. Counsel for the respondent, Mr Anthony Radevsky, called Mr John K Willson BSc MRICS, a consultant in the Birmingham office of Lambert Smith Hampton, as his expert witness. I inspected the appeal properties, accompanied by representatives of the parties, on the
C afternoon of the second day of the hearing.

[5] I was told that there are outstanding appeals to this tribunal in respect of approximately 20 other flats in Edgbaston, which have been stayed pending the decision on this appeal.

Facts

[6] In the light of the evidence and my inspection, I find the following facts. The appeal properties are on the Calthorpe Estate, comprising around 1,500 acres in Edgbaston and Harborne, Birmingham. At its nearest point, the estate is around 1.2km from the centre of Birmingham. It contains approximately 7,000 properties, including offices, shops, hotels, leisure and recreational facilities as well as residential properties.
D There are around 2,000 flats on the estate, most of which are purpose-built. A scheme of management to which all freehold properties are subject was granted to the respondent in 1974, pursuant to section 19 of the Leasehold Reform Act 1967 (1967 Act).

[7] The appeal properties are located within the Edgbaston conservation area, in an established residential area on the south side of Carpenter Road at the junction with Church Road. The Kelton Court development comprises 36 purpose-built flats, erected in the 1970s and located in six three-storey blocks. The accommodation in each flat comprises an entrance hall, one double bedroom, one single bedroom, living/dining room, kitchen, bathroom and separate WC. Each of the
E appeal properties includes a single garage in a separate block.

[8] Each flat is held on an underlease for a term expiring on 26 September 2071 at current ground rents of £75 pa, increasing to £100 in September 2022 and £125 in September 2047. The landlord is responsible for insuring the premises and recovers the appropriate premium from each lessee as part of the service charge. The landlord's responsibilities include maintenance of the structure and the common areas of the flats and garages, the private roads and amenity areas. The lessees contribute to a service charge to meet the cost of these items. The use of each flat is restricted to that of a private residence of a single family only, such family to comprise not more than four persons.

F [9] It is agreed that, for the purposes of valuing all the flats except no 33, the unexpired lease term is 64 years. The notice of claim for flat 33 was served on 28 January 2008, when there were approximately 63.75 years' unexpired. The headlease was surrendered on 19 May 1981, since when the appellants have held their properties directly from the respondents.

Evidence

[10] Mr Rutledge has worked in Birmingham and the West Midlands for approximately 30 years and is regularly involved in inspections and valuations of a broad range of residential, commercial and industrial properties. He is increasingly involved in leasehold reform work in London, Hertfordshire and other regions in England and Wales. Mr Rutledge said that, in *Earl Cadogan* v *Sportelli* G
[2007] 1 EGLR 153, the Lands Tribunal (LT) (Mr George Bartlett QC, president, HH Judge Michael Rich QC and Mr Paul Francis FRICS) had determined the deferment rate for flats in prime central London (PCL) at 5%. This was arrived at as follows: risk-free rate (2.25%), minus real growth rate (2%), plus risk premium (4.5%), plus increased management risk for flats (0.25%). He considered that two matters were relevant when considering the appropriate risk premium for Birmingham, namely obsolescence/deterioration and long-term prospects of capital growth. When considering the potential for obsolescence/deterioration, the relationship between capital value and the cost of maintenance, repairs and refurbishment should be taken H
into account. He produced photographs of several of the properties that were considered in *Sportelli*. He said that the majority were within a conservation area, with significantly higher market values than Kelton Court. He also produced statistical information that suggested that, in 2007, building costs in London were 21.87% higher than in the West Midlands. The capital values of flats in the PCL were much greater than the values of the flats in Kelton Court. On the other hand, the difference in the cost of maintaining buildings in the two areas was small.

[11] It would therefore remain cost-effective to rebuild, repair or refurbish high-value PCL properties for an almost infinitely longer period than flats such as those at Kelton Court. Kelton Court was a J
development on the site of the old Royal Institute for the Blind. This was just one example of the cycle of redevelopment of old properties in the area. Such redevelopment was common throughout Birmingham and, indeed, in most parts of the country outside the PCL. Other examples were a number of properties on Hagley Road on the Calthorpe Estate, which had been derelict and awaiting redevelopment for several years, and a 1960s development of flats at Montague Road, Edgbaston, just outside the Calthorpe Estate but only around a mile from Kelton Court, where the state of deterioration was very marked. These were good examples of what happened when the cost of building works and repair was very high compared to the market value of the existing properties. K

[12] Turning to changes in property values, Mr Rutledge said that he had tried to find reliable, long-term statistics relating to the PCL and Birmingham or the West Midlands. It was difficult to find information and statistics split into localities going back 40 or 50 years that would be the ideal. He had researched information available from Nationwide Building Society, HBOS (previously Halifax Building Society), the Land Registry, the Valuation Office Agency and others.

[13] He produced a graph, incorporating house-price information obtained from various published statistics, together with the available information on sale prices in Kelton Court. He said that this graph demonstrated a significant difference in property prices between L
Kensington and Chelsea (in the PCL) and the West Midlands and the UK as a whole and Kelton Court in particular. It also showed a significant difference in the rate of growth in the value of properties in these areas. The trend in Kelton Court was a little below that of both the West Midlands and the UK national average in the periods plotted.

[14] Mr Rutledge also compared the attributes of investments in flats and houses. He said that the liquidity of the freehold interest in any flats development had always been lower and more of a problem than the equivalent freehold investment in houses. The right of first refusal granted by the Landlord and Tenant Act 1987 had been in place for more than 20 years and those procedural requirements had had a M
clear effect on the ability of any freehold investor to deal freely with its assets.

[15] In addition, the gradual introduction of the various sections of the Commonhold and Leasehold Reform Act 2002 (the 2002 Act) had placed more significant management burdens on investors in flats and maisonettes. The need to consult lessees on qualifying long-term agreements, administration charges and major expenditure had increased the risk that landlords could be tripped up and challenged. Freeholders then faced the real prospect of having to pay for certain significant items of expenditure without any chance of recovering the cost from lessees. Such legislation applied equally to property in the PCL and in Kelton Court, but Mr Rutledge believed that it was a

A material factor that would widen the gap between the yields for houses compared to flats and maisonettes. In short, he felt that the 0.25% difference arrived at in *Sportelli* no longer realistically reflected the differences between the two forms of investment.

[16] The number of cases brought before LVTs in connection with disputed service charges, and in particular the consultation requirements contained in section 20 of the Landlord and Tenant Act 1985 (as amended by the 2002 Act), were very significant. He produced statistics obtained from the Residential Property Tribunal Service website that showed that, in 2007, there were 411 service charge disputes in London, almost all of which related to flats and maisonettes. This compared with B 232 cases in 2005 and 27 in 2004.

[17] In Mr Rutledge's opinion, there should be added to the 5% starting point for flats adopted in *Sportelli* between 0.5% and 0.75% for the higher rate of obsolescence/deterioration at Kelton Court compared to the PCL; between 0.5% and 1% to reflect investors' concern as to whether the growth rate in capital values would be as high in Birmingham as in the PCL, and 0.25% to reflect the increased gap between houses and flats resulting from recent legislative changes. Cumulatively, this produced a minimum deferment rate of 6.25% and a maximum of 7%. In his view, the appropriate rate for the appeal properties should be 6.5%. This assessment did not include a specific C allowance for increased volatility. In the course of his evidence-in-chief, Mr Rutledge said that although he had suggested at the LVT hearing that volatility was greater in the West Midlands than in the PCL, on reflection he no longer considered that to be the case. In cross-examination, he accepted that this change of opinion should be reflected by reducing the deferment rate from 6.5% to 6.25%.

[18] Between 1984 and 1997, Mr Willson was employed by the respondents, dealing with a range of commercial and residential estate-management matters, including responsibility for all freehold sales. He worked as a surveyor in the Birmingham office of Lambert Smith Hampton between February 1998 and February 1999, dealing D with a variety of professional work and property management, including Leasehold Reform Act work and sales, on a voluntary basis. In his present position, he specialises in leasehold enfranchisement work and voluntary sales of freeholds and grants of extended leases. Mr Willson has personally dealt with around 250 lease extensions under the 1993 Act and more than 1,000 claims under the 1967 Act. He considered that the expectations of future growth were reflected in current capital values. He disagreed with Mr Rutledge's view that property values in the West Midlands had grown at a slower rate than in the PCL. He said that there were now many indices that attempted to measure changes in the housing market. Some sought to track national or regional E trends and others considered different types of residential property. The compilation of such indices involved complex sampling and mathematical modelling. There were considerable variations between indices. An individual index would mask major differences in changes in prices between property types and locations.

[19] The difficulties of using indices were illustrated in the average prices data compiled by Communities and Local Government and UK National Statistics. These data were broken down by property type, including purpose-built flats. The data was, however, compiled by region and presumably included a diverse range of properties, both of type and location. Two separate tables were compiled, with some F variation in the make-up of the regions and different start dates. The first table showed the changes in average prices for purpose-built flats in different parts of the UK between 1983 and 2007. The second showed the corresponding changes between 1986 and 2007. Although there were differences between the extent of some of the regions in the two tables, the extent of the London and the West Midlands regions was the same in both. The standard West Midlands economic region included Herefordshire, Shropshire, Staffordshire, Warwickshire and Worcestershire, plus the metropolitan districts making up the old West Midlands county. The region was therefore both large and diverse. The only difference between the two sets of data, as far as London and the West Midlands were concerned, was the start date. Table 1 showed growth in London that was higher than in any other region of the UK and higher than in any of the other eight regions. In table 2, with a start G date three years later, the growth rate in London was very close to that in the country as a whole, whereas growth in the West Midlands was higher than in London. This showed how a relatively small difference in the period for which data had been compiled could lead to a radically different result.

[20] The two tables did not show real growth. In order to strip out the effects of inflation, Mr Willson deducted the inflation rate, based on the retail price index, from the annual growth rate for the West Midlands and London. Between 1983 and 2007, the average real growth rate was 6.43% pa for London and 5.88% pa for the West Midlands. Between 1986 and 2007, the figures were 5.49% and 6.16% respectively. For H both areas over these periods, the average level of real growth was significantly greater than 2%.

[21] Any attempt to use indices to determine rates of growth in different areas presented considerable difficulties. Such evidence as was available did not generally establish any sustained differences in growth rates over the long term. Moreover, such indices showed only past performance. Although they might assist a potential investor in deciding the likely rate of future growth, it was far from certain that they would be a decisive influence on the price it would be prepared to pay for the reversion.

[22] Mr Willson did not agree with Mr Rutledge's view that J obsolescence and deterioration would be greater at Kelton Court than in properties in the PCL. He said that repair and obsolescence should not be confused. Parts of buildings broke or wore out over time and needed to be repaired or replaced. Other parts needed regular maintenance. In most purpose-built flats, including Kelton Court, the landlord was responsible for the external and structural repairs, but recovered these costs from the tenants via a service charge. When the leases expired, the building should be handed back in repair. Obsolescence was considered in detail by the tribunal in *Daejan Investments Ltd* v *The Holt (Freehold) Ltd* LRA/133/2006 2 May 2008*. Here, a building that was "tired and shabby" ([85]) was not considered to be at greater risk K of obsolescence than the norm.

[23] Mr Willson considered that the relatively high level of repair costs in London was significant. Moreover, costs in the PCL area were probably higher than in London postal districts generally. Buildings were often taller than in the West Midlands. Four-, five- and six-storey buildings, especially where the site coverage was extensive and the buildings close to the highway, meant that access for building works was difficult, with potentially significant cost consequences. A much higher proportion of flats in the PCL area were in buildings converted to use as flats and this could result in higher maintenance costs.

[24] The state of repair did not determine whether a building was L obsolete. Obsolescence had more to do with function than condition. A building became obsolete because it no longer met the requirements for which it was built or because it no longer met the needs of occupiers or because the use for which it was designed no longer existed or was no longer appropriate. Nearly all buildings, even new ones, had a degree of obsolescence. Many houses in the PCL area, for example, were built before the era of almost universal car ownership and had no parking spaces and nowhere to create one. They therefore had a degree of obsolescence that could normally not be remedied without redevelopment. As a result, properties with off-street parking usually sold for higher prices than comparable properties without parking. M The degree of obsolescence in this case was reflected in the vacant possession value of the property. Mr Willson considered that that was the correct approach.

[25] The extensive redevelopment of properties in Birmingham, and particularly on the Calthorpe Estate, had to be viewed in its historical context. In the immediate post-war period, the estate trustees were under considerable pressure to increase the density of development on their land. There was a fear that, otherwise, large areas of the estate could be compulsorily purchased by the local authority for

* Editor's note: Reported at [2008] PLSCS 146

redevelopment. It was therefore decided to instruct a local architect to prepare a plan for the redevelopment of three residential sectors within the estate. This was published in 1958 and was designed to raise the density of the estate to 30 persons to the acre by 2000. It would have been possible to achieve this by clearing large areas and building traditional suburban houses at eight to 10 to the acre. It was decided, however, to consider the characteristics of individual sites and preserve landscape features, trees and boundary walls. Various sites were developed with a mixture of high- and low-rise flats, town houses and traditional low-density housing. The Ampton Green Development, of which Kelton Court forms part, comprises 36 houses and 36 flats. There are relatively large four-bedroom detached houses at a low density, with the flats serving to increase the overall density. Whether the redeveloped buildings had all been truly obsolete was not a major factor leading to their redevelopment.

[26] Paradoxically, many properties that would have been thought to be obsolete in the past had been retained. When Kelton Court was built, many older houses in the area were converted to flats, bedsits and hostels. Most have now been restored to single family use. Although not expensive by the standards of some parts of the PCL area, some of the most valuable houses in the West Midlands were close to Kelton Court. Moreover, Kelton Court was now within a conservation area and there were therefore greater planning restrictions on redevelopment.

[27] The nature of property was such that it would deteriorate over time. Provided that leases were adequate to ensure suitable standards of repair, this should be mitigated. Both building quality and design had an effect on the speed with which a building became obsolete and would potentially have an effect on condition. Factors such as type and quality of construction, however, were reflected in the vacant possession value of the property. This approach was endorsed by the LT in *Sportelli*.

[28] Mr Willson then considered the distinction between the treatment of houses and flats. In *Arbib* v *Earl Cadogan* [2005] 3 EGLR 139, the LT concluded that there should be a general addition to the deferment rate to reflect the differences between the two classes of property. This was largely because of a perceived difference in management requirements. However, provided that leases were properly drafted, were comprehensive in their requirements and allowed for the full recovery of management costs, the actual cost to the landlord of a block of flats should be no greater than for any other property investment. All residential property has been subject to increased statutory intervention in the relationship between landlord and tenant over the years. Changes to consultation requirements were introduced by the 2002 Act and came into force prior to the decision in *Sportelli*. The requirement for consultation applied to houses as well as flats. Mr Willson did not consider that there was any reason why the effect in Birmingham should be different from that in the PCL. There was probably some justification for an addition for flats, but this should not exceed the figure of 0.25% adopted in both *Arbib* and *Sportelli*.

[29] In Mr Willson's opinion, the length of the lease and the extent of the landlord's management obligations could influence the value of the landlord's reversion and lead to an adjustment to the deferment rate. Matters such as expectations of future growth, location, obsolescence and condition, however, would influence the value of all interests in the property, not just the freehold reversion. Mr Willson considered that such factors were already reflected in the vacant possession value. Only in the most exceptional cases would they justify a change in the deferment rate.

[30] In a supplementary report issued in the light of comments made in [33] of my decision in *Re Mansal Securities Ltd* [2009] 20 EG 104*, Mr Willson made the following additional points. He was not aware of any general evidence to suggest that there were widespread problems in recovering service charge contributions. It did not follow from the fact that people lived in high-value properties that they had substantial incomes from which to meet repair costs. Pensioners, for example, might be "asset rich" but still have modest disposable incomes. A number of factors needed to be borne in mind when attempting to use house-price data to assess variations in growth rates. In *Hildron Finance Ltd* v *Greenhill Hampstead Ltd* [2008] 1 EGLR 179, the LT (HH Judge Reid QC and Mr Norman J Rose FRICS) suggested that periods in the region of 50 years, with different starting dates, should be looked at in order to obtain a reliable indication of the long-term movement in residential values. This analysis could not be undertaken in practice. The Nationwide UK House Price Index was the only one that covered a 50-year period and this looked only at the national picture. In the 1970s, the Nationwide and the Halifax started to produce indices showing changes in house prices on a regional basis. They used the standard West Midlands economic region, used regularly in the production of economic statistics. It covered an area of 12,998 sq km (5,019 sq miles), ranging from Stoke-on-Trent in the north to Hereford in the south and from the Welsh border to Rugby. It contained a wide range of properties in many very different locations. There was therefore likely to be considerable variation, at least in the short term, in the rate of change in prices within the region. Local factors were likely to be masked by the wider regional picture. A number of indices had recently been compiled at a more local level, usually based on local authority areas. Even at this level, variations could be hidden, particularly in large local authority areas such as Birmingham. Moreover, most of these indices went back only to 2000, when HM Land Registry started to make sale prices available. Ideally, long-term data would be available for local areas. Even this would create problems, since if the pool of data was too small, individual transactions started to skew the results.

[31] Mr Willson drew attention to the fact that the way in which the Nationwide UK House Price Index was calculated had been changed four times. It was not clear whether this made any difference, but it highlighted the difficulties of using such data, particularly for non-statisticians. The LT had expressed a desire to see a comparison of different indices over different periods. It was difficult to make that comparison, even when the indices existed, because it was not clear how they had been compiled and what data had been used. The Nationwide and Halifax West Midlands indices showed very different rates of change for the same region and property types.

[32] Mr Willson concluded that all these factors tended to undermine the usefulness of such statistical information. At best, they presented an unreliable picture of the past and were not a useful guide to future performance.

[33] Dealing with Mr Rutledge's reference to changes in legislation, Mr Willson said that the Residential Property Tribunal Service website indicated that there had been only six service charge disputes in Birmingham in 2007. Given that Birmingham was significantly larger than any of the London boroughs, this did not suggest that the problem was more serious in Birmingham.

[34] In his supplementary report, Mr Rutledge said that he had spent much time seeking an experienced expert witness who might be willing to give further evidence before the tribunal in respect of the PCL investment market. In preliminary informal conversations with senior partners/directors of many of the largest firms in the country, it was clear that they shared his general view that the property investment market in the PCL was unlike that in any other part of the country. However, in every case the individual concerned had been unable or unwilling to appear before the tribunal because of conflicts of interest with clients owning properties outside London.

[35] Mr Rutledge said that he had made further extensive enquiries to discover whether there were other relevant sources of statistical information. He had obtained clarification from the compilers of the Nationwide and Halifax statistics of the different methodologies employed. It was clear from his enquiries of one of Nationwide's senior analysts that, contrary to the suggestion in *Mansal*, the Nationwide did not include studio flats in their calculations. Their main regional indices were weighted according to transactions in each area and covered all properties, subject to their "cleansing" criteria used to strip out unusual cases. This system of weighting allowed them to adjust the mix of properties in the sample in each quarter, with weights being updated periodically to reflect changes in the types of property

* Editor's note: Also reported at [2009] 2 EGLR 87

sold. This process ensured that the series was adjusted to remove any inconsistencies. The purpose of these adjustments was to achieve a consistent series that could be used for long-run analysis. Mr Rutledge also produced a copy of the Halifax House Price Index technical details and background. Halifax, too, used methods of standardisation and disaggregation of prices into their constituent parts to enable it to track the value of a typical property over time on a like-for-like basis (that is, with the same characteristics). The Halifax database started in 1983. Its technical details section emphasised that the size of the database was exceptionally large because of the number of transactions financed by the Halifax each month and because the scope of its data collection was more extensive than anything hitherto available in the UK.

[36] Although Mr Rutledge was not an expert statistician, he felt that the striking feature of the statistics summarised in his first expert report, and as illustrated in his comparative property price graph, was that, in the long-term, both the Halifax and Nationwide tables produced remarkably similar statistics. A comparison of the Nationwide regional table from 1973 for the West Midlands and the Halifax West Midlands table for all properties from 1983 showed little variance.

[37] Mr Rutledge pointed out that the LVT had found that "the growth rate in PCL probably has exceeded that in the West Midlands". Any investor would be interested in the prospects for growth, as well as the security of the investment and the anticipated income. His professional experience led him to conclude that an investor would not pay the same price for a property in the West Midlands as for one in the PCL, which was demonstrably likely to appreciate at a greater rate in the long term. On the balance of probabilities, therefore, he would have to advise an investor that an investment in the PCL would always, in the long term, be expected to produce more growth than a comparable investment in the West Midlands, and this would be reflected in the level of the investor's bid.

[38] Mr Rutledge said that he had given further careful thought to the comments of the LT in *Hildron* and, in particular, in [39]. He did not know why a 50-year period had been highlighted as being particularly important when looking at statistics for movements in property prices. There was no comprehensive set of figures or statistics for property prices going back 50 years from the valuation date to 1958, except for the Nationwide UK House Price Index. That overall UK index was started in 1952 and there did not appear to be anything else that distinguished between the PCL and the West Midlands going back that far.

[39] Looking at the available information, whether one started with figures in 1952 or any other date over that decade, the conclusion was very much the same. The first point at which different figures were available for London as compared to other parts of the country appeared to be in around 1973, when the Nationwide regional table was created. A noticeable divergence between that and other parts of the country opened up from that date gradually but inexorably, as could be seen from the comparative property price graph.

Conclusions

[40] *Sportelli* was concerned with residential properties in the PCL. As Carnwath LJ pointed out in the course of his judgment in the appeal, [2007] EWCA Civ 1042; [2008] 1 WLR 2142*, at [102]:

> The issues within the PCL were fully examined in a fully contested dispute between directly interested parties. The same cannot be said in respect of other areas. The judgment that the same deferment rate should apply outside the PCL area was made, and could only be made, on the evidence then available. That must leave the way open to the possibility of further evidence being called by other parties in other cases directly concerned with different areas. The deferment rate adopted by the tribunal will no doubt be the starting point; and their conclusions on the methodology, including the limitations of market evidence, are likely to remain valid. However, it is possible to envisage other evidence being called, for example, on issues relevant to the risk premium for residential property in different areas. That will be a matter for those advising future parties, and for the tribunals, to consider as such issues arise.

* Editor's note: Also reported at [2008] 1 EGLR 137

[41] Thus, the starting point when considering the appropriate deferment rate to be adopted in this appeal is the 5% determined in *Sportelli* for flats in the PCL. This rate was arrived at as follows:

Risk free rate	2.25%, minus
Real growth rate	2.00%, plus
Risk premium	4.5%, plus
Increased management risks for flats	0.25%

[42] In its decision dated 7 September 2009 in *Culley* v *Daejan Properties Ltd* LRA/163/2007 [2009] UKUT 168 (LC)*, the tribunal was required to determine the appropriate deferment rate for a block of flats in Eastcote, Middlesex. The members (Mr George Bartlett QC, president, and Mr Paul Francis FRICS) quoted the remarks of Carnwath LJ in [102]. They continued:

> 35 Thus evidence of valuers as to whether a higher risk premium should be taken because of the features of the property under consideration, including its location, is of undoubted relevance, and if a tribunal is satisfied on the evidence before it that such features justify the application of a higher deferment rate then, of course, it ought to apply such higher rate. In determining whether a higher rate is appropriate it will need to bear in mind the considerations that led the Tribunal in *Sportelli* to adopt the approach that it did, and the primary question will always be whether there are particular features that are not fully reflected in the vacant possession value and thus should be reflected in a higher risk premium. Moreover — and this is a matter that may not, or may not sufficiently, emerge from the Tribunal's post-*Sportelli* decisions — what matters is the view that the market, properly informed on relevant factual matters, would take on such features (the prospective movement of house prices in the area, for instance, or the potential obsolescence of the property) in considering an investment in the reversion. On this the expert opinion of the valuer is likely to be important.

[43] Mr Willson considered that there was no justification for a departure from any of the *Sportelli* elements of the deferment rate when valuing a block of flats such as Kelton Court, located in Edgbaston in the West Midlands. Mr Rutledge, on the other hand, was of the view that the risk premium should be increased to reflect a greater risk of deterioration and obsolescence and the risk that the real growth rate of 2% pa assumed in *Sportelli* would not be achieved in Edgbaston over the long term. He also thought that *Sportelli*'s 0.25% addition for management was now insufficient in view of the increased market awareness of the financial risk to landlords of flats, resulting from the coming into effect of various sections of the 2002 Act. My conclusions on each of these factors follow.

Deterioration and obsolescence conclusions

[44] In Mr Rutledge's opinion, by comparison with the much more valuable properties in the PCL, there was a greater risk that flats in Edgbaston would become obsolete over the term of the existing leases or that the extent of deterioration would be so great that the flats would no longer be worth repairing. Mr Willson disagreed. He said that it was inevitable that parts of buildings would need to be repaired or replaced over time and other parts would need regular maintenance. In most purpose-built blocks of flats, including Kelton Court, the buildings should be handed back in repair when the leases expired. As for obsolescence, Mr Willson said that this was a factor that was common to most buildings. As in the case of the "tired and shabby" building considered by the LT in *Daejan,* Kelton Court was no more at risk of obsolescence than the norm.

[45] In *Sportelli*, the LT's conclusion on obsolescence and condition was set out in [91] in these terms:

> As with location, while we do not rule out the possible need to adjust the deferment rate to take account of such matters as obsolescence and condition, we think that it would only exceptionally be the case that such factors were not fully reflected in the vacant possession value and the risk premium. Evidence would be needed to establish that they were not fully reflected in this way.

[46] Mr Rutledge produced a schedule showing that the values of the properties considered in *Sportelli* were of a different order of magnitude

* Editor's note: Reported at [2009] 3 EGLR 165

from those of the flats in Kelton Court. Of the six PCL buildings considered in *Sportelli*, the LVT decisions provided the necessary information to enable the value psf of four to be calculated. These values ranged from £740 to £1,100 psf and the relevant valuation dates were between December 2003 and July 2005. By contrast, the value of Kelton Court as at September/October 2007 was only £198.50 psf. I do not consider that the fact that there has been extensive redevelopment in Birmingham proves that the previously existing buildings had become obsolete. Nor does the fact that most of the *Sportelli* properties are in a conservation area mean that they cannot become obsolete. Nevertheless, as Mr Rutledge observed, the difference between the value of flats in Kelton Court and those considered in *Sportelli* is striking. Although building costs were somewhat higher in London than in Edgbaston, I accept Mr Rutledge's opinion that it is likely to remain economically viable to repair high-value properties in the PCL for considerably longer than it will for similar-sized flats in Kelton Court. As a result, while the individual flats might be leased on full repairing terms, there is a greater risk of deterioration at Kelton Court than in PCL properties, but this is not reflected in the respective vacant possession values. I find that a purchaser of the freehold reversion to Kelton Court would have required an increase of 0.25% in the risk premium to 4.75% to compensate for this difference.

Prospect of future growth conclusions

[47] Mr Rutledge produced statistical evidence that, he said, demonstrated why a potential purchaser of a reversion secured on Kelton Court would be concerned that the long-term growth rate of 2% adopted by the LT in *Sportelli* might not be achieved. Mr Willson disagreed, on the basis that the deficiencies in the available statistics meant that it was difficult to draw reliable conclusions from them on long-term growth rates in the past and that, even if these difficulties could be overcome, historic information would not provide a useful guide to future performance.

[48] In *Hildron*, the LT said, in [39]:

> In order to provide a reliable indication of the long-term movement in residential values so as to justify a departure from the *Sportelli* starting point, we consider that a period in the region of 50 years should be looked at, and that a series of statistics with different starting dates should be considered in order to ensure that an unrepresentative period is not relied upon.

[49] Both experts in the present case have taken the trouble to search for all the available statistics illustrating residential property movements in the West Midlands and in the PCL over the past 50 years. As a result, it appears that the statistical information that the LT indicated in *Hildron* would be required is simply not available. Contrary to the impression that I obtained in *Mansal*, the Nationwide statistics do not provide a reliable indication of the relative change in house prices in the West Midlands and Inner London between 1952 and 2007. Nationwide did not provide separate statistics for the West Midlands until 1973. The information referred to in [30] of *Mansal* was based on an attempt by a surveyor giving evidence at an earlier LVT hearing to extrapolate back from statistics relating to the West Midlands, rather than using actual base data.

[50] Despite their deficiencies, the patterns of price movements shown in the Nationwide Regional Index West Midlands for all residential properties from the fourth quarter of 1973 and the corresponding Halifax Regional Index from the first quarter of 1983 are similar. The trend of increase illustrated by both indices and also by the history of price increases in Kelton Court is significantly slower than that shown by the Knight Frank index for Kensington and Chelsea (part of the PCL) since 1976. Notwithstanding the statistical evidence to which he referred in his written report (see [20] above), Mr Willson accepted in cross-examination that information similar to that shown on Mr Rutledge's graph would have been provided to an investor seeking to compare the likely growth rates of residential properties in the West Midlands and the PCL. In his view, the graph showed that, over the periods covered by the available statistics, London had a slight advantage, but:

> we do not know whether they show the long-term position. All you can do is give the available information to the investor and he has to take the decision [on which location offers better growth prospects] which cannot be based on hard evidence.

[51] In my judgment, in spite of its undoubted limitations, the available statistical information demonstrates that the difference between past rates of long-term price increases in the PCL and in the West Midlands has been not slight but considerable. I accept Mr Rutledge's opinion that this information would persuade an investor that it could reasonably anticipate significantly slower long-term growth from residential properties in the West Midlands generally than in the PCL and I find that there was no reason to suppose that the position would be significantly different in the case of Kelton Court.

[52] In arriving at its conclusion in [72] of *Sportelli*, that a real growth rate of 2% pa should be assumed in calculating the deferment rate, the LT said:

> we think a realistic, or neutral, assumption would be 2%, the one made by Mr Cullum, with any concern on the part of the investor that this rate might not be achieved being reflected in the risk premium.

[53] As I have said, the 5% deferment rate determined in *Sportelli* for flats in the PCL is the starting point for calculating the appropriate rate for Kelton Court. Since, as I have found, an investor considering long-term growth prospects in Kelton Court would not be confident that the PCL growth rate would be achieved (or, put another way, would be less confident that the real growth rate of 2% would be achieved in the West Midlands than in the PCL), it would reduce its bid for Kelton Court accordingly. The appropriate way to assess that reduction, in my view, is by further increasing the risk premium by 0.5% to 5.25%.

Allowance for flats

[54] In *Sportelli*, the LT's conclusions on the difference between the deferment rate for houses and flats were contained in [95], as follows:

> We think, however, that an adjustment needs to be made to reflect the greater management problems [associated with flats], although we do not consider it appropriate to differentiate between flats that are the subject of headleases and those that are not. Nor do we think that the management concerns are necessarily so much less for a single flat than for a block to warrant a different adjustment. Even where flats are efficiently managed, service charge and repairs problems inevitably occur, and the management exercise in itself is, we feel, sufficiently more complex to warrant a generalised 0.25% addition for flats. We do not consider that any fine-tuning below this percentage is justified.

[55] Mr Rutledge relied primarily on the introduction of the provisions of the 2002 Act to support his view that the risks to landlords of flats has increased since *Sportelli*. He had in mind in particular the Service Charges (Consultation Requirements)(England) Regulations 2003 (the 2003 Regulations), which impose strict procedural requirements, failure to comply with which could leave landlords unable to recover a large proportion of the costs of repairs. Although these regulations had come into force nearly three years before the *Sportelli* hearing, the potential seriousness of their effect has become increasingly widely known and this has reduced the relative attraction of investments secured on blocks of flats. Mr Willson replied that there have not been any changes to legislation post-*Sportelli* that would have made a difference. There was no reason why the effect in Birmingham should be different from that in the PCL.

[56] The provisions of the 2003 Regulations are potentially extremely serious for landlords: see the LT decision dated 11 April 2008 (Mr George Bartlett QC, president, and Mr Norman J Rose FRICS) in *Camden London Borough Council* v *Leaseholders of 37 flats at 30-40 Grafton Way* LRX/185/2006 30 June 2008*. I accept Mr Rutledge's evidence that although LVTs have heard only a limited number of service charge appeals relating to properties in Birmingham, there have been other examples of landlords of such properties agreeing to bear part of the cost of disputed items without the need for a tribunal hearing. The 2003 Regulations came into force on 31 October 2003. I am satisfied that, by September 2007, the first date with which I am concerned, the market was more aware of the

* Editor's note: Reported atRe 30-40 Grafton Way [2008] PLSCS 198

A dangers posed by the regulations than was the case in *Sportelli*, where the properties fell to be valued between two-and-a-quarter and three-and-three-quarter years earlier. I conclude that, in the 11 cases with which I am concerned, investors would have required an addition of 0.5% to reflect the greater management problems associated with flats than with houses. In reaching this conclusion, I have borne in mind that the subject flats are no longer subject to the original headlease. Had that headlease still been in existence, I would not have considered it appropriate to depart from the *Sportelli* uplift of 0.25%.

Summary

B [57] The LVT reached the conclusion, based on the evidence before it, that the appellants had not made out their case for a departure from the *Sportelli* starting point of a 5% deferment rate. In the light of the more extensive evidence produced before this tribunal, I have concluded that the risk premium of 4.5% determined in *Sportelli* should be increased to 5.25%, and that the 0.25% management allowance for flats should be increased to 0.5%, resulting in a deferment rate of 6%. The appeal is allowed. The premium payable by the leaseholder of flat 33 is £8,150. The premium payable by the respective leaseholder or leaseholders of each of the remaining 10 flats is £8,100. The calculations producing these figures appear in the appendix to this decision.

C [58] I would add that my conclusion in this appeal differs from the one reached by the LT in *Culley*, namely that there was no justification for a departure from the *Sportelli* starting point. In *Culley*, as in the present appeal, the basis of the LT's decision was the expert valuation evidence adduced by the parties.

[59] The tribunal's power to order the payment of costs is extremely limited in cases of this nature. I make no order as to costs.

Appendix

Flats 1,2,5,6,19,21,31,33,34,35 and 36 Kelton Court
Carpenter Road, Edgbaston, Birmingham
Valuation by Lands Tribunal

D All properties except flat 33

(i) Diminution in the freeholder's interest	
Term:	
Current rent: £75.00 per year	
Years' purchase: 15 years @ 6% = 9.712	
£75.00 x 9.712 = £728.40	£728.40
Rent (from 2022): £100.00 per year	
Years' purchase: 25 years @ 6% @ 12.783	
PV £1 in 15 years @ 6% = 0.417	
£100.00 x 12.783 x 0.417 = £533.05	£533.05
Rent (from 2047): £125.00 per year	
Years' purchase: 24 years @ 6% = 12.550	
E PV £1 in 40 years @ 6% = 0.097	
£125.00 x 12.550 x 0.097 = £152.17	£152.17
	£1,413.62
F *Reversion*:	
Open market value with extended lease: £158,025	
(excluding tenant's improvements)	
PV £1 in 64 years @ 6%: 0.024	
£158,025 x 0.024 = £3,792.60	£3,792.60
	£5,206.22
(ii) Marriage value	
Open market value with extended lease:	£158,025.00
(excluding tenant's improvements)	
Less Freehold interest	5,206.22
Leasehold interest	£147,000.00
	£5,818.78
G (iii) Premium payable	
Freehold interest:	£5,206.22
Marriage value (£5,818.78) x 50% =	£2,909.39
	£8,115.61
Say	£8,100

Flat 33

(i) Diminution in the freeholder's interest	
Term:	
Current rent: £75.00 per year	
Years' purchase: 14.75 years @ 6% = 9.608	
£75.00 x 9.608 = £720.60	£720.60
Rent (from 2022): £100.00 per year	
H Years' purchase: 25 years @ 6% = 12.783	
PV £1 in 14.75 years @ 6% = 0.423	
£100.00 x 12.783 x0.423 = £540.72	£540.72
Rent (from 2047): £125.00 per year	
Years' purchase: 24 years @ 6% = 12.550	
PV £1 in 39.75 years @ 6% = 0.098	
£125.00 x 12.550 x 0.098 = £153.74	£153.74
	£1,415.06
Reversion:	
Open market value with extended lease: £158,025	
(excluding tenant's improvements)	
PV £1 in 63.75 years @ 6%: 0.0244	
J £158,025 x 0.0244 = £3,855.81	£3,855.81
	£5,270.87
(ii) Marriage value	
Open market value with extended lease:	£158,025.00
(excluding tenant's improvement)	
Less Freehold interest	£5,270.87
Leasehold interest	£147,000.00
	£5,754.13
(iii) Premium payable	
Freehold interest:	£5,270.87
Marriage value (£5,754.13) x 50% =	£2,877.07
	£8,147.94
K *Say*	£8,150

Appeal allowed.

For Product Safety Concerns and Information please contact our EU representative GPSR@taylorandfrancis.com Taylor & Francis Verlag GmbH, Kaufingerstraße 24, 80331 München, Germany

Batch number: 10731598

Printed by Printforce, the Netherlands